Calcium Signals

From single molecules to physiology

Online at: https://doi.org/10.1088/978-0-7503-2009-2

Biophysical Society–IOP Series

About the Series

The Biophysical Society and IOP Publishing have forged a new publishing partnership in biophysics, bringing the world-leading expertise and domain knowledge of the Biophysical Society into the rapidly developing IOP ebooks program.

The program publishes textbooks, monographs, reviews, and handbooks covering all areas of biophysics research, applications, education, methods, computational tools, and techniques. Subjects of the collection will include: bioenergetics; bioengineering; biological fluorescence; biopolymers *in vivo*; cryo-electron microscopy; exocytosis and endocytosis; intrinsically disordered proteins; mechanobiology; membrane biophysics; membrane structure and assembly; molecular biophysics; motility and cytoskeleton; nanoscale biophysics; and permeation and transport.

A full list of titles published in this series can be found here: https://iopscience.iop.org/bookListInfo/iop-series-in-biophysical-society.

Calcium Signals

From single molecules to physiology

Edited by
Leslie S Satin
University of Michigan Medical School, Brehm Diabetes Center, Kellogg Eye Center, Ann Arbor, Michigan, MI, USA

Manu Ben-Johny
Department of Physiology and Cellular Biophysics, Columbia University, New York, USA

Ivy E Dick
Physiology Department, School of Medicine, University of Maryland, Baltimore, MD, USA

IOP Publishing, Bristol, UK

ISBN 978-0-7503-2009-2 (ebook)
ISBN 978-0-7503-2007-8 (print)
ISBN 978-0-7503-2010-8 (myPrint)
ISBN 978-0-7503-2008-5 (mobi)

DOI 10.1088/978-0-7503-2009-2

Version: 20230301

IOP ebooks

British Library Cataloguing-in-Publication Data: A catalogue record for this book is available from the British Library.

Published by IOP Publishing, wholly owned by The Institute of Physics, London

IOP Publishing, No.2 The Distillery, Glassfields, Avon Street, Bristol, BS2 0GR, UK

US Office: IOP Publishing, Inc., 190 North Independence Mall West, Suite 601, Philadelphia, PA 19106, USA

This book is dedicated to our teachers, mentors and students.

Contents

Part II Calcium signaling on channels and other effectors

Part III Physiological roles of calcium

Part IV Calcium channels in organelles

Part V Calcium channels and disease

Acknowledgement

We would like to thank our colleague Dr Andrea Meredith of the University of Maryland for initiating this project. Research in Dr Satin's laboratory is supported by NIH RO1 DK46409 and U01 DK127747.

Editor biographies

Leslie S Satin

Dr Leslie S Satin is the Joanne I. Moore Endowed Research Professor of Pharmacology of the University of Michigan Medical School. He obtained his PhD from UCLA in the laboratory of D Junge and conducted postdoctoral work at Stony Brook University with P R Adams and at the University of Washington, Seattle, with D Cook. His research has focused on calcium signaling, ion channels, oscillations, and stimulus-secretion coupling in pancreatic beta cells.

Manu Ben-Johny

Manu Ben-Johny is an Assistant Professor in the Department of Physiology and Cellular Biophysics at Columbia University. He received a BS in biomedical engineering and mathematics from Saint Louis University, and subsequently completed his PhD in biomedical engineering at Johns Hopkins University under the tutelage of David Yue. His lab focuses on the biophysics and physiology of voltage-gated sodium and calcium ion channels that shape cardiac and neuronal action potentials and in developing nature-inspired approaches to manipulate ion channel function.

Ivy E Dick

Ivy E Dick is an assistant professor in the Physiology Department at the University of Maryland, School of Medicine. She began her career under the direction of Drs Charles Cohen and Owen McManus at Merck Research Labs and subsequently earned her doctoral degree under the direction of Dr David Yue in the Biomedical Engineering department at Johns Hopkins University, where she studied the mechanisms underlying calmodulin regulation of voltage-gated calcium channels. Dr Dick's current research focuses on understanding the mechanisms of voltage-gated calcium channel regulation, and how genetic mutations disrupt channel function.

List of contributors

Enrique Balderas Angeles
Nora Eccles Harrison Cardiovascular Research and Training Institute, University of Utah, Salt Lake City, UT, USA

Dipayan Chaudhuri
Nora Eccles Harrison Cardiovascular Research and Training Institute, University of Utah, Salt Lake City, UT, USA

Ivy E Dick
Physiology Department, School of Medicine, University of Maryland, Baltimore, MD, USA

David Eberhardt
Nora Eccles Harrison Cardiovascular Research and Training Institute, University of Utah, Salt Lake City, UT, USA

Patrick A Fletcher
Laboratory of Biological Modeling, NIDDK, National Institutes of Health, Bethesda, MD, USA

Marc Freichel
Pharmakologisches Institut, Universität Heidelberg, Heidelberg, Germany

Teresa Giraldez
Department of Basic Medical Sciences and Institute of Biomedical Technologies, School of Medicine, University of La Laguna, Tenerife, Spain

Alberto J Gonzalez-Hernandez
Department of Basic Medical Sciences and Institute of Biomedical Technologies, School of Medicine, University of La Laguna, Tenerife, Spain

Erick O Hernández-Ochoa
Department of Biochemistry and Molecular Biology, University of Maryland School of Medicine, Baltimore, MD, USA

Kevin G Herold
Department of Physiology, University of Maryland School of Medicine, Baltimore, MD, USA

John W Hussey
Department of Physiology, University of Maryland School of Medicine, Baltimore, MD, USA

Gary J Iacobucci
Jacobs School of Medicine and Biomedical Sciences, University at Buffalo, Buffalo New York, USA

Alec Kittredge
Department of Pharmacology, Columbia University, New York, New York, USA

Aravind Kshatri
Department of Basic Medical Sciences and Institute of Biomedical Technologies, School of Medicine, University of La Laguna, Tenerife, Spain

Amy Lee
Department of Neuroscience, The University of Texas-Austin, Austin, TX, USA

Sandra Lee
Nora Eccles Harrison Cardiovascular Research and Training Institute, University of Utah, Salt Lake City, UT, USA

Gilbert Q Martinez
AGC Biologics, Bothell, WA, USA

Sohum Mehta
Biomedical Research Facility 2, Department of Pharmacology, University of California, San Diego, La Jolla, CA, USA

Aaron P Owji
Department of Pharmacology, Columbia University, New York, New York, USA

Gabriela K Popescu
Jacobs School of Medicine and Biomedical Sciences, University at Buffalo, Buffalo New York, USA

Sara Ragi
Jonas Children's Vision Care, and Bernard & Shirlee Brown Glaucoma Laboratory, Departments of Ophthalmology and Pathology & Cell Biology, Edward S. Harkness Eye Institute, Columbia Stem Cell Initiative, New York Presbyterian Hospital/Columbia University, New York, USA

Anjali M Rajadhyaksha
Department of Pediatric Neurology, Pediatrics, Feil Family Brain and Mind and Research Institute, Weill Cornell Autism Research Project, Weill Cornell Medicine, New York, USA

Leslie S Satin
University of Michigan Medical School, Brehm Diabetes Center, Kellogg Eye Center, Ann Arbor, Michigan, MI, USA

Marin F Schneider
Department of Biochemistry and Molecular Biology, School of Medicine, University of Maryland, Baltimore, MD, USA

Eric N Senning
Department of Neuroscience, The University of Texas-Austin, Austin, TX, USA

Arthur S Sherman
Laboratory of Biological Modeling, NIDDK, National Institutes of Health, Bethesda, MD, USA

Salah Sommakia
Nora Eccles Harrison Cardiovascular Research and Training Institute, University of Utah, Salt Lake City, UT, USA

Herie Sun
Department of Pediatric Neurology, Pediatrics, Feil Family Brain and Mind and Research Institute, Weill Cornell Autism Research Project, Weill Cornell Medicine, New York, USA

Jessica R Thomas
Colbran Laboratory, Department of Molecular Physiology and Biophysics, Vanderbilt University, Nashville, TN, USA

Stephen H Tsang
Jonas Children's Vision Care, and Bernard & Shirlee Brown Glaucoma Laboratory, Departments of Ophthalmology and Pathology & Cell Biology, Edward S. Harkness Eye Institute, Columbia Stem Cell Initiative, New York Presbyterian Hospital/Columbia University, New York, USA

Richard W Tsien
Department of Neuroscience and Physiology at NYU Grossman School of Medicine, New York, USA

Volodymyr Tsvilovsky
Pharmakologisches Institut, Universität Heidelberg, Heidelberg, Germany

Hristo Varbanov
Institute of Neurophysiology, Hannover Medical School, Hannover, Germany

Christian Wahl-Schott
Institut für Kardiovaskuläre Physiologie und Pathophysiologie, Biomedizinisches Zentrum, Ludwig-Maximilians-Universität München, Muenchen, Germany

Brittany Williams
Department of Cell Biology and Physiology, University of North Carolina, Chapel Hill, NC, USA and Neuroscience Center, University of North Carolina, Chapel Hill, NC, USA

Tingting Yang
Department of Ophthalmology, Columbia University, New York, New York, USA

Jin Zhang
Biomedical Research Facility II, University of California, San Diego, La Jolla, CA, USA

Yu Zhang
Department of Ophthalmology, Columbia University, New York, New York, USA

Yanghao Zhong
Department of Pharmacology, University of California, San Diego, La Jolla, CA, USA

IOP Publishing

Calcium Signals
From single molecules to physiology
Leslie S Satin, Manu Ben-Johny and Ivy E Dick

Chapter 1

Introduction

Les Satin

The importance of Ca^{2+} as an intracellular signaling molecule *par excellence* is undisputed today. Being critical for diverse physiological processes ranging from muscle contraction, hormone secretion, learning and memory to the control of gene expression, Ca^{2+} is the veritable king of the cellular signaling world due to its ubiquitous presence, the mechanisms that have evolved to control its free concentration and its precise localization not only signals critical messages to various sites within the cell, but also induces death when elevated inappropriately in the cytosol (Cheng *et al* 2006).

The history of calcium signaling research was initiated by the work of Sidney Ringer, who first showed that the isolated heart required calcium for contraction. More recent research on calcium regulation of protein conformation and control of gene expression has been covered by a number of excellent recent reviews (Patel 2019, Carafoli 2002, Clapham 2007). The gist of the story is that calcium ions are special compared to related ions, such as sodium and magnesium, because of calcium's special coordination chemistry (Carafoli and Krebs 2016), the evolution of E–F hand Ca^{2+} binding proteins, which have the necessary flexibility but also the requisite specificity (Kretsinger and Nockolds 1973, Carafoli and Krebs 2016, Clapham 2007). In addition, cells were subjected to evolutionary pressures favoring the establishment of low cytosolic Ca^{2+} levels so that they could avoid forming insoluble calcium phosphate complexes once they turned to using phosphate as an energy source (Carafoli 2004). The intricate Ca^{2+} buffering and decoding mechanisms of certain proteins, and the occurrence of specialized membrane-localized calcium pumping systems to eject calcium from cells or sequester it at the expense of ATP hydrolysis (Carafoli and Krebs 2016) have ensured that changes in Ca^{2+} can occur rapidly within highly localized domains within the cell (Brini and Carafoli 2000).

Our goal in creating the present volume was to get some of the leading researchers in the field to discuss key aspects of calcium signaling in cells. The wide variety of molecules that mediate Ca^{2+} influx in cells are represented by chapters describing voltage-gated Ca^{2+} channels and their unique permeation properties (Tsien), the structure and function of TRP channels, which transduce a host of environmental stimuli, including odors, changes in pH or changes in temperature (Martinez and Senning), and a description of the glutamate-activated Ca^{2+} influx mechanisms of neurons (Iacobucci and Popescu).

Once Ca^{2+} enters the cytosol, it can activate other ion channels, such as Ca^{2+}-activated K^+ channels (Gonzalez-Hernandez) and Ca^{2+}-activated Cl channels (Kittredge and Yang). The dynamics of Ca^{2+} concentration changes within the cytosol and endoplasmic reticulum, due to Ca^{2+} influx and Ca^{2+} uptake, respectively, and the dependence of Ca^{2+} transduction patterns on spatial compartmentalization in endocrine cells are discussed (Fletcher and Sherman), while the ways Ca^{2+} effectors interact to help Ca^{2+} signal, and how biosensors are used to measure Ca^{2+} experimentally are covered by Zhong and Zhang.

In muscle cells, intracellular Ca^{2+} release is critical for excitation–contraction coupling, which is discussed by Hernández-Ochoa and Schneider for skeletal muscle. Williams and Lee discuss how Ca^{2+} entry triggers neurotransmitter release upon activation of voltage-gated Ca^{2+} channels in synapses, while Dick presents an up-to-date summary of channelopathies involving Ca^{2+}, and Sun and Rajadhyaksha discuss the role of dysfunctional calcium regulation in neuropsychiatric disorders.

As expected of an up-to-date book on Ca^{2+} channels and Ca^{2+} signaling in cells, most chapters emphasize the molecular structures of the key proteins involved in Ca^{2+}-influx and Ca^{2+} signaling in cells as well as their function.

Acknowledgements

I am grateful to Drs Arthur Sherman of NIH and Richard Bertram of Florida State University for reading an earlier draft of this chapter and providing me with valuable input.

Getting busy researchers to commit to writing a book chapter in this day and age, and during a pandemic to boot, is a big challenge. The editors are thus deeply grateful to all of our wonderful colleagues who contributed chapters to this book. We also thank Dr Andrea Meredith of the University of Maryland who had the original idea for the book and who inspired us to work towards making it happen. Thanks, Andrea!

References

Brini M and Carafoli E 2000 Calcium signaling: a historical account, recent developments and future perspectives *Cell. Mol. Life Sci.* **57** 354–70

Carafoli E and Krebs J 2016 Why calcium? How calcium became the best communicator *J. Biol. Chem.* **291** 20849–57

Carafoli E 2004 Calcium signaling: a historical account *Biol. Res.* **37** 497–505

Carafoli E 2002 Calcium signaling: a tale for all seasons *Proc. Natl Acad. Sci. USA* **99** 1115–22

Cheng H-P, Wei S, Wei L-P and Verkhratsky A 2006 Calcium signaling in physiology and pathophysiology *Acta Pharmacol. Sin.* **7** 767–72

Clapham D 2007 Calcium signaling *Cell* **131** 1047–58

Kretsinger R H and Nockolds C E 1973 Carp muscle calcium-binding protein. II. Structure determination and general description *J. Biol. Chem.* **248** 3313–26

Patel S 2019 The secret life of calcium in cell signaling *Elements in Biochemistry* (London: Portland Press) pp 34–7 (online)

Part I

Molecules mediating calcium influx

IOP Publishing

Calcium Signals
From single molecules to physiology
Leslie S Satin, Manu Ben-Johny and Ivy E Dick

Chapter 2

Calcium channel selectivity and permeation: function meets 3D structure

Richard W Tsien

2.1 Introduction

In a remarkable erratum of a scientific paper, Sydney Ringer pointed out that his earlier report on the frog heart's contraction had overlooked the trace level of calcium in London tap water (Ringer 1883). This amount of calcium, less than 1 millimolar, proved to be essential for supporting the heartbeat; distilled water-based saline solution would not do the job unless a pinch of calcium salt was added. Thus, the field of calcium signaling was born. At the core of Ringer's finding was the voltage-gated calcium channel and its ability to select for Ca^{2+} rather than Mg^{2+} or the hundred-fold more abundant sodium ions in extracellular fluid. Understanding calcium channel selectivity and permeation is a worthy topic because it relies jointly on electrophysiological analysis, molecular manipulations and 3D channel structures. Together, these approaches provide a satisfying account of how Ca^{2+} channels prove to be more than a thousand-fold selective for Ca^{2+} over Na^+, ions of almost identical crystal radius (~1 Å), yet capable of fluxing 10 million Ca^{2+} ions per second. Spoiler alert: Ca^{2+} channels meet these challenges by use of a selectivity mechanism significantly different from potassium or sodium channels, yet illuminate the evolutionary relationship between each of these essential channel types. Vive les différences: Ca^{2+}, Na^+ and K^+ channels each use their own distinctive ion selectivity to support the action potentials and excitation–response coupling, the basis of information transfer within and among excitable cells. Leading benefits beyond the heartbeat (Noble 1979) are neural functions ranging from the simplest reflex behavior (Tsien and Barrett 2011) up to the most sophisticated aspects of sensation, action and memory (Mack *et al* 2021).

A full century after Ringer's 1883 erratum, electrophysiologists were applying new and improved methods for studying voltage-gated, calcium-selective channels,

doi:10.1088/978-0-7503-2009-2ch2

having settled an acrimonious debate about whether 'calcium currents' were an artifact of poor voltage clamp (Johnson and Lieberman 1971, Reuter 1973). Methods in multicellular preparations improved (Kass *et al* 1979) but suction pipette methods for whole-cell recordings from single cells (Lee *et al* 1980, Hamill *et al* 1981, Kostyuk *et al* 1983) and cell attached patch recordings (Hamill *et al* 1981) were the big step forward.

At the very least, unitary recordings allowed an easy distinction between three multiplicative factors ($N \times p \times i$) that determined whole-cell current: N, the number of functional channels, p, the probability that channels were open and thus allowing ion flux, and i, the flux through the open channel. This factorization welcomed investigations of conceptually different aspects of channel function. For example, cell biologists and biochemists could study N and how it was determined by gene expression, membrane trafficking and phosphorylation, leaving others to tackle the voltage-dependent factors $p(V)$ and $i(V)$. Hodgkin and Huxley had inspired studies of gating by describing the rate constants for channel opening and closing that in turn govern $p(V)$. Their treatment of $i(V)$ effectively opted for a simple ohmic relationship, reversing at the Nernst potential for a particular ionic species. This left room for others to study the determinants of *selectivity*—the choice of which ion species was allowed to traverse the pore—and of *permeation*, their rate of passage in ions per unit time. Introductions to the mathematical descriptions and underlying energetics can be found in Hille's foundational book (Hille 2007). The idea of a *selectivity filter* was most clearly developed for sodium channels (Hille 1971, 1972) but has relevance for all ion channels.

Back in the 1980s, studies of Ca^{2+} channel selectivity and permeation still faced the special issue of Ca^{2+} as a minority ion, far outnumbered by Na^{+}. As methods for studying Ca^{2+} channel currents continued to improve, researchers with a biophysical bent could now focus on the dynamic interaction between Ca^{2+} channels and permeant and/or blocking ions (Hagiwara and Byerly 1981, Kostyuk *et al* 1983) and generate provocative models (Kostyuk *et al* 1983) (figure 2.1(A)). In 1984, studies from Yale (Hess and Tsien 1984) and the University of Washington (Almers and McCleskey 1984) independently proposed that Ca^{2+} ions moved through the channel in single file, and that micromolar Ca^{2+} sufficed to block Na^{+} flux but millimolar Ca^{2+} favored multi-ion occupancy and self-facilitated influx. The basic experimental findings have been repeatedly confirmed (figure 2.2(A)), but the interpretation continues to stir up lively debate (e.g. Shuba 2014, Nonner *et al* 1998). To test the basics of the 1984 selectivity hypothesis, unitary recordings resolved interactions of single Ca^{2+} ions with individual Ca^{2+} channels, thus providing a quantitative look at the first step in Ca^{2+} permeation, as rapid as diffusion limits would allow (Lansman *et al* 1986). Cloning of the skeletal muscle Ca^{2+} channel ($Ca_V1.1$) (Tanabe *et al* 1987) and cardiac muscle Ca^{2+} channel ($Ca_V1.2$) (Mikami *et al* 1989), both L-type channels (Nowycky *et al* 1985), enabled many amino acid modifications aimed at deciphering how Ca^{2+} channel selectivity works. Lacking direct structural information about the 3D anatomy of the channel pore, the next investigations relied on electrophysiology and molecular manipulations to build a picture of Ca^{2+} selectivity and permeation (reviewed in Tsien *et al*

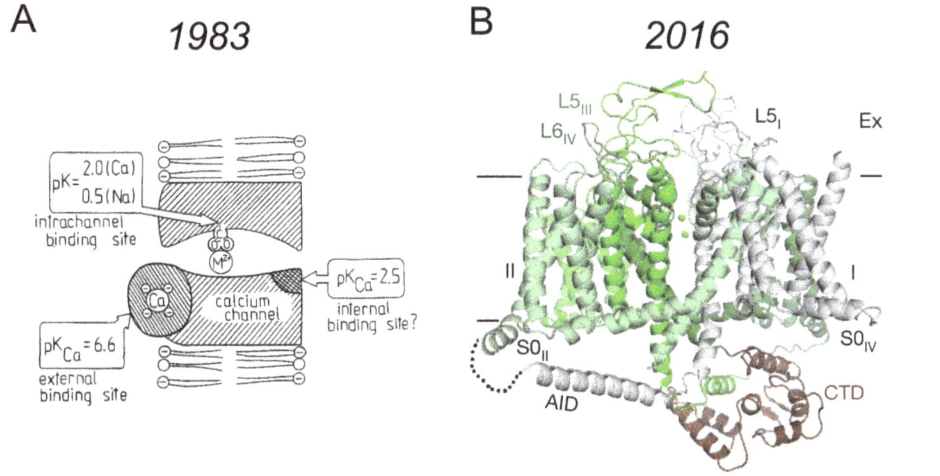

Figure 2.1. Advances in understanding Ca^{2+} channel selectivity and permeation over the third of a century from 1983 to 2016. (A) Kostyuk's 1983 model, invoking a low-affinity binding site for a single Ca^{2+} ($K_{1/2}$ ~10 mM) to explain how unitary flux saturates with increasing $[Ca^{2+}]_o$. An external binding site of high affinity ($K_{1/2}$ ~0.3 μM) is postulated to regulate transition from divalent to monovalent cation selectivity. A refinement of this model has been put forward (Kostyuk et al 1983, copyright 1983, Springer-Verlag with permission of Springer). (B) Cryo-EM structure of skeletal muscle Ca^{2+} channel, determined in presence of 10 mM $[Ca^{2+}]_o$, suggests pore occupancy by two closely spaced Ca^{2+} ions (Wu *et al* 2016, copyright 2016, Macmillan Publishers Limited, part of Springer Nature, all rights reserved, with permission of Springer).

1987, Sather and McCleskey 2003). This led to a revised model in which: (1) the ion permeation pathway is much wider than the crystal diameter of Ca^{2+} or Na^+ ions (both ~2 A); (2) selectivity arises because monovalent ions like Na^+, Li or K^+ cannot leapfrog ever-present Ca^{2+} ion(s); and (3) acidic side chains, be they glutamates or aspartates, protrude into the pore and provide a flexible complex to closely coordinate one or more Ca^{2+} ions when these are present; (4) rapid throughput of ions is possible because a newly incoming Ca^{2+} can loosen the affinity for an already present Ca^{2+} further along the permeation pathway (Yang *et al* 1993, Ellinor *et al* 1995); (5) the EEEE cluster constitutes a single locus, aptly called a selectivity filter, that can accommodate multiple divalent cations (see also Armstrong and Neytonc 1991), thus mediating key ion–ion interactions among multiple divalent ions; (6) the same glutamate side chains constitute a binding site for polyvalent metal ions (Hagiwara and Byerly 1981, Lansman *et al* 1986) that act as environmental toxins; (7) the complex of carboxylates presents a high-affinity titration site for hydrogen ions with a pKa near neutral pH, thus supporting physiological pH-regulation of Ca^{2+} entry (Chen *et al* 1996, Chen and Tsien 1997).

Today, nearly four decades after the proposals from Yale and UW in 1984, a set of 3D structures for mammalian voltage-gated calcium channels have been published. These cryo-EM structures all come from the amazing work of Nieng Yan and colleagues (Wu *et al* 2015, 2016, Zhao *et al* 2019). Two structures, at increasingly high resolution, describe the purified $Ca_V1.1$ l-type channel from

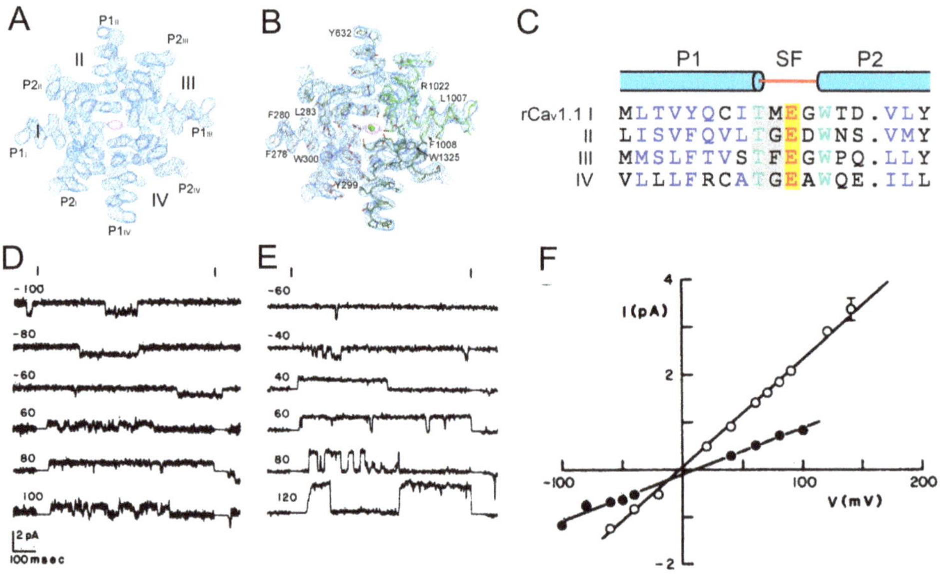

Figure 2.2. Structural overview of $Ca_V1.1$ (skeletal muscle) selectivity filter, sequence and comparison of single-channel flux with $Ca_V1.2$ (cardiac muscle). (A) and (B) Overview of structure of Ca^{2+} channel selectivity filter (Wu *et al* 2016, copyright 2016, Macmillan Publishers Limited, part of Springer Nature, all rights reserved, with permission of Springer). (A) Top-down view of the EM map of the selectivity filter and the supporting P1 and P2 helices for repeats I through IV, labeled $P1_i$ and $P2_i$ etc, determined in presence of 10 mM $[Ca^{2+}]_o$. (B) Side-chain assignment was assisted by bulky residues in P1 and P2 helices as labeled. Two green spheres depict assignment of intrapore density to two Ca^{2+} ions, appearing nearly concentric in this top-down perspective. (C) Amino acid sequence of PI and P2 helices and key linker region between them that governs the presentation of liganding oxygen atoms for Ca^{2+} channel selectivity (from Wu *et al* (2015) reprinted with permission from AAAS). (D)–(F) Highly homologous L-type channels, studied under identical experimental conditions, in planar bilayers with symmetrical ion concentrations (Rosenberg *et al* 1986, reprinted with permission from AAAS). Dihydropyridine agonist Bay K 8644 included to prolong channel openings. (D) Ca^{2+} channel purified from skeletal muscle T-tubules ($Ca_V1.1$). (E) Ca^{2+} channels from cardiac muscle $Ca_V1.2$. (F) open channel current versus voltage relationships obtained with equal permeant ion concentrations (100 mM Ba^{2+}, 50 mM Na^+) on both sides of the planar bilayer show significant differences in single-channel conductance between the $Ca_V1.2$ l-type channels of heart (22.7 pS, open circles) and the $Ca_V1.1$ l-type channels of skeletal muscle (10.6 pS, filled circles).

skeletal muscle (Wu *et al* 2015, 2016), the same subtype studied by Almers' group (Almers and McCleskey 1984). These were characterized in the presence of 0.5 mM and 10 mM external Ca^{2+}, a range over which the unitary Ca^{2+} current increases the most (Hess *et al* 1986). The other structures are for a cloned T-type channel, $Ca_V3.1$ (Zhao *et al* 2019), also with 10 mM Ca^{2+} present, and a cloned N-type channel, with 2 mM Ca^{2+} present (Gao *et al* 2021). Disease-generating modifications and binding sites for therapeutic drugs are rapidly being mapped (Wu *et al* 2017, Zhao *et al* 2019, Gao and Yan 2021, Gao *et al* 2021). The advent of high-resolution 3D structures also provides an opportunity to reassess the biophysical measurements and structural conclusions derived from twenty years of research from 1983 to 2003, elegantly summarized by Sather and McCleskey (2003). The following sections briefly review the basic cornerstones of those conclusions with simple questions in mind. How valid were the earlier inferences about principles of calcium channel selectivity and permeation? Can the exquisitely detailed but static structures be leveraged to reframe dynamic studies of ion-channel and ion–ion interactions, now decades-old? How would our general picture of what makes a Ca^{2+} channel a Ca^{2+} channel be sharpened and revised, and do the same principles extend to other Ca^{2+}-selective channels important for non-excitable cells? In turn, what insights can we gain about K^+, Ca^{2+}, and Na^+ channels, how members of this superfamily are evolutionarily related and perform their essential functions?

Although the frog heart's Ca^{2+} channel hasn't been cloned yet, we already have a multiplicity of mammalian Ca^{2+} channels, well-described elsewhere in this book. Here, some background and justification might be helpful before embarking on a review of data and interpretation. The voltage-gated Ca^{2+} channels considered here, $Ca_V1.1$ (skeletal), $Ca_V1.2$ (cardiac and neuronal), $Ca_V2.2$ and $Ca_V3.1$ (neuronal) are virtually identical in overall body plan (figures 2.1(B) and 2.2(A and B)) and are highly homologous in their amino acid sequence, especially in a common locus of acidic amino acids, generally known as the EEEE cluster or 'selectivity filter' (figure 2.2(C)). The abundance of $Ca_V1.1$ has facilitated pioneering biochemistry (Catterall 2000), and in turn, 3D structural studies (Wu *et al* 2017), while the anatomical accessibility and large conductance of $Ca_V1.2$ has enabled the bulk of the real-time biophysical studies on selectivity and permeation. Incorporating native skeletal and cardiac Ca2+ channels into planar lipid bilayers puts $Ca_V1.1$ and $Ca_V1.2$ into the same recording system (figures 2.2(D)–(F)). Electrophysiology can pick up interesting differences, observed as open channel unitary current–voltage (*i–V*) relationship with the same ion concentrations on both sides of the bilayer. Interestingly, the *i–V* relationships for both channel subtypes are both symmetric and linear even though their slopes are different (figure 2.2(F)). Such single-channel studies have capitalized on two experimental stratagems: first, the use of a dihydropyridine agonist, Bay K 8644, to modify channel gating and prolong the open state, enabling a clearer view of ion-open pore interactions; second, the use of Ba^{2+} as a divalent charge carrier because it often shows double the open channel flux rate seen with equimolar Ca^{2+} (Hagiwara and Byerly 1981, Hess *et al* 1986).

2.1.1 Conclusion #1. The Ca^{2+} channel pore has an intrinsically large diameter

$Ca_V1.1$ calcium channels display a large minimal pore diameter in the absence of Ca^{2+}; tetramethylammonium and other quaternary ammonium compounds as large as 6 Å across can permeate and carry significant current (Almers *et al* 1984). This weighs strongly against selectivity by a snug and rigid fit of dehydrated Ca^{2+} within a fixed hole, the scenario favored for dehydrated K^+ ions in potassium channels (Doyle *et al* 1998) and monohydrated Na^+ ions in sodium channels (Hille 1971). Working with $Ca_V1.2$, Cataldi, Perez-Reyes and I confirmed the $Ca_V1.1$-based conclusions of Almers *et al* (1984) and went on to probe the newly cloned representatives of T-type channels, Ca_V3 (Cataldi *et al* 2002). This was of interest because the acidic residues of the selectivity filter of $Ca_V3.1$ are Glu Glu Asp Asp (EEDD) rather than (EEEE). If pore size were based solely on side chain length, the pore size of Ca_V3 channels should be larger than $Ca_V1.1$, all other factors being equal. Surprisingly, we found the opposite based on organic cation permeation (Cataldi *et al* 2002) (figures 2.3(A) and (B)). The key finding was obtained with trimethylammonium (TriMA), which fails to pass through Ca_V3 channels yet permeates $Ca_V1.2$ channels. Contrary to expectation, Ca_V3 channels' inferred pore size was *smaller* than that of $Ca_V1.1$ (figure 2.3(A)). These inferences were recently confirmed by the new 3D structures for $Ca_V1.1$ (Wu *et al* 2016) (figures 2.2(C) and (D)) and $Ca_V3.1$ (Zhao *et al* 2019) (figures 2.2(E) and (F)). The pore diameters estimated by the HOLE program (Smart *et al* 1996) corroborated the predicted differential between $Ca_V1.1$ and $Ca_V3.1$ (figure 2.3(D)). The agreement in relative pore sizes is particularly reassuring because the nature of the experimental evidence is so different. Note that absolute pore diameters predicted from 3D structure falls short of those predicted if organic cations were hard spheres. This may be expected because the 3D structures were obtained with Ca^{2+} present, whereas organic cation permeation was studied with Ca^{2+} absent, perhaps allowing carboxylate side chains to relax to a 'wheels-up' position.

2.1.2 Conclusion #2. At sub-millimolar external Ca^{2+}, the pore's selectivity filter (EEEE locus) is continually occupied by a single Ca^{2+} ion

This conclusion fits with the prevailing selectivity mechanism at the turn of the century: rejection of Na^+ by ongoing occupation of the single-file pore by a single Ca^{2+} ion, blocking the permeation pathway. Quantitative analysis had suggested 1:1 Ca^{2+} binding to an intrapore binding site with K_D of ~1 μM. This scenario had superseded previous proposals of a single low-affinity (~10 mM) binding site within the pore (figure 2.1(A)), based on the $K_{1/2}$ of the saturating relationship between external Ca^{2+} concentration and channel flux (Hagiwara and Byerly 1981, Kostyuk *et al* 1983). While there is no dispute about block of Na^+ flux through the channel at micromolar Ca^{2+}, the idea of an intrapore site of low affinity is still under consideration (Shuba 2014). However, this would be inconsistent with the presence of an intrapore Ca^{2+} ion density even at 0.5 mM external Ca^{2+} (Wu *et al* 2015) (figure 2.3).

2.1.3 Conclusion #3. The selectivity filter is supported by 4 carboxyl side chains, provided by glutamate residues from each of the 4 repeats, protruding into the aqueous pathway (EEEE locus); no additional high-affinity Ca^{2+} binding site are uncovered by looking beyond it or splitting it up

Structure–function analyses indicated that amino side chains protrude into the pore lumen and contribute carboxyl groups, likely carboxylates at physiological pH, to make up the high affinity site. For example, various studies mutated elements of the EEEE locus in L-type calcium channels (Mikala *et al* 1993, Tang *et al* 1993, Yang *et al* 1993, Yatani *et al* 1994). Neutralizing any one of the glutamates to a glutamine

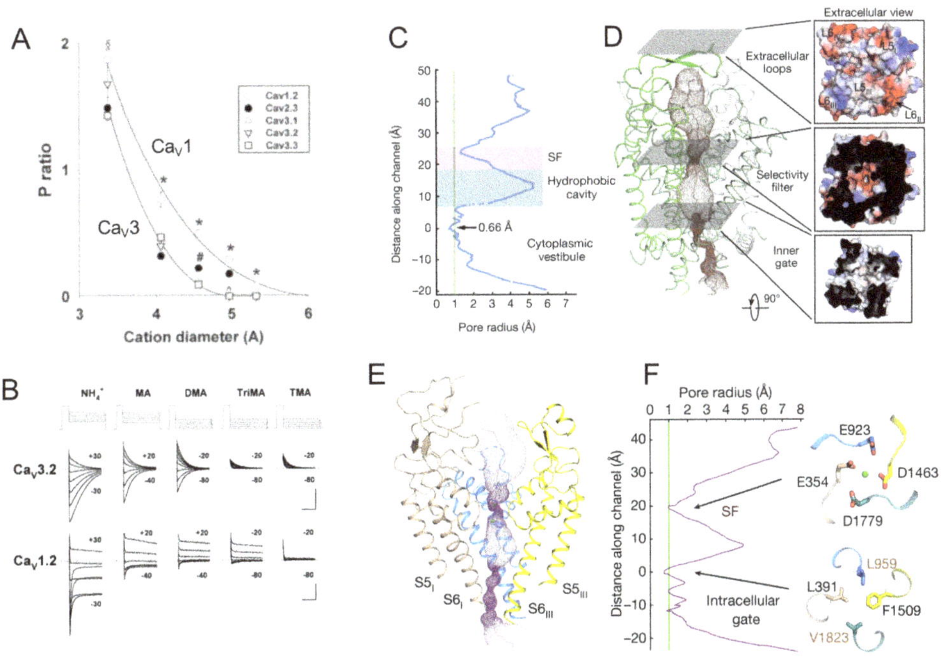

Figure 2.3. Conduction pathway and pore diameter of Ca_V1 (L-type) and Ca_V3 (T-type) Ca^{2+} channels. Minimal pore diameter of various Ca^{2+} channels, studied from different perspectives. (A) Dependence of the permeability ratio on cation diameter. Permeability ratios of ammonium (NH_4), methylammonium (MA), dimethylammonium (DMA), trimethylammonium (TriMA), and tetramethylammonium (TMA) versus Li^+ in $Ca_V1.2$ (○), $Ca_V2.3$ (●), $Ca_V3.1$ (Δ), $Ca_V3.2$ (∇), and $Ca_V3.3$ (■) plotted vertically against mean cation diameter, derived from corresponding cation volumes (Sun *et al* 1997). Estimated pore diameter ~1 Å smaller for various Ca_V3 (T-type) channels than $Ca_V1.2$ (L-type) (reprinted from Cataldi *et al* 2002, copyright (2002), with permission from Elsevier). (B) Source data of tail currents carried by NH_4 and its methyl-substituted derivatives in oocytes expressing $Ca_V3.2$ IVA channels (*upper row*) and $Ca_V1.2$ HVA channels (*lower row*). Tail currents were elicited by depolarizing the membrane potential from a holding potential of −80 mV to +70 mV for 5 ms to maximally open the calcium channels and then stepping down to potential levels ranging from −30 to +30 mV in 10 mV increments. Key difference between Ca_V3 and Ca_V1 channels detected with TriMA permeation, fourth column (reprinted from Cataldi *et al* 2002, copyright (2002), with permission from Elsevier). (C) and (D) The ion permeation path of $Ca_V1.1$ Ca^{2+} channels (skeletal L-type), (from Wu *et al* 2016, copyright 2016, Macmillan Publishers Limited, part of Springer Nature, all rights reserved, with permission of Springer). The pore radii along the length of the pore are tabulated in (C), corresponding to the ion conducting passage calculated by HOLE (Smart *et al* 1996), depicted by brown dots in the middle panel. The cut-open extracellular views of the electrostatic potentials calculated in PyMol are shown for the indicated layers. (E and F) The permeation path of $Ca_V3.1$ Ca^{2+} channels (T-type) (from Zhao *et al* (2019, copyright 2019, The Author, under exclusive licence to Springer Nature Limited, with permission of Springer). The ion-conducting passage is demarcated by purple dots (E), and the calculated pore radii are shown (F). Two constriction sites, the entrance to the selectivity filter, enclosed by the EEDD motif and the intracellular gate, are shown in extracellular views (F, right). The EEDD locus defines the narrowest aperture, one significantly narrower than for $Ca_V1.1$ Ca^{2+} channels (C and D).

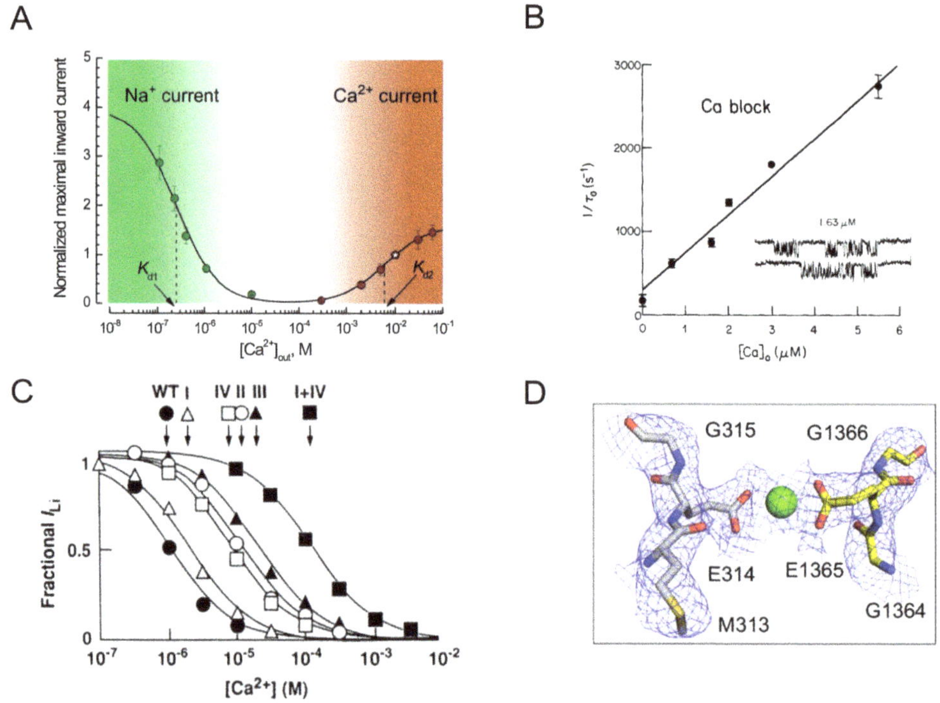

Figure 2.4. Pore interaction with single Ca^{2+} ions: electrophysiological monitoring and 3D structural basis. (A) Ca^{2+} as both a regulator and charge carrier in voltage-gated Ca^{2+} channels of snail neurons, from Shuba (2014, copyright 2014, Springer-Verlag Berlin Heidelberg, with permission of Springer). Data in orange area depicts typical dependence of maximal inward whole-cell current through VGCCs on extracellular Ca^{2+} concentration ($[Ca^{2+}]_{out}$), eventually saturating with $K_{d2} = 6$ mM. Measured current amplitudes at each $[Ca^{2+}]_{out}$ were normalized to the amplitude at $[Ca^{2+}]_{out} = 10$ mM (white open symbol). Data in green area shows Ca^{2+} block of Na^+ current with a half-blocking concentration $K_{d1} = 250$ nM. (B) Block by Ca^{2+} of unitary Li current through an L-type Ca channel, evoked by stepping the patch membrane potential was from −80 to −40 mV (inset). Unitary currents carried by Li influx with 150 mM LiCl and 12.5 μM TTX in the patch pipette. The Ca concentration in the patch pipette (abscissa) was buffered with HEDTA. From Lansman *et al* (1986), originally published in *The Journal of General Phsyiology*. (C) Dose–response relationships for Ca^{2+} block of inward Li+ currents. From Yang *et al* (1993), copyright 1993, Nature Publishing Group, with permission of Springer). (D) Electron microscopy densities for the bound Ca^{2+} ion and surrounding residues in the selectivity filter of the $Ca_V2.2$ (N-type) pore. Structure obtained in the presence of 2 mM $CaCl_2$ (Gao *et al* 2021, copyright 2021, The Authors, under exclusive licence to Springer Nature Limited, with permission of Springer).

shifted the block of monovalent current to higher Ca^{2+} concentrations, thereby demonstrating the role of these carboxylate groups in high affinity Ca^{2+} binding (figure 2.4(C)). The extent of the shift differed for glutamates in the different repeats, but was always significant; neutralizing multiple glutamate residues shifted the blocking concentration by as much as 1000-fold (Ellinor *et al* 1995). The inference was that the four glutamate side chains help form a binding site for a single Ca^{2+} ion. Only at low, submicromolar concentrations, when Ca^{2+} is absent from the site, can Na^+ permeate.

The protrusion of amino acid side chains has now been validated by the 3D structures (figure 2.4(D)): side chain carboxyl groups of the four glutamates do indeed contribute to coordination of the bound Ca^{2+}, depicted as a single green sphere in structures obtained in 0.5 Ca^{2+} (Wu *et al* 2015) or 2 mM Ca^{2+} (Gao *et al* 2021).

At the same time, the permeation path lacked any additional divalent cation binding site distant from the EEEE locus. This was concluded in two complementary ways, by systematic replacement of glutamate side chains at the EEEE locus (Ellinor *et al* 1995), and by accessing the selectivity filter with Ca^{2+} ions from the cytoplasmic side (Kuo and Hess 1993). In both cases, a second, high-affinity Ca^{2+} binding site was not found.

2.1.4 Conclusion #4. It is the EEEE locus itself that becomes occupied by multiple divalent ions at higher divalent concentrations

The capability of the Ca^{2+} channel pore to accommodate multiple divalent cations was a cornerstone of the 1984 papers. The first experimental argument was the anomalous mole fraction effect (AMFE), visible in whole-cell currents (Almers and McCleskey 1984, Hess and Tsien 1984), and importantly, in unitary currents (Friel and Tsien 1989, Nonner *et al* 1998, Rodriguez-Contreras *et al* 2002), especially at divalent concentrations near the $K_{1/2}$. In a kinetic approach, the departure of blocking divalent ions was hastened by increasing the concentration of permeant divalent ions, indicative of intrapore ion–ion interactions (Lansman *et al* 1986) (figure 2.6(A)). This kind of competitive interaction was modified by amino acid substitutions at the EEEE selectivity filter (figure 2.6(B)), consistent with this locus operating as the site of ion–ion competition (Ellinor *et al* 1995). This reasoning was extended by Cloues *et al* (2000) who showed the EEEE-based effects on blocker on-rate or off-rate depend on valence of the permeant ion. These conclusions from electrophysiological analysis are now vindicated by cryo-EM structures obtained in the presence of 10 mM $[Ca]_o$ (Wu *et al* 2016) (figure 2.6(C)). They show densities in the permeation path that are interpreted as two Ca^{2+} ions, one right at the EEEE locus and another slightly further along, near the carbonyls at the −1 position. This provides a satisfying structural vindication of the earlier but less direct inferences from electrophysiology. The closeness of the two bound Ca^{2+} ions confirms the idea of a single pore locus. The EEEE locus thus provides functional flexibility, strongly stabilizing a single Ca^{2+} ion at submicromolar $[Ca^{2+}]_o$, while also coordinating two Ca^{2+} ions in tandem when $[Ca^{2+}]_o$ is elevated to levels at which Ca^{2+} influx would occur. Thus, there is neither need for nor structural indication of another high-affinity Ca^{2+} binding site closer to the cytoplasm. A similar conclusion was reached for the $Ca_V3.1$ structure (figure 2.6(D)), supplemented by a careful discussion of the attribution of intrapore density near the EEDD to two Ca^{2+} ions (Zhao *et al* 2019). It is reasonable to think that the Ca^{2+}-in-carbonyl ring complex is less energetically favorable than the singly bound Ca^{2+}, in tune with the notion of energetic stairsteps (figure 2.5(E3)) (Dang and McCleskey 1998, Sather and McCleskey 2003). The axial

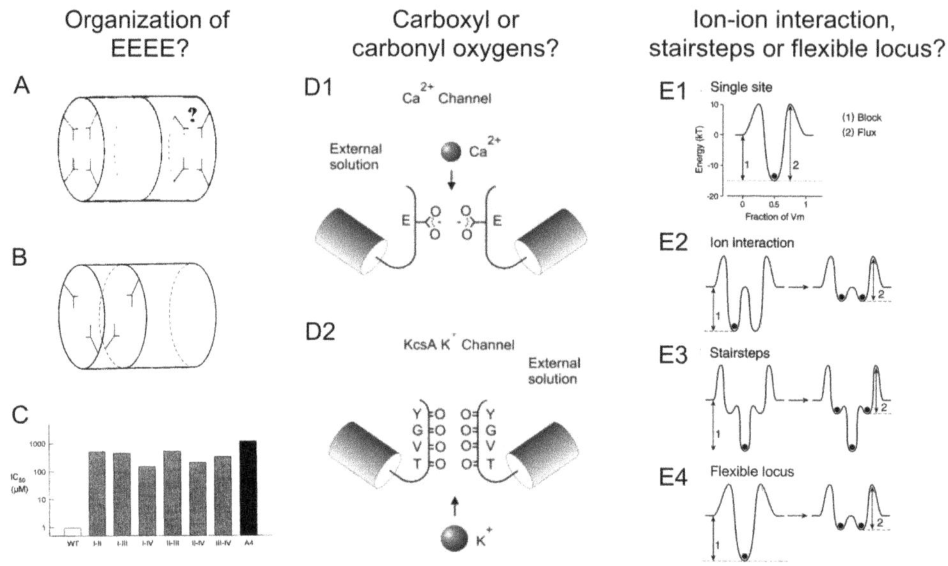

Figure 2.5. Issues about intrapore interactions seeking insight from 3D structure. (A) Hypothesis that another high-affinity Ca^{2+} binding locus co-exists with the EEEE locus as summarized by Ellinor *et al* (1995). (B) Hypothesis that the four P-region glutamates form two distinct high-affinity Ca^{2+} binding sites as considered by Ellinor *et al* (1995). (C) Quadruple-alanine and all double-alanine substitutions strongly attenuate the potency of Ca^{2+} block of monovalent (Li^+) current. IC50 values for 6 double-alanine mutants (shaded bars, roman numerals indicate the pair of repeats bearing substitutions) and quadruple-alanine mutant (A4, solid bar). Reprinted from Ellinor *et al* 1995, copyright (1995), with permission from Elsevier. (D) Comparison of models of pore structure for Ca and K channels. Reproduced with permission from Sather and McCleskey (2003), copyright ESO. Cylinders indicate predicted pore helices; for simplicity, only two of four pore-lining regions are shown. D1, in Ca channels, carboxylate oxygen atoms from glutamate side chains of the EEEE locus project into the pore to form a high electronegativity region. D2, in K^+ channels, carbonyl oxygen atoms from the polypeptide backbone project into the pore to form a series of high-affinity K^+-binding sites. (E) Rate theory descriptions of pores that bind Ca^{2+} tightly. E1–E3 reproduced with permission from from Sather and McCleskey (2003), copyright ESO. (E1) A single-site pore alone predicts low flux because of the high energy barrier for exit. Horizontal axis: fraction of the electric field through the pore. Vertical axis: the chemical potential energy for Ca^{2+} as it traverses the voltage drop. Arrow 1 corresponds to binding energy for the micromolar site, while arrow 2 indicates the highest energy barrier that limits the rate of Ca^{2+} exit from the pore. Various hypothetical mechanisms for rapid permeation (E2–E4). (E2) In multi-ion pores, ion–ion interactions (either electrostatic or chemical) are hypothesized to lower the exit energy barrier. (E3) Stairsteps generated by low-affinity binding sites flanking the selectivity site. Modified from Dang and McCleskey (1998). (E4) Additional hypothesis of a flexible locus wherein a single high-affinity Ca^{2+} binding site morphs into a locus capable of coordinating two Ca^{2+} ions at once.

position of the singly bound Ca^{2+} ion is intermediate between the positions of the pair of Ca^{2+} ions, consistent with a flexible locus (figure 2.5(E4)).

Thus, structure affirms that Ca^{2+} channel selectivity is not simply reliant on coordination by carbonyl oxygens by strict extension of the pore structure of potassium channels (Doyle *et al* 1998) (figure 2.5(D2)), as was speculated when the K^+ channel structure first appeared. On the other hand, carbonyl groups from the neighboring amino acids participate in coordination of multiple ions. This was not

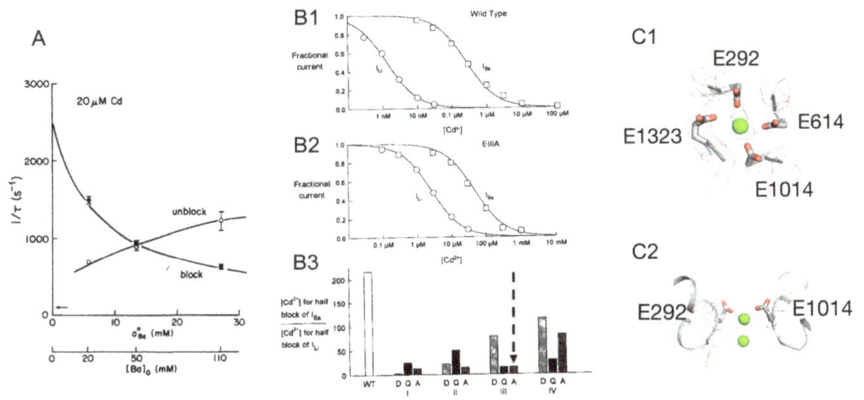

Figure 2.6. Multi-ion interactions in the permeation pathway, localized to EEEE locus. (A) Early evidence for multiple divalent cations within the pore: Cd^{2+} unblock rate speeds up as $[Ba]_o$ is elevated along with decrease in block rate (copyright Lansman *et al* 1986, originally published in *The Journal of General Physiology*). (B) Further localization of the interaction among permeating and pore-blocking ions to the sphere of influence of EEEE mutations. (B1) Concentration-dependence of Cd^{2+} block of Ba^{2+} and Li^+ currents through cloned wild-type $Ca_V1.2$ channels studied in oocytes. (B2) Similar information shown for a EEEE glutamate point mutant, EIIIA. (B3) Comparison of the relative potencies in B1 and B2 indicates that the mutation has altered the interaction of permeant and blocking ions within the pore (downward arrow). Similar alterations in the Ba^{2+}–Cd^{2+} antagonism were found for aspartate (D), glutamine (Q) and alanine (A) substitutions at each position in the EEEE locus (Ellinor *et al* 1995, copyright (1995), with permission from Elsevier). (c) cryo-EM structure of the selectivity filter vestibule of $Ca_V1.1$. Extracellular view (C1) and side view (C2) of the EEEE locus. Green spheres depict EM densities assigned to two Ca^{2+} ions positioned along permeation pathway, the outer Ca^{2+} coordinated by carboxylate oxygen atoms (red spheres), the inner Ca^{2+} partially coordinated by backbone carbonyl oxygens. Data from Wu *et al* (2016), represented in Zhao *et al* (2019, copyright 2019, The Author, under exclusive licence to Springer Nature Limited, with permission of Springer). Doubly occupied EEEE locus can be compared to singly occupied locus in $Ca_V1.1$ structure obtained with 0.5 mM Ca^{2+} (Wu *et al* 2015) or in $Ca_V2.2$ structure generated with 2 mM Ca^{2+} (Gao *et al* 2021).

tested in our functional approach to structure and was discussed with appropriate caution: 'The participation of backbone carbonyl oxygens and/or water molecules remains an open possibility for the Ca^{2+} channel pore' (Ellinor *et al* 1995).

2.1.5 Conclusion #5. The rate of Ca^{2+} arrival at the EEEE locus approaches the diffusion limit

While 3D structures are by nature static and cannot measure reaction rates, they can provide a valuable framework to understand how such rates might be achieved. Using Li^+ as a charge carrier, we estimated the on-rate for Ca^{2+} block of the pore as 4.5×10^8 M^{-1} s^{-1}, approaching the diffusion rate, and fast enough not to be the main limitation on ion transfer (Lansman *et al* 1986) (figure 2.4(B)). How is this rapid on-rate achieved? For years, physiologists have envisioned the outer mouth of the Ca^{2+} channel as a dish antenna, a large negatively charged funnel, to attain this diffusion-limited rate (e.g. Nonner *et al* 1998). Now the 3D structure teaches us that the approach to the selectivity filter displays abundant negative charged residues but is not shaped like a funnel. The permeating ion must navigate passage through a dome made up of extracellular domains from all four repeats. In this case, information from single-channel recordings and cryo-EM structures needs to be integrated and specific hypotheses tested. One could propose that negative charges surrounding the dome passageway could begin the process of Ca^{2+} dehydration and thereby lower the energy barrier for access to the EEEE locus. Alternatively, the same negatively charged region could serve as a shallow external energy well, stabilizing the binding of external monovalents to prevent Ca^{2+} ions from moving backwards (Kuo and Hess 1993, Dang and McCleskey 1998).

2.1.6 Conclusion #6. Ca_V1 channels exemplify principles found in other Ca^{2+}-selective pores

Principles derived from studies of voltage-gated Ca^{2+} channels may be partly or largely generalizable to other Ca^{2+}-selective pores (figure 2.7). A compelling example is the Ca^{2+}-release activated (CRAC) channel, an important conduit for Ca^{2+}-entry in non-excitable cells (Lewis 2007). The pore-forming subunit of CRAC channels, known as orai or orai1, is highly selective for Ca^{2+} but in the absence of Ca^{2+} allows permeation of Na^+. CRAC channels achieve exquisite Ca^{2+} selectivity by high-affinity binding of Ca^{2+} to a ring of six glutamates, one from each of the six subunits, Glu178 in *Drosophila* Orai or Glu106 in Orai1 (Hou *et al* 2012). Mutation of the glutamic acid to aspartic acid dramatically increases channel permeation by monovalent cations (Prakriya *et al* 2006, Vig *et al* 2006, Yeromin *et al* 2006,

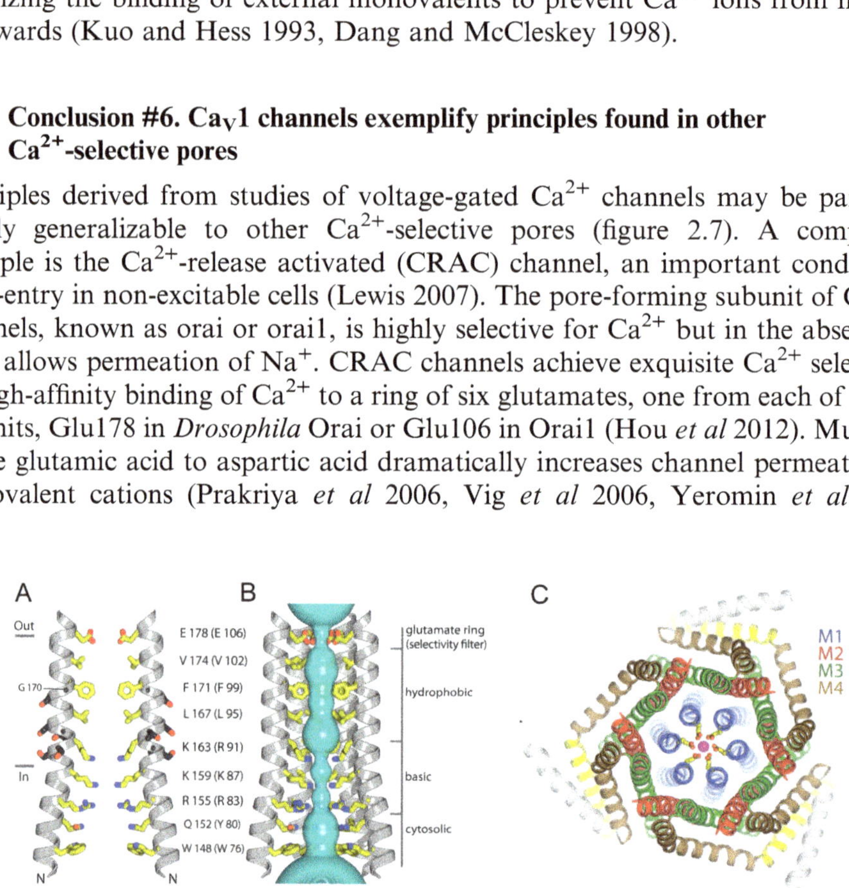

Figure 2.7. Extension of general principles to other Ca^{2+}-selective ion channels, exemplified by Orai and Orai1. (A) and (B) Architecture of pore-forming helices of Orai channel of *Drosophila*, determined by x-ray crystallography to show 6-fold symmetry. (A) Two M1 helices drawn (four omitted for clarity), showing amino acid side chains lining the pore (yellow). The permeation pathway extends well beyond the membrane-spanning region (horizontal lines). Labeled are key amino acids, numbered for fly Orai and closely corresponding residues in human Orai1 (parentheses). (B) View of the permeation pathway, showing four of six helices and minimal radial distance from the central axis to the nearest van der Waals protein contact (teal). Note constriction at the glutamate ring (selectivity filter). (C) Orthogonal view of the channel from the extracellular side showing a bound Ca^{2+} ion (magenta sphere) and the nearby Glu178 side chains (yellow sticks) with carboxylate oxygens (red balls) along with six M1 helices (blue). From Hou *et al* (2012), with permission from AAAS.

McNally *et al* 2009). The same externally exposed EEEEEE locus is responsible for high-affinity blockage of Orai by the trivalent lanthanide Gd^{3+}. Hou *et al* obtained crystals of Orai and determined x-ray structures with Ca^{2+}, Ba^{2+} or Gd^{3+} bound (figure 2.7) (Hou *et al* 2012). They did not observe a second bound Ca^{2+} ion, even in crystals soaked in 50 mM $CaCl_2$, but speculate that a second Ca^{2+} binding site, lying a bit further along the transmembrane voltage gradient, just below the glutamate ring, might account for the observed voltage-dependent Ca^{2+} block (Bakowski and Parekh 2002, Prakriya and Lewis 2006). Thus, Orai pores behave remarkably like voltage-gated Ca^{2+} channels despite having six-fold rather than four-fold architecture. Similar biophysical experiments were performed as for voltage-gated Ca_V channels and were comparably accurate in anticipating the illuminating 3D structures that followed.

2.1.7 Conclusion #7. Ca_V channels employ a selectivity mechanism distinct from K^+ and Na^+ channels, yet illuminating channel evolution

While K^+, Ca^{2+} and Na^+ channels all undergo opening upon depolarization, each channel type needs its own distinctive selectivity to perform its job. Sharply different selectivity reflects variations on the same basic architecture in the P region. Thus, understanding the various ion selectivity mechanisms calls for more comparisons between functional properties and 3D structures and investigating how the channels evolved to their present state.

The prevailing wisdom about K^+, Ca^{2+} and Na^+ channels' evolutionary relationship envisions Ca^{2+} channels as a four-fold concatenation of an ancestor of modern single-repeat K^+ channels (Hille 2007). Gene duplication and gene fusion would participate. After mutation to or insertion of a single glutamic acid in the P loop, the EEEE locus would emerge because of two rounds of two-fold duplication. In turn, Na^+ channels are presumed to arise later (<7 M years ago) by modification of a 4-repeat ancestor of modern Ca^{2+} channels. One key alteration would be transformation of the EEEE locus of Ca_V1 channels to the DEKA motif of Na^+ channels. In this evolutionary framework, 4-fold symmetrical rings of carbonyls arising from successive duplications of the peptide backbone could provide supportive channel-ion interactions for all three cations. Carbonyl rings serve as the dominant motif of tetrameric K^+ channels (Doyle *et al* 1998), and provide an contributory component of Ca^{2+} pores as well (figure 2.6(C2)) (Wu *et al* 2015, 2016). Less attention has been paid to this motif for pseudo-tetrameric Na^+ channels, but its existence is implied by the ability to convert Na^+ channels into quasi-Ca^{2+} channels by the reverse transformation of DEKA→EEEE (Heinemann *et al* 1992).

2.1.8 Conclusion #8. Studies of Ca^{2+} channel selectivity and permeation exemplify advantages of studying biological function with multiple approaches

Early electrophysiological studies that distinguished calcium channel types and proposing mechanisms for selectivity and permeation (Almers and McCleskey 1984, Hess and Tsien 1984, Nowycky *et al* 1985) paved the way for discovery of selective pharmacological blockers (McCleskey *et al* 1987, Mintz *et al* 1992, Newcomb *et al* 1998) and aided molecular cloning of the major neuronal subtypes (Mikami *et al* 1989, Dubel *et al* 1992, Ellinor *et al* 1993, Soong *et al* 1993). This in turn enabled detailed structure–function studies that highlighted the molecular basis for subtype identity, selectivity, permeation, pharmacological blockade and modulation. Now that all three classes of voltage-gated calcium channels have yielded exemplar 3D structures by cryo-EM, a remarkable tour-de-force by one group (Wu *et al* 2016, Zhao *et al* 2019, Gao *et al* 2021), what is left to do? Beyond taking stock of progress, a rational response is to integrate the information obtained by multiple approaches, stitching these together with help of molecular dynamic modeling and further experiments. Understanding single-channel flux—the functional output of the channel—must remain a primary objective. How is it that $Ca_V1.1$ has only about half the single-channel conductance of $Ca_V1.2$ under identical ionic conditions (figure 2.2(F) (Rosenberg *et al* 1986)? Domain swapping provides a partial answer: determinants in the IS5-IS6 linker (Dirksen *et al* 1997)—now one may find out how the molecular differences influence ion motion. Likewise, single-channel conductances of $Ca_V2.1$ and $Ca_V1.2$ differ to a similar degree, leading to the prescient finding of the importance of neighboring residues just N-terminal to the pore glutamates (Williamson and Sather 1999). As providers of carbonyl oxygens, how do these '−1' amino acids wind up impacting pore geometry and divalent cation dynamics? An outstanding challenge for a future generation of ion channel biologists is to use molecular dynamics to reconstruct the current versus voltage relationship of single open channels, knowing some critical determinants. One way that structural studies can contribute further is by determining the 3D structure of the experimental platform for much early work on selectivity and permeation (Sather and McCleskey 2003)—$Ca_V1.2$ channels with Bay K 8644 bound (Hess *et al* 1986, Lansman *et al* 1986)—for comparison with other family members. Combining the exquisite yet static detail of 3D structure with the rich but ambiguous dynamics of single-channel recordings will test our mettle as channel biologists.

Acknowledgments

This work was supported by funding from NIH (Heart, Lung and Blood Institute; Neurological Disorders and Stroke Institute) and the Mathers Foundation, the Burnett Family Foundation and George and Camilla Smith.

RWT wishes to thank each of the excellent colleagues who contributed to our studies of calcium channel selectivity, permeation and block and to whom I owe so much. In alphabetical order, these co-workers include: Ilya Bezprozvanny, Mario Cataldi, Xiao-Hua Chen, Patrick Ellinor, David Friel, Peter Hess, Jeffry Lansman, Esther Lee, Kai Lee, Edwin McCleskey, Edward Perez-Reyes, Robert Rosenberg, William Sather, Jian Yang, Ji-Fang Zhang.

References

Almers W and McCleskey E W 1984 Non-selective conductance in calcium channels of frog muscle: calcium selectivity in a single-file pore *J. Physiol.* **353** 585–608

Almers W, McCleskey E W and Palade P T 1984 A non-selective cation conductance in frog muscle membrane blocked by micromolar external calcium ions *J. Physiol.* **353** 565–83

Armstrong C M and Neytonc J 1991 Ion permeation through calcium channels: a one-site model *Ann. N.Y. Acad. Sci.* **635** 18–25

Bakowski D and Parekh A B 2002 Monovalent cation permeability and Ca^{2+} block of the store-operated Ca^{2+} current I(CRAC) in rat basophilic leukemia cells *Pflugers Arch.* **443** 892–902

Cataldi M, Perez-Reyes E and Tsien R W 2002 Differences in apparent pore sizes of low and high voltage-activated Ca^{2+} channels *J. Biol. Chem.* **277** 45969–76

Catterall W A 2000 Structure and regulation of voltage-gated Ca^{2+} channels *Annu. Rev. Cell Dev. Biol.* **16** 521–55

Chen X-H and Tsien R W 1997 Aspartate substitutions establish the concerted action of P-region glutamates in repeats I and III in forming the protonation site of L-type Ca^{2+} channels *J. Biol. Chem.* **272** 30002–8

Chen X H, Bezprozvanny I and Tsien R W 1996 Molecular basis of proton block of L-type Ca^{2+} channels *J. Gen. Physiol.* **108** 363–74

Cloues R K, Cibulsky S M and Sather W A 2000 Ion interactions in the high-affinity binding locus of a voltage-gated Ca^{2+} channel *J. Gen. Physiol.* **116** 569–86

Dang T X and McCleskey E W 1998 Ion channel selectivity through stepwise changes in binding affinity *J. Gen. Physiol.* **111** 185–93

Dirksen R T, Nakai J, Gonzalez A, Imoto K and Beam K G 1997 The S5-S6 linker of repeat I is a critical determinant of L-type Ca^{2+} channel conductance *Biophys. J.* **73** 1402–9

Doyle D A, Morais Cabral J, Pfuetzner R A, Kuo A, Gulbis J M, Cohen S L, Chait B T and MacKinnon R 1998 The structure of the potassium channel: molecular basis of K^+ conduction and selectivity *Science* **280** 69–77

Dubel S J, Starr T V, Hell J, Ahlijanian M K, Enyeart J J, Catterall W A and Snutch T P 1992 Molecular cloning of the alpha-1 subunit of an omega-conotoxin-sensitive calcium channel *Proc. Natl Acad. Sci. USA* **89** 5058–62

Ellinor P T, Yang J, Sather W A, Zhang J-F and Tsien R W 1995 Ca^{2+} channel selectivity at a single locus for high-affinity Ca^{2+} interactions *Neuron* **15** 1121–32

Ellinor P T, Zhang J F, Randall A D, Zhou M, Schwarz T L, Tsien R W and Horne W A 1993 Functional expression of a rapidly inactivating neuronal calcium channel *Nature* **363** 455–8

Friel D and Tsien R 1989 Voltage-gated calcium channels: direct observation of the anomalous mole fraction effect at the single-channel level *Proc. Natl Acad. Sci.* **86** 5207–11

Gao S and Yan N 2021 Structural basis of the modulation of the voltage-gated calcium ion channel Ca_V1. 1 by Dihydropyridine compounds *Angew. Chem.* **133** 3168–74

Gao S, Yao X and Yan N 2021 Structure of human $Ca_V2.2$ channel blocked by the painkiller ziconotide *Nature* **596** 143–7

Hagiwara S and Byerly L 1981 Calcium channel *Annu. Rev. Neurosci.* **4** 69–125

Hamill O P, Marty A, Neher E, Sakmann B and Sigworth F J 1981 Improved patch-clamp techniques for high-resolution current recording from cells and cell-free membrane patches *Pflugers Arch.* **391** 85–100

Heinemann S H, Terlau H, Stühmer W, Imoto K and Numa S 1992 Calcium channel characteristics conferred on the sodium channel by single mutations *Nature* **356** 441–3

Hess P, Lansman J B and Tsien R W 1986 Calcium channel selectivity for divalent and monovalent cations. voltage and concentration dependence of single channel current in ventricular heart cells *J. Gen. Physiol.* **88** 293–319

Hess P and Tsien R W 1984 Mechanism of ion permeation through calcium channels. *Nature* **309** 453–6

Hille B 1971 The permeability of the sodium channel to organic cations in myelinated nerve *J. Gen. Physiol.* **58** 599–619

Hille B 1972 The permeability of the sodium channel to metal cations in myelinated nerve *J. Gen. Physiol.* **59** 637–58

Hille B 2007 *Ion Channels of Excitable Membranes* (Sunderland, MA: Sinauer)

Hou X, Pedi L, Diver M M and Long S B 2012 Crystal structure of the calcium release–activated calcium channel Orai *Science* **338** 1308–13

Johnson E A and Lieberman M 1971 Heart: excitation and contraction *Annu. Rev. Physiol.* **33** 479–532

Kass R S, Siegelbaum S A and Tsien R W 1979 Three-micro-electrode voltage clamp experiments in calf cardiac Purkinje fibres: is slow inward current adequately measured? *J. Physiol.* **290** 201–25

Kostyuk P, Mironov S and Shuba Y M 1983 Two ion-selecting filters in the calcium channel of the somatic membrane of mollusc neurons *J. Membr. Bio.* **76** 83–93

Kuo C-C and Hess P 1993 Characterization of the high-affinity Ca^{2+} binding sites in the L-type Ca^{2+} channel pore in rat phaeochromocytoma cells *J. Physiol.* **466** 657–82

Lansman J B, Hess P and Tsien R W 1986 Blockade of current through single calcium channels by Cd^{2+}, Mg^{2+}, and Ca^{2+} voltage and concentration dependence of calcium entry into the pore *J. Gen. Physiol.* **88** 321–47

Lee K S, Akaike N and Brown A M 1980 The suction pipette method for internal perfusion and voltage clamp of small excitable cells *J. Neurosci. Methods* **2** 51–78

Lewis R S 2007 The molecular choreography of a store-operated calcium channel *Nature* **446** 284–7

Mack S, Kandel E R, Koester J D, Siegelbaum S A and McGraw H 2021 *Principles of Neural Science* (New York: McGraw Hill)

McCleskey E W, Fox A P, Feldman D H, Cruz L J, Olivera B M, Tsien R W and Yoshikami D 1987 Omega-conotoxin: direct and persistent blockade of specific types of calcium channels in neurons but not muscle *Proc. Natl Acad. Sci. USA* **84** 4327–31

McNally B A, Yamashita M, Engh A and Prakriya M 2009 Structural determinants of ion permeation in CRAC channels *Proc. Natl Acad. Sci. USA* **106** 22516–21

Mikala G, Bahinski A, Yatani A, Tang S and Schwartz A 1993 Differential contribution by conserved glutamate residues to an ion-selectivity site in the L-type Ca^{2+} channel pore *FEBS Lett.* **335** 265–69

Mikami A, Imoto K, Tanabe T, Niidome T, Mori Y, Takeshima H, Narumiya S and Numa S 1989 Primary structure and functional expression of the cardiac dihydropyridine-sensitive calcium channel *Nature* **340** 230–3

Mintz I M, Venema V J, Swiderek K M, Lee T D, Bean B P and Adams M E 1992 P-type calcium channels blocked by the spider toxin omega-aga-IVA *Nature* **355** 827–9

Newcomb R *et al* 1998 Selective peptide antagonist of the class E calcium channel from the venom of the tarantula hysterocrates gigas *Biochemistry* **37** 15353–62

Noble D 1979 *The Initiation of the Heartbeat* (Oxford; New York: Clarendon Press; Oxford University Press)

Nonner W, Chen D P and Eisenberg B 1998 Anomalous mole fraction effect, electrostatics, and binding in ionic channels *Biophys. J.* **74** 2327–34

Nowycky M C, Fox A P and Tsien R W 1985 Three types of neuronal calcium channel with different calcium agonist sensitivity *Nature* **316** 440–3

Prakriya M, Feske S, Gwack Y, Srikanth S, Rao A and Hogan P G 2006 Orai1 is an essential pore subunit of the CRAC channel *Nature* **443** 230–3

Prakriya M and Lewis R S 2006 Regulation of CRAC channel activity by recruitment of silent channels to a high open-probability gating mode *J. Gen. Physiol.* **128** 373–86

Reuter H 1973 Divalent cations as charge carriers in excitable membranes *Prog. Biophys. Mol. Biol.* **26** 1–43

Ringer S 1883 A further contribution regarding the influence of the different constituents of the blood on the contraction of the heart *J. Physiol.* **4** 29–42

Rodriguez-Contreras A, Nonner W and Yamoah E N 2002 Ca2+ transport properties and determinants of anomalous mole fraction effects of single voltage-gated Ca^{2+} channels in hair cells from bullfrog saccule *J. Physiol.* **538** 729–45

Rosenberg R L, Hess P, Reeves J P, Smilowitz H and Tsien R W 1986 Calcium channels in planar lipid bilayers: insights into mechanisms of ion permeation and gating *Science* **231** 1564–6

Sather W A and McCleskey E W 2003 Permeation and selectivity in calcium channels *Annu. Rev. Physiol.* **65** 133–59

Shuba Y M 2014 Models of calcium permeation through T-type channels *Pflügers Arch.-Eur. J. Physiol.* **466** 635–44

Smart O S, Neduvelil J G, Wang X, Wallace B A and Sansom M S 1996 HOLE: a program for the analysis of the pore dimensions of ion channel structural models. *J. Mol. Graph.* **14** 354–60

Soong T W, Stea A, Hodson C D, Dubel S J, Vincent S R and Snutch T P 1993 Structure and functional expression of a member of the low voltage-activated calcium channel family *Science* **260** 1133–6

Sun Y M, Favre I, Schild L and Moczydlowski E 1997 On the structural basis for size-selective permeation of organic cations through the voltage-gated sodium channel. effect of alanine mutations at the DEKA locus on selectivity, inhibition by Ca^{2+} and H^+, and molecular sieving *J. Gen. Physiol.* **110** 693–715

Tanabe T, Takeshima H, Mikami A, Flockerzi V, Takahashi H, Kangawa K, Kojima M, Matsuo H, Hirose T and Numa S 1987 Primary structure of the receptor for calcium channel blockers from skeletal muscle *Nature* **328** 313–8

Tang S, Mikala G, Bahinski A, Yatani A, Varadi G and Schwartz A 1993 Molecular localization of ion selectivity sites within the pore of a human L-type cardiac calcium channel *J. Biol. Chem.* **268** 13026–9

Tsien R W and Barrett C F 2011 A brief history of calcium channel discovery *Voltage-gated Calcium Channels* ed G W Zamponi (New York: Springer)

Tsien R W, Hess P, McCleskey E W and Rosenberg R L 1987 Calcium channels: mechanisms of selectivity permeation, and block *Annu. Rev. Biophys. Biophys. Chem.* **16** 265–90

Vig M *et al* 2006 CRACM1 multimers form the ion-selective pore of the CRAC channel *Curr. Biol.* **16** 2073–9

Williamson A V and Sather W A 1999 Nonglutamate pore residues in ion selection and conduction in voltage-gated Ca^{2+} channels *Biophys. J.* **77** 2575–89

Wu J, Yan N and Yan Z 2017 Structure–function relationship of the voltage-gated calcium channel $Ca_v1.1$ complex *Adv. Exp. Med. Biol.* **981** 23–39

Wu J, Yan Z, Li Z, Qian X, Lu S, Dong M, Zhou Q and Yan N 2016 Structure of the voltage-gated calcium channel $Ca_v1.1$ at 3.6 Å resolution *Nature* **537** 191–6

Wu J, Yan Z, Li Z, Yan C, Lu S, Dong M and Yan N 2015 Structure of the voltage-gated calcium channel $Ca_v1.1$ complex *Science* **350** aad2395

Yang J, Ellinor P T, Sather W A, Zhang J F and Tsien R W 1993 Molecular determinants of Ca^{2+} selectivity and ion permeation in L-type Ca^{2+} channels *Nature* **366** 158–61

Yatani A, Bahinski A, Mikala G, Yamamoto S and Schwartz A 1994 Single amino acid substitutions within the ion permeation pathway alter single-channel conductance of the human L-type cardiac Ca^{2+} channel *Circ. Res.* **75** 315–23

Yeromin A V, Zhang S L, Jiang W, Yu Y, Safrina O and Cahalan M D 2006 Molecular identification of the CRAC channel by altered ion selectivity in a mutant of Orai *Nature* **443** 226–9

Zhao Y, Huang G, Wu J, Wu Q, Gao S, Yan Z, Lei J and Yan N 2019 Molecular basis for ligand modulation of a mammalian voltage-gated Ca^{2+} channel *Cell* **177** 1495–506

Zhao Y, Huang G, Wu Q, Wu K, Li R, Lei J, Pan X and Yan N 2019 Cryo-EM structures of apo and antagonist-bound human $Ca_v3.1$ *Nature* **576** 492–7

IOP Publishing

Calcium Signals
From single molecules to physiology
Leslie S Satin, Manu Ben-Johny and Ivy E Dick

Chapter 3

Structure and function of TRP channels

Gilbert Q Martinez and Eric N Senning

Transient receptor potential (TRP) channels encompass a diverse class of non-selective cation channels that are primarily grouped by sequence homology to an ion channel discovered in the light transduction pathway of fly photoreceptors. Emphasis on gene sequence and protein domains rather than a functional classification scheme is necessary given the broad array of physiological roles for these channels as well as their modes of activation. To date, approximately 30 mammalian orthologues of the original fly TRP have been identified, and these have been further classified into six sub-groups: TRPC, TRPV, TRPA, TRPM, TRPP and TRPML. Functional characterization by electrophysiology and mutational or chimeric analysis have been the cornerstone of advancing our knowledge about TRP channels. A recent wave of TRP channel structures has confirmed many of the keenest observations from electrophysiology experiments and provided fertile starting ground for a new wave of functional analysis. This chapter aims to consolidate an understanding of TRP channel function with the growing number of TRP channel structures that are being solved at an incredible pace. In these structures, 4 TRP subunits assemble into a pore-forming ion channel, and each subunit is defined by six transmembrane segments separated into a pore domain and voltage sensor-like domain, common to the voltage-gated ion channel super-family. A high degree of structural similarity in the transmembrane region will be the backdrop to a discussion on general principles of gating, lipid interactions, Ca^{2+} dependent modulation and cell signaling in the TRP channel family.

3.1 What's in a name?

The story of TRP channels begins with a mutant *Drosophila melanogaster* phenotype that exhibited an aberrant electrophysiological response in retinal cells. Cosens and Manning discovered a phenotype that becomes temporarily blind after prolonged exposure to bright light, and after further investigation of the cumulative electrical responses from the eye, they deduced that the mutation impaired normal

doi:10.1088/978-0-7503-2009-2ch3

photoreceptor function (Cosens and Manning 1969). Technically challenging experiments were done by investigators from various groups to pinpoint the exact location of this mutation within the phototransduction cascade. A prevailing view at the time was that the photopigment regeneration time was impaired, but Baruch Minke was able to show that in this regard there was no difference between wild-type and mutants (Minke 1982). An agreed upon name for the Cosens and Manning mutation was then put forward as 'Transient Receptor Potential' to address the characteristic transient response to extended periods of bright light.

Following the discovery that the *trp* mutation could be traced to a gene product with numerous transmembrane domains that is expressed in the rhabdomeres, which is the site of phototransduction processes in the photoreceptor cell, evidence began to accumulate that the mutation resided in an ion channel (Montell and Rubin 1989). A critical finding came from the research of Hochstrate, who showed that the addition of La^{3+} in the extracellular media of blowfly electrophysiological preparations could mimic the *trp* phenotype (Hochstrate 1989). Since La^{3+} is known to inhibit cellular Ca^{2+} handling, new and innovative approaches became necessary to tease apart the differences between photoreceptor ionic currents in wild-type and *trp* mutant flies. Whole-cell patch-clamp recordings were introduced to the study of fly photoreceptor by Hardie, Ranganathan, Harris, Stevens and Zuker (Hardie 1991, Ranganathan *et al* 1991). Single photoreceptor electrophysiology permitted the observation of a highly selective Ca^{2+} conductance in response to light. Further biophysical characterization of the currents in wild-type and *trp* photoreceptors also revealed the existence of two distinct currents with different reversal potentials, indicating that there could be a second channel in the phototransduction pathway (Hardie and Minke 1992). At the same time the Kelly group had cloned a close homologue of the *trp* gene, which they termed *trpl* (*trp*-like), proposing that the gene product was a channel with similar properties to *trp* (Phillips *et al* 1992). Thus, the identification of the TRP and TRPL channels as fundamental components of the phototransduction pathway in fly stretched across more than 2 decades in time between Cosens and Manning's mutant and the electrophysiological characterizations[1].

Once the origin of the trp phenotype could be traced to one or more ion channels, questions turned to the mechanism of TRP channel pore gating and regulation. Genetic evidence for the involvement of phosphoinositide signaling in the phototransduction pathway came from the *norpA* mutant of drosophila, which had a null response to light (Deland and Pak 1973). The gene product was identified as a beta-class PLC, and expression of the functional *norpA* gene in a transgenic fly could rescue the *norpA* mutant from all its phenotypes (McKay *et al* 1995). Phosphoinositide signaling via PLC is strongly coupled to intracellular Ca^{2+} dynamics, which would account for differences in the electroretinogram responses between *trp* flies with a much lower Ca^{2+} influx than wild-type (Peretz *et al* 1994,

[1] This retelling of the *trp* and *trpl* account is distilled from a more intricate history, and the reader is kindly directed to the commentaries of Hardie and Minke for a more complete description of events (Hardie 2011, Minke 2010).

Ranganathan *et al* 1994). The molecular details of how the PLC signaling pathway intersects with TRP and TRPL channels are still being worked out (Hardie and Juusola 2015). However, there is a clear involvement of Ca^{2+}-dependent regulation and PLC in the phototransduction of flies. The initial search criteria that identified the *trpl* gene were aimed toward identifying ion channels with calmodulin binding motifs (Phillips *et al* 1992). There are two calmodulin binding motifs in the *trpl* sequence, and the sequences of both *trp* and *trpl* gave the first picture of TRP channel architecture: An N-terminal domain (consisting of ankyrin repeat domains for TRP and TRPL), a voltage sensor-like domain and pore domain (similar to the family of voltage-gated channels), and a short sequence just C-terminal of the sixth transmembrane segment called the TRP box (figure 3.1).

The serendipitous discovery of *trpl* and its sequence similarity to *trp* led to an intensive search through genetic data for other TRP family members. Soon other members of the TRP family turned up in vertebrate genetic sequence screens, and these were termed the TRPC (canonical) sub-family since they shared the architecture of their fly orthologues (Wes *et al* 1995, Zhu *et al* 1995). In addition to TRP channels that could be identified through sequence homology, there were also successful expression cloning screens in combination with bioinformatic approaches that identified members of the TRPV, TRPM and TRPA sub-families (Caterina *et al* 1997, McKemy *et al* 2002, Peier *et al* 2002, Story *et al* 2003). All told, there are now about 30 TRP channel genes belonging to 6 sub-families in vertebrates: TRPC, TRPV, TRPA, TRPM, TRPP and TRPML.

3.2 Sequence similarity in TRP channels

As previously mentioned, TRPL was identified by sequence similarity as a close relative to TRP, and the surge of genetic sequence data in the 1990s would drive a successful search for further TRP channels, beginning with TRPC1 (Wes *et al* 1995, Zhu *et al* 1995). The general features of TRPs are: (1) a recognizable 6-transmemberane (6-TM) region at the core of the sequence that resembles the functional voltage sensor domain (VSD) and pore domain (PD) of the voltage-gated ion channel super-family (figure 3.1(b)), and (2) a TRP domain that contains the TRP box amino acid sequence (EWKFAR), with E, F, and A being wobble positions (figure 3.1(a)) (Montell 2005). Certain structural features beyond the channel core may be shared outside the sub-family such as in the case of ankyrin repeat domains on the N-terminal side of the 6-TM domain (TRP, TRPL, TRPCs, TRPVs, TRPA1). However, the prevailing characteristic of the N-terminal and C-terminal regions when comparing across sub-families is an increased divergence in comparison to the common aspects of the 6-TM core (figure 3.1(c)). The flexibility of these channels to interface with various aspects of physiological function is well documented, and TRP channel variation to allow for this diversity in signaling is evident in the differentiation of the N- and C-termini (see review by Ramsey *et al* 2006).

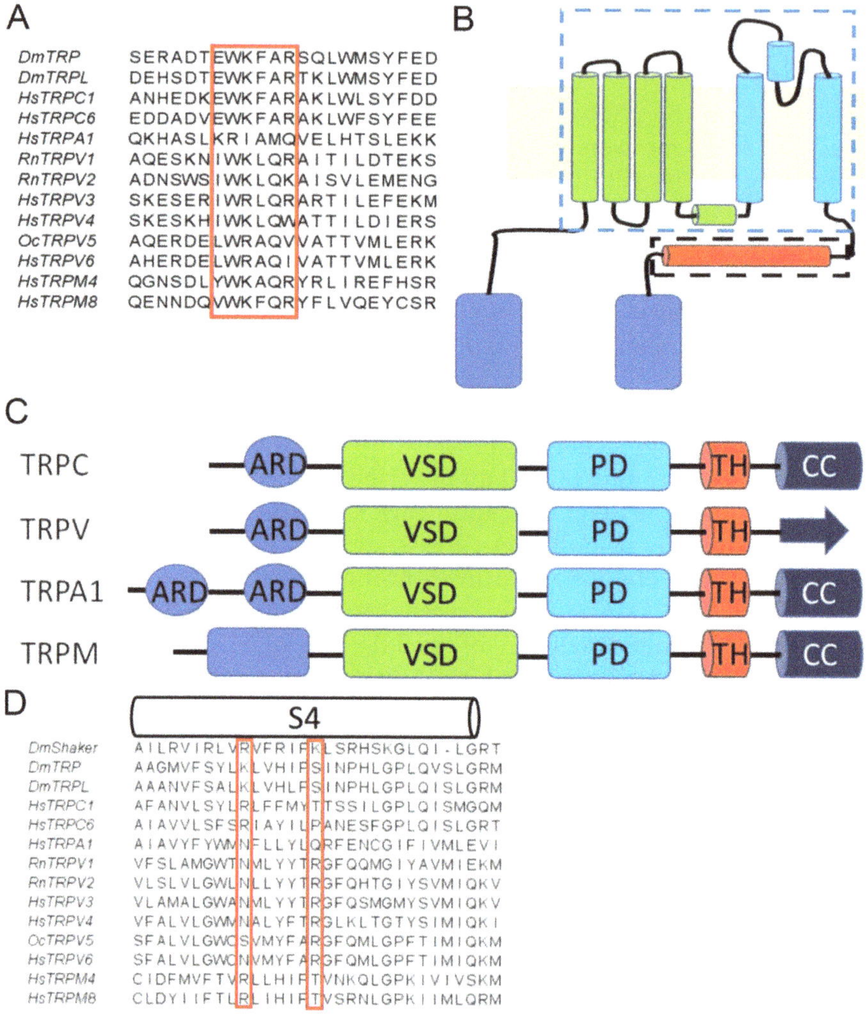

Figure 3.1. Topology and conserved sequence motifs of TRP channels. (a) Sequence alignment of TRP box across several TRP family members. (b) Topology of generic TRP channel with N- and C-termini (blue) projecting into the cellular space. The 6-TM core of the TRP channel contains the VSLD (green) and PD (light blue). The TRP helix is shown in red. (c) Alignment of domains across several TRP members. Topological conservation extends from VSLD (green) through PD (light blue) and TRP domain. Variability in N- and C-termini is shown as ARDs (blue ovals), coiled coil domains (purple cylinder), beta-sheet (purple arrow). (d) Sequence alignments of the S4 transmembrane helix from Shaker and several TRPs. The red boxes outline positions along TRP channel sequences where an arginine or lysine residue occasionally correlates with a conserved K_V channel residue. Species are designated by the first two letters in each name (Hs: *Homo sapiens*; Dm: *Drosophila melanogaster*; Rn: *Rattus norvegicus*; Oc: *Oryctolagus cuniculus*).

3.3 The conserved 6-TM core of TRP channels

Channels in the TRP family are largely non-selective cation channels. Conductances through TRP channels at physiological resting potential depolarize the cell and bring it to a more excitable state. The influx of Ca^{2+} has the added effect of signaling by second messenger. However, the selectivity of divalents over monovalents can vary strongly even within the same sub-family such as the TRPVs (see the review by Gees *et al* for selectivity tables) (Gees *et al* 2012). TRPV1 and TRPV2 have a P_{Ca2+}/P_{Na+} ~ 10 and ~3, respectively, which is easily distinguishable from TRPV5 and TRPV6 with P_{Ca2+}/P_{Na+} ~ 100. Mutagenesis experiments in the fly TRP channel at a site where a putative selectivity filter is expected revealed the importance of an Asp621 residue in the PD and validated the site as important for ion selectivity (Liu *et al* 2007). The single amino acid substitution from Aspartate to Glycine at position 621 profoundly affected the selectivity for Ca^{2+} over Na^{+} in the fly TRP channel. In TRPV1, 3, and 4, single amino acid substitutions of an aspartic acid near a selectivity filter sequence similarly altered the unitary conductance or ionic selectivity of these channels (Liu *et al* 2009, Voets *et al* 2002, Xiao *et al* 2008). These functional studies supported the sequence identity of a selectivity filter in the PD that was positioned slightly upstream of the S6 helix in a manner that closely resembled the structure of the pore in KcsA (Doyle *et al* 1998). A critical aspartic acid residue was identified in TRPV5/6 (D542/541) approximately in the location of a selectivity filter, and it affected ion selectivity to a similar extent as did aspartic acid residue replacement in the selectivity filter of fly TRP (Nilius *et al* 2001).

The first detailed examination of single-channel currents through TRPV1 revealed a pronounced voltage dependence of the unitary conductance, showing the channel's outward rectification at the single-channel level (Premkumar *et al* 2002). However, the voltage dependence of gating for different TRP channels is minimal compared to Ca_V, Na_V or K_V channels, and the transmembrane S1–S4 segments of TRP channels are therefore referred to as a voltage sensor-like domain (VSLD). TRPV1 and TRPM8 have been studied extensively for their voltage dependence, and an approximate total gating charge of less than $1e_o$ for both channels has been described by different groups, which is small compared to the ~10 e_o found in K_V channels (Brauchi *et al* 2004, Fernández *et al* 2011, Liu *et al* 2003, Voets *et al* 2007). A single lysine and arginine found in the fourth trans-membrane segment (S4) of TRPV1 and TRPM8, respectively, align with one of the numerous arginines and lysines found in the S4 voltage sensor of the Shaker K_V channel (figure 3.1(d)). This correspondence of positively charged amino acids in the S4 of K_V channels and TRP channels has directed research to investigate the S4 of TRPs as a voltage sensor. However, the voltage activated gating of TRPV1 and TRPM8 is small in comparison to the energetics of ligand and temperature gating of these channels (figure 3.2) (Brauchi *et al* 2004, Yao *et al* 2010). Although single point amino acid substitutions for the S4 arginines in TRPV1 and TRPM8 are known to impact the voltage dependent gating of both channels, there are also mutagenesis experiments done with charged residues in the S4–S5 linker that disrupted voltage dependent gating to a similar extent (Boukalova *et al* 2010, Voets *et al* 2007).

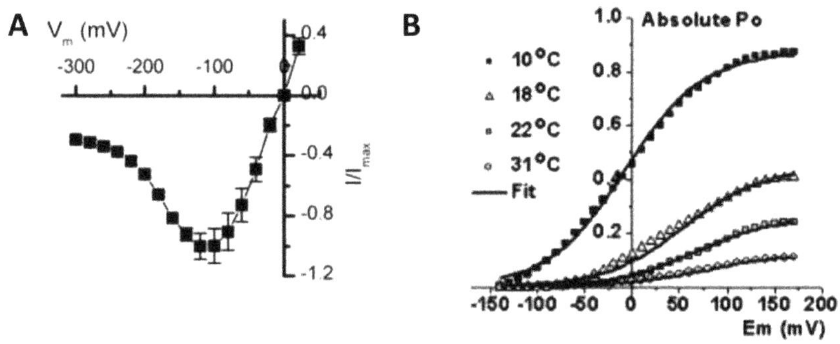

Figure 3.2. Relative energetics of ligand and temperature gating in TRPV1 and TRPM8, respectively, compared to voltage activation. (a) *I–V* plot of TRPV1 in the presence of saturating (10 mM) capsaicin. Persistent current at strong hyperpolarizing potentials (−300 mV) demonstrates that TRPV1 gates by capsaicin independent of voltage (adapted from Yao *et al* (2010), copyright 2010 with permission from Elsevier). (b) Po versus voltage plot of TRPM8 shows the significant increase in Po that temperatures below 22 °C achieve at extreme depolarizations of the membrane (adapted with permission from Brauchi *et al* (2004), copyright 2004 National Academy of Sciences, USA).

TRP channels have a prominent role in physiology because of their spectacular range of activation modalities. They have been documented as, to name just a few: thermoreceptors for hot and cold; mechanoreceptors; chemoreceptors to oxidation and acidity; calcium sensors; lipid sensors; and ligand-gated receptors. These modalities address many of the different receptors in the TRP channel class, but individual TRP channels such as TRPV1, TRPA1 and TRPM8 are also well known as polymodal receptors, with the sensory capabilities to detect multiple sensory inputs. Although TRP channel research has made important strides in uncovering the myriad ways that these channels activate and conduct, the mechanisms of how these signals integrate at the core of the channel to gate the pore are poorly understood. Fortunately, recent technical advances in single-particle cryo-EM reconstruction are now able to provide a wealth of information about TRP channel structure. The Cheng and Julius labs published the structure of TRPV1 in 2013, and in a few years the number of TRP channel structures has skyrocketed (Cao *et al* 2013, Liao *et al* 2013). During the publication of this chapter, there will no doubt be many more additions. This rapid increase in structural data now requires systematic analysis and careful interpretation.

3.4 TRP channel structures

Single molecule cryo-EM opened the proverbial floodgate for TRP channel structures and numerous other integral membrane proteins (Cheng 2018). Since the publication of a TRPV1 structure in 2013, there have been over 30 other TRP channel structures resolved to less than 4.5 Å, which span the sub-families of TRPs (figure 3.3). The remarkable persistence of the 6-TM core sequence across all TRP channels is clear in the structures of the TRPs. TRPs assemble as tetrameric channels with the pore domain defined by S5 and S6 transmembrane helices and a peripheral VSLD positioned in a domain swapped configuration nearest to its neighboring

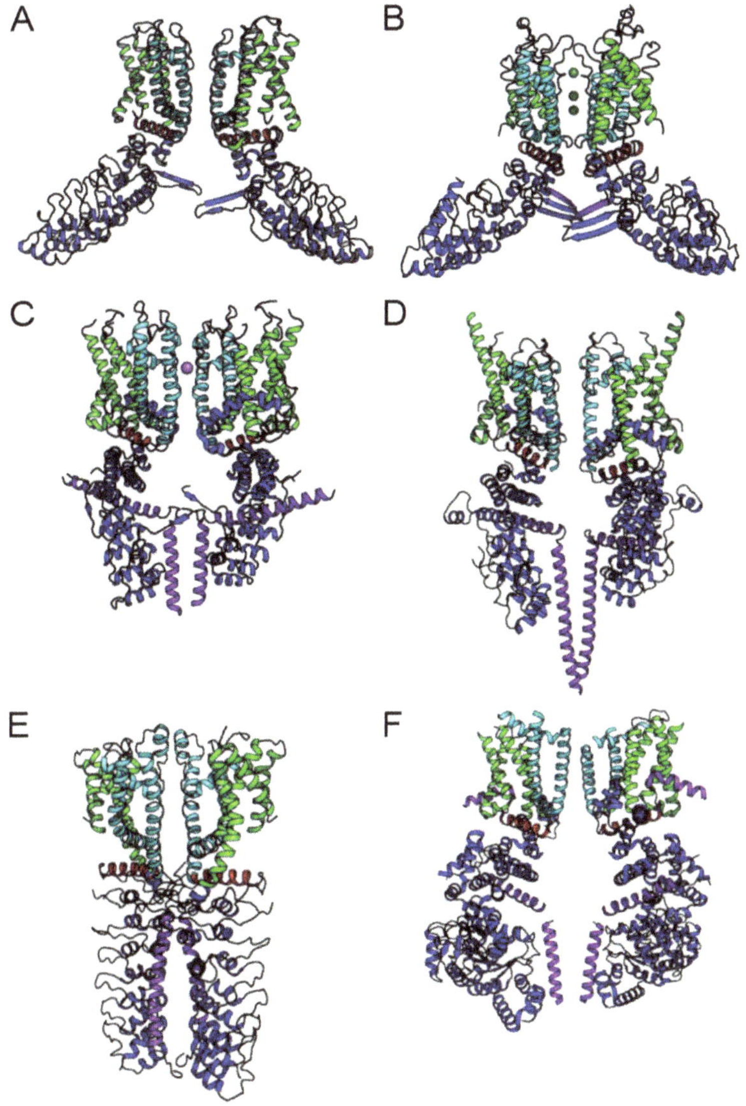

Figure 3.3. TRP channel structures viewed in side profile with blue and purple intracellular domains directed towards the bottom of the page. Instead of 4 subunits, each structure is shown as two, diagonally opposed subunits for the sake of clarity. All models were reconstructions from Cryo-EM data apart from TRPV6, which was solved from x-ray crystallography data. (a) TRPV1 (PDBID: 3J5P) (b) TRPV6 (PDBID: 5WO7) (c) TRPC4 (PDBID: 5Z96) (d) TRPC6 (PDBID: 5YX9) (e) TRPA1 (PDBID: 3J9P) (f) TRPM8 (PDBID: 6NR3). Coloring follows the scheme of figure 3.1(b): N-terminus, blue; VSLD, green; PD, light blue; TRP domain, red; C-terminus, purple.

subunit's PD. Arrangement of the VSLDs in tight bundles around the 'inverted teepee' assemblage of the pore domain highly resembles the K_V channel structure (Long *et al* 2005). If it can be said that there is any disappointment to the plentitude of structural data that has come forward in the TRP field, it would only be that the regions outside of the 6-TM core, which comprise part of each channel's uniqueness, are frequently of lower resolution. In the case of TRPV1, a prior x-ray crystallography structure of the free TRPV1 ankyrin repeat domain aided in the reconstruction of the first TRPV1 map (Lishko *et al* 2007). Perhaps the most exciting aspect to having numerous TRP structures was the anticipated resolving of different conformational states bound by channel agonists and modulators.

3.5 TRP channel ligand binding

TRPV1 is potently activated by the exogenous ligand capsaicin, which is one of the pungent compounds derived from chili peppers. It appears that capsaicin may have evolved in plants to favor consumption of the fruits by birds over mammals in order to more widely distribute seeds (Levey *et al* 2006, Tewksbury and Nabhan 2001). In the first set of TRPV1 structures, one of which contained capsaicin, the ligand is thought to be situated between the VSLD and PD but the capsaicin molecule is poorly resolved. To force the channel into an open state a structure of TRPV1 was solved with resinoferotoxin (RTX) and double-knot toxin (DkTx) derived from a tarantula venom (Bae *et al* 2012, 2016, Bohlen *et al* 2010, Gao *et al* 2016). This structure showed a clearer density between the VSLD and PD with the RTx forming contacts with amino acids in S3, S4, and S4–S5 linker (Cao *et al* 2013, Gao *et al* 2016). The side-chain of Y511 in the S3 becomes reoriented with the hydroxyl group directed towards the ligand when it is present, and in the absence of the agonist, the side-chain of Y511 is contrarily directed away from the ligand (figure 3.4(a)). The RTX structure provides a more convincing placement of RTX in the ligand pocket and demonstrates how the interactions of RTX with the channel induce a conformational change on the S4–S5 linker that reorients the side-chain of I679 to make a path for ion conduction through the pore. As the only site along the ion conduction pathway that shows a convincing change from an open to closed state between the apo- and agonist bound structures of TRPV1, I679 has been defined as a putative gate.

The molecular details offered through the structure of TRPV1 have informed studies of the ligand-channel interaction. The orientation and placement of capsaicin in the binding pocket was carefully examined by Yang *et al* using an elegant electrophysiological approach (Yang *et al* 2015). Their work strengthened the argument for a pulling mechanism along the S4–S5 linker to induce channel opening. Systematic mutagenesis of the side-chains in the binding pocket and introduction of different chemical groups on the capsaicin molecule augmented the structural information by specifically defining the interactions between the head, neck, tail of the capsaicin molecule with the channel. Their detailed analysis of the ligand interaction explained the decreased sensitivity of rabbit TRPV1 and loss of sensitivity of avian TRPV1 to capsaicin. It was deduced that the sequential loss of two hydrogen bonding interactions between TRPV1 at T551 and E571 with the neck and head of the capsaicin molecule are the underlying cause for the loss in sensitivity.

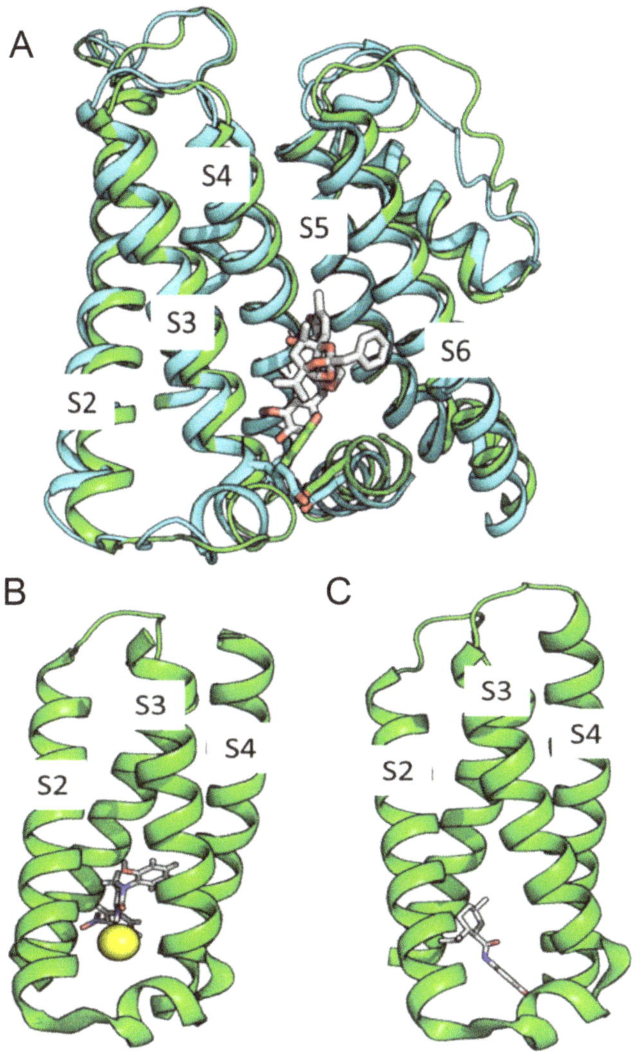

Figure 3.4. ligand binding pocket of TRPV1 and TRPM8. (a) Overlaid structures of TRPV1 without ligand (PDBID: 5IRZ; cyan) or the agonist Resiniferatoxin (RTx) and double-knot toxin (DkTx) (PDBID: 5IRX; green). Only the RTx molecule is shown in the binding pocket for clarity's sake. The side-chain of Y511 in the S3 is shown to be in an upwards position in the presence of RTx but directed away from the ligand binding pocket in the apo-structure. TRPV1 agonists are situated in a pocket between the VSLD of one subunit and the S5 and S6 of a neighboring subunit. (b) The VSLD of TRPM8, showing the location of icillin and a Ca^{2+} ion (yellow) in the ligand binding pocket (PDBID: 6NR3). (c) The VSLD of TRPM8, showing the location of the menthol analogue, WS-12, in the ligand binding pocket (PDBID: 6NR2).

Structural data of TRPM8 shows that TRPV1 is not alone in binding to its ligand at a site within the channel core. Yin *et al* found that the menthol analog WS-12 and icilin bind to TRPM8 between the VSLD and the TRP domain (figure 3.4(b) and (c)) (Yin *et al* 2019). The structure of TRPM8 and icilin is part of a complex that includes the lipid phosphatidylinositol 4,5-bisphosphate ($PI(4,5)P_2$). The presence of $PI(4,5)P_2$ is accounted for since it was a reagent used in the experimental preparation for cryo-EM imaging, and functional experiments on TRPM8 make a strong case for its inclusion as a necessary factor for channel opening (Rohács *et al* 2005).

Conservation of the gating mechanism within TRP sub-families was explored through artificial engineering of the TRPV1 RTx ligand binding site into other members of the TRPV sub-family (Yang *et al* 2016, Zhang *et al* 2016, 2019). By introducing 4 point mutations into the TRPV2 channel that mapped onto the binding site of RTX for TRPV1, researchers could potently open the TRPV2 channel with RTX. The Swartz group went on to introduce similar mutations into TRPV3, which allowed RTX to open this channel as well. Opening TRPV2 and TRPV3 with RTX through minimal mutagenesis to install an RTX binding pocket showed that the TRPV1-3 mutants shared a common gating mechanism. Experiments such as these, which explored the gating mechanism of TRP channels, establish that the VSLDs of TRP channels are part of a separately evolved gating system that is unlike the canonical voltage sensor in a specialized role as a ligand binding domain (Cabezas-Bratesco *et al* 2022).

3.6 Endogenous TRP ligands

As the first vertebrate TRP channels identified with the highest sequence homology to the fly TRPs, TRPC channels have been at the focus of TRP channel research for a long time. TRPC4 and TRPC5 have a lengthier story arch due to their similarity and the importance of TRPC5 in growth cone development of hippocampal neurons (Greka *et al* 2003). It has been shown that both channels gate via a PLC dependent $G\alpha q/11$ based mechanism. Although it was put forward that the lipid diacylglyrcerol (DAG) is the channel agonist, experiments that implemented a direct application of DAG revealed that DAG was insufficient to activate the channels and that application of high intracellular Ca^{2+} was also necessary (Schaefer *et al* 2000, Strübing *et al* 2001). Similarly, TRPC3,6, and 7 open through a $G\alpha q/11$ activation pathway (Estacion *et al* 2004, Hofmann *et al* 1999, Okada *et al* 1999). Structures for TRPC3,4, and 6 have been published, and the TRPC3 structure was determined, revealing lipid densities (Duan *et al* 2018a, Fan *et al* 2018, MacGregor *et al* 2002, Tang *et al* 2018). The closed-state structure of TRPC3 deserves special consideration because of lipids that are observed in the channel reconstruction. Although the identity of the lipids is unknown, one lipid is situated between the VSLD and the PD of the channel core and makes numerous contacts between the S4–S5 linker and the S4 helix. The authors describe the arrangement of the S1–S4 bundle in TRPC3 as much more similar to the VSLD of TRPV1, which opens up a pocket for lipids to take up residence, than that of TRPA1 and TRPM4 (Fan *et al* 2018). The placement of the lipid binding pocket has an unmistakable similarity to the ligand binding pockets of the TRPV channels, but it is yet to be shown by functional studies whether the TRPC3 lipid sites are of a significant nature.

Lipid interactions with TRP channels have also been noted in the cryo-EM structures of the TRPV, TRPM, TRPP families that were determined in lipid

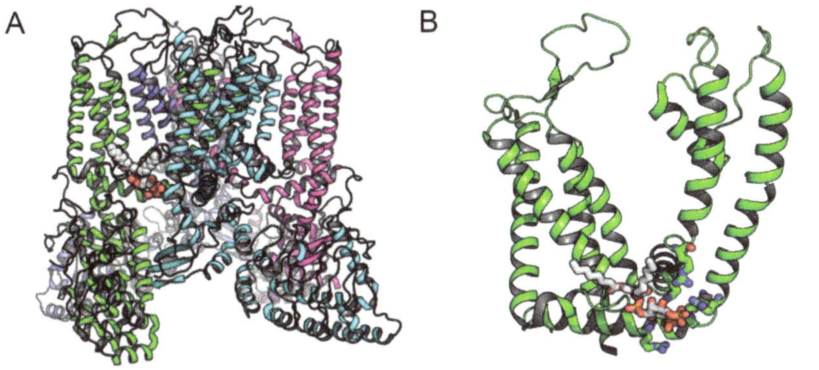

Figure 3.5. Structure of TRPV5 in complex with $PI(4,5)P_2$ (PDBID: 6DMU). (a) Side profile of TRPV5 with a single $PI(4,5)P_2$ molecule shown proximal to the green subunit of the channel. (b) Detailed depiction of the interactions between the head group of $PI(4,5)P_2$ and positively charged side-chains of the S4–S5 linker, S5, and S6.

nanodiscs, and in some cases it is possible to identify the lipid more specifically (Hughes *et al* 2018, Yin *et al* 2019). In the rabbit TRPV5 structure determined in nanodiscs, numerous annular lipids are obvious in the EM density map of the structure, and a $PI(4,5)P_2$ molecule can be identified, residing in a position between the VSLD and the PD (Hughes *et al* 2018). An important distinction to be made between the interaction of $PI(4,5)P_2$ and TRPV5 versus the ligand interaction observed in TRPV1-3 structures is the separation between the location of the $PI(4,5)P_2$ inositol ring in TRPV5 and the RTX density observed in TRPV structures (Fluck *et al* 2022). The head group of PIP(4,5)2 makes numerous contacts with positive charges on the S4–S5 linker and elbow between the S6 and TRP helix. The lipid tails are splayed out with one branch directed along the VSLD and the second branch reaching up the S4–S5 linker (figure 3.5). The functional importance of the $PI(4,5)P_2$ interaction with TRPV5 was previously known since there is a significant decrease in current through TRPV5 following the loss of $PI(4,5)P_2$ in the membrane, and the current can be recovered by reintroducing $PI(4,5)P_2$ to the inner leaflet in an excised patch electrophysiological experiment (Lee *et al* 2005). The $PI(4,5)P_2$ head group is positioned where it can interact with side-chains from the bend at the end of S6 where the TRP helix begins. Thus, $PI(4,5)P_2$ interacts with a site in TRPV5 that is an appreciable >15 Å from the analogous, capsaicin sensitive T551 residue in S4 of TRPV1 (Yang *et al* 2015). The TRPM8 structure also showed a lipid binding site that is different from ligand binding site. It is an open question whether or not other TRPVs have a distinct binding site for $PI(4,5)P_2$ than that of TRPV5 (Yin *et al* 2019).

3.7 Ca^{2+}-dependent modulation

Electrophysiology studies of the fly TRP channels signified the early importance of Ca^{2+} in the signaling roles of TRP channels. The high Ca^{2+} permeability across the TRPs places some of these channels under tight feedback constraints that may be exerted directly or indirectly through Ca^{2+} sensors such as calmodulin (CaM). The earliest TRP channel structure to portray the indirect role that Ca^{2+} has on the channel through CaM is TRPV5 (de Groot *et al* 2011). A close interaction between the side-chain of K116 of C-terminal CaM and the W583 side-chain of TRPV5 led to a functional investigation of these pair of amino acids, and mutating W583 to leucine disrupted the CaM-dependent inhibition of currents by Ca^{2+} (figure 3.6) (Hughes *et al* 2018). A C-terminal helix of the channel binds the N-terminus of CaM and the C-terminus of CaM binds with W583 near the pore (Dang *et al* 2019).

Ca^{2+} need not be a regulator of negative feedback, and in the structure of TRPM2 a direct association of Ca^{2+} with the channel shows it in an open state. TRPM2 was solved in three structural states: As an apo-protein; bound by the activator adenosine diphosphate ribose (ADPR); and bound by ADPR and Ca^{2+} (Wang *et al* 2018). A large structural rearrangement of the cytosolic domains occurs with the addition of ADPR, and the addition of Ca^{2+} to the ADPR-associated channel gates the pore of the channel along with further rearrangements in the cytosolic domains. Since TRPM2 will not open in the presence of ADPR and absence of Ca^{2+}, the ADPR structure was termed the 'primed' structure of TRPM2 (Csanády and Törocsik 2009). Transitioning from the apo-state of the channel to the primed state, results in multiple domains experiencing disruptions of their interfaces, including one interface that bridges subunits. By comparison, the structural changes observed in going from the 'primed' ADPR bound state to the ADPR/ Ca^{2+} bound state are much smaller. Side-chains from acidic residues in S2 and S3 of the VSLD and also side-chains from the TRP helix form numerous chelating contacts with the Ca^{2+}. These interactions are thought to induce movement in the TRP helix that promotes conformational changes in the channel pore, which lead to channel opening. Ca^{2+} binding to the VSLD is, therefore, a potent allosteric modulator of TRPM2, eliciting conformational changes that gate the pore.

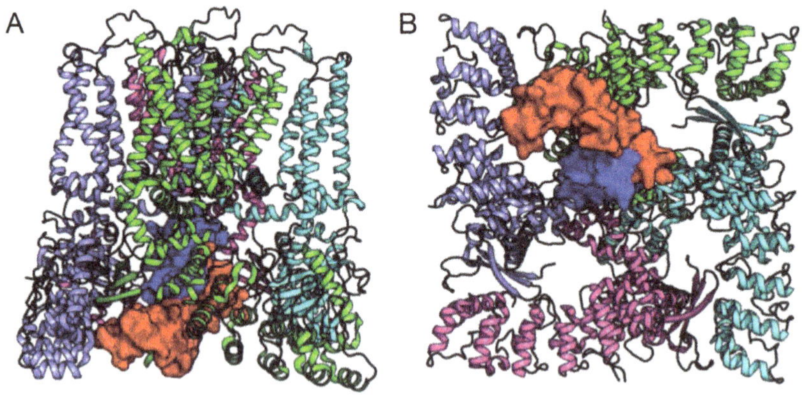

Figure 3.6. Structure of TRPV5 in complex with CaM (PDBID: 6DMW). (a) Side profile of TRPV5, showing each subunit of the channel as the polypeptide backbone in a different color. A single CaM molecule is depicted as a surface density map with the N-terminus colored red and the C-terminus in blue. (b) The representation in panel (a) as viewed from beneath at the cytosolic mouth of TRPV5. The coloring scheme is maintained from panel (a).

A recent TRPM8 cryo-EM structure revealed direct interactions between Ca^{2+} and TRPM8 in data collected from a sample that included the strong agonist icillin and Ca^{2+} (Yin *et al* 2019). Icillin requires the presence of Ca^{2+} in order to activate the channel, and the icillin binding site is located immediately adjacent to the Ca^{2+} ion (figure 3.4(b) and (c)). TRPM8 coordinates Ca^{2+} with the side-chains of several amino acids, including an asparagine at position 799 (N799) and aspartate at position 802 (D802), which were previously shown to be important for activation by icillin (Chuang *et al* 2004). Moreover, the salt bridge observed between D802 and arginine 841 (R841) in the apo-TRPM8 structure must be broken to form the binding pocket for Ca^{2+}, and the resulting accommodation of Ca^{2+} and icillin widens the internal cavity of the VSLD. Although the placement of Ca^{2+} within the binding pocket necessitates breaking the D802/R841 salt bridge between S3 and S4, it is unclear if Ca^{2+} binding to its site precedes icillin interactions in the binding pocket. Importantly, the TRPM8 structure with icillin–Ca^{2+} exhibits large conformational changes that distinguish the channel structure from the apo-state. However, a close examination of the pore domain suggests that the channel is in a non-conductive state, and the authors surmise that the icillin–Ca^{2+} structures capture either a state just prior to opening or a state that follows after this.

3.8 Thermosensitivity in TRP channels

The most debated topic of TRP channels is their activity as thermoreceptors. The TRPV sub-family has several members that activate as temperatures exceed homeostatic temperatures for mammals at around 40°, and, in an unusual twist of circumstance, the TRPM8 channel activates at temperatures below 18°. Both TRPV1 and TRPM8 have been the focus of many studies as the representative ion channels for hot and cold thermoreceptors, respectively, and it is not an exaggeration to say that we know slightly more now than when we started investigating these modalities of activation. One aspect of the problem in addressing thermosensitivity in an ion channel is the inherent difference between how heat is sensed and how modalities such as electric fields or chemical reactivity are detected. In the cases of the latter, a physical location within the protein is responsive to the sensory modality as in the case of a voltage sensor, ligand binding site, or sulfide reactive side-chains on cysteine residues. Upon activation, these allosteric sites containing the sensory apparatus would then induce conformational changes of the channel that open a channel pore. The sensory apparatus for heat, on the other hand, can be either localized or broadly distributed as a thermodynamic property of the ion channel protein.

Initial searches for the TRP temperature sensor were based on mutagenesis techniques with the hope of discovering a similar domain within TRPV1 or TRPM8. However, it soon became evident that disparate sites on the ion channels could disrupt or enhance their apparent temperature sensitivity. For TRPV1, sites that have been studied for their unique contribution to temperature gating include the pore turret, between the S6 and the pore helix, the pore domain itself, and the n-terminus of the channel (Cui *et al* 2012, Laursen *et al* 2016, Yao *et al* 2011,

Zhang *et al* 2018). An important consideration given to all these temperature sensing deficient constructs is that the capsaicin dependent activity remains nearly intact. Comparative physiology between extremophiles such as 13-lined grown squirrels (*Ictidomys tridecemlineatus*) or Bactrian camels (*Camelus ferus*) and model organisms like the rat have revealed surprising aspects about temperature gating of TRPV1 (Laursen *et al* 2016). Both squirrel and camel TRPV1 exhibit a significantly higher threshold of activation by heat compared to rat TRPV1. Although the three variants of TRPV1 share a high degree of sequence identity, a pair of single point amino acid changes in the squirrel TRPV1 n-terminus to the corresponding amino acids in rat TRPV1 (N126S, E190Q) shifts the threshold of activation dramatically towards the lower activation temperature of rat TRPV1 (Laursen *et al* 2016). It is bewildering to attempt to understand thermosensitivity in TRPs given the numerous sites that have been proposed as a temperature sensing module for TRPV1 and TRPM8.

To account for temperature sensing being altered by changes throughout the TRP channel sequences, a theory of temperature sensing in ion channels was proposed that relegates temperature dependent pore opening to a distributed property of the channel (Clapham and Miller 2011). A thermodynamic description of cold- and hot-sensitive gating has the distinction of not relying heavily on the localized temperature sensor model. If channel opening were to induce rearrangements within the 6-TM domain to expose hydrophobic side-chains to water molecules, a consequent, negative heat capacity change of channel gating would favor the open state in a colder environment (Makhatadze and Privalov 1990, Privalov and Makhatadze 1992, 1993). By virtue of whether the heat capacity change has either a positive or negative sign, this mechanism describes either cold- or hot-sensing channels. Experimental evidence for the heat-capacity model of temperature sensing has been limited, but at least one research group has attempted to test this idea. By changing the hydrophobicity of amino acid side-chains of the Shaker K_V channel at sites that are buried in the membrane, researchers were able to engineer temperature activation into the channel, which was consistent with a change in the heat capacity due to solvation of these sites (Chowdhury *et al* 2014). The distributed temperature sensing model of Clapham and Miller has the benefit of describing both cold and hot sensing by a change in sign of the heat capacity contribution to the enthalpy. In a similar turn, Jara-Oseguera and Islas have proposed that a cooperativity model based on the mechanism of binding by Monod, Wyman and Changeaux could rely on a single type of temperature sensor (ie. only responsive to heat) but couple gating to anti-cooperative or cooperative interactions with the end effect of achieving cold or heat dependent channel opening (Jara-Oseguera and Islas 2013).

3.9 Selectivity filter and pore

The similarity in sequence between TRPV1 and KcsA near the pore helix is mirrored in the structural motif of a selectivity filter identifiable in the extracellular mouth of both ion channels. A conserved pair of glycines are separated by methionine in TRPV1, which is followed by an aspartic acid in both ion channels. Structures of members in the TRPV family show the greatest consistency with potassium channel filters, and an x-ray crystallography structure of TRPV6 shows the aspartic acid residue, which immediately follows the G-x-G motif, in a configuration that coordinates a heavy atom (figure 3.3(b)) (Saotome *et al* 2016). As had been previously known about the aspartic acid residue in an identical site of fly TRP, site-directed mutagenesis of this amino acid drastically affected Ca^{2+} selectivity of the channel, and the coordinating action of the aspartic acids provided the molecular evidence for the importance of these sites in ion selectivity. Indeed, the crystal structures of an engineered calcium selective ion channel showed the critical nature of acidic aspartic acid amino acids for calcium selectivity (Tang *et al* 2014). One aspect, worth considering in the engineered calcium selective channel, however, is that direct ion coordination resulted in block rather than permeation.

The structure of TRPV1 gave us our first glimpse of a TRP channel pore. Comparison of the apo- and capsaicin-bound structures reveals that TRPV1 exhibits a dilation of the pore near the cytosolic mouth of the channel at position I679. The 'open' conformation structure, which is in the presence of capsaicin, has a pore diameter at this constriction which is double that of the apo-TRPV1 structure (radius apo: 1 Å; radius capsaicin: 2 Å). A complicating concern about I679 being designated as a channel gate is that the selectivity filter never exceeds an open radius of 1 Å in either the apo- or the capsaicin-bound structure. Only in the presence of the more potent agonist RTX and the tarantula toxin DkTx is a structure resolved with ample space at both the selectivity filter (radius of opening at narrowest point > 2 Å) and the conventional lower gate I679 (radius of opening at narrowest point >3 Å) for cations to pass unencumbered. Functional studies with ion channel blockers that preceded the structural work on TRPV1 had presented the earliest evidence that there could be two gates in these channels (Jara-Oseguera *et al* 2008).

Since the first TRPV1 structures have been published, a nomenclature has evolved, distinguishing a lower gate near I679 and an upper gate in the selectivity filters for TRP channels. Compelling evidence for either gate being the primary site of channel constriction is still forthcoming. However, a sweeping review of TRP channel structures and sequence conservation has identified an important structural feature known as the 'π'-helix in the S6 segment of the pore, and a nearby, highly conserved asparagine exhibits a large displacement away from the pore cavity in the closed state (Kasimova *et al* 2018). The 'π'-helix behaves as a meta-stable structure within the channel pore that undergoes a large rearrangement between the opening and closing of the channel. Because the conserved asparagine lies at the end of the 'π'-helix in S6, its side-chain is susceptible to any changes transmitted through a rearrangement of the 'π'-helix. Once the asparagine is directed into the pore cavity it serves to 'wet' the pore, thereby increasing the permeation of cations. The authors of the study posit that the structural analysis and sequence conservation of the 'π'-helix and asparagine side-chain rotation argue in favor of a gate being closer to the cytosolic mouth of the channel. Nevertheless, there remains the important task of providing experimental evidence for these hypotheses that were developed with molecular dynamics simulations and bioinformatics tools.

Further evidence to stoke the debate over the existence of lower and upper gates in TRPV channels comes from differences observed in apo-TRPV2 structures. Several published structures of TRPV1 and TRPV2 necessitated the excision of an extracellular loop between the S5 and pore helix, known as the pore turret, in order to collect high quality structural data (Cao *et al* 2013, Gao *et al* 2016, Zubcevic *et al* 2016). Along with an earlier reconstruction of full-length TRPV2 resolved at ~5 Å, more recent TRPV2 structures included the pore turret, and were determined with less than 4 Å resolution (Huynh *et al* 2016, Pumroy *et al* 2019, Zubcevic *et al* 2019). In contrast to apo-TRPV1 and apo-TRPV2 structures without the pore turret, the upper gate in some of the full-length apo-TRPV2 structures exhibits an open state. A distinguishing feature of the full-length apo-TRPV2 structures solved in lipid nanodiscs is that both open and closed selectivity filters could be observed in reconstructions derived from the same data set (figure 3.7). In refining the reconstruction of a cryo-EM structure, images from a single data set may be distributed according to different 3D classes, which improves the final resolution of a structure. It so happens that the apo-TRPV2 lipid nanodisc data collected by Pumroy *et al* separated into two 3D classes with differences in the final structure, which, in part, could be attributed to the width of the selectivity filter. The lower gate of both apo-TRPV2 structures with variable selectivity filters remained constricted, rendering both channels in a non-conductive state. Although the resolution of turret-inclusive TRPV2 structures is variable, evidence for two unique selectivity filter widths in the unliganded channel signifies the

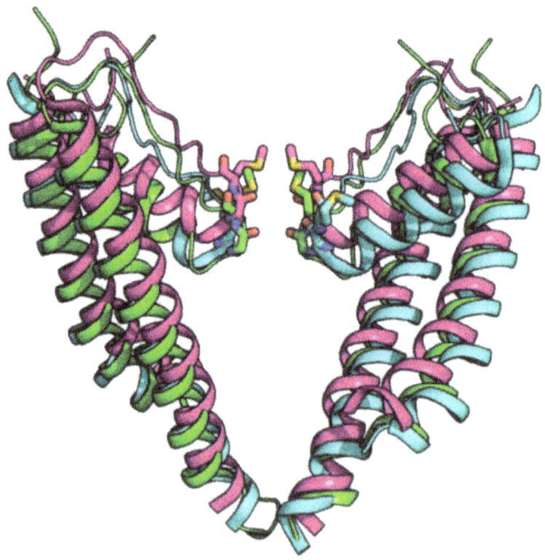

Figure 3.7. Overlay of oppositely facing pore domains from full-length apo-TRPV2 structures in side profile. Front and back pore domains are removed for clarity's sake. Alignments of two full-length rat apo-TRPV2 structures with closed (green, PDBID: 6U84) and open (cyan, PDBID: 6U86) selectivity filters, which were reconstructed as separate 3D classes from the same image data set. The side-chain of the methionine at position 607 and backbone carbonyl of G606 are shown where a constriction forms in the selectivity filter of the green apo-TRPV2 structure. A third full-length apo-TRPV2 structure from rabbit also shows the constriction of the selectivity filter at the backbone carbonyl of G604 and the side-chain of M605 (magenta, PDBID: 6OO4).

importance of structures derived from full-length channels, and higher resolution structures of unliganded, full-length TRPV1 and TRPV2 channels could be revelatory (Kwon *et al* 2021, Nadezhdin *et al* 2021, Zhang *et al* 2021).

There is a compelling need to find out if the closed selectivity filter is a native conformation of TRP channels. In four independent structural studies of the TRPM4 channel, reconstructions show the selectivity filter in a conductive state although the lower gate is constricted (Autzen *et al* 2018, Duan *et al* 2018b, Guo *et al* 2017, Winkler *et al* 2017). Structural studies of TRPML3 with the lower gate in either open or closed states also exhibits an open, and presumably, conductive selectivity filter, which places further emphasis on the significance of the lower gate and the importance of retaining the pore turret in obtaining structural data on the TRPVs (Hirschi *et al* 2017, Zhou *et al* 2017).

3.10 Cytsolic domains of TRP channels

The resolution of an ion channel structure solved by single-particle cryo-EM is typically defined by the Fourier shell correlation (FSC) of the three-dimensional reconstruction. In most cases the reported resolution is the spatial frequency at which the FSC measurement drops below a value of 0.143, and it is important to recognize that this resolution is distilled from the full volume of the three-dimensional structure as an ensemble measure (Lyumkis 2019). Furthermore, knowledge of the 4-fold symmetry of tetrameric channels, including TRPs, is frequently incorporated within the averaging algorithms of the reconstruction process in order to increase the sample number in the average. The improved resolution due to the symmetry operation may then come at the expense of non-symmetrical differences across the channel subunits. Resolution within a single structure also varies across the entirety of the structure. The central pore axis and surrounding transmembrane segments of TRP channel structures are invariably captured with the highest resolution, and as the scope of the structure extends beyond these regions, a trend to lower resolution becomes apparent (Jakobi *et al* 2017). In addition to a loss in resolution due to increasing the distance from the pore axis, reconstruction of the cytosolic domains may also suffer from a greater amount of flexibility. Although most TRP channel structures have a well-defined reconstruction of the 6-TM domains, there are varying degrees of success at depicting an atomistic reconstruction of the N-terminal and C-terminal domains with the available electron density information. Two cytosolic domains, showing consistency across independent reconstructions made of the same TRP channel and appearing across several members within a TRP sub-family are the ankyrin repeat domains (ARD) of TRPV and TRPC channels and the coiled coil domains (CCD) of TRPA1, TRPM, and TRPC channels.

3.11 Ankyrin repeat domains

The ankyrin repeat in an ARD refers to a 33-residue motif identifiable as a genetic sequence that is repeated a minimum of four times. Pattern recognition methods and database sequences have recognized ankyrin repeats across almost all phyla and present in various types of proteins, ranging from enzymes to transcription factors (Bork 1993). As a protein–protein interaction module, ARDs have been adapted to recognize diverse macromolecular targets, and the frequent appearance of ARDs in TRP channels, especially amongst TRPC, TRPV and TRPA sub-families, as well as known channelopathies arising from mutations in the ARDs has garnered them a fair share of attention in TRP channel studies (Kang *et al* 2012, Sedgwick and Smerdon 1999). The ARDs from TRPV1 and TRPV4 have been structurally resolved in isolation from the rest of the channel (Inada *et al* 2012, Jin *et al* 2006, Lishko *et al* 2007, McCleverty *et al* 2006, Phelps *et al* 2008). In the case of TRPV4, a rearrangement of a loop extending between two fingers of the ARD is observed in the presence of ATP, and, therefore, consequential changes to the interaction surfaces of ARD could be interpreted from the differences in the structures. Successful experiments on the structural and biochemical properties of TRP channel ARDs have shown that ion channel domains can be studied in isolation to further our functional understanding of the full ion channel.

Evidence for the protein–protein interaction properties of the N-terminal ARD of TRPV1 derives from biochemical pull-down experiments done with a fragment of TRPV1(1-432) and the regulatory domain of PI3 kinase (p85) (Stein *et al* 2006). Through subsequent rounds of protein truncation, the necessary portion of p85 for a successful pull-down experiment was found to be mediated by SH2 domains. In a more recent study it was shown that cellular expression of the soluble N-terminal TRPV1 fragment, which includes the ARD, potentiates PI3 kinase activity (Stratiievska *et al* 2018). Association of the regulatory domain of PI3 kinase (p85) and the TRPV1 ARD could, therefore, be implicated in the cellular activation of PI3K by overexpression of the TRPV1 n-terminus. When considered in the context of the full TRPV1 channel structure, the ARD of each TRPV1 subunit has a high degree of solvent accessibility to allow for protein–protein interactions to occur.

The arrangement of ARDs in TRPVs, which project outward, contrasts with the arrangement of ARDs in TRPA1 and TRPC channels, which project straight downward from the pore of the channel (figure 3.3). With insight provided through the cryo-EM reconstruction of TRPV1 in amphipols, distance measurements between different positions along the ARD of TRPV1 and the inner leaflet of the plasma membrane were made in cellular preparations with the fluorescence method known as tmFRET. Although no detectable changes in the distance between the ARD and the inner leaflet of the membrane were observed with the addition of capsaicin, the separation between different amino acid positions and the inner leaflet trended according to the attributes of the ARD in the TRPV1 structure (Zagotta *et al* 2016). With the spatial detail afforded through reconstructed TRP channels, FRET methods may corroborate structural details of ion channels in the context of a more physiological environment as well as report on dynamic changes occurring through channel activation.

Although the placement of the ARDs in TRPV1 differs significantly from those of TRPA1 and TRPC channels, they do share the feature of forming direct interactions with the C-termini. For TRPV1, the N-terminal ARDs have interactions with a C-terminal beta-sheet, while TRPA1 and TRPC have interactions with a C-terminal coiled coil domain (figures 3.1 and 3.3). Indeed, extensive interactions between N- and C-termini are also conserved in TRPM channels in which the N-terminus forms interactions with a C-terminal coiled coil (figure 3.3). Interactions between the N- and C-terminus of TRP channels demonstrate how the different domains in the channel are strongly coupled and are amenable to allosteric properties.

3.12 Coiled coil domains

A coiled coil is a protein structural domain formed by the parallel or anti-parallel bundling of between 2 and 7 alpha helices. As would be expected for the tetrameric arrangement of subunits in TRP channels, the coiled coil domains which have been identified in the structures of TRPC and TRPM channels as well as TRPA1 assemble as a tetrameric left-hand super helix. The coiled coil domains for all TRP channels are C-terminal of the TRP domain and project along the axis of the pore on the intracellular face of the channel (figure 3.3). As an example of how the intracellular N- and C-terminal regions of TRP channels interact, the coiled coil of TRPA1 is in contact with a significant section of the N-terminal ARD, and the report accompanying the structure posits that an IP_6 molecule, which was an essential co-factor for TRPA1 purification, is positioned between the coiled coil and the ARD (Paulsen *et al* 2015). Loss of an endogenous co-factor during excised patch-clamp experiments in cells expressing human TRPA1 causes loss of channel function. However, function of human TRPA1 is restored when polyphosphates are reintroduced to the intracellular solution (Kim and Cavanaugh 2007). In a study looking at the biochemical properties of the TRPA1 coiled coil, it was determined that the isolated coiled coil assembles in a concentration dependent manner, but assembly in the cell is likely mediated through additional interactions provided through the full channel. However, the formation of the coiled coil occurred spontaneously and did not depend on the presence of IP_6 (Martinez and Gordon 2019). An important finding in this study was that temperature factors into the stability of the coiled coil and the interaction between purified TRPA1 coiled coil and purified TRPA1 ARD. This demonstrated that temperatures above 37 °C could influence the interaction across the N- and C-terminus of TRPA1 and may lead to ion channel conformational changes, a hypothesis first put forward by the Minor lab (Arrigoni *et al* 2016).

Cryo-EM reconstructions of the TRPC channels have successfully modeled the coiled coil domains of numerous different members in this sub-family. TRPC3–6 structures from various vertebrates have been resolved at <4 Å (Duan *et al* 2018a, 2019, Fan *et al* 2018, Tang *et al* 2018, Vinayagam *et al* 2018). Apart from their ability to assemble as homomeric ion channels, several members of the TRPC subfamily are believed to heteromerically assemble into functional ion channels *in vivo*. Several experiments have shown that TRPC4 pairs with TRPC5 and TRPC1, and it is thought that TRPC3 pairs with TRPC6. Biochemical evidence points to a strict adherence in each, determining that members of one group will not oligomerize with members of the other (Cheng *et al* 2010, Hofmann *et al* 2002). Although the structures represent the homomeric TRPC channels, the configuration of the coiled coil domain is highly similar within the pairs of TRPC channels that are believed to form heteromeric channels. Indeed, the structures of TRPC4 and TRPC6, which are not believed to form heteromeric channels, are significantly different when compared to the structural similarity of TRPC6 and TRPC3. Structures reported on members of the TRPM sub-family also contain resolved coiled coil domains. However, there are few reports on the heteromerization of these channels apart from TRPM6 and TRPM7 (Duan *et al* 2018b, Li *et al* 2006).

3.13 Post-translational modifications

Numerous TRP channels are the targets of phosphorylation pathways as a mechanism of cellular regulation over channel activity. In the context of TRPV1 pain signaling, phosphorylation by PKC, PKA CAMKII and Src kinase through at least three sites, S116, S502, and S800 have a regulatory effect on channel function (Bhave *et al* 2003, Jin *et al* 2004, Jung *et al* 2004, Mandadi *et al* 2006, Numazaki *et al* 2002). There is a considerable feedback effect of channel regulation through kinases since the Ca^{2+} permeability through TRPV1 will modulate the activity of kinases in the cell. Unfortunately, there is no structural information about how the phosphorylation of different sites in TRPV1 will affect channel activity. The general absence of phosphorylated sites is not unique to TRPV1 structures but rather to all TRP structures that have been solved to date. This may be due to a limitation attributed to the resolution of TRPV1 reconstructions, a matter of sample heterogeneity in the phosphorylated state of the protein, which is lost in the averaging steps of structure determination, or simply that the protein preparations are not conducive to significant phosphorylation. Aside from the problems that arise due to sample variation in cryo-EM reconstruction, this method also has reduced sensitivity to negative charges relative to positive charges in the molecule under observation, which could factor into detecting phosphorylated side-chains (Wang and Moore 2017).

Covalent modifications resulting from an interaction with an activating compound have been closely studied in TRPA1. This channel is activated by chemical electrophiles such as alpha, beta-unsaturated aldehydes and isothiocyanates, which are found in plant derived irritants and as endogenous products to oxidative stress. TRPA1 expressing neurons generate a nociceptive response upon encountering these compounds because they open the channel upon reacting with cysteine residues in the cytoplasmic N-terminus (Bandell *et al* 2004, Bautista *et al* 2006, Hinman *et al* 2006, Macpherson *et al* 2007). The cysteine residues form an 'allosteric nexus' that has been characterized through site-directed mutagenesis and structurally resolved in a reconstruction of TRPA1 (Bahia *et al* 2016, Paulsen *et al* 2015). Modifications of the cysteines causes a conformational change at the interface of the N-terminal cytosolic domain and the 6-TM core of TRPA1, and the rearrangement of the channel is allosterically coupled to opening of the gate.

Glycosylations are another post-translational modification of integral membrane proteins that signal maturation steps in cellular trafficking pathways. Asparagine or arginine amino acids on TRP channels are occasionally N-glycosylated, meaning that the nitrogen on the side-chain becomes covalently linked to a carbohydrate known as a glycan. Unfortunately, many TRP channel protein preparations use cells that have been engineered to limit glycosylation. However, glycosylation of an extracellular site was recently reported in a TRPM4 structure solved at 3.7 Å, and the glycosylated asparagine residue is located at the extracellular mouth of the channel with the potential to influence the permeation of ions (Duan *et al* 2018b). How the glycosylation of a TRP channel affects function must ultimately be determined by electrophysiology. In the case of TRPV1 there is a glycosylated asparagine in the wild-type channel that can be mutated (N604S) without any dilatory affect to channel function (Rosenbaum *et al* 2002).

3.14 Summary

Structural data on TRP channels is currently being published at an incredible pace because of the recent technical advances in single-particle cryo-EM reconstruction. In addition to published structures representing many of the members in each of the TRP sub-families, there are examples of the same channel appearing in the apo- or liganded configuration. Although the resolution of most cryo-EM TRP channel structures is greater than 3 Å, which is not fine enough to provide atomic detail, the knowledge emerging from TRP structural studies are an important building block to a mechanistic understanding of ion channels. The wealth of information available to TRP channel researchers is nearly overwhelming at this point, and through careful evaluation, it is hopeful that these data provide critical information to ion channel biophysicists as they design experiments to elucidate function. Perhaps one of the more meaningful aspects to collecting so many TRP channel structures in the past few years is the broad representation of the many different sub-families. There is a clear diversity apparent in the sequences and overall architecture across the TRP family. However, the conserved elements in the 6-TM domain and cytosolic ARDs and CC domains invite us to ask whether general mechanisms are shared between all or some of the TRPs. At this time there is also a serendipitous alignment between the power of molecular dynamics and the structural detail of large integral membrane proteins such as ion channels. We are undeniably near an intersection of structural insight, computational prowess and biophysical experimentation where great discoveries await.

References

Arrigoni C, Rohaim A, Shaya D, Findeisen F, Stein R A, Nurva S R, Mishra S, Mchaourab H S and Minor D L 2016 Unfolding of a temperature-sensitive domain controls voltage-gated channel activation *Cell* **164** 922–36

Autzen H E, Myasnikov A G, Campbell M G, Asarnow D, Julius D and Cheng Y 2018 Structure of the human TRPM4 ion channel in a lipid nanodisc *Science* **359** 228–32

Bae C *et al* 2016 Structural insights into the mechanism of activation of the TRPV1 channel by a membrane-bound tarantula toxin *Elife* **5** e11273

Bae C, Kalia J, Song I, Yu J, Kim H H, Swartz K J and Kim J I 2012 High yield production and refolding of the double-knot toxin, an activator of TRPV1 channels *PLoS One* **7** e51516

Bahia P K, Parks T A, Stanford K R, Mitchell D A, Varma S, Stevens S M and Taylor-Clark T E 2016 The exceptionally high reactivity of Cys 621 is critical for electrophilic activation of the sensory nerve ion channel TRPA1 *J. Gen. Physiol.* **147** 451–65

Bandell M, Story G M, Hwang S W, Viswanath V, Eid S R, Petrus M J, Earley T J and Patapoutian A 2004 Noxious cold ion channel TRPA1 is activated by pungent compounds and bradykinin *Neuron* **41** 849–57

Bautista D M, Jordt S E, Nikai T, Tsuruda P R, Read A J, Poblete J, Yamoah E N, Basbaum A I and Julius D 2006 TRPA1 mediates the inflammatory actions of environmental irritants and proalgesic agents *Cell.* **124** 1269–82

Bhave G, Hu H J, Glauner K S, Zhu W, Wang H, Brasier D J, Oxford G S and Gereau R W 2003 Protein kinase C phosphorylation sensitizes but does not activate the capsaicin receptor transient receptor potential vanilloid 1 (TRPV1) *Proc. Natl Acad. Sci. USA* **100** 12480–5

Bohlen C J, Priel A, Zhou S, King D, Siemens J and Julius D 2010 A bivalent tarantula toxin activates the capsaicin receptor, TRPV1, by targeting the outer pore domain *Cell* **141** 834–45

Bork P 1993 Hundreds of ankyrin-like repeats in functionally diverse proteins: mobile modules that cross phyla horizontally? *Proteins* **17** 363–74

Boukalova S, Marsakova L, Teisinger J and Vlachova V 2010 Conserved residues within the putative S4–S5 region serve distinct functions among thermosensitive vanilloid transient receptor potential (TRPV) channels *J. Biol. Chem.* **285** 41455–62

Brauchi S, Orio P and Latorre R 2004 Clues to understanding cold sensation: thermodynamics and electrophysiological analysis of the cold receptor TRPM8 *Proc. Natl Acad. Sci. USA* **101** 15494–9

Cabezas-Bratesco D, Mcgee F A, Colenso C K, Zavala K, Granata D, Carnevale V, Opazo J C and Brauchi S E 2022 Sequence and structural conservation reveal fingerprint residues in TRP channels *Elife* **11** e73645

Cao E, Liao M, Cheng Y and Julius D 2013 TRPV1 structures in distinct conformations reveal activation mechanisms *Nature* **504** 113–8

Caterina M J, Schumacher Ma, Tominaga M, Rosen Ta, Levine J D and Julius D 1997 The capsaicin receptor: a heat-activated ion channel in the pain pathway *Nature* **389** 816–24

Cheng W, Sun C and Zheng J 2010 Heteromerization of TRP channel subunits: extending functional diversity *Protein Cell* **1** 802–10

Cheng Y 2018 Membrane protein structural biology in the era of single particle cryo-EM *Curr. Opin. Struct. Biol.* **52** 58–63

Chowdhury S, Jarecki B W and Chanda B 2014 A molecular framework for temperature-dependent gating of ion channels *Cell* **158** 1148–58

Chuang H H, Neuhausser W M and Julius D 2004 The super-cooling agent icilin reveals a mechanism of coincidence detection by a temperature-sensitive TRP channel *Neuron* **43** 859–69

Clapham D E and Miller C 2011 A thermodynamic framework for understanding temperature sensing by transient receptor potential (TRP) channels *Proc. Natl Acad. Sci. USA* **108** 19492–7

Cosens D J and Manning A 1969 Abnormal electroretinogram from a *Drosophila* mutant *Nature* **224** 285–7

Csanády L and Törocsik B 2009 Four Ca^{2+} ions activate TRPM2 channels by binding in deep crevices near the pore but intracellularly of the gate *J. Gen. Physiol.* **133** 189–203

Cui Y, Yang F, Cao X, Yarov-Yarovoy V, Wang K and Zheng J 2012 Selective disruption of high sensitivity heat activation but not capsaicin activation of TRPV1 channels by pore turret mutations *J. Gen. Physiol.* **139** 273–83

Dang S, van Goor M K, Asarnow D, Wang Y, Julius D, Cheng Y and van der Wijst J 2019 Structural insight into TRPV5 channel function and modulation *Proc. Natl Acad. Sci. USA* **116** 8869–78

de Groot T, Kovalevskaya N V, Verkaart S, Schilderink N, Felici M, van der Hagen E A, Bindels R J, Vuister G W and Hoenderop J G 2011 Molecular mechanisms of calmodulin action on TRPV5 and modulation by parathyroid hormone *Mol. Cell. Biol.* **31** 2845–53

Deland M C and Pak W L 1973 Reversibly temperature sensitive phototransduction mutant of *Drosophila melanogaster Nat. New Biol.* **244** 184–86

Doyle D A, Morais Cabral J, Pfuetzner R A, Kuo A, Gulbis J M, Cohen S L, Chait B T and MacKinnon R 1998 The structure of the potassium channel: molecular basis of K^+ conduction and selectivity *Science* **280** 69–77

Duan J *et al* 2019 Cryo-EM structure of TRPC5 at 2.8-Å resolution reveals unique and conserved structural elements essential for channel function *Sci. Adv.* **5** eaaw7935

Duan J, Li J, Zeng B, Chen G L, Peng X, Zhang Y, Wang J, Clapham D E, Li Z and Zhang J 2018a Structure of the mouse TRPC4 ion channel *Nat. Commun.* **9** 3102

Duan J, Li Z, Li J, Santa-Cruz A, Sanchez-Martinez S, Zhang J and Clapham D E 2018b Structure of full-length human TRPM4 *Proc. Natl Acad. Sci. USA* **115** 2377–82

Estacion M, Li S, Sinkins W G, Gosling M, Bahra P, Poll C, Westwick J and Schilling W P 2004 Activation of human TRPC6 channels by receptor stimulation *J. Biol. Chem.* **279** 22047–56

Fan C, Choi W, Sun W, Du J and Lü W 2018 Structure of the human lipid-gated cation channel TRPC3 *Elife* **7** e36852

Fernández J A, Skryma R, Bidaux G, Magleby K L, Scholfield C N, McGeown J G, Prevarskaya N and Zholos A V 2011 Voltage- and cold-dependent gating of single TRPM8 ion channels *J. Gen. Physiol.* **137** 173–95

Fluck E C, Yazici A T, Rohacs T and Moiseenkova-Bell V Y 2022 Structural basis of TRPV5 regulation by physiological and pathophysiological modulators *Cell Rep.* **39** 110737

Gao Y, Cao E, Julius D and Cheng Y 2016 TRPV1 structures in nanodiscs reveal mechanisms of ligand and lipid action *Nature* **534** 347–51

Gees M, Owsianik G, Nilius B and Voets T 2012 TRP channels *Compr. Physiol.* **2** 563–608

Greka A, Navarro B, Oancea E, Duggan A and Clapham D E 2003 TRPC5 is a regulator of hippocampal neurite length and growth cone morphology *Nat. Neurosci.* **6** 837–45

Guo J, She J, Zeng W, Chen Q, Bai X C and Jiang Y 2017 Structures of the calcium-activated, non-selective cation channel TRPM4 *Nature* **552** 205–9

Hardie R C 1991 Whole-cell recordings of the light-induced current in *Drosophila* photoreceptors: evidence for feedback by calcium permeating the light sensitive channels *Proc. R. Soc. Lond B.* **245** 203–10

Hardie R C 2011 A brief history of trp: commentary and personal perspective *Pflügers Archiv Eur. J. Physiol.* **461** 493–8

Hardie R C and Juusola M 2015 Phototransduction in *Drosophila Curr. Opin. Neurobiol.* **34** 37–45

Hardie R C and Minke B 1992 The *trp* gene is essential for a light-activated Ca^{2+} channel in *Drosophila* photoreceptors *Neuron* **8** 643–51

Hinman A, Chuang H H, Bautista D M and Julius D 2006 TRP channel activation by reversible covalent modification *Proc. Natl Acad. Sci. USA* **103** 19564–8

Hirschi M, Herzik M A, Wie J, Suo Y, Borschel W F, Ren D, Lander G C and Lee S Y 2017 Cryo-electron microscopy structure of the lysosomal calcium-permeable channel TRPML3 *Nature* **550** 411–4

Hochstrate P 1989 Lanthanum mimicks the trp photoreceptor mutant of *Drosophila* in the blowfly *Calliphora J. Comp. Physiol. A.* **166** 179–87

Hofmann T, Obukhov A G, Schaefer M, Harteneck C, Gudermann T and Schultz G 1999 Direct activation of human TRPC6 and TRPC3 channels by diacylglycerol *Nature* **397** 259–63

Hofmann T, Schaefer M, Schultz G and Gudermann T 2002 Subunit composition of mammalian transient receptor potential channels in living cells *Proc. Natl Acad. Sci. USA* **99** 7461–6

Hughes T E T *et al* 2018 Structural insights on TRPV5 gating by endogenous modulators *Nat. Commun.* **9** 4198

Huynh K W, Cohen M R, Jiang J, Samanta A, Lodowski D T, Zhou Z H and Moiseenkova-Bell V Y 2016 Structure of the full-length TRPV2 channel by cryo-EM *Nat. Commun.* **7** 11130

Inada H, Procko E, Sotomayor M and Gaudet R 2012 Structural and biochemical consequences of disease-causing mutations in the ankyrin repeat domain of the human TRPV4 channel *Biochemistry* **51** 6195–206

Jakobi A J, Wilmanns M and Sachse C 2017 Model-based local density sharpening of cryo-EM maps *Elife* **6** e27131

Jara-Oseguera A and Islas L D 2013 The role of allosteric coupling on thermal activation of thermo-TRP channels *Biophys. J.* **104** 2160–9

Jara-Oseguera A, Llorente I, Rosenbaum T and Islas L D 2008 Properties of the inner pore region of TRPV1 channels revealed by block with quaternary ammoniums *J. Gen. Physiol.* **132** 547–62

Jin X, Morsy N, Winston J, Pasricha P J, Garrett K and Akbarali H I 2004 Modulation of TRPV1 by nonreceptor tyrosine kinase, c-Src kinase *Am. J. Physiol. Cell Physiol.* **287** C558–63

Jin X, Touhey J and Gaudet R 2006 Structure of the N-terminal ankyrin repeat domain of the TRPV2 ion channel *J. Biol. Chem.* **281** 25006–10

Jung J, Shin J S, Lee S Y, Hwang S W, Koo J, Cho H and Oh U 2004 Phosphorylation of vanilloid receptor 1 by Ca^{2+}/calmodulin-dependent kinase II regulates its vanilloid binding *J. Biol. Chem.* **279** 7048–54

Kang S S, Shin S H, Auh C K and Chun J 2012 Human skeletal dysplasia caused by a constitutive activated transient receptor potential vanilloid 4 (TRPV4) cation channel mutation *Exp. Mol. Med.* **44** 707–22

Kasimova M A, Yazici A T, Yudin Y, Granata D, Klein M L, Rohacs T and Carnevale V 2018 A hypothetical molecular mechanism for TRPV1 activation that invokes rotation of an S6 asparagine *J. Gen. Physiol.* **150** 1554–66

Kim D and Cavanaugh E J 2007 Requirement of a soluble intracellular factor for activation of transient receptor potential A1 by pungent chemicals: role of inorganic polyphosphates *J. Neurosci.* **27** 6500–509

Kwon D H, Zhang F, Suo Y, Bouvette J, Borgnia M J and Lee S Y 2021 Heat-dependent opening of TRPV1 in the presence of capsaicin *Nat. Struct. Mol. Biol.* **28** 554–63

Laursen W J, Schneider E R, Merriman D K, Bagriantsev S N and Gracheva E O 2016 Low-cost functional plasticity of TRPV1 supports heat tolerance in squirrels and camels *Proc. Natl Acad. Sci. USA* **113** 11342–7

Lee J, Cha S K, Sun T J and Huang C L 2005 PIP2 activates TRPV5 and releases its inhibition by intracellular Mg^{2+} *J. Gen. Physiol.* **126** 439–51

Levey D J, Tewksbury J J, Cipollini M L and Carlo T A 2006 A field test of the directed deterrence hypothesis in two species of wild chili *Oecologia* **150** 61–8

Li M, Jiang J and Yue L 2006 Functional characterization of homo- and heteromeric channel kinases TRPM6 and TRPM7 *J. Gen. Physiol.* **127** 525–37

Liao M, Cao E, Julius D and Cheng Y 2013 Structure of the TRPV1 ion channel determined by electron cryo-microscopy *Nature* **504** 107–12

Lishko P V, Procko E, Jin X, Phelps C B and Gaudet R 2007 The ankyrin repeats of TRPV1 bind multiple ligands and modulate channel sensitivity *Neuron* **54** 905–18

Liu B, Hui K and Qin F 2003 Thermodynamics of heat activation of single capsaicin ion channels VR1 *Biophys. J.* **85** 2988–3006

Liu B, Yao J, Wang Y, Li H and Qin F 2009 Proton inhibition of unitary currents of vanilloid receptors *J. Gen. Physiol.* **134** 243–58

Liu C H, Wang T, Postma M, Obukhov A G, Montell C and Hardie R C 2007 *In vivo* identification and manipulation of the Ca^{2+} selectivity filter in the *Drosophila* transient receptor potential channel *J. Neurosci.* **27** 604–15

Long S B, Campbell E B and Mackinnon R 2005 Crystal structure of a mammalian voltage-dependent Shaker family K^+ channel *Science* **309** 897–903

Lyumkis D 2019 Challenges and opportunities in cryo-EM single-particle analysis *J. Biol. Chem.* **294** 5181–97

MacGregor G G, Dong K, Vanoye C G, Tang L, Giebisch G and Hebert S C 2002 Nucleotides and phospholipids compete for binding to the C terminus of KATP channels *Proc. Natl Acad. Sci. USA* **99** 2726–31

Macpherson L J, Dubin A E, Evans M J, Marr F, Schultz P G, Cravatt B F and Patapoutian A 2007 Noxious compounds activate TRPA1 ion channels through covalent modification of cysteines *Nature* **445** 541–5

Makhatadze G I and Privalov P L 1990 Heat capacity of proteins. I. Partial molar heat capacity of individual amino acid residues in aqueous solution: hydration effect *J. Mol. Biol.* **213** 375–84

Mandadi S, Tominaga T, Numazaki M, Murayama N, Saito N, Armati P J, Roufogalis B D and Tominaga M 2006 Increased sensitivity of desensitized TRPV1 by PMA occurs through PKCepsilon-mediated phosphorylation at S800 *Pain* **123** 106–16

Martinez G Q and Gordon S E 2019 Multimerization of *Homo sapiens* TRPA1 ion channel cytoplasmic domains *PLoS One* **14** e0207835

McCleverty C J, Koesema E, Patapoutian A, Lesley S A and Kreusch A 2006 Crystal structure of the human TRPV2 channel ankyrin repeat domain *Protein Sci.* **15** 2201–6

McKay R R, Chen D M, Miller K, Kim S, Stark W S and Shortridge R D 1995 Phospholipase C rescues visual defect in norpA mutant of *Drosophila melanogaster J. Biol. Chem.* **270** 13271–6

McKemy D D, Neuhausser W M and Julius D 2002 Identification of a cold receptor reveals a general role for TRP channels in thermosensation *Nature* **416** 52–8

Minke B 1982 Light-induced reduction in excitation efficiency in the trp mutant of *Drosophila* *J. Gen. Physiol.* **79** 361–85

Minke B 2010 The history of the *Drosophila* TRP channel: the birth of a new channel superfamily *J. Neurogenet.* **24** 216–33

Montell C 2005 The TRP superfamily of cation channels *Sci STKE* **2005** re3

Montell C and Rubin G M 1989 Molecular characterization of the *Drosophila* trp locus: a putative integral membrane protein required for phototransduction *Neuron* **2** 1313–23

Nadezhdin K D, Neuberger A, Nikolaev Y A, Murphy L A, Gracheva E O, Bagriantsev S N and Sobolevsky A I 2021 Extracellular cap domain is an essential component of the TRPV1 gating mechanism *Nat. Commun.* **12** 2154

Nilius B, Vennekens R, Prenen J, Hoenderop J G, Droogmans G and Bindels R J 2001 The single pore residue Asp542 determines Ca^{2+} permeation and Mg^{2+} block of the epithelial Ca^{2+} channel *J. Biol. Chem.* **276** 1020–5

Numazaki M, Tominaga T, Toyooka H and Tominaga M 2002 Direct phosphorylation of capsaicin receptor VR1 by protein kinase Cepsilon and identification of two target serine residues *J. Biol. Chem.* **277** 13375–8

Okada T *et al* 1999 Molecular and functional characterization of a novel mouse transient receptor potential protein homologue TRP7. Ca^{2+}-permeable cation channel that is constitutively activated and enhanced by stimulation of G protein-coupled receptor *J. Biol. Chem.* **274** 27359–70

Paulsen C E, Armache J P, Gao Y, Cheng Y and Julius D 2015 Structure of the TRPA1 ion channel suggests regulatory mechanisms *Nature* **525** 552

Peier A M *et al* 2002 A TRP channel that senses cold stimuli and menthol *Cell* **108** 705–15

Peretz A, Suss-Toby E, Rom-Glas A, Arnon A, Payne R and Minke B 1994 The light response of *Drosophila* photoreceptors is accompanied by an increase in cellular calcium: effects of specific mutations *Neuron* **12** 1257–67

Phelps C B, Huang R J, Lishko P V, Wang R R and Gaudet R 2008 Structural analyses of the ankyrin repeat domain of TRPV6 and related TRPV ion channels *Biochemistry* **47** 2476–84

Phillips A M, Bull A and Kelly L E 1992 Identification of a *Drosophila* gene encoding a calmodulin-binding protein with homology to the trp phototransduction gene *Neuron* **8** 631–42

Premkumar L S, Agarwal S and Steffen D 2002 Single-channel properties of native and cloned rat vanilloid receptors *J. Physiol.* **545** 107–17

Privalov P L and Makhatadze G I 1992 Contribution of hydration and non-covalent interactions to the heat capacity effect on protein unfolding *J. Mol. Biol.* **224** 715–23

Privalov P L and Makhatadze G I 1993 Contribution of hydration to protein folding thermodynamics. II. The entropy and Gibbs energy of hydration *J. Mol. Biol.* **232** 660–79

Pumroy R A, Samanta A, Liu Y, Hughes T E, Zhao S, Yudin Y, Rohacs T, Han S and Moiseenkova-Bell V Y 2019 Molecular mechanism of TRPV2 channel modulation by cannabidiol *Elife* **8** e48792

Ramsey I S, Delling M and Clapham D E 2006 An introduction to TRP channels *Annu. Rev. Physiol.* **68** 619–47

Ranganathan R, Bacskai B J, Tsien R Y and Zuker C S 1994 Cytosolic calcium transients: spatial localization and role in *Drosophila* photoreceptor cell function *Neuron* **13** 837–48

Ranganathan R, Harris G L, Stevens C F and Zuker C S 1991 A *Drosophila* mutant defective in extracellular calcium-dependent photoreceptor deactivation and rapid desensitization *Nature* **354** 230–2

Rohács T, Lopes C M, Michailidis I and Logothetis D E 2005 PI(4,5)P2 regulates the activation and desensitization of TRPM8 channels through the TRP domain *Nat. Neurosci.* **8** 626–34

Rosenbaum T, Awaya M and Gordon S E 2002 Subunit modification and association in VR1 ion channels *BMC Neurosci.* **3** 4

Saotome K, Singh A K, Yelshanskaya M V and Sobolevsky A I 2016 Crystal structure of the epithelial calcium channel TRPV6 *Nature* **534** 506–11

Schaefer M, Plant T D, Obukhov A G, Hofmann T, Gudermann T and Schultz G 2000 Receptor-mediated regulation of the nonselective cation channels TRPC4 and TRPC5 *J. Biol. Chem.* **275** 17517–26

Sedgwick S G and Smerdon S J 1999 The ankyrin repeat: a diversity of interactions on a common structural framework *Trends Biochem. Sci.* **24** 311–6

Stein A T, Ufret-Vincenty Ca, Hua L, Santana L F and Gordon S E 2006 Phosphoinositide 3-kinase binds to TRPV1 and mediates NGF-stimulated TRPV1 trafficking to the plasma membrane *J. Gen.Physiol.* **128** 509–22

Story G M *et al* 2003 ANKTM1, a TRP-like channel expressed in nociceptive neurons, is activated by cold temperatures *Cell* **112** 819–29

Stratiievska A, Nelson S, Senning E N, Lautz J D, Smith S E and Gordon S E 2018 Reciprocal regulation among TRPV1 channels and phosphoinositide 3-kinase in response to nerve growth factor *Elife* **7** e38869

Strübing C, Krapivinsky G, Krapivinsky L and Clapham D E 2001 TRPC1 and TRPC5 form a novel cation channel in mammalian brain *Neuron* **29** 645–55

Tang L, Gamal El-Din T M, Payandeh J, Martinez G Q, Heard T M, Scheuer T, Zheng N and Catterall W A 2014 Structural basis for Ca^{2+} selectivity of a voltage-gated calcium channel *Nature* **505** 56–61

Tang Q, Guo W, Zheng L, Wu J X, Liu M, Zhou X, Zhang X and Chen L 2018 Structure of the receptor-activated human TRPC6 and TRPC3 ion channels *Cell Res.* **28** 746–55

Tewksbury J J and Nabhan G P 2001 Seed dispersal. Directed deterrence by capsaicin in chilies *Nature* **412** 403–04

Vinayagam D, Mager T, Apelbaum A, Bothe A, Merino F, Hofnagel O, Gatsogiannis C and Raunser S 2018 Electron cryo-microscopy structure of the canonical TRPC4 ion channel *Elife* **7** e36615

Voets T, Owsianik G, Janssens A, Talavera K and Nilius B 2007 TRPM8 voltage sensor mutants reveal a mechanism for integrating thermal and chemical stimuli *Nat. Chem. Biol.* **3** 174–82

Voets T, Prenen J, Vriens J, Watanabe H, Janssens A, Wissenbach U, Bödding M, Droogmans G and Nilius B 2002 Molecular determinants of permeation through the cation channel TRPV4 *J. Biol. Chem.* **277** 33704–10

Wang J and Moore P B 2017 On the interpretation of electron microscopic maps of biological macromolecules *Protein Sci.* **26** 122–9

Wang L, Fu T M, Zhou Y, Xia S, Greka A and Wu H 2018 Structures and gating mechanism of human TRPM2 *Science* 362

Wes P D, Chevesich J, Jeromin A, Rosenberg C, Stetten G and Montell C 1995 TRPC1, a human homolog of a *Drosophila* store-operated channel *Proc. Natl Acad. Sci. USA* **92** 9652–656

Winkler P A, Huang Y, Sun W, Du J and Lü W 2017 Electron cryo-microscopy structure of a human TRPM4 channel *Nature* **552** 200–4

Xiao R, Tang J, Wang C, Colton C K, Tian J and Zhu M X 2008 Calcium plays a central role in the sensitization of TRPV3 channel to repetitive stimulations *J. Biol. Chem.* **283** 6162–74

Yang F, Vu S, Yarov-Yarovoy V and Zheng J 2016 Rational design and validation of a vanilloid-sensitive TRPV2 ion channel *Proc. Natl Acad. Sci. USA* **113** E3657–66

Yang F, Xiao X, Cheng W, Yang W, Yu P, Song Z, Yarov-Yarovoy V and Zheng J 2015 Structural mechanism underlying capsaicin binding and activation of the TRPV1 ion channel *Nat. Chem. Biol.* **11** 518–24

Yao J, Liu B and Qin F 2010 Kinetic and energetic analysis of thermally activated TRPV1 channels *Biophys. J.* **99** 1743–53

Yao J, Liu B and Qin F 2011 Modular thermal sensors in temperature-gated transient receptor potential (TRP) channels *Proc. Natl Acad. Sci. USA* **108** 11109–14

Yin Y, Le S C, Hsu A L, Borgnia M J, Yang H and Lee S Y 2019 Structural basis of cooling agent and lipid sensing by the cold-activated TRPM8 channel *Science* **363**

Zagotta W N, Gordon M T, Senning E N, Munari M A and Gordon S E 2016 Measuring distances between TRPV1 and the plasma membrane using a noncanonical amino acid and transition metal ion FRET *J. Gen. Physiol.* **147** 201–16

Zhang F, Hanson S M, Jara-Oseguera A, Krepkiy D, Bae C, Pearce L V, Blumberg P M, Newstead S and Swartz K J 2016 Engineering vanilloid-sensitivity into the rat TRPV2 channel *Elife* **5** e16409

Zhang F, Jara-Oseguera A, Chang T H, Bae C, Hanson S M and Swartz K J 2018 Heat activation is intrinsic to the pore domain of TRPV1 *Proc. Natl Acad. Sci. USA* **115** E317–324

Zhang F, Swartz K J and Jara-Oseguera A 2019 Conserved allosteric pathways for activation of TRPV3 revealed through engineering vanilloid-sensitivity *Elife* **8** e42756

Zhang K, Julius D and Cheng Y 2021 Structural snapshots of TRPV1 reveal mechanism of polymodal functionality *Cell* **184** 5138–50

Zhou X, Li M, Su D, Jia Q, Li H, Li X and Yang J 2017 Cryo-EM structures of the human endolysosomal TRPML3 channel in three distinct states *Nat. Struct. Mol. Biol.* **24** 1146–54

Zhu X, Chu P B, Peyton M and Birnbaumer L 1995 Molecular cloning of a widely expressed human homologue for the *Drosophila trp* gene *FEBS Lett.* **373** 193–8

Zubcevic L, Herzik M A, Chung B C, Liu Z, Lander G C and Lee S Y 2016 Cryo-electron microscopy structure of the TRPV2 ion channel *Nat. Struct. Mol. Biol.* **23** 180–6

Zubcevic L, Hsu A L, Borgnia M J and Lee S Y 2019 Symmetry transitions during gating of the TRPV2 ion channel in lipid membranes *Elife* **8** e45779

Chapter 4

Glutamate-gated calcium currents in the central nervous system: structural determinants, regulatory mechanisms, and biological functions

Gary J Iacobucci and Gabriela K Popescu

Glutamate is the main neurotransmitter in the mammalian central nervous system. It gates calcium-rich excitatory postsynaptic currents by binding to and opening ionotropic glutamate receptors (iGluRs). Although iGluRs are generally non-selective, cation-permeable channels, and thus all mediate some calcium influx, NMDA receptors are the main sources of glutamatergic calcium in postsynaptic compartments. NMDA receptor-mediated calcium fluxes are responsible for essential physiologic processes such as synapse formation, maintenance, and for the activity-dependent plasticity of synaptic connections, which underlies learning and memory. They also initiate the toxic actions of glutamate associated with several acute and chronic neuropsychiatric disorders. Given the critical processes set in motion by NMDA receptor-mediated calcium fluxes, several mechanisms are in place to modulate and control them. Here, we review these mechanisms and their impact on critical functions of the central nervous tissue.

4.1 Introduction

Glutamate mediates fast neurotransmission at the vast majority of mammalian synapses in the central nervous system (CNS). The first clue to its powerful excitatory action came from the observation that when injected into primate brain *in vivo*, it produced immediate muscular spasms (Hayashi 1954). Additionally, when applied for a longer period to *ex vivo* neuronal tissue, it was strongly neurotoxic (Lucas and Newhouse 1957, Olney 1969). Over the intervening decades, it has become clear that both the depolarizing and toxic actions of glutamate are the result of its binding to an assortment of ionotropic glutamate receptors (iGluRs) and of the

doi:10.1088/978-0-7503-2009-2ch4

cationic currents thus produced (Curtis *et al* 1959, Johnston *et al* 1974, McCulloch *et al* 1974). Notably, a substantial fraction of the glutamate-gated current is carried by calcium ions and this feature is critical for the cellular outcome of glutamate exposure (Wigstrom *et al* 1979, Lynch *et al* 1983, Choi 1985, 1987, Olney *et al* 1986).

Molecular cloning has documented a family of 18 homologous mammalian genes encoding iGluR subunits, which are widely expressed in central neurons and some glial cells (Hollmann and Heinemann 1994, Traynelis *et al* 2010). Of these, GluA1-4, GluK1-6, and GluN1-3 subunits combine within each subclass to form tetrameric glutamate-gated channels named for their sensitivity to the eponymous glutamate analogues: AMPA (α-amino-3-hydroxy-5-methyl-4-isoxazolepropionic acid), kainate, and NMDA (N-methyl-D-aspartic acid) receptors, respectively. Most CNS cells express several iGluR subunits contemporaneously and the excitatory current elicited by glutamate often consists of a mixture of AMPA-sensitive and NMDA-sensitive components (Traynelis *et al* 2010). The presence of iGluRs in the plasma membrane is dynamic and highly regulated. Receptors are capable of clustering into tight postsynaptic configurations, often located directly vis à vis presynaptic release sites (Scheefhals and MacGillavry 2018), and can move in and out of postsynaptic regions in a controlled manner (Tovar and Westbrook 2002, Borgdorff and Choquet 2002). For these reasons, the exact molecular composition of native iGluRs, at postsynaptic sites or elsewhere, remains unknown beyond pharmacologic dissection of the glutamate-elicited current. Similarly, the calcium content and the precise molecular origins of the excitatory postsynaptic current (EPSC) remain to be determined.

Much of the present knowledge on the biophysical properties of glutamate-gated channels, including information about their ability to flux calcium, and their sensitivity to extracellular and intracellular calcium levels originates from experiments with mixed native receptors or with recombinant receptors expressed in heterologous cells. A systematic investigation of the calcium-passing properties of iGluRs is currently lacking, and the diversity of preparations and experimental conditions for which reported results exist allow only general conclusions. Even so, it has become clear that the vast majority of glutamate-gated calcium currents in CNS emanate from NMDA receptor activations. Therefore, this review will focus mainly on mechanisms that mediate and regulate calcium fluxes through NMDA receptor channels.

Ample evidence demonstrates that calcium ions and NMDA receptors interact in complex ways. Calcium permeates open NMDA receptors (Dingledine 1983, MacDermott *et al* 1986), and thus NMDA receptor activations produce calcium transients with critical consequences for cellular functions. In addition, calcium ions can change NMDA receptor activity with several mechanisms, each observable within distinct concentration ranges and developing with distinct kinetics (Mayer and Westbrook 1985, Ascher and Nowak 1988, MacDonald *et al* 1989, Zorumski *et al* 1989). As with other modulators, the many aspects of NMDA receptor interactions with calcium have come into view incrementally, often boosted by critical discoveries and technological advances. For example, glycine became recognized as an obligatory co-agonist for NMDA receptor activation in the late

1980s (Kleckner and Dingledine, 1988, Johnson and Ascher 1987). For this reason, prior experiments were done with no added glycine, such that the uncontrolled levels of 'background' glycine introduced unknown variability. Similarly, the temporal control of ligand application, the resolution of current measurement, and the ability to record currents from central neurons required new technologies such as the patch-clamp technique (Hamill *et al* 1981), the development calcium imaging dyes (Grynkiewicz *et al* 1985), methods for the *in vitro* culture of central neurons (Beaujouan *et al* 1982), and many others. In particular, the cloning of NMDA receptor subunits provided a molecular basis for the observed multiplicity of NMDA receptor functions and made available critical tools for more rigorous investigations (Moriyoshi *et al* 1991, Sucher *et al* 1996, Monyer *et al* 1992).

To understand the complex relationships between fluctuations in calcium levels and iGluR currents, and how these changes affect CNS functions, research has focused on delineating how calcium-dependent mechanisms control the amplitude and time course of the glutamate-gated calcium flux. Among these, the present literature documents effects of calcium on the kinetics of current gating and the effects of allosteric modulators, the biophysical mechanisms of channel permeability and block, and more recently into the three-dimensional structures that underlie these functional properties.

4.2 Functional aspects of NMDA receptor gating, conductance, and permeation

The impact of NMDA receptors on biology rests primarily on their ability to generate a glutamate-gated, calcium-rich, excitatory current. Therefore, biophysical properties such as gating, conductance, and permeation are fundamental to the biology they control. Further, the impact of these intrinsic biophysical capacities of NMDA receptors is critically influenced by the pattern of stimulation and by the intracellular machinery that will further decode the biological significance of the calcium signals generated. Therefore, the experimental settings in which biophysical properties of ligand-gated ion channels, such as NMDA receptors, can be examined in a rigorous and mechanistic manner require control of as many variables as possible and as such they scarcely ever reproduce the *in situ* environment under which natural receptors operate. This reality requires continuous and consistent integration of information obtained with different approaches, especially as new technologies become available.

4.2.1 Characteristic features of the macroscopic current

The portion of the excitatory postsynaptic potential (EPSP) mediated by NMDA receptors can be observed reliably by recording excitatory postsynaptic currents (EPSC) in the presence of controlled concentrations of glycine in Mg^{2+}-free solutions (Ault *et al* 1980, Johnson and Ascher 1987), and in the presence of a non-NMDA receptor antagonist, such as CNQX (Blake *et al* 1988, Honore *et al* 1988). Groundbreaking work revealed that the NMDA receptor-EPSC is a slow-rising and surprisingly long-lasting current, clearly different from the fast rising short-lived current mediated by non-NMDA receptors (figure 4.1(a)). In hippocampal neurons, the rise times observed for NMDA receptor currents (10–50 ms) are orders of magnitude longer relative to those observed for AMPA receptors (0.2–0.4 ms), and the current decline is also much slower for NMDA receptors (50–500 ms) relative to AMPA receptors (~2 ms) (Lester *et al* 1990). These characteristic kinetics, which are largely dictated by intrinsic differences between the reaction mechanisms of the two receptor types (Clements *et al* 1992), imply that when activated, NMDA receptors are poised to pass substantially more charge than non-NMDA receptors. Further, given their generally higher affinity for glutamate, incomplete desensitization, and larger single-channel conductance it is highly probable that NMDA receptors mediate the majority of glutamate-evoked currents at extrasynaptic locations NMDA (figure 4.1(b)).

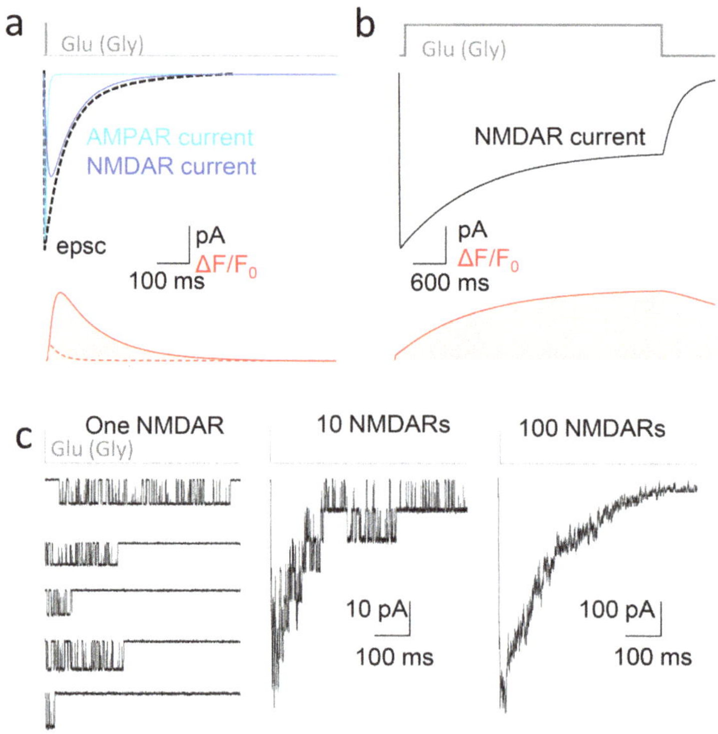

Figure 4.1. NMDA receptors generate the majority of glutamate-gated calcium current in CNS. (a) *Top*, excitatory postsynaptic current (dashed) is the ensemble response of AMPA (AMPAR, blue) and NMDA (NMDAR, red) receptors. The relative abundance of the two receptor types varies across preparations. *Bottom*, change in intracellular calcium concentration recorded as the fluorescence waveform generated by AMPA (dashed line) and NMDA (solid line) receptors. (b) Whole-cell current (*top*) recorded from NMDA receptors during chronic exposure to glutamate (1 mM) and the corresponding elevation of intracellular calcium concentration (*bottom*). (c) Current recorded from an excised patch containing one NMDA receptor following a series of synaptic-like glutamate pulses (1 ms, 1 mM) (*left*), and the combined response from 10 (*middle*) and 100 (*right*) simultaneously stimulated channels.

Aside from electrophysiological observations of NMDA receptor activity, the availability of calcium-sensitive dyes and more recently of genetically encoded calcium sensors has allowed direct visualization of NMDA receptor-mediated calcium fluxes (MacDermott *et al* 1986, Schneggenburger *et al* 1993). These approaches have confirmed that NMDA receptors are the main calcium sources at postsynaptic sites and that a substantial portion of the NMDA receptor current is carried by calcium (figures 4.1(a) and (b)). Notably, currents recorded from recombinant NMDA receptors expressed in heterologous systems closely reproduce those obtained from neuronal preparations (Stern *et al* 1992, Borschel *et al* 2012), suggesting that auxiliary subunits, if any exist, have little influence on gating and permeation, and therefore on the NMDA receptor-mediated calcium current. This is in contrast to the drastic effects produced by auxiliary subunits on the biophysical and pharmacological properties of non-NMDA receptors (Tomita and Castillo 2012, Jackson and Nicoll 2011).

In considering these broad generalizations, it will be important to keep in mind that both NMDA and non-NMDA receptors exist as multiple molecular entities with a range of kinetic and permeation properties, which can further change in response to endogenous modulators. Importantly, rigorous studies exist for only a fraction of the naturally occurring iGluRs. Therefore, as information continues to accumulate about the characteristic features of individual receptors in this family, very likely a more nuanced picture will emerge describing the depolarizing effects of and calcium signals produced by glutamate in the CNS.

4.2.2 Characteristic features of the single-molecule current

The characteristic features of the ensemble responses observed for both NMDA and non-NMDA receptors originate from their distinct gating and permeation characteristics, which can be investigated in mechanistic detail by examining currents generated by individual receptors. Generally, NMDA receptors have substantially higher equilibrium open probabilities (P_o) relative to AMPA and kainate receptors, even if this property, which is sensitive to a variety of regulatory mechanisms, manifests *in situ* for all iGluR representatives as a range of values. Generally, the lower P_o of non-NMDA receptors reflect shorter openings, shorter bursts of openings in response to a single glutamate pulse, and longer desensitized intervals (Howe 1996, Swanson *et al* 1996, 1997). By contrast, the higher activity of NMDA receptors reflect longer openings, longer activations, and shorter desensitized periods (Howe *et al* 1988, 1991, Gibb and Colquhoun 1991).

Similarly, the unitary conductance properties of NMDA and non-NMDA receptors are distinct. Single-channel currents recorded from non-NMDA receptors reveal up to four conductance levels ranging from 9 to 45 pS (Swanson *et al* 1997, Tomita *et al* 2005, Banke *et al* 2000, Smith *et al* 2000). This feature imparts substantial complexity onto the microscopic signal and has limited its comprehensive kinetic modeling (Zhang *et al* 2008, Poon *et al* 2010, 2011). Both AMPA and kainate receptors can pass mixed cationic currents and their measured fractional Ca^{2+} currents appear similar to the bulk Ca^{2+} content in the extracellular space (2%). Even if small, it is highly likely that this calcium current is of biological importance because it is tightly regulated. Notably, certain AMPA receptor subunits can be modified post-transcriptionally to render receptors completely impermeable to calcium (Hollmann *et al* 1991, Hume *et al* 1991, Verdoorn *et al* 1991, Burnashev *et al* 1992a).

By contrast, NMDA receptors have large and relatively uniform unitary conductance (Nowak *et al* 1984, Jahr and Stevens 1987, Howe *et al* 1991), which can be observed for long periods of time under conditions of continuous stimulation. Importantly, summation of unitary activations obtained with a series of synaptic-like activations recapitulates responses recorded in excised patches with brief glutamate application, and approximates the shape of NMDA receptor-EPSCs (figure 4.1(c)). This feature, together with the analytical tractability of their microscopic kinetics, have allowed mechanistic investigations of NMDA receptor gating and reliable extrapolation of mechanisms inferred from microscopic observations to macroscopic, synaptically-relevant behaviors (Iacobucci and Popescu 2017a).

All NMDA receptor subtypes are highly permeable to calcium and calcium ions may carry as much as 20% of the total charge. This observation suggests that NMDA receptors have the ability to enrich the Ca^{2+} content of their current (P_{Ca}/P_{Na} ~ 10) (Burnashev *et al* 1995, Mayer and Westbrook 1987, Ascher and Nowak 1988, Iino *et al* 1990, Schneggenburger 1996) to produce a substantial calcium flux (Schneggenburger *et al* 1993, Wollmuth and Sakmann 1998, Jahr and Stevens 1993). Combined with their relatively large unitary conductance, and prolonged high P_o activations, this feature of NMDA receptors ensures that in addition to depolarizing the postsynaptic membrane, their activations also produce considerable calcium entry. These speculations have been supported by calcium imaging of dendritic spines following focal glutamate uncaging, which demonstrated that NMDA receptors represent the main source of postsynaptic calcium and that, depending on endogenous calcium-binding proteins, a single activation can elevate the bulk amplitude of calcium in spines up to 10 μM (Sabatini *et al* 2002, Higley and Sabatini 2012). Therefore, in addition to the type of simulation (phasic or tonic), and the intrinsic properties of the iGluRs present, the final intracellular calcium elevation will also depend on the cellular calcium buffers, effectors, and extrusion mechanisms present in each preparation.

4.2.3 Functional aspects of NMDA receptor interactions with calcium

The precise quantification of calcium currents produced by NMDA receptors across physiological and pathological conditions still faces many difficulties. Aside from the poorly characterized composition of receptors responding to glutamate *in situ* and the lack of systematic investigation of gating and permeability across NMDA receptor representatives, physiologic Ca^{2+} concentrations affect both the conductance and the activation kinetics of NMDA receptors, with direct and indirect mechanisms, and become apparent over characteristic concentration and time domains.

Relative to sodium-only conditions, physiological levels of extracellular Ca^{2+} reduce substantially the macroscopic NMDA receptor currents (Mayer and

Westbrook 1987, Ascher *et al* 1988, Ascher and Nowak 1988). Ascher and Nowak were first to show that as the extracellular calcium concentration increases, the unitary NMDA receptor conductance decreases linearly from a high value (γ_{Na}, 70 pS), when the sodium ions carry all the current, to a much lower value (17 pS), when calcium ions make up all the inward current (Ascher and Nowak 1988). They referred to this phenomenon as 'calcium block', although the effect is voltage independent and therefore unlikely to reflect a classic open-channel block mechanism (Neher and Steinbach 1978) (figure 4.2(a)). In addition to reducing channel

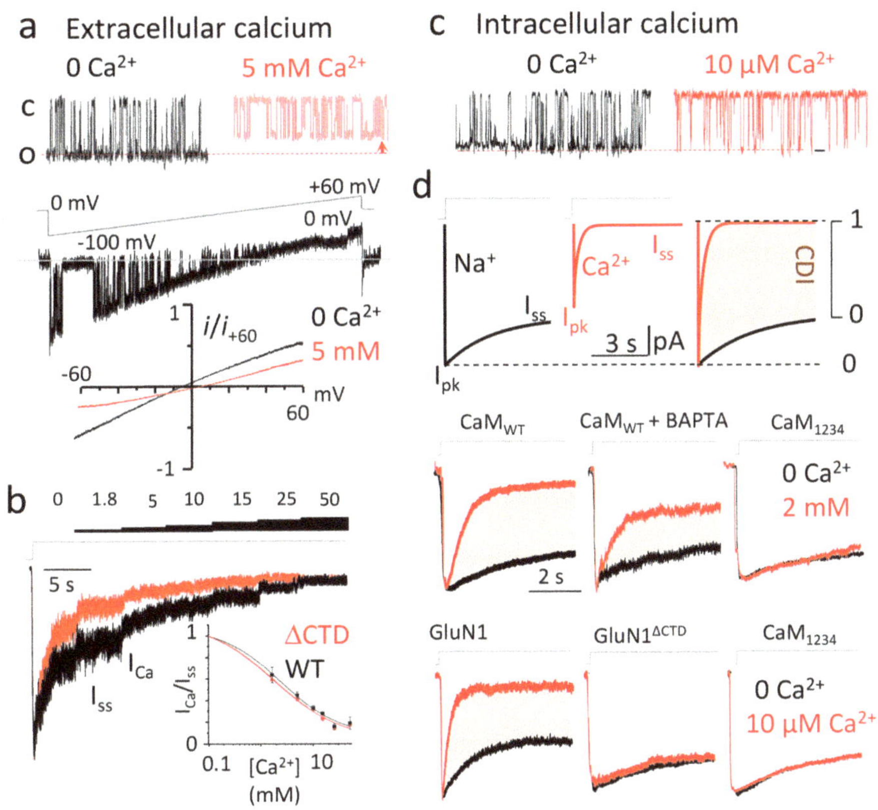

Figure 4.2. Calcium reduces NMDA receptor currents by multiple mechanisms. (a) Extracellular calcium reduces the unitary current amplitude in cell attached patches and shifts the current reversal potential to more positive values, indicative of voltage-independent block, and Ca^{2+} permeability, respectively. (b) Extracellular calcium concentrations reduce whole-cell currents (IC_{50}, 2.4 mM) independently of the receptors' intracellular domain (WT, wild type; ΔCTD, CTD truncation) (adapted from Maki and Popescu (2014), originally published in *Journal of General Physiology*). (c) Intracellular calcium, fluxed through ionophore applied in bath, shortens openings and extends closures of single-channel inward sodium currents in cell-attached patches, without affecting unitary conductance (Iacobucci and Popescu 2019a). (d) *Top*, protocol used to quantify CDI (by intracellular Ca^{2+}) independently of effects of extracellular Ca^{2+}, and of intrinsic differences in kinetics and conductance across receptor preparations. *Middle*, CDI produced by NMDA receptor-fluxed Ca^{2+} is reduced by intracellular BAPTA and abolished by CaM mutants that cannot bind Ca^{2+} (CaM_{1234}). *Bottom*, CDI produced by direct Ca^{2+} dialysis requires intact GluN1 CTD and Ca^{2+} binding to CaM (reprinted from Iacobucci and Popescu (2017b), copyright (2017), with permission from Elsevier).

conductance, physiologic levels of extracellular calcium reduce NMDA receptor currents with a direct allosteric mechanism (figure 4.2(b)). This inhibition is fast and promptly reversible, with a half-maximal dose close to physiological calcium concentrations (2.4 mM, residual current 8%) (Maki and Popescu 2014).

In contrast, increasing concentrations of intracellular calcium reduce NMDA receptor activity with a purely allosteric and indirect mechanism (figure 4.2(c)). This inhibition manifests as an increase in macroscopic desensitization and for this reason has been referred to in the literature as calcium-dependent desensitization or calcium-dependent inactivation (CDI) (Mayer and Westbrook 1985, Legendre *et al* 1993, Ehlers *et al* 1996). CDI is independent from the external effects of calcium, and it can be elicited by calcium fluxed by NMDA receptors themselves, as well as by dialyzing calcium experimentally into cells (Iacobucci and Popescu 2017b) (figure 4.2(d)). CDI of NMDA receptors occurs with an indirect mechanism, requiring both calcium binding to calmodulin (CaM) and binding of CaM to NMDA receptors, more specifically to the membrane-proximal C0 segments of the GluN1 intracellular tails (Ehlers *et al* 1996, Zhang *et al* 1998, Akyol *et al* 2004). Therefore, CDI is an indirect allosteric mechanism that can be initiated by calcium fluxed by NMDA receptors or by vicinal calcium sources (Iacobucci and Popescu 2019b, Rozov and Burnashev 2016). Relative to the direct effects of extracellular calcium, intracellular calcium ions reduce NMDA currents at much lower concentrations (half-maximal dose ~5 μM, residual current ~20%) and with slower time scale (0.5 s onset, 10 s recovery) (Iacobucci and Popescu 2017b).

On longer time domains, intracellular calcium concentrations modulate NMDA receptor currents by controlling in a reversible manner the phosphorylation status of NMDA receptors (Lieberman and Mody 1994, Wang and Salter 1994). Similarly, calcium can initiate the internalization of receptors from the plasma membrane, producing a slow irreversible decrease in the ensemble current, often referred to as run-down (Zorumski *et al* 1989, Rosenmund and Westbrook 1993, Wechsler and Teichberg 1998, Krupp *et al* 1999). These latter two mechanisms involve the activation of cellular signaling cascades and will not be discussed here.

4.3 Structural aspects of NMDA receptor interactions with calcium

The bidirectional relationship between glutamate receptors and Ca^{2+} is determined both by the biophysical properties of the receptor (gating kinetics, conductance, and permeation), which set the maximal probability of calcium entry into the cell, as well as by the mechanisms that control the effects of Ca^{2+} on receptor function. Recent approaches link quantitatively the kinetic attributes of NMD receptors with atomic-resolution structural models and afford unprecedented opportunities to understand how the unique biophysical properties of this class of proteins arise from structure and how they are controlled by mutations and modulators. Lastly, *in vivo* manipulations of receptors' primary sequence through modern genetic approaches, and a rapidly increasing number of naturally occurring disease-related mutations offer new means to investigate how changes in protein structure relate with changes in kinetic mechanism and therefore in electrophysiological signal and ultimately with animal behavior (Burnashev and Szepetowski 2015, Myers *et al* 2019).

All iGluR representatives have similar overall architecture (figure 4.3(a)). Within each class, functional channels assemble as tetrameric combinations of homologous subunits and each subunit consists of several modules or domains. Along the longitudinal axis, modules are arranged in string-like fashion starting with two stacked extracellular domains, which are globular: a membrane-distal N-terminal domain and a membrane-proximal ligand-binding domain (LBD). The LBD is connected through three short linkers to transmembrane helices (M1, M3, and M4) embedded within a cylindrical, transmembrane domain (TMD), which continues intracellularly with a disordered C-terminal domain (CTD). The NTD binds several diffusible small molecules such as protons, zinc, and ifenprodil, which act as allosteric modulators and control the receptor's kinetics. The LBD binds the native

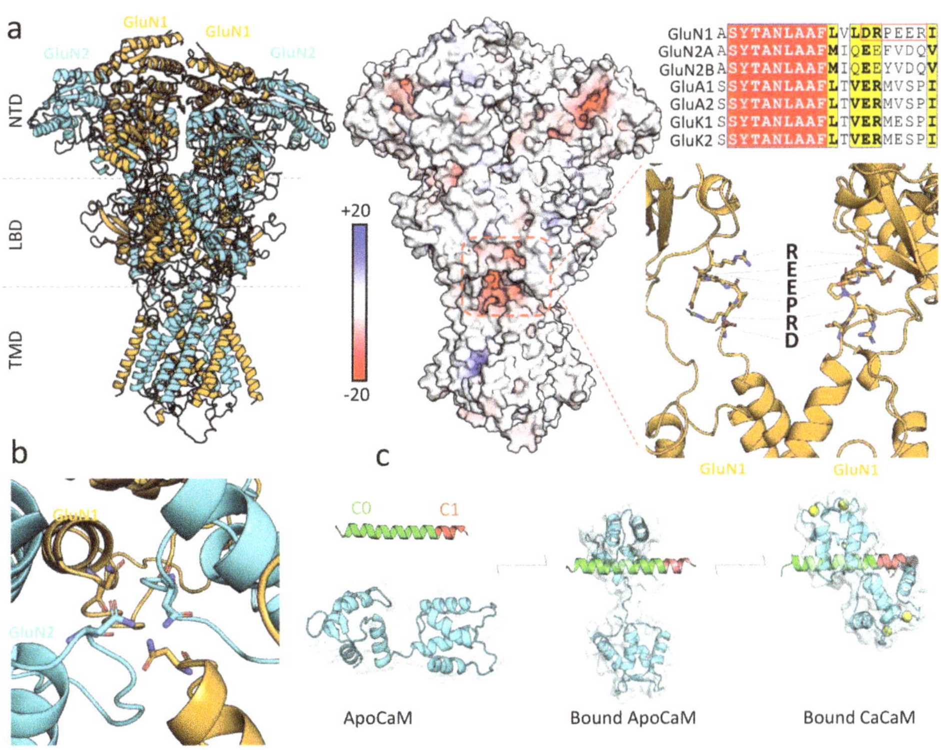

Figure 4.3. Structural determinants mediating NMDA receptor sensitivity to Ca^{2+}. (a) *Left*, structural model of NMDA receptors derived from x-ray crystallographic data (homology model from PDB 4PE5). The intracellular CTD is unresolved. *Middle*, calculated electrostatic surface potential reveals a cluster of negative charge density immediately external to the pore aperture, which includes the GluN1-specific DRPEER segment. (b) A ring of asparagine residues located at the apex of the M2 segment mediates voltage-dependent bock by Mg^{2+} and MK-801 (c) Apo-calmodulin (PDB 2HQW) binds to the C0 segment of the GluN1 CTD via its C-lobe (homology model from PDB 6MBA) and primes receptors for CDI (homology model from PDB 6MC9).

co-agonists, glutamate and glycine or d-serine, or a series of antagonists and in so doing controls the activation gate. In addition to the three transmembrane helices, the TMD also includes a re-entrant P-loop (M2). Close to the external portion of the membrane, the intersection of the four M3 helices forms a hydrophobic ring, whose aperture is governed by the LBD and thus represents the agonist-controlled gate. Deeper within the membrane, the narrowest portion of the pore is lined by M2 residues, specifically a ring of asparagine residues, which control the channel's sensitivity to voltage-dependent cationic blockers such as magnesium, ketamine, and memantine. Lastly the CTD, whose sequence is least conserved across subunits and lacks specific evolutionary origin, is responsible for interactions with cytoskeleton molecules and contains modulatory sites controlled by post-translational modifications or interactions with effector molecules (Hansen *et al* 2018, Zhu and Gouaux 2017, Regan *et al* 2015).

Except for the CTD, whose three-dimensional arrangement remains unknown and is presumed unstructured, several atomic-resolution models are available for NTD and LBD modules of several iGluR subunits and more recently for CTD-truncated tetrameric receptors. Overall, these models confirmed domain-specific properties demonstrated previously with functional approaches, such as binding sites for agonists, antagonists and modulators. Notably, they produced substantial evidence for an activation mechanism where agonist binding between the two mobile lobes of each LBD, controls the position of the M3 helices and thus toggle receptors between impermeable (closed) and permeable (open) conformations (Furukawa and Gouaux 2003, Furukawa *et al* 2005, Jin *et al* 2009, Regan *et al* 2015). In addition, movements in the NTD and CTD modulate the ability of agonist-bound LBD to control the position of the M3 helices and, therefore, the receptor's kinetics (Dolino *et al* 2015, Sirrieh *et al* 2013).

Still unresolved and, therefore, the subject of current research are the structural details of the allosteric cross-talk between domains both within a single subunit, and across subunits; the identity of structural features that determine subtype-specificity of receptor functions at the macroscopic and microscopic levels; and the mechanisms by which disease-related mutations cause receptor dysfunction. Specifically, the molecular determinants responsible for this receptor's high calcium permeability and the mechanisms by which calcium ions reduce channel amplitude and kinetics remain insufficiently understood.

4.3.1 Pore determinants of calcium-dependent modulatory mechanisms

Evolutionarily, iGluRs have ionic permeation pathways that, when inverted, resemble those made famous by potassium channel structures (Doyle *et al* 1998). As for potassium channels, the iGluR pore is lined partly by four transmembrane helices (M3 of each subunit), whose intersection forms the gate, and partly by four re-entrant loops (M2 of each subunit), which form the selectivity filter (Sutcliffe *et al* 1996, Blanpied *et al* 1997, Kuner *et al* 1996, Jahr and Stevens 1987, Premkumar and Auerbach 1996). However, the mechanisms that control ionic flux and selectivity in iGluR are necessarily distinct from those delineated for potassium channels. Notably, in contrast to the sophisticated design of the potassium channel filter, which forms a series of stacked high-affinity potassium binding sites and confers exquisite selectivity (P_K/P_{Na}~1000), the corresponding region in iGluRs appears substantially less structured (Karakas and Furukawa 2014, Lee *et al* 2014, Song *et al* 2018) and affords no selectivity for monovalent cations ($P_{Na}/P_K = 1$) (Jahr and Stevens 1987, Hablitz and Langmoen 1982). In the presently available iGluR structures, this region is insufficiently resolved to offer clues to the mechanism by which these residues participate to permeation and/or block, however, functional measurements showed that residues situated at the apex of the re-entrant M2 loop control the effective pore diameter. In AMPA and kainate receptor subunits, a glutamine (Q) residue occupies this key position. Changing this residue into arginine (Q/R), which occurs naturally and is physiologically controlled, renders receptors impermeable to calcium (Kohler *et al* 1993, Sommer *et al* 1991). Therefore, modifications at this site simply deny calcium access and turn off completely the small and non-specific calcium permeability of these channels (figure 4.3(b)). In NMDA receptor subunits, these residues are also critically involved in calcium permeability and voltage-dependent block (Wollmuth *et al* 1996, Burnashev *et al* 1992b, Premkumar and Auerbach 1996, Jatzke *et al* 2002). However, the effects of modifications at this site are more complex and remain insufficiently delineated.

In the GluN1 and GluN2 subunits, which form the majority of glutamatergic NMDA receptors, the position corresponding to the Q/R site identified in non-NMDA receptors is occupied by an asparagine (N) (Burnashev *et al* 1992a, Sommer *et al* 1991). The *N* residues of the four constituent subunits form a ring structure that is critical for the receptor's physiological function. Naturally occurring substitutions at this site can cause neuropsychiatric disorders, although the exact mechanism of the impairment is unknown (Marwick *et al* 2019, Endele *et al* 2010). Experimental substitutions at this positon identified it as an important determinant of the receptor's conductance (Premkumar *et al* 1997), calcium permeability, and voltage-dependent Mg^{2+} block (Burnashev *et al* 1992b, Wollmuth *et al* 1996, 1998). Notably, in GluN3 subunits, whose incorporation renders channels impermeable to calcium, this position is occupied by a glycine residue (Perez-Otano *et al* 2001).

Residues other than those situated at the apex of the M2 segment are important determinants of both channel conductance and calcium selectivity. Even though all GluN2 subunits have a conserved *N* at this site, channels containing GluN2A and/or GluN2B subunits have conductance and calcium permeability properties that are distinct from channels containing GluN2C and/or GluN2D subunits (Glasgow *et al* 2015, Siegler Retchless *et al* 2012). Further, residues situated on the extracellular portion of the receptor also play critical roles in setting permeation properties of NMDA receptors, as summarized in the following section.

4.3.2 Extracellular determinants of calcium-dependent modulatory mechanisms

Prior to the surge in structural data on NMDA receptors, the vast majority of evidence regarding the physical pathway by which cations permeate NMDA receptors emerged necessarily from functional studies. Two important observations prompted the development of new and original hypotheses regarding the role of extracellular residues in controlling calcium fluxes (Ascher and Nowak 1988). First, the observation that the calcium-dependent reduction in current amplitude is independent of membrane voltage implied that this property of NMDA receptors relies on structural determinants located outside of the membrane field, likely in a vestibule guarding the pore entrance. Second, the observation that increasing the ionic strength of the external solution facilitated the outward flow of current through NMDA receptors fits the hypothesis that several external residues collaborate to produce a negatively charged electrical field, which influences ionic flow through the channel, in both directions. Over the past 30 years, data that substantiate these pioneering hypotheses have continued to accumulate.

Critical information necessary to begin even to imagine the contour of the permeation pathway emerged from accessibility studies with MTS reagents. Scanning mutagenesis of residues on M3 and M4 segments coupled with functional evaluation the receptor's permeation properties identified residues most likely to line the pore (Beck *et al* 1999, Kuner *et al* 1996). Further, the hypothesis that negative charges external to the membrane field likely influence the channel's permeability motivated the examination of residues situated further along M3 helix, which likely protruded from the membrane (Watanabe *et al* 2002). Mutagenesis of residues in this region identified a sequence of charged residues, dubbed the DRPEER motif that critically influences the channel's calcium permeability. This motif is unique to the GluN1 subunit, which being intrinsic to all NMDA receptor proteins is ideally positioned confer properties that distinguish NMDA receptors from other iGluRs. Further, its strong negative electrostatic character and its proximity to the pore provide additional features necessary to influence the flow of ions through the channel. Functional measurements strongly support these hypotheses and demonstrate that the negative charges provided by residues in this motif are required for the characteristically high fraction of calcium ions present in the currents gated by NMDA receptors (Watanabe *et al* 2002).

These features also distinguish the mechanisms by which NMDA receptors flux calcium from those governing calcium permeation in voltage-gated calcium channels (VGCC). In contrast to the calcium currents produced by NMDA receptors, where calcium ions carry up to 15%–20% of the charge, VGCCs strongly select calcium ions from physiological solutions producing virtually pure calcium currents (Simms and Zamponi, 2014, Nanou and Catterall 2018). Size-filtration is an unlikely mechanism for this calcium preference for either channel type, because calcium and sodium ions are relatively similar in size; yet, permeability ratios for calcium relative to sodium are substantial for both NMDA receptors (5–10) and VGCCs (1000) (Mayer and Westbrook 1987, Watanabe *et al* 2002, Wollmuth and Sakmann 1998). In VGCC's a ring of aspartate residues (EEEE locus) bind calcium with high affinity and practically exclude monovalent ions from passing across this true selectivity filter (Tang *et al* 2016, Wu *et al* 2016). In NMDA receptors, the apposition of DRPEER residues on GluN1 subunits may also provide some preference for the passage of calcium ions (Karakas and Furukawa 2014,

Lee *et al* 2014) (figure 4.3(a)). This segment may harbor a true calcium-binding site that affects both calcium permeability and the receptor's opening kinetics (Maki and Popescu 2014, Watanabe *et al* 2002) (figure 4.2(a)).

4.3.3 Intracellular determinants of calcium-dependent modulatory mechanisms

In addition to directly producing elevations in intracellular calcium levels, NMDA receptors are themselves targets of modulation by fluctuations in intracellular calcium ions. In contrast to the direct effects of calcium ions on NMDA receptor functions, which engage external and pore-lining residues, intracellular calcium ions change receptor functions with indirect mechanisms. Among these, investigations into calmodulin-mediated allosteric modulation (CDI) and phosphorylation-dependent regulatory mechanisms have provided the bulk of evidence, so far. Both these processes rely necessarily on residues located on the cytoplasmic portion of NMDA receptors.

The NMDA receptor intracellular domain is characteristically large. It represents one third of the receptor's mass and serves vital roles in receptor expression, trafficking, activity modulation, and function. Despite its functional significance, and aggressive investigations, the structure of NMDA receptor cytoplasmic domain remains unknown. The prevailing hypothesis is that this portion of the receptor lacks an intrinsic stable arrangement, being dependent instead on myriad interaction with cellular structures that provide both anchorage and a communication conduit with signaling cascades (Choi *et al* 2011, Frank and Grant 2017).

The GluN1 subunit includes several alternative splicing sites, which organize this sequence into distinct regions, or cassettes. The C0 cassette is common to al GluN1 subunits, and depending on which two of the other cassettes are present, four primary structures of the GluN1 CTD are possible, corresponding to subunits designated GluN1-1 through GluN1-4 (Zukin and Bennett 1995). Developmental, regional, and activity-related factors determine which GluN1 subunit will be present in any preparation. Each cassette contains unique protein-recognition sites and these sensitize NMDA receptors to distinct regulatory mechanisms. Specifically, calmodulin-binding sites, and serine/threonine residues susceptible to modification by calcium-dependent protein kinases/phosphatases control the access of these calcium-dependent effectors to NMDA receptors (Chen and Roche 2007).

The constitutive C0 cassette is immediately contiguous with the M4 transmembrane helix, hosts a calmodulin-binding site, and is required and sufficient for calcium-dependent inactivation of NMDA receptors. When bound to calmodulin, the C0 segment displays clear helical structure (figure 4.3(c)) (Akyol *et al* 2004). Yet, it is unclear whether this structure persists when alpha-actinin displaces calmodulin from this site (Wyszynski *et al* 1997, Krupp *et al* 1999, Leonard *et al* 2002, Merrill *et al* 2007). The C1 cassette also binds calmodulin and adopts a helical structure when calmodulin occupies this site (Merrill *et al* 2007). It is unclear whether the CaM-binding site on C1 mediates calcium-dependent regulation of NMDA receptor currents *per se* or whether it permits/prevents separate regulatory mechanisms. To this point, CaM binding to C1 enhances C0-mediated CDI but cannot induce CDI in the absence of C0 (Ehlers *et al* 1996, Rafiki *et al* 1997). Similarly, because the C1 segment hosts several phosphorylation sites, its CaM-binding properties may impart control the ability of the responsible protein kinase to access this site, and therefore to change the activity of NMDA receptor in a calcium-dependent manner (Tingley *et al* 1993).

NMDA receptors are direct targets of protein kinases and protein phosphatases and dynamic changes in the phosphorylation status of NMDA receptor residues correlate with dramatic changes in the receptor's activity levels (Chen and Roche 2007, Lieberman and Mody 1994, Wang and Salter 1994, Tingley *et al* 1993, Tong *et al* 1995, Swope *et al* 1999). In turn, intracellular calcium levels control the activity of several protein kinases/phosphatases, and their access to their target sites on NMDA receptors (Zheng *et al* 1997). Therefore, calcium-dependent phosphorylation of NMDA receptors represents a major mechanism by which intracellular calcium levels control NMDA receptor currents. Generally, phosphorylation modulates NMDA receptor currents with an allosteric mechanism, such that phosphorylated receptors have higher activities relative to their phosphate-free counter parts (Lieberman and Mody 1994, Zheng *et al* 1997, Wang and Salter 1994). However, evidence indicate that the phosphorylation status of specific residues can also change the calcium content of the currents carried by NMDA receptors.

Protein kinase A (PKA), a cAMP-dependent protein kinase, phosphorylates several residues on NMDA receptors, and calcium, through calmodulin and adenylate cyclase, controls PKA activity (Tingley *et al* 1997, Raman *et al* 1996, Leonard and Hell 1997, Skeberdis *et al* 2006). Consistent with a potentiation effect of phosphorylation on NMDA receptor currents, PKA blockers reduce the NMDA receptor-mediated calcium influx with profound effects on cellular plasticity and animal behavior (Lau *et al* 2009, Murphy *et al* 2014, Aman *et al* 2014. Skeberdis *et al* 2006). Experiments with recombinant receptors suggest that PKA phosphorylation of NMDA receptors enhances receptor activity and be necessary for its high-calcium permeability. The same experiments suggest that the observed modulatory effects of PKA on NMDA receptor gating and permeation occur through separate residues (Aman *et al* 2014). Given the multiplicity of PKA sensitive sites, and their selective presence on NMDA receptor genetic isoforms and splice variants, the regulation of NMDA receptor currents by PKA and its dependency on intracellular calcium levels is likely more complex.

Protein kinase C (PKC) a calcium- and phospholipid-dependent protein kinase, phosphorylates several NMDA receptor residues and regulates multiple facets of NMDA receptor function with mechanisms that remain poorly understood (Chen and Roche 2007, Leonard and Hell 1997, Tingley *et al* 1997, Lan *et al* 2001). Increasing PKA activity produces sustained facilitation of NMDA receptors in neurons (Chen and Huang 1991, Mori *et al* 1993, Zheng *et al* 1997). Yet in other preparations, PKA potentiation suppresses NMDA receptor currents (Markram and Segal 1992). These apparently contradictory results may reflect the multiplicity of mechanisms by which PKC-dependent increase in NMDA receptor phosphorylation status affects the receptor's activity (Chen and Huang 1992, Wagner and

Leonard 1996, 1999, Liao *et al* 2001, Kelso *et al* 1992, Logan *et al* 1999). In addition, PKA also affects NMDA receptor trafficking and surface retention (Scott *et al* 2003, Lan *et al* 2001).

Calcineurin, a calcium- and CaM-dependent protein phosphatase, also known as PP2B or as protein phosphatase 3, decreases NMDA receptor currents (Tong *et al* 1995, Krupp *et al* 2002). Specifically, this regulation is important in developmentally controlled reduction in NMDA receptor currents (Shi *et al* 2000, Townsend *et al* 2004). The calcineurin effect likely involves residues on the GluN2 CTD (Krupp *et al* 2002, Maki *et al* 2013). However, more information is necessary to understand its mechanism, especially given the biological significance of this regulation.

4.4 Biological functions of calcium-dependent regulatory mechanisms

Several lines of evidence demonstrate that specifically, the calcium fluxed by NMDA receptors controls fundamental brain processes such as neurogenesis, synaptic formation, maturation, plasticity, and even pruning (Lau *et al* 2009, Clapham 2007, Berridge *et al* 2003, Butt 2006). Further, these changes in cellular function and morphology support critical brain function such as learning, memory, cognition, and influence behaviors in myriad ways. Given the wide range and importance of these phenomena for human health and wellbeing, it is important to understand the precise mechanisms by which NMDA receptors produce calcium fluxes, and how in turn, intracellular calcium transients affect the activity of NMDA receptors in real time. Kinetic modeling of NMDA receptor fluxes, both in physiological and in disease-related preparations, can facilitate such quantitative and understanding of the dynamics of NMDA receptor-mediated calcium transients.

Using reaction mechanisms derived from kinetic modeling of individual NMDA receptors in the absence and presence of physiological levels of external calcium ions helps to understand how the layered temporal installation of various calcium-dependent mechanisms affects the postsynaptic calcium transient (figure 4.4(a)) (Iacobucci and Popescu 2019a). Simulations with one such model illustrate the differential influence of calcium-dependent mechanisms on fluxes elicited with brief pulses of glutamate, such as during phasic synaptic transmission (figures 4.4(b) and (c)) or with sustained stimulation (figure 4.4(c)).

Congruent with their critical roles in fundamental brain functions, defective or dysregulated NMDA receptor currents produce pernicious neuropsychiatric pathologies and conditions. Schizophrenia, cognitive delays or impairments, including autism spectrum disorders, often correlate with NMDA receptor hypofunction. In contrast, NMDA receptor hyperfunction results in hyperexcitability and associates with epilepsies, pathologic plasticity related to central sensitization, such as chronic pain, and addictive behaviors, and in neurotoxic phenotypes resulting from acute or chronic neurodegenerative pathologies, such as ischemic stroke, Alzheimer's, Parkinson's, and Huntington's diseases. In the majority of studied cases, an abnormal NMDA receptor-generated calcium transient is an offending or aggravating event. Therefore, to make the promise of rational drug design a reality for NMDA

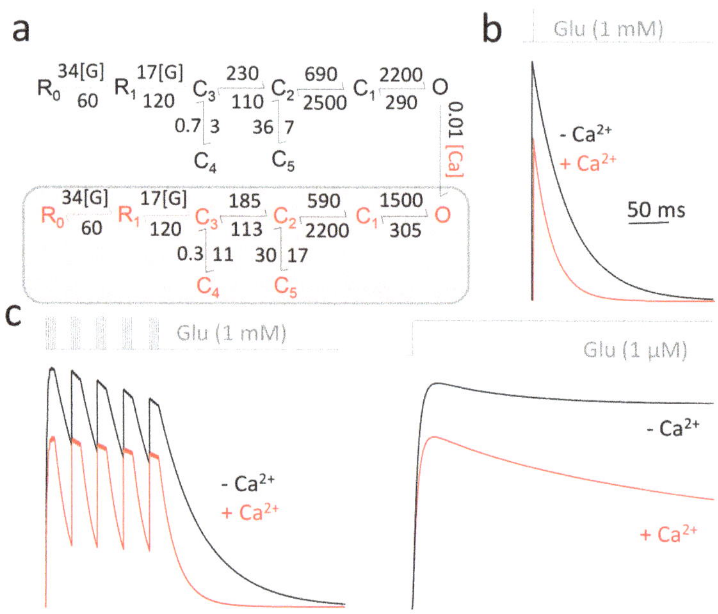

Figure 4.4. Predicted effects of combined Ca^{2+} regulatory mechanisms on NMDA receptor current responses to physiologic stimulation patterns. (a) Tiered kinetic model used for simulations includes a Ca^{2+}-free arm (top) and a Ca^{2+}-bound arm (bottom) was derived for GluN1/GluN2A receptors. R_0 and R_1 represent receptor states un- or mono-liganded with glutamate (resting); C and O, represent glutamate-bound states that are either non-conducting or conducting, respectively. (b) Simulated current responses from 100 channels (10 pA in Ca^{2+}-free, or 8 pA, in 1.8 mM Ca^{2+}). During synaptic-like phasic stimulation (1 ms of 1 mM Glu) Ca^{2+} block and allosteric inhibition by external Ca^{2+} predominate. (c) *Left*, during high frequency stimulation of synaptic receptors (theta-like bursts), fluxed Ca^{2+} engages CaM to further reduce responses through CDI. *Right*, during tonic, low-amplitude stimulation, such as following glutamate spillover into extrasynaptic regions, sustained Ca^{2+} flux elicits full expression of CDI and may prepare receptors for internalization (observed functionally as run-down).

receptor-mediated neuropsychiatric disorders it will be important to delineate the complex interactions between calcium ions and NMDA receptor residues, and how these interactions affect the channel's calcium permeability, ionic conductance, and kinetics.

References

Akyol Z, Bartos J A, Merrill M A, Faga L A, Jaren O R, Shea M A and Hell J W 2004 Apo-calmodulin binds with its C-terminal domain to the N-methyl-D-aspartate receptor NR1 C0 region *J. Biol. Chem.* **279** 2166–75

Aman T K, Maki B A, Ruffino T J, Kasperek E M and Popescu G K 2014 Separate intramolecular targets for protein kinase a control NMDA receptor gating and Ca^{2+} permeability *J. Biol. Chem.* **289** 18805–17

Ascher P, Bregestovski P and Nowak L 1988 NMDA-activated channels of mouse central neurones in magnesium-free solutions *J. Physiol.* **399** 207–26

Ascher P and Nowak L 1988 The role of divalent cations in the N-methyl-D-aspartate responses of mouse central neurones in culture *J. Physiol.* **399** 247–66

Ault B, Evans R H, Francis A A, Oakes D J and Watkins J C 1980 Selective depression of excitatory amino acid induced depolarizations by magnesium ions in isolated spinal cord preparations *J. Physiol.* **307** 413–28

Banke T G, Bowie D, Lee H, Huganir R L, Schousboe A and Traynelis S F 2000 Control of GluR1 AMPA receptor function by cAMP-dependent protein kinase *J. Neurosci.* **20** 89–102

Beaujouan J C, Torrens Y, Herbet A, Daguet M C, Glowinski J and Prochiantz A 1982 Specific binding of an immunoreactive and biologically active 125I-labeled substance P derivative to mouse mesencephalic cells in primary culture *Mol. Pharmacol.* **22** 48–55

Beck C, Wollmuth L P, Seeburg P H, Sakmann B and Kuner T 1999 NMDAR channel segments forming the extracellular vestibule inferred from the accessibility of substituted cysteines *Neuron* **22** 559–70

Berridge M J, Bootman M D and Roderick H L 2003 Calcium signalling: dynamics, homeostasis and remodelling *Nat. Rev. Mol. Cell Biol.* **4** 517–29

Blake J F, Brown M W and Collingridge G L 1988 CNQX blocks acidic amino acid induced depolarizations and synaptic components mediated by non-NMDA receptors in rat hippocampal slices *Neurosci. Lett.* **89** 182–6

Blanpied T A, Boeckman F A, Aizenman E and Johnson J W 1997 Trapping channel block of NMDA-activated responses by amantadine and memantine *J. Neurophysiol.* **77** 309–23

Borgdorff A J and Choquet D 2002 Regulation of AMPA receptor lateral movements *Nature* **417** 649–53

Borschel W F, Myers J M, Kasperek E M, Smith T P, Graziane N M, Nowak L M and Popescu G K 2012 Gating reaction mechanism of neuronal NMDA receptors *J. Neurophysiol.* **108** 3105–15

Burnashev N, Monyer H, Seeburg P H and Sakmann B 1992a Divalent ion permeability of AMPA receptor channels is dominated by the edited form of a single subunit *Neuron* **8** 189–98

Burnashev N, Schoepfer R, Monyer H, Ruppersberg J P, Gunther W, Seeburg P H and Sakmann B 1992b Control by asparagine residues of calcium permeability and magnesium blockade in the NMDA receptor *Science* **257** 1415–9

Burnashev N and Szepetowski P 2015 NMDA receptor subunit mutations in neurodevelopmental disorders *Curr. Opin. Pharmacol.* **20** 73–82

Burnashev N, Zhou Z, Neher E and Sakmann B 1995 Fractional calcium currents through recombinant GluR channels of the NMDA, AMPA and kainate receptor subtypes *J. Physiol.* **485** 403–18

Butt A M 2006 Neurotransmitter-mediated calcium signalling in oligodendrocyte physiology and pathology *Glia* **54** 666–75

Chen B S and Roche K W 2007 Regulation of NMDA receptors by phosphorylation *Neuropharmacology* **53** 362–8

Chen L and Huang L Y 1991 Sustained potentiation of NMDA receptor-mediated glutamate responses through activation of protein kinase C by a mu opioid *Neuron* **7** 319–26

Chen L and Huang L Y 1992 Protein kinase C reduces Mg^{2+} block of NMDA-receptor channels as a mechanism of modulation *Nature* **356** 521–3

Choi D W 1985 Glutamate neurotoxicity in cortical cell culture is calcium dependent *Neurosci. Lett.* **58** 293–7

Choi D W 1987 Ionic dependence of glutamate neurotoxicity *J. Neurosci.* **7** 369–79

Choi U B, Xiao S, Wollmuth L P and Bowen M E 2011 Effect of Src kinase phosphorylation on disordered C-terminal domain of NMDA receptor subunit GluN2B protein *J. Biol. Chem.* **286** 29904–12

Clapham D E 2007 Calcium signaling *Cell* **131** 1047–58

Clements J D, Lester R A, Tong G, Jahr C E and Westbrook G L 1992 The time course of glutamate in the synaptic cleft *Science* **258** 1498–501

Curtis D R, Phillis J W and Watkins J C 1959 Chemical excitation of spinal neurones *Nature* **183** 611–2

Dingledine R 1983 NMDA activates voltage-dependent calcium conductance in rat hippocampal pyramidal cells *J. Physiol.* **343** 385–405

Dolino D M, Cooper D, Ramaswamy S, Jaurich H, Landes C F and Jayaraman V 2015 Structural dynamics of the glycine-binding domain of the N-methyl-D-aspartate receptor *J. Biol. Chem.* **290** 797–804

Doyle D A, Morais Cabral J, Pfuetzner R A, Kuo A, Gulbis J M, Cohen S L, Chait B T and MacKinnon R 1998 The structure of the potassium channel: molecular basis of K^+ conduction and selectivity *Science* **280** 69–77

Ehlers M D, Zhang S, Bernhadt J P and Huganir R L 1996 Inactivation of NMDA receptors by direct interaction of calmodulin with the NR1 subunit *Cell* **84** 745–55

Endele S *et al* 2010 Mutations in GRIN2A and GRIN2B encoding regulatory subunits of NMDA receptors cause variable neurodevelopmental phenotypes *Nat. Genet.* **42** 1021–6

Frank R A and Grant S G 2017 Supramolecular organization of NMDA receptors and the postsynaptic density *Curr. Opin. Neurobiol.* **45** 139–47

Furukawa H and Gouaux E 2003 Mechanisms of activation, inhibition and specificity: crystal structures of the NMDA receptor NR1 ligand-binding core *EMBO J.* **22** 2873–85

Furukawa H, Singh S K, Mancusso R and Gouaux E 2005 Subunit arrangement and function in NMDA receptors *Nature* **438** 185–92

Gibb A J and Colquhoun D 1991 Glutamate activation of a single NMDA receptor-channel produces a cluster of channel openings *Proc. R. Soc. Lond. B: Biol. Sci.* **243** 39–45

Glasgow N G, Siegler Retchless B and Johnson J W 2015 Molecular bases of NMDA receptor subtype-dependent properties *J. Physiol.* **593** 83–95

Grynkiewicz G, Poenie M and Tsien R Y 1985 A new generation of Ca^{2+} indicators with greatly improved fluorescence properties *J. Biol. Chem.* **260** 3440–50

Hablitz J J and Langmoen I A 1982 Excitation of hippocampal pyramidal cells by glutamate in the guinea-pig and rat *J. Physiol.* **325** 317–31

Hamill O P, Marty A, Neher E, Sakmann B and Sigworth F J 1981 Improved patch-clamp techniques for high-resolution current recording from cells and cell-free membrane patches *Pflugers Arch.* **391** 85–100

Hansen K B, Yi F, Perszyk R E, Furukawa H, Wollmuth L P, Gibb A J and Traynelis S F 2018 Structure, function, and allosteric modulation of NMDA receptors *J. Gen. Physiol.* **150** 1081–105

Hayashi T 1954 Convulsant effects on monkeys following of L-glutamate injections in brains *Keio J. Med.* **3** 183

Higley M J and Sabatini B L 2012 Calcium signaling in dendritic spines *Cold Spring Harb. Perspect. Biol.* **4** a005686

Hollmann M, Hartley M and Heinemann S 1991 Ca^{2+} permeability of KA-AMPA-gated glutamate receptor channels depends on subunit composition *Science* **252** 851–3

Hollmann M and Heinemann S 1994 Cloned glutamate receptors *Annu. Rev. Neurosci.* **17** 31–108

Honore T, Davies S N, Drejer J, Fletcher E J, Jacobsen P, Lodge D and Nielsen F E 1988 Quinoxalinediones: potent competitive non-NMDA glutamate receptor antagonists *Science* **241** 701–3

Howe J R 1996 Homomeric and heteromeric ion channels formed from the kainate-type subunits GluR6 and KA2 have very small, but different, unitary conductances *J. Neurophysiol.* **76** 510–9

Howe J R, Colquhoun D and Cull-Candy S G 1988 On the kinetics of large-conductance glutamate-receptor ion channels in rat cerebellar granule neurons *Proc. R. Soc. Lond. B: Biol. Sci.* **233** 407–22

Howe J R, Cull-Candy S G and Colquhoun D 1991 Currents through single glutamate receptor channels in outside-out patches from rat cerebellar granule cells *J. Physiol. (Lond.)* **432** 143–202

Hume R I, Dingledine R and Heinemann S F 1991 Identification of a site in glutamate receptor subunits that controls calcium permeability *Science* **253** 1028–31

Iacobucci G J and Popescu G K 2017a NMDA receptors: linking physiological output to biophysical operation *Nat. Rev. Neurosci.* **18** 236–49

Iacobucci G J and Popescu G K 2017b Resident calmodulin primes NMDA receptors for Ca^{2+}-dependent inactivation *Biophys. J.* **113** 2236–48

Iacobucci G J and Popescu G K 2019a Ca^{2+}-dependent inactivation of GluN2A and GluN2B NMDA receptors occurs by a common kinetic mechanism *Biophys. J.* **118** 798–812

Iacobucci G J and Popescu G K 2019b Spatial coupling tunes NMDA receptor responses via Ca^{2+} diffusion *J. Neurosci.* **39** 8831–44

Iino M, Ozawa S and Tsuzuki K 1990 Permeation of calcium through excitatory amino acid receptor channels in cultured rat hippocampal neurones *J. Physiol.* **424** 151–65

Jackson A C and Nicoll R A 2011 The expanding social network of ionotropic glutamate receptors: TARPs and other transmembrane auxiliary subunits *Neuron* **70** 178–99

Jahr C E and Stevens C F 1987 Glutamate activates multiple single channel conductances in hippocampal neurons *Nature* **325** 522–5

Jahr C E and Stevens C F 1993 Calcium permeability of the N-methyl-D-aspartate receptor channel in hippocampal neurons in culture *Proc. Natl Acad. Sci. USA* **90** 11573–7

Jatzke C, Watanabe J and Wollmuth L P 2002 Voltage and concentration dependence of Ca^{2+} permeability in recombinant glutamate receptor subtypes *J. Physiol.* **538** 25–39

Jin R, Singh S K, Gu S, Furukawa H, Sobolevsky A I, Zhou J, Jin Y and Gouaux E 2009 Crystal structure and association behaviour of the GluR2 amino-terminal domain *EMBO J.* **28** 1812–23

Johnson J W and Ascher P 1987 Glycine potentiates the NMDA response in cultured mouse brain neurons *Nature* **325** 529–31

Johnston G A, Curtis D R, Davies J and McCulloch R M 1974 Spinal interneurone excitation by conformationally restricted analogues of L-glutamic acid *Nature* **248** 804–5

Karakas E and Furukawa H 2014 Crystal structure of a heterotetrameric NMDA receptor ion channel *Science* **344** 992–7

Kelso S R, Nelson T E and Leonard J P 1992 Protein kinase C-mediated enhancement of NMDA currents by metabotropic glutamate receptors in *Xenopus* oocytes *J. Physiol.* **449** 705–18

Kleckner N W and Dingledine R 1988 Requirement for glycine in activation of NMDA-receptors expressed in *Xenopus* oocytes *Science* **241** 835–7

Kohler M, Burnashev N, Sakmann B and Seeburg P H 1993 Determinants of Ca^{2+} permeability in both TM1 and TM2 of high affinity kainate receptor channels: diversity by RNA editing *Neuron* **10** 491–500

Krupp J J, Vissel B, Thomas C G, Heinemann S F and Westbrook G L 1999 Interactions of calmodulin and alpha-actinin with the NR1 subunit modulate Ca^{2+}-dependent inactivation of NMDA receptors *J. Neurosci.* **19** 1165–78

Krupp J J, Vissel B, Thomas C G, Heinemann S F and Westbrook G L 2002 Calcineurin acts via the C-terminus of NR2A to modulate desensitization of NMDA receptors *Neuropharmacology* **42** 593–602

Kuner T, Wollmuth L P, Karlin A, Seeburg P H and Sakmann B 1996 Structure of the NMDA receptor channel M2 segment inferred from the accessibility of substituted cysteines *Neuron* **17** 343–52

Lan J Y, Skeberdis V A, Jover T, Grooms S Y, Lin Y, Araneda R C, Zheng X, Bennett M V and Zukin R S 2001 Protein kinase C modulates NMDA receptor trafficking and gating *Nat. Neurosci.* **4** 382–90

Lau C G, Takeuchi K, Rodenas-Ruano A, Takayasu Y, Murphy J, Bennett M V and Zukin R S 2009 Regulation of NMDA receptor Ca^{2+} signalling and synaptic plasticity *Biochem. Soc. Trans.* **37** 1369–74

Lee C H, Lu W, Michel J C, Goehring A, Du J, Song X and Gouaux E 2014 NMDA receptor structures reveal subunit arrangement and pore architecture *Nature* **511** 191–7

Legendre P, Rosenmund C and Westbrook G L 1993 Inactivation of NMDA channels in cultured hippocampal neurons by intracellular calcium *J. Neurosci.* **13** 674–84

Leonard A S, Bayer K U, Merrill M A, Lim I A, Shea M A, Schulman H and Hell J W 2002 Regulation of calcium/calmodulin-dependent protein kinase II docking to N-methyl-D-aspartate receptors by calcium/calmodulin and alpha-actinin *J. Biol. Chem.* **277** 48441–8

Leonard A S and Hell J W 1997 Cyclic AMP-dependent protein kinase and protein kinase C phosphorylate N-methyl-D-aspartate receptors at different sites *J. Biol. Chem.* **272** 12107–15

Lester R A, Clements J D, Westbrook G L and Jahr C E 1990 Channel kinetics determine the time course of NMDA receptor-mediated synaptic currents *Nature* **346** 565–7

Liao G Y, Wagner D A, Hsu M H and Leonard J P 2001 Evidence for direct protein kinase-C mediated modulation of N-methyl-D-aspartate receptor current *Mol. Pharmacol.* **59** 960–4

Lieberman D N and Mody I 1994 Regulation of NMDA channel function by endogenous Ca^{2+}-dependent phosphatase *Nature* **369** 235–9

Logan S M, Rivera F E and Leonard J P 1999 Protein kinase C modulation of recombinant NMDA receptor currents: roles for the C-terminal C1 exon and calcium ions *J. Neurosci.* **19** 974–86

Lucas D R and Newhouse J P 1957 The toxic effect of sodium L-glutamate on the inner layers of the retina *AMA Arch. Ophthalmol.* **58** 193–201

Lynch G, Larson J, Kelso S, Barrionuevo G and Schottler F 1983 Intracellular injections of EGTA block induction of hippocampal long-term potentiation *Nature* **305** 719–21

MacDermott A B, Mayer M L, Westbrook G L, Smith S J and Barker J L 1986 NMDA-receptor activation increases cytoplasmic calcium concentration in cultured spinal cord neurones *Nature* **321** 519–22

MacDonald J F, Mody I and Salter M W 1989 Regulation of N-methyl-D-aspartate receptors revealed by intracellular dialysis of murine neurones in culture *J. Physiol.* **414** 17–34

Maki B A, Cole R and Popescu G K 2013 Two serine residues on GluN2A C-terminal tails control NMDA receptor current decay times *Channels* **7** 126–32

Maki B A and Popescu G K 2014 Extracellular Ca^{2+} ions reduce NMDA receptor conductance and gating *J. Gen. Physiol.* **144** 379–92

Markram H and Segal M 1992 Activation of protein kinase C suppresses responses to NMDA in rat CA1 hippocampal neurones *J. Physiol.* **457** 491–501

Marwick K F M, Hansen K B, Skehel P A, Hardingham G E and Wyllie D J A 2019 Functional assessment of triheteromeric NMDA receptors containing a human variant associated with epilepsy *J. Physiol.* **597** 1691–704

Mayer M L and Westbrook G L 1985 The action of N-methyl-D-aspartic acid on mouse spinal neurones in culture *J. Physiol.* **361** 65–90

Mayer M L and Westbrook G L 1987 Permeation and block of N-methyl-D-aspartic acid receptor channels by divalent cations in mouse cultured central neurones *J. Physiol.* **394** 501–27

McCulloch R M, Johnston G A, Game C J and Curtis D R 1974 The differential sensitivity of spinal interneurones and renshaw cells to Kainate and N-methyl-D-aspartate *Exp. Brain Res.* **21** 515–8

Merrill M A, Malik Z, Akyol Z, Bartos J A, Leonard A S, Hudmon A, Shea M A and Hell J W 2007 Displacement of alpha-actinin from the NMDA receptor NR1 C0 domain By Ca^{2+}/calmodulin promotes CaMKII binding *Biochemistry* **46** 8485–97

Monyer H, Sprengel R, Schoepfer R, Herb A, Higuchi M, Lomeli H, Burnashev N, Sakmann B and Seeburg P H 1992 Heteromeric NMDA receptors: molecular and functional distinction of subtypes *Science* **256** 1217–21

Mori H, Yamakura T, Masaki H and Mishina M 1993 Involvement of the carboxyl-terminal region in modulation by TPA of the NMDA receptor channel *Neuroreport* **4** 519–22

Moriyoshi K, Masu M, Ishii T, Shigemoto R, Mizuno N and Nakanishi S 1991 Molecular cloning and characterization of the rat NMDA receptor *Nature* **354** 31–7

Murphy J A *et al* 2014 Phosphorylation of Ser1166 on GluN2B by PKA is critical to synaptic NMDA receptor function and Ca^{2+} signaling in spines *J. Neurosci.* **34** 869–79

Myers S J, Yuan H, Kang J Q, Tan F C K, Traynelis S F and Low C M 2019 Distinct roles of GRIN2A and GRIN2B variants in neurological conditions *F1000Res.* **8** 1–14

Nanou E and Catterall W A 2018 Calcium channels, synaptic plasticity, and neuropsychiatric disease *Neuron* **98** 466–81

Neher E and Steinbach J H 1978 Local anaesthetics transiently block currents through single acetylcholine-receptor channels *J. Physiol.* **277** 153–76

Nowak L, Bregestovski P, Ascher P, Herbet A and Prochiantz A 1984 Magnesium gates glutamate-activated channels in mouse central neurones *Nature* **307** 462–5

Olney J W 1969 Brain lesions, obesity, and other disturbances in mice treated with monosodium glutamate *Science* **164** 719–21

Olney J W, Price M T, Samson L and Labruyere J 1986 The role of specific ions in glutamate neurotoxicity *Neurosci. Lett.* **65** 65–71

Perez-Otano I, Schulteis C T, Contractor A, Lipton S A, Trimmer J S, Sucher N J and Heinemann S F 2001 Assembly with the NR1 subunit is required for surface expression of NR3A-containing NMDA receptors *J. Neurosci.* **21** 1228–37

Poon K, Ahmed A H, Nowak L M and Oswald R E 2011 Mechanisms of modal activation of GluA3 receptors *Mol. Pharmacol.* **80** 49–59

Poon K, Nowak L M and Oswald R E 2010 Characterizing single-channel behavior of GluA3 receptors *Biophys. J.* **99** 1437–46

Premkumar L S and Auerbach A 1996 Identification of a high affinity divalent cation binding site near the entrance of the NMDA receptor channel *Neuron* **16** 869–80

Premkumar L S, Qin F and Auerbach A 1997 Subconductance states of a mutant NMDA receptor channel kinetics, calcium, and voltage dependence *J. Gen. Physiol.* **109** 181–9

Rafiki A, Gozlan H, Ben-Ari Y, Khrestchatisky M and Medina I 1997 The calcium-dependent transient inactivation of recombinant NMDA receptor-channel does not involve the high affinity calmodulin binding site of the NR1 subunit *Neurosci. Lett.* **223** 137–9

Raman I M, Tong G and Jahr C E 1996 Beta-adrenergic regulation of synaptic NMDA receptors by cAMP-dependent protein kinase *Neuron* **16** 415–21

Regan M C, Romero-Hernandez A and Furukawa H 2015 A structural biology perspective on NMDA receptor pharmacology and function *Curr. Opin. Struct. Biol.* **33** 68–75

Rosenmund C and Westbrook G L 1993 Rundown of N-methyl-D-aspartate channels during whole-cell recording in rat hippocampal neurons: role of Ca^{2+} and ATP *J. Physiol.* **470** 705–29

Rozov A and Burnashev N 2016 Fast interaction between AMPA and NMDA receptors by intracellular calcium *Cell Calcium* **60** 407–14

Sabatini B L, Oertner T G and Svoboda K 2002 The life cycle of Ca^{2+} ions in dendritic spines *Neuron* **33** 439–52

Scheefhals N and MacGillavry H D 2018 Functional organization of postsynaptic glutamate receptors *Mol. Cell. Neurosci.* **91** 82–94

Schneggenburger R 1996 Simultaneous measurement of Ca^{2+} influx and reversal potentials in recombinant NMDA receptor channels *Biophys. J.* **70** 2165–74

Schneggenburger R, Zhou Z, Konnerth A and Neher E 1993 Fractional contribution of calcium to the cation current through glutamate receptor channels *Neuron* **11** 133–43

Scott D B, Blanpied T A and Ehlers M D 2003 Coordinated PKA and PKC phosphorylation suppresses RXR-mediated ER retention and regulates the surface delivery of NMDA receptors *Neuropharmacology* **45** 755–67

Shi J, Townsend M and Constantine-Paton M 2000 Activity-dependent induction of tonic calcineurin activity mediates a rapid developmental downregulation of NMDA receptor currents *Neuron* **28** 103–14

Siegler Retchless B, Gao W and Johnson J W 2012 A single GluN2 subunit residue controls NMDA receptor channel properties via intersubunit interaction *Nat. Neurosci.* **15** S1–2

Simms B A and Zamponi G W 2014 Neuronal voltage-gated calcium channels: structure, function, and dysfunction *Neuron* **82** 24–45

Sirrieh R E, MacLean D M and Jayaraman V 2013 Amino-terminal domain tetramer organization and structural effects of zinc binding in the N-methyl-D-aspartate (NMDA) receptor *J. Biol. Chem.* **288** 22555–64

Skeberdis V A *et al* 2006 Protein kinase a regulates calcium permeability of NMDA receptors *Nat. Neurosci.* **9** 501–10

Smith T C, Wang L Y and Howe J R 2000 Heterogeneous conductance levels of native AMPA receptors *J. Neurosci.* **20** 2073–85

Sommer B, Kohler M, Sprengel R and Seeburg P H 1991 RNA editing in brain controls a determinant of ion flow in glutamate-gated channels *Cell* **67** 11–9

Song X, Jensen M Ø, Jogini V, Stein R A, Lee C-H, McHaourab H S, Shaw D E and Gouaux E 2018 Mechanism of NMDA receptor channel block by MK-801 and memantine *Nature* **556** 515–9

Stern P, Behe P, Schoepfer R and Colquhoun D 1992 Single-channel conductances of NMDA receptors expressed from cloned cDNAs: comparison with native receptors *Proc. Biol. Sci.* **250** 271–7

Sucher N J, Awobuluyi M, Choi Y B and Lipton S A 1996 NMDA receptors: from genes to channels *Trends Pharmacol. Sci.* **17** 348–55

Sutcliffe M J, Wo Z G and Oswald R E 1996 Three-dimensional models of non-NMDA glutamate receptors *Biophys. J.* **70** 1575–89

Swanson G T, Feldmeyer D, Kaneda M and Cull-Candy S G 1996 Effect of RNA editing and subunit co-assembly single-channel properties of recombinant kainate receptors *J. Physiol.* **492** 129–42

Swanson G T, Kamboj S K and Cull-Candy S G 1997 Single-channel properties of recombinant AMPA receptors depend on RNA editing, splice variation, and subunit composition *J. Neurosci.* **17** 58–69

Swope S L, Moss S I, Raymond L A and Huganir R L 1999 Regulation of ligand-gated ion channels by protein phosphorylation *Adv. Second Messenger Phosphoprotein Res.* **33** 49–78

Tang L, Gamal El-Din T M, Swanson T M, Pryde D C, Scheuer T, Zheng N and Catterall W A 2016 Structural basis for inhibition of a voltage-gated Ca^{2+} channel by Ca^{2+} antagonist drugs *Nature* **537** 117–21

Tingley W G, Ehlers M D, Kameyama K, Doherty C, Ptak J B, Riley C T and Huganir R L 1997 Characterization of protein kinase a and protein kinase C phosphorylation of the N-methyl-D-aspartate receptor NR1 subunit using phosphorylation site-specific antibodies *J. Biol. Chem.* **272** 5157–66

Tingley W G, Roche K W, Thompson A K and Huganir R L 1993 Regulation of NMDA receptor phosphorylation by alternative splicing of the C-terminal domain *Nature* **364** 70–3

Tomita S, Adesnik H, Sekiguchi M, Zhang W, Wada K, Howe J R, Nicoll R A and Bredt D S 2005 Stargazin modulates AMPA receptor gating and trafficking by distinct domains *Nature* **435** 1052–8

Tomita S and Castillo P E 2012 Neto1 and Neto2: auxiliary subunits that determine key properties of native kainate receptors *J. Physiol.* **590** 2217–23

Tong G, Shepherd D and Jahr C E 1995 Synaptic desensitization of NMDA receptors by calcineurin *Science* **267** 1510–2

Tovar K R and Westbrook G L 2002 Mobile NMDA receptors at hippocampal synapses *Neuron* **34** 255–64

Townsend M, Liu Y and Constantine-Paton M 2004 Retina-driven dephosphorylation of the NR2A subunit correlates with faster NMDA receptor kinetics at developing retinocollicular synapses *J. Neurosci.* **24** 11098–107

Traynelis S F, Wollmuth L P, McBain C J, Menniti F S, Vance K M, Ogden K K, Hansen K B, Yuan H, Myers S J and Dingledine R 2010 Glutamate receptor ion channels: structure, regulation, and function *Pharmacol. Rev.* **62** 405–96

Verdoorn T A, Burnashev N, Monyer H, Seeburg P H and Sakmann B 1991 Structural determinants of ion flow through recombinant glutamate receptor channels *Science* **252** 1715–8

Wagner D A and Leonard J P 1996 Effect of protein kinase-C activation on the Mg^{2+}-sensitivity of cloned NMDA receptors *Neuropharmacology* **35** 29–36

Wagner D A and Leonard J P 1999 Protein kinase C potentiation of currents from mouse zeta1/epsilon2 NMDA receptors expressed in *Xenopus* oocytes depends on f-actin/g-actin cycling *Neurosci. Lett.* **272** 187–90

Wang Y T and Salter M W 1994 Regulation of NMDA receptors by tyrosine kinases and phosphatases *Nature* **369** 233–5

Watanabe J, Beck C, Kuner T, Premkumar L S and Wollmuth L P 2002 DRPEER: a motif in the extracellular vestibule conferring high Ca^{2+} flux rates in NMDA receptor channels *J. Neurosci.* **22** 10209–16

Wechsler A and Teichberg V I 1998 Brain spectrin binding to the NMDA receptor is regulated by phosphorylation, calcium and calmodulin *EMBO J.* **17** 3931–9

Wigstrom H, Swann J W and Andersen P 1979 Calcium dependency of synaptic long-lasting potentiation in the hippocampal slice *Acta Physiol. Scand.* **105** 126–8

Wollmuth L P, Kuner T and Sakmann B 1998 Intracellular Mg^{2+} interacts with structural determinants of the narrow constriction contributed by the NR1-subunit in the NMDA receptor channel *J. Physiol.* **506** 33–52

Wollmuth L P, Kuner T, Seeburg P H and Sakmann B 1996 Differential contribution of the NR1- and NR2A-subunits to the selectivity filter of recombinant NMDA receptor channels *J. Physiol.* **491** 779–97

Wollmuth L P and Sakmann B 1998 Different mechanisms of Ca^{2+} transport in NMDA and Ca^{2+}-permeable AMPA glutamate receptor channels *J. Gen. Physiol.* **112** 623–36

Wu J, Yan Z, Li Z, Qian X, Lu S, Dong M, Zhou Q and Yan N 2016 Structure of the voltage-gated calcium channel Ca(v)1.1 at 3.6 A resolution *Nature* **537** 191–6

Wyszynski M, Lin J, Rao A, Nigh E, Beggs A H, Craig A M and Sheng M 1997 Competitive binding of alpha-actinin and calmodulin to the NMDA receptor *Nature* **385** 439–42

Zhang S, Ehlers M D, Bernhardt J P, Su C T and Huganir R L 1998 Calmodulin mediates calcium-dependent inactivation of N-methyl-D-aspartate receptors *Neuron* **21** 443–53

Zhang W, Cho Y, Lolis E and Howe J R 2008 Structural and single-channel results indicate that the rates of ligand binding domain closing and opening directly impact AMPA receptor gating *J. Neurosci.* **28** 932–43

Zheng X, Zhang L, Wang A P, Bennett M V and Zukin R S 1997 Ca^{2+} influx amplifies protein kinase C potentiation of recombinant NMDA receptors *J. Neurosci.* **17** 8676–86

Zhu S and Gouaux E 2017 Structure and symmetry inform gating principles of ionotropic glutamate receptors *Neuropharmacology* **112** 11–5

Zorumski C F, Yang J and Fischbach G D 1989 Calcium-dependent, slow desensitization distinguishes different types of glutamate receptors *Cell. Mol. Neurobiol.* **9** 95–104

Zukin R S and Bennett M V 1995 Alternatively spliced isoforms of the NMDAR1 receptor subunit *Trends Neurosci.* **18** 306–13

Part II

Calcium signaling on channels and other effectors

Chapter 5

Monitoring the dynamics of calcium signaling effectors using genetically encoded fluorescent biosensors

Yanghao Zhong, Sohum Mehta and Jin Zhang

5.1 Introduction

Cells need to quickly respond and adapt to changes in the surrounding environment in order to survive and proliferate. While signals from the extracellular environment are usually transduced and propagated by networks of protein effectors, small molecules known as 'second messengers' play a vital role in regulating the dynamics of this signal transduction network. As the simplest second messenger, the calcium ion (Ca^{2+}) regulates exquisite and complex cellular processes including muscle contraction, cell proliferation and synaptic plasticity [1, 2]. Ca^{2+} concentrations vary dramatically between the extracellular (~mM) and intracellular space (~100 nM), and even within intracellular compartments (nM range in the cytoplasm versus mM range in the endoplasmic reticulum). Cells maintain these large differences in Ca^{2+} concentrations between cellular compartments by investing much of their energy and resources into ion exchangers and Ca^{2+} channels [3, 4], which cooperatively maintain a low Ca^{2+} concentration in the cytoplasm. Ca^{2+}-mediated signal transduction is usually initiated by an increase in the cytoplasmic Ca^{2+} concentration, which triggers different downstream pathways that control many specific biological functions in parallel.

To achieve this specificity, the spatiotemporal dynamics of Ca^{2+} signaling are very precisely regulated, and the presence of Ca^{2+} microdomains has been examined through both *in silico* mathematical modeling [5] and live-cell imaging [6]. Local Ca^{2+} elevations can be generated around the mouths of open Ca^{2+} channels, and such localized Ca^{2+} then acts on nearby downstream effector proteins to further trigger specific biological processes. Besides localized signaling by Ca^{2+} itself, a variety of

doi:10.1088/978-0-7503-2009-2ch5

downstream effector proteins, including calmodulin (CaM), Ca^{2+}/CaM-dependent protein phosphatase calcineurin (CaN), Ca^{2+}/CaM-dependent protein kinase II (CaMKII), and protein kinase C (PKC), enable the simple calcium ion to achieve sophisticated biological functions. Generally, Ca^{2+} activates downstream effector proteins through direct binding (CaM, PKC) or indirect activation via another effector protein (CaN, CaMKII). Positively charged Ca^{2+} binds to effector proteins containing C2 domains or EF-hand motifs [1], changing the conformation of the effector protein or inducing interactions with specific signaling proteins to switch on the subsequent signal transduction.

The idea that Ca^{2+} signaling is compartmentalized within living cells could not be directly tested until the development of optical techniques to visualize cellular Ca^{2+} signaling events. Earlier techniques for measuring cytosolic Ca^{2+} levels, such as those that require microinjection of the luminescent photoprotein aequorin or the absorbance dye arsenazo III, were often limited to use in giant cells. A family of fluorescent indicators were later generated based on the well-known Ca^{2+}-selective chelator EGTA (ethylene glycol tetra-acetic acid), which can be easily loaded into a broad range of cell types through the use of hydrolyzable esters [7]. Such easy-to-use chemical Ca^{2+} indicators, including quin-2 [8], fluo-3 [9] and fura-2 [10], became popular in the research community and are still widely used to study intracellular Ca^{2+} signaling. Despite the great achievements made using chemical fluorescent indicators, these indicators have certain key limitations: they can neither be used to probe Ca^{2+} signaling at subcellular compartments easily, nor can they be utilized to look at the dynamics of Ca^{2+} effector proteins. The discovery of *Aequorea victoria* green fluorescent protein (GFP), along with the identification and engineering of numerous GFP derivatives [11], led to the development of genetically encoded fluorescent Ca^{2+} indicators (GECIs). These genetically encoded indicators can be constructed using recombinant DNA techniques and can be specifically targeted to desired cellular regions by incorporating well-defined subcellular targeting motifs [12, 13] or fusion with proteins that natively localize to specific cellular compartments [14]. This relatively straightforward and universal targeting strategy enables the direct and specific examination of Ca^{2+} signaling at subcellular compartments. Indeed, since the development of the very first GECI, 'cameleon' [15], a wide variety of genetically encoded fluorescent reporters have been developed to investigate different signaling molecules besides Ca^{2+} itself, including those to monitor many effector proteins downstream of Ca^{2+} signaling.

5.1.1 General design strategies of biosensors for calcium effectors

Owing to the extensive efforts of the research community, over the years, there have been many different ways to engineer a fluorescent biosensor since the success of the original 'cameleon design', and readers can refer to other reviews for a complete list [16]. In general, however, all genetically encoded fluorescent biosensors share a modular design containing a sensing unit that detects a biochemical input corresponding to a specific signaling process, and a reporting unit that transforms the biochemical input into optical readout. This modular, 'sensing-and-reporting' type of biosensor design enables us to examine very specific biological changes in a cell.

The sensing unit in a biosensor usually functions as a 'molecular switch' that undergoes a conformational change when the specific signaling is turned on or off. Such conformational changes are rapidly transmitted to the reporting unit, which is physically linked to the sensing unit, generating a measurable optical signal reporting the real-time status of the biological signal. The molecular switch components are often derived from native cellular proteins or protein fragments whose specific function determines the type of biochemical event a biosensor can detect. For example, biosensors containing domains that undergo a conformational change upon second messenger binding often serve as 'indicators' for these small molecules. Similarly, the dynamics of signaling enzymes can be monitored using 'activation' sensors, in which a full-length target protein whose conformation changes between 'active' and 'inactive' states is used as the molecular switch, and 'activity' sensors, in which the molecular switch functions as a surrogate substrate for the enzyme of interest and undergoes a conformational change in response to the target catalytic activity, such as post-translational modification. While indicators are useful for probing Ca^{2+} itself, activation/activity sensors are required for monitoring the actions of Ca^{2+} effector proteins.

The reporting unit usually consists of one or more fluorescent proteins (FPs) that change their spectral properties in response to the conformational change in the molecular switch, and the readout from the reporting unit serves as a dynamic, quantitative representation of the biological event being recorded. Depending on how the reporting unit is constructed, there are two most commonly used biosensor designs to visualize Ca^{2+} effector signaling: fluorescence resonance energy transfer (FRET)-based biosensors and single-fluorescent protein-based biosensors. Before introducing each of the genetically encoded biosensors that are designed to report Ca^{2+} downstream effector signaling and answer specific biological questions, we will first briefly overview these two types of biosensors.

5.1.2 FRET-based biosensors

The most popular and versatile design for genetically encoded fluorescent biosensors is based on FRET [17]. FRET is a photophysical process in which a 'donor' fluorophore that initially absorbs energy and reaches its electronic exited state, transfers part of the energy to a neighboring 'acceptor' fluorophore through nonradiative dipole–dipole coupling [18]. Therefore, when FRET happens, even when the donor fluorophore is illuminated at its characteristic excitation wavelength, its fluorescence is quenched, and the acceptor fluorophore is excited through nonradiative energy transfer. The energy transfer efficiency, i.e., FRET efficiency, between two fluorophores mainly depends on the distance, orientation and spectral overlap of the donor and acceptor [19]. In FRET-based biosensor designs, the molecular switch is typically sandwiched between a pair of FPs between which FRET can happen. Upon detecting the specific cellular signaling event of interest, the molecular switch will undergo a conformational change that will alter the relative distance and orientation between the attached FPs, resulting in changes in FRET efficiency that can be measured by fluorescence microscopy. This design is exemplified by the FRET-based Ca^{2+} indicator 'yellow cameleon', wherein the molecular switch is composed of the Ca^{2+} sensor CaM and a CaM-binding peptide from myosin light chain kinase (M13) sandwiched between the FRET pair cyan fluorescent protein (CFP) and yellow fluorescent protein (YFP). The binding of cellular Ca^{2+} triggers CaM to form a complex with the M13 peptide, inducing a conformational change that alters the relative proximity of CFP and YFP and changes the FRET signal [15]. Over the years, this initial design has been adapted to probe a wide range of targets through the engineering and incorporation of new molecular switches [16]. While many FRET-based biosensors feature a unimolecular design, where the molecular switch and the FRET pair are physically combined in a single protein chain, biosensors can also exist in a bimolecular form, in which specific signaling events can induce binding between two separate parts of the biosensor, leading to increases in FRET.

The response of a FRET-based biosensor is frequently read out either by monitoring changes in the ratio between the donor fluorescence and sensitized emission from the acceptor or by monitoring changes in the excited-state fluorescence lifetime of the donor FP using fluorescence lifetime imaging microscopy (FLIM) [19]. The latter case often utilizes a subclass of FRET-based biosensors, hereafter referred to as FLIM-FRET biosensors, that essentially follow the same design principles as ratiometric FRET sensors but must also meet certain special requirements for compatibility with FLIM, such as utilizing a FRET donor whose excited-state lifetime undergoes a single exponential decay [20]. Furthermore, because only the donor fluorescence intensity is recorded to calculate fluorescence lifetime changes when using FLIM-FRET biosensors, FLIM-FRET biosensors can be designed using non-fluorescent, or dark, FRET acceptors that reduce or eliminate acceptor fluorescence emission, which can minimize spectral contamination during FRET imaging and also serve as single-color fluorescent biosensors for enhanced multiplexed imaging [21].

5.1.3 Single-fluorescent protein-based biosensors

In contrast to FRET-based biosensors, which rely on an FP pair, an alternative design involves directly inserting the molecular switch into a single FP. These single-FP biosensors utilize conformational changes in the molecular switch to distort the FP barrel and modulate fluorescence. For example, the popular GCaMP series of Ca^{2+} indicators contain CaM and the M13 peptide fused to the N- and C-termini of circularly permuted EGFP (cpEGFP), wherein Ca^{2+} binding alters the sensor conformation and increases green fluorescence intensity [22–26]. As illustrated with GCaMP, one of the important advantages of single-FP biosensors is their often higher signal-to-noise ratio and larger dynamic range compared with FRET-based biosensors, which enables investigation of subtle changes in cell signaling. Although initially confined to monitoring Ca^{2+}, the popularity and development of single-FP biosensors has exploded over the years to yield numerous small-molecule indicators and also recently expanded to monitor enzymatic activities [27]. Single-FP biosensors occupy less spectral space than FRET-based biosensors, and are thus more conveniently applied to multiplexed imaging that utilizes color variants of different biosensors to study dynamic signaling propagation and crosstalk in living cells.

5.2 Monitoring calcium effector proteins in living cells

Ca^{2+} is a versatile second messenger that impacts nearly every aspect of biology. During a typical Ca^{2+} signaling event, intracellular Ca^{2+} concentrations are increased upon the activation of various receptors (G-protein coupled receptors or receptor tyrosine kinases) and ion channels (Orai or voltage-dependent Ca^{2+} channels) [1]. Direct Ca^{2+}-activated downstream effectors such as CaM and PKC are activated immediately after binding to Ca^{2+}, and active CaM can further bind to and activate indirect effectors including CaMKII and CaN. When activated, both direct and indirect effectors will then exert their enzymatic activities on various target proteins to regulate myriad cellular processes. Thus, developing biosensors to monitor the dynamic behaviors of downstream Ca^{2+} effectors in living cells is crucial in order to understand the molecular logic of Ca^{2+} signaling. Among these biosensors, CaM biosensors were the first ever constructed to study the basis of dynamic Ca^{2+} signaling in live cells, while CaMKII and CaN biosensors have frequently been used to study Ca^{2+}-mediated neuronal activities, and PKC biosensors have helped elucidate the role of PKC as a tumor suppressor. Below, we will discuss the development and applications of genetically encoded biosensors for each of these Ca^{2+} effector proteins.

5.2.1 CaM

CaM is an EF-hand motif-containing protein that is highly conserved across different species and is a key transducer of intracellular Ca^{2+} signals. CaM generally does not bind to downstream target proteins on its own, but once bound to Ca^{2+} (Ca^{2+}/CaM), it becomes active and can bind and activate a variety of targets, including enzymes, ion pumps, channels and cytoskeletal proteins [28]. Ca^{2+}/CaM also has different binding affinities towards its targets, with K_d values ranging from high ($K_d \leqslant 2$ nM) and intermediate (10 nM $\leqslant K_d \leqslant$ 100 nM) to low ($K_d \geqslant$ 100 nM). Thus, understanding the kinetics of Ca^{2+}/CaM and its interactions with different targets inside cells is of great importance to link the changes of cellular Ca^{2+} to diverse biological processes. Romoser and colleagues developed the first FRET-based fluorescent biosensor to monitor Ca^{2+}/CaM levels in live cells [29]. Their CaM biosensor was constructed by linking a FRET pair consisting of blue-shifted GFP (BGFP) [30] and red-shifted GFP (RGFP) [31] via a Ca^{2+}/CaM-binding sequence from avian smooth muscle myosin light chain kinase and was named fluorescent indicator protein–calmodulin binding (FIP–CB). FIP–CB was microinjected into human embryonic kidney cells (HEK293), and upon Ca^{2+}/CaM-binding, FIP–CB exhibited a decrease in FRET signal, providing a view of free Ca^{2+}/CaM (active CaM) levels in living cells. This study revealed that physiological levels of free Ca^{2+}/CaM are very low in cells, indicating that small changes in the binding affinity of Ca^{2+}/CaM targets should significantly alter the levels of their activity. To further evaluate the relationship between Ca^{2+} and Ca^{2+}-CaM, Persechini *et al* first improved the FIP–CB biosensor by replacing the original FRET pair with enhanced CFP (ECFP) and enhanced YFP (EYFP), which is a better FRET pair [15]. The advantage of this improvement is that the ECFP donor fluorophore can be excited at a longer wavelength (430 nm), thereby reducing phototoxicity in live-cell imaging. Furthermore, they inserted CaM-binding motifs with different affinities between the FRET pair to construct a series of FIP–CB sensors. In fractional saturation experiments, data from the biosensor containing a high-affinity motif suggested tight-binding targets outnumbered CaM by approximately a factor of 2 and that high-affinity CaM targets are efficiently activated throughout the cell. In contrast, low-affinity CaM targets only become activated in regions with locally high concentrations of free Ca^{2+}/CaM, supporting the compartmentalization of Ca^{2+} signaling.

In another application of the FIP–CB biosensor, Chew and colleagues modified the probe design in an effort to monitor the dynamics of myosin light chain kinase (MLCK) activation during cell migration and mitosis [32]. MLCK is a Ca^{2+}/CaM-dependent kinase that can phosphorylate the regulatory light chain (RLC), which regulates non-muscle myosin II activity, and MLCK is implicated in non-muscle contraction, cytokinesis, stress fiber formation, and motility [33–36]. Chew *et al* fused full-length MLCK to the N-terminus of the FIP–CB biosensor, generating an MLCK activation biosensor. Whenever Ca^{2+}/CaM binds to and activates MLCK in the cell, Ca^{2+}/CaM will also bind the FIP–CB biosensor linked to the MLCK molecule. By monitoring the FIP–CB signal, Chew *et al* were thus able to observe localized MLCK activation in different cellular processes. This MLCK–FIP sensor reported the recruitment of MLCK to various cytoskeletal structures during contraction, motility and cytokinesis, revealing the spatiotemporal dynamics of MLCK activation (figure 5.1). While this biosensor provides a way of visualizing localized MLCK activation in live cells, some limitations should also be noted. First, activation of MLCK is indirectly represented by the activity of CaM bound to the FIP–CB part of the biosensor; other regulators such as protein kinase A (PKA) that suppress MLCK activation [37] may lead to confounding results. Second, although overexpression of MLCK fused to FIP–CB may not interfere with the global kinase activity of MLCK, one cannot rule out the possibility that the fusion protein behaves differently in more localized environments, as reported previously [38].

5.2.2 CaMKII

CaMKII belongs to the CAMK2 family and was one of the first Ca^{2+}/CaM-regulated kinases to be reported. There are four CaMKII isoforms, α, β, γ and δ, encoded by separate genes. The four isoforms are highly homologous, with isoform α being the most brain-specific. CaMKII consists of an N-terminal kinase domain, followed by a regulatory domain that can be bound by CaM, and a C-terminal association domain which allows CaMKII to form a dodecameric holoenzyme. The CaM-binding site in the regulatory domain overlaps with an autoinhibitory region;

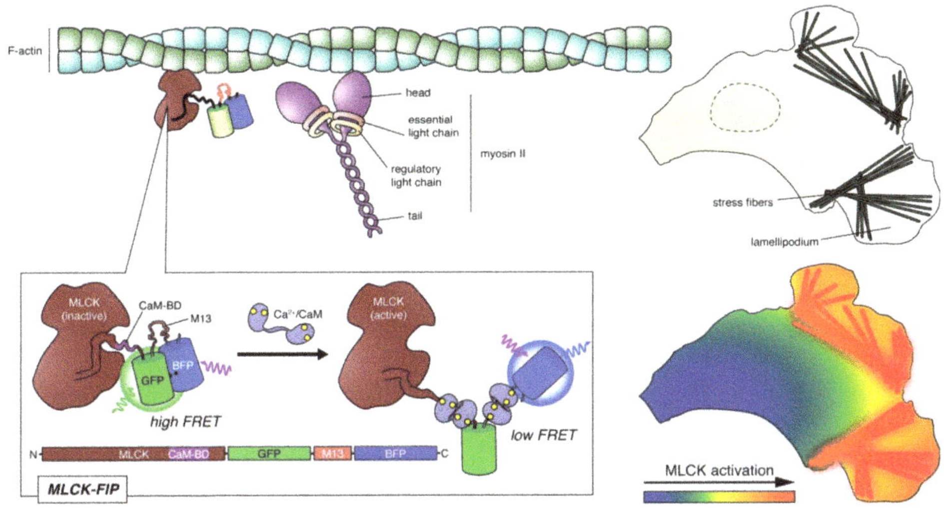

Figure 5.1. FIP–CB is used to monitor MLCK activation. To visualize MLCK activation, full-length MLCK is fused to the N-terminus of the FIP–CB CaM biosensor (lower left). In this design, binding and activation of MLCK by Ca^{2+}/CaM coincides with the binding of Ca^{2+}/CaM to the FIP–CB module, which induces a FRET signal change. MLCK natively localizes to the actin cytoskeleton, where it regulates phosphorylation of the non-muscle myosin II regulatory light chain (upper left), and the distribution of FIP–CB fluorescence can also indicate the subcellular localization of MLCK; the activation of localized MLCK can be monitored through the FRET signal change. For example, Chew *et al* [32] observed MLCK to be highly activated at stress fibers near lamellipodium (right). CaM–BD, CaM-binding domain.

therefore, CaM binding to CaMKII releases the autoinhibition and activates CaMKII [39]. CaMKII is highly enriched in excitatory synapses in the central nervous system and plays a crucial role in regulating neuronal functions such as synaptic plasticity, learning and memory [40]. Ca^{2+} enters postsynaptic neurons through N-methyl-D-aspartate (NMDA) receptors and activates CaM, which further triggers CaMKII activation, followed by autophosphorylation events catalyzed by neighboring subunits within the dodecameric CaMKII holoenzyme. The phosphorylated CaMKII can remain activated in the absence of Ca^{2+} and eventually lead to the insertion of α-amino-3-hydroxy-5-methyl-4-isoxazolepropionic acid (AMPA) receptors at the synapse, a persistent increase in synaptic efficacy known as long-term potentiation (LTP) [41]. Due to the strong interest in the research community to understand how CaMKII contributes to neuronal activities and the lack of information on spatiotemporal patterns of CaMKII activation in living cells, many efforts over the years have been devoted to developing and improving CaMKII biosensors.

The first CaMKII biosensor, named 'Camuiα', was designed to monitor CaMKII activation and was constructed by flanking the entire CaMKIIα protein with YFP and CFP [42]. CaMKIIα undergoes a conformational change resulting from Ca^{2+}/CaM binding and subsequent release of the autoinhibitory domain [39], allowing full-length CaMKIIα to act as a molecular switch to modulate FRET between CFP and YFP. Camuiα exhibits a decrease in FRET signal upon CaMKII activation and has been used to detect autophosphorylation-dependent CaMKII activation in living neurons at single-dendrite and spine resolution, and this prototype CaMKIIα activation sensor has laid the foundation for many follow-up studies investigating the spatiotemporal modulation of CaMKII in synaptic plasticity and learning [43–45]. Camuiα has also been used in various fields beyond neuroscience. For example, Erickson *et al* adapted the design of Camuiα to visualize the spatiotemporal activation of CaMKIIδ in intact cardiomyocytes [46], finding that CaMKIIδ activation increased with Ca^{2+} pacing frequency and intensity, demonstrating the acute regulation of CaMKIIδ activation in cardiomyocytes.

It is not surprising that the first generation of a biosensor did not perform optimally; thus, Lam *et al* modified Camuiα using a better FRET pair to improve its dynamic range and sensitivity [47]. Specifically, they engineered Clover and mRuby2, which were the brightest green- and red-emitting FPs developed to date, and used them to replace the original CFP and YFP in Camuiα to generate Camuiα-CR. The resultant Camuiα-CR probe showed lower photobleaching, lower phototoxicity and a higher signal-to-noise ratio compared with Camuiα. Fujii *et al* also sought to optimize the biosensor based on studies of the CaMKII holoenzyme structure. In order to optimize the proximity of the donor and acceptor FPs in the resting state, they repositioned the FRET pair so that one FP was located within the N-terminal domain and the other within an internal variable region, both of which are exposed to the surface in the holoenzyme structure. They also generated a series of CaMKII activation biosensor color variants, which allowed them to perform co-imaging with a CaN biosensor [45], as will be discussed in a later section.

While ratiometric FRET measurements can produce robust biosensor readouts, lifetime-based measurements offer several specific advantages. FLIM-FRET biosensors measure FRET-induced changes in the nanosecond-order fluorescence decay time of the FRET donor, and because this fluorescence decay constant is independent of the biosensor concentration, FRET-FLIM sensors can circumvent some of the difficulties associated with ratiometric FRET imaging, such as changes in tissue scattering or absorption [21]. However, the FRET pair most frequently used in ratiometric FRET imaging, CFP and YFP, is not suitable for FLIM-FRET because CFP has a complex fluorescence decay time and is not bright enough to minimize the spectral bleed-through effect that reduces sensor dynamic range [19]. Therefore, alternative FRET pairs are often utilized in FLIM-FRET biosensors. Along these lines, Kwok *et al* found that GFP and the monomeric RFP (mRFP) DsRed are a suitable FLIM-FRET pair, which they used to construct a FLIM-FRET version of Camuiα, Camuiα-GFP/mRFP [20]. They also tested various spectral variants of the GFP/mRFP pair and generated a panel of CaMKII activation biosensors suitable for FLIM-FRET. Although attempts to utilize nonradiative YFP variants as the FLIM-FRET acceptor were not successful in this study, Lee *et al* later managed to develop a FLIM-FRET CaMKII activation biosensor called green-Camuiα using a nonradiative YFP derivative REACh (Resonance Energy Accepting Chromoprotein) as the acceptor and mEGFP as the donor [48]. Combining two-photon glutamate uncaging techniques and FLIM-FRET imaging using their improved CaMKII biosensor, which exhibited higher sensitivity and brightness compared to the original Camuiα, Lee and colleagues found that CaMKII activation by specific channels is restricted to the stimulated spines during LTP, suggesting a high degree of compartmentalization, and that the channel-specificity of CaMKII signaling allows for spatiotemporal regulation of CaMKII in synaptic plasticity. Furthermore, Murakoshi *et al* developed another dark acceptor, ShadowG, to replace REACh in green-Camuiα, generating ShadowG-Camuiα. ShadowG has a very low quantum yield which removes the residual fluorescence contamination in green-Camuiα and shows low cell-to-cell variation, enabling more precise measurements of individual cell responses [49]. Furthermore, because both GFP/YFP and GFP/RFP versions of FLIM-FRET CaMKII biosensors occupy a broad spectral band and are thus incompatible with multiplexed imaging, Nakahata *et al* developed a red-shifted FLIM-FRET pair using mRuby2 and dark-mCherry to occupy a narrower part of the spectrum and facilitate dual observation with GFP and other blue-to-green light based optogenetic tools [50].

Notably, because enzyme activation biosensors contain a full-length protein as the sensing unit, their use entails overexpressing the target enzyme. In contrast, activity biosensors function as surrogate substrates and are thus able to monitor the enzymatic activity of the target protein at its endogenous level. Recently, Ardestani and colleagues developed the first activity biosensor for measuring endogenous CaMKII kinase activity [51]. This CaMKII activity reporter, known as FRESCA (FRET-based Sensor for CaMKII Activity), was constructed by sandwiching a Thr-containing version of the synthetic CaMKII substrate syntide (e.g., syntide-FRESCA) and a forkhead associated (FHA2) phosphoamino acid binding domain (PAABD) between a CFP/YFP FRET pair. Upon phosphorylation by CaMKII, syntide-FRESCA will bind to FHA2, resulting in a conformational change which can be measured as a ratiometric FRET change. In this study, the authors examined the differential responses of Camuiα and FRESCA in parallel in mouse eggs and found that the Camuiα response was delayed or terminated earlier compared with the FRESCA response, suggesting that FRESCA is more suitable for capturing long-term CaMKII activity dynamics, which are important during fertilization. It is anticipated that future studies will use FRESCA or its derivatives to look at endogenous CaMKII activities in other systems such as neurons and cardiomyocytes.

5.2.3 CaN

CaN is a serine/threonine protein phosphatase that regulates diverse physiological processes, including cell proliferation, cardiac development and differentiation, immune system activation, and neuronal plasticity. The only known Ca^{2+}/CaM-dependent protein phosphatase, CaN is comprised of two subunits, a catalytic A subunit (CaNA) and CaM-like regulatory B subunit (CaNB) that binds Ca^{2+}. CaNA contains a catalytic domain and a regulatory arm containing the CaNB-binding domain, Ca^{2+}/CaM-binding domain, and an autoinhibitory domain (AID) that blocks the active site under basal Ca^{2+} concentrations [52]. The phosphatase activity of CaN is dependent on Ca^{2+}/CaM, which binds to CaNA and releases the autoinhibitory domain from the active site. Given the prominent role of CaN in broad cellular functions under physiological and pathological conditions [53], it is of great importance to develop biosensors to monitor CaN activation and activity with spatiotemporal precision in living cells.

Currently, two FRET-based CaN activity biosensors, CaN activity reporter 1 (CaNAR1) and an improved second-generation version (CaNAR2) have been reported to date [13, 54]. CaNAR1 was constructed by flanking a portion of the N-terminal regulatory domain of NFAT (Nuclear Factor of Activated T-cells), a well-studied CaN substrate, with CFP and YFP; CaN-mediated dephosphorylation of this NFAT domain induces a conformational change that results in a FRET signal change in CaNAR1 [54]. As has been illustrated above, modifying the FRET pair can significantly increase the dynamic range of a FRET-based biosensor. In this case, the brighter cyan- and yellow-emitting FPs Cerulean3 and YPet were used to replace the original CFP/YFP pair in CaNAR1 to generate CaNAR2, yielding an approximately threefold increase in dynamic range. Using CaNAR2 to monitor subcellular CaN activity, we previously found that Ca^{2+} oscillations in pancreatic β-cells induced distinct CaN activity across different cellular compartments: cytosolic/plasma membrane CaN activity showed an integrating, step-like pattern, whereas mitochondrial/endoplasmic reticulum (ER) CaN activity exhibited a more reversible, oscillatory pattern. To examine if CaN was differentially activated in these different compartments, we also built a CaN activation biosensor named CaNARi (CaN Activation Ratiometric indicator) by sandwiching the catalytic CaNA subunit between CFP and YFP. Close inspection of subcellular CaNARi responses suggested weaker mitochondrial/ER CaN activation compared with the cytosol and plasma membrane. Further investigation using the CaM biosensor BSCaM-2 (a specific version of the FIP–CB sensor [55], see section 5.2.1) revealed lower free CaM concentrations near the ER surface, which likely contribute to weaker CaN activity that is more susceptible to antagonism by local kinase activity, thereby generating CaN activity oscillations near the ER surface [13].

In the nervous system, multiple signaling pathways are responsible for modulating synaptic properties, such as the morphology of dendritic spines, in order to adapt to changes in neuronal activity [56]. Along with CaMKII, CaN is implicated as a key player in controlling the direction and extent of these modulations, and CaN activation biosensors are thus especially valuable tools when combined with CaMKII biosensors to dissect neuronal activity. For instance, when Fujii *et al* were trying to answer the question of how information encoded in glutamate release rates at individual synapses is converted into biochemical activation patterns of postsynaptic enzymes, they built a system called dual FRET with optical manipulation (dFOMA) that allowed them to perform co-imaging of CaMKII and CaN activation during glutamate uncaging in a single spine [45]. As mentioned above (see section 5.2.2), to improve the performance of Camuiα, the authors tested different FRET pairs inserted at different positions in the biosensor to optimize the donor–acceptor proximity of Camuiα based on the CaMKII holoenzyme structure, ultimately selecting the NH_2 terminal domain and an internal variable domain as optimal insertion points. Similarly, they used the structure of CaN to construct a CaN activation FRET biosensor to monitor activation-associated conformational changes by tagging a FRET pair to the N- and C-termini of the CaNA subunit. Among the many color variants tested, the authors chose RS-K2α (mCherry/Sapphire-Camuiα) and RY-CaN (mCherry/YFP-CaN), and simultaneously measured the activation dynamics of both CaMKII and CaN using the dFOMA system. Interestingly, they found that CaMKIIα and CaN showed differential sensitivity under different glutamate uncaging frequencies and amplitudes: CaMKIIα sensed both higher frequencies and input amplitudes, acting as an input frequency/number decoder, whereas CaN sensed input number with little frequency dependence, thus serving only as an input number decoder (figure 5.2). These results provided

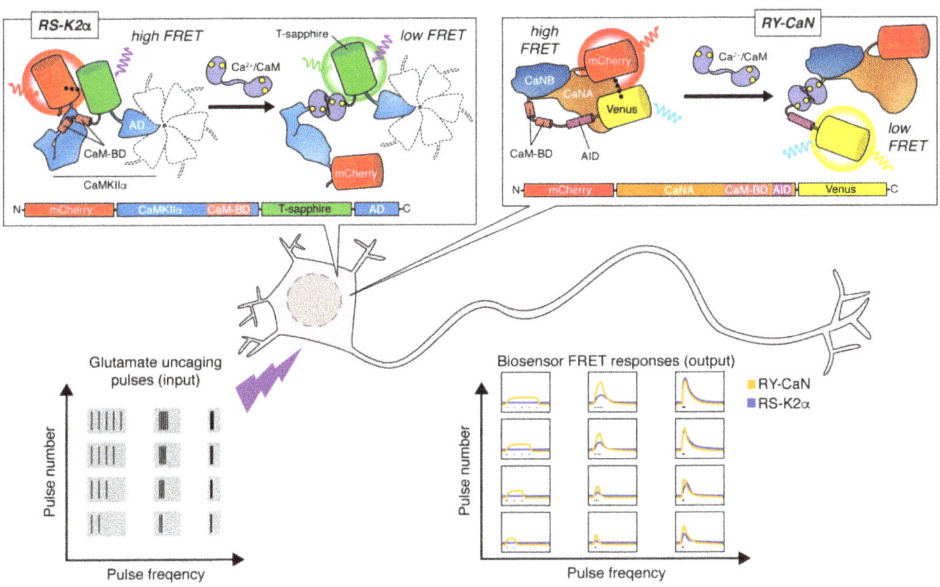

Figure 5.2. dFOMA system to study decoding and representation of neuronal input information by CaMKIIα and Calcineurin. (upper left) In the RS-K2α CaMKIIα activation biosensor, which is an improved version of Camuiα designed to optimize donor–acceptor proximity in the high-FRET sensor 'off' state, the binding of Ca^{2+}/CaM to the CaM-binding domain (CaM-BD) induces an open conformation that decreases FRET between T-sapphire and mCherry. The association domain (AD) allows RS-K2α to incorporate into endogenous CaMKIIα holoenzymes. (upper right) The RY-CaN Calcineurin activation biosensor is constructed by tagging mCherry and Venus to the N- and C-termini of the CaNA subunit of Calcineurin. Binding of Ca^{2+}/CaM to the CaM-BD induces a conformational change that displaces the Calcineurin autoinhibitory domain (AID) and leads to a decrease in FRET. (lower) In Fujii *et al* [45], light-controlled glutamate uncaging was applied with different pulse numbers and frequencies to hippocampal neurons, and the FRET responses from RS-K2α and RY-CaN were recorded simultaneously. CaMKIIα and CaN showed differential sensitivity under different glutamate uncaging frequencies and pulse counts: CaMKIIα sensed both higher frequencies and counts, thus acting as an input frequency/number decoder, whereas CaN sensed input number with little frequency dependence, thus serving only as an input number decoder.

evidence for the finely tuned neuronal signaling processing functions of these two critical downstream Ca^{2+} effectors.

CaN activation also plays a key role in the heart, where physiological CaN signaling is essential for normal cardiac development and cardiac myocyte differentiation, while dysregulation of CaN signaling by depleting the CaNB1 subunit specifically in the heart results in abnormal cardiac growth, impaired systolic and diastolic functions, and mortality [57–59]. Yet while CaN signaling has been extensively studied in both neonatal and adult cardiomyocytes, the traditional biochemical methods used often lack spatial and temporal precision. Thus, Bazzazi *et al* developed both bimolecular (DuoCaN) and unimolecular (UniCaN) CaN activation biosensors, which they used to monitor Ca^{2+} downstream effector dynamics in neonatal rat ventricular myocytes (NRVMs) and adult guinea pig ventricular myocytes (aGPVMs) [60]. Specifically, in DuoCaN, the cyan- and yellow-emitting FPs Cerulean and Venus were fused to the N-terminus of CaNB and C-terminus of CaNA, respectively, to yield a bimolecular FRET biosensor. UniCaN was generated following the same design, except that the two subunits from DuoCaN were joined together by a linker sequence, as the unimolecular design is more suitable for subcellular targeting. Notably, in order to correlate CaN activation with CaN phosphatase activity, they also measured the translocation of NFAT to the nucleus upon dephosphorylation by CaN, corroborating the readout of the CaN biosensor. Using these CaN activation biosensors, the authors observed differential CaN activation in neonatal and adult cardiomyocytes: whereas CaN was readily activated in response to individual Ca^{2+} pulses in neonatal myocytes, in adult myocytes, CaN seemed to only be activated at accumulative Ca^{2+} frequencies.

5.2.4 PKC

There are 10 major mammalian protein kinase C (PKC) isozymes that share highly conserved catalytic domains but vary in their regulatory domain, with only conventional PKC family members being responsive to Ca^{2+} signaling. Conventional PKC isozymes (α, βI, βII, and γ) contain an autoinhibitory pseudo-substrate sequence in the regulatory domain that allosterically tunes access of substrates to the catalytic site. Additionally, the regulatory domain also contains a C1 domain that binds to diacylglycerol (DAG) and phorbol esters, and a C2 domain that binds to Ca^{2+}; thus, conventional PKC isozymes are Ca^{2+} sensitive and are primed by Ca^{2+} binding to translocate to the plasma membrane to be fully activated. PKC function is dynamic and exquisitely controlled in live cells, and dysregulated PKC activity is involved in many pathological conditions, including cancer [61]. To dissect how PKC signaling is spatiotemporally regulated in living cells, various fluorescent biosensors have been created and applied to study the role of PKC in cancer biology.

The Newton lab developed the first PKC activity reporter (CKAR), which is composed of monomeric CFP (mCFP) and monomeric YFP (mYFP) flanking a PKC substrate sequence and FHA2 domain [62]. PKC-mediated phosphorylation of the biosensor results in a conformational change such that decreases in the FRET signal reflect enhanced PKC kinase activity in cells. Subcellular imaging by tethering CKAR to the plasma membrane revealed phase-locked oscillations between PKC activity and Ca^{2+}, supporting spatiotemporal regulation of the PKC signaling pathway [62]. A PKC activation biosensor has also been generated by the Newton lab: Kinameleon was engineered by fusing CFP and YFP to the N- and C-termini, respectively, of the wildtype PKCβII [63], an approach that has been previously validated with PKCδ [64]. Kinameleon is able to sense the intricate intramolecular conformational change that PKC undergoes during maturation, activation and downregulation. Further experiments to correlate PKC translocation kinetics with DAG dynamics using a FRET-based DAG sensor [62] revealed that the ligand-binding surface of the C1 domain is masked during maturation, highlighting the careful tuning of PKC signaling.

Considerable efforts of the community have since been invested in generating better CKAR variants, including the development of KCP-1 [65], which does not require a PAABD and instead utilizes the endogenous PKC substrate pleckstrin to reduce interference from signaling crosstalk by avoiding unwanted interactions between endogenous proteins and the PAABD, and the dual PKA and PKC reporter KCAP-1, though KCAP-1 is not suitable for cases where PKC and PKA are activated simultaneously. Komatsu *et al* also developed an improved version of CKAR by incorporating their universal EV linker for FRET biosensors [66], which is designed to reduce basal FRET efficiency and thus increase dynamic range. This new PKC activity reporter was localized to the plasma membrane through a plasma membrane targeting sequence derived from the Kras protein and named Eevee-PKC-pm, which showed responses to tetradecanoylphorbol 13-acetate (TPA) stimulation in Hela cells. Meanwhile, Ross *et al* developed CKAR2, which follows the original design of CKAR but contains mCerulean3 (a CFP variant) and cpVenus (a YFP variant) flanking an FHA1 domain and a modified PKC substrate sequence and shows four times larger dynamic range than the original CKAR [67]. An even greater improvement was also achieved with the development of the single-FP biosensor ExRai-CKAR, in which the PKC substrate sequence and FHA1 domain flank cpEGFP [27]. Upon phosphorylation, ExRai-CKAR undergoes a conformational change that results in a change in the cpEGFP excitation spectrum, with the response being read out as the ratio of fluorescence intensities at two different excitation wavelengths (i.e., excitation ratio), largely increasing the dynamic range of the biosensor. This generalizable design was also used to generate color variants such as blueCKAR and sapphireCKAR using cpBFP and cp-T-sapphire, respectively, for multiplexed activity imaging to study signal transduction and crosstalk.

PKC isozymes have long been implicated in synaptic plasticity, but whether PKC is activated in dendritic spines and how PKC encodes synaptic plasticity remains unknown [68, 69]. While CKAR is able to monitor PKC activity, it is not straightforward to identify which PKC isozyme is responsible for catalyzing substrate phosphorylation in a given situation due to the poor isozyme selectivity of PKC inhibitors and the fact that the preferred substrate sequences of PKC isozymes only exhibit subtle differences [61]. While isozyme-specific CKARs have been developed, these are selective for novel PKCs, which do not require Ca^{2+} [70, 71]. Therefore, to investigate isozyme-specific signaling by conventional PKCs, Colgan *et al* developed two new FRET-FLIM biosensors, ITRACK (Isozyme-specific TRAnslocation of C Kinase) and IDOCKS (Isozyme-specific Docking of C Kinase Substrate), to monitor the activation of individual PKC isozymes [72]. Both ITRACK and IDOCKS feature bimolecular designs where a specific PKC isozyme is directly tagged with EGFP as the FRET donor, while mCherry is either targeted directly to the plasma membrane (ITRACK) or fused to the PKC pseudosubstrate domain (IDOCKS) as the FRET acceptor. PKC translocation to the plasma membrane (ITRACKS), or translocation followed by binding to the pseudosubstrate domain (IDOCKS), leads to increased proximity between EGFP and mCherry, resulting in decreased EGFP fluorescence lifetime (figure 5.3). Using ITRACK and IDOCKS to

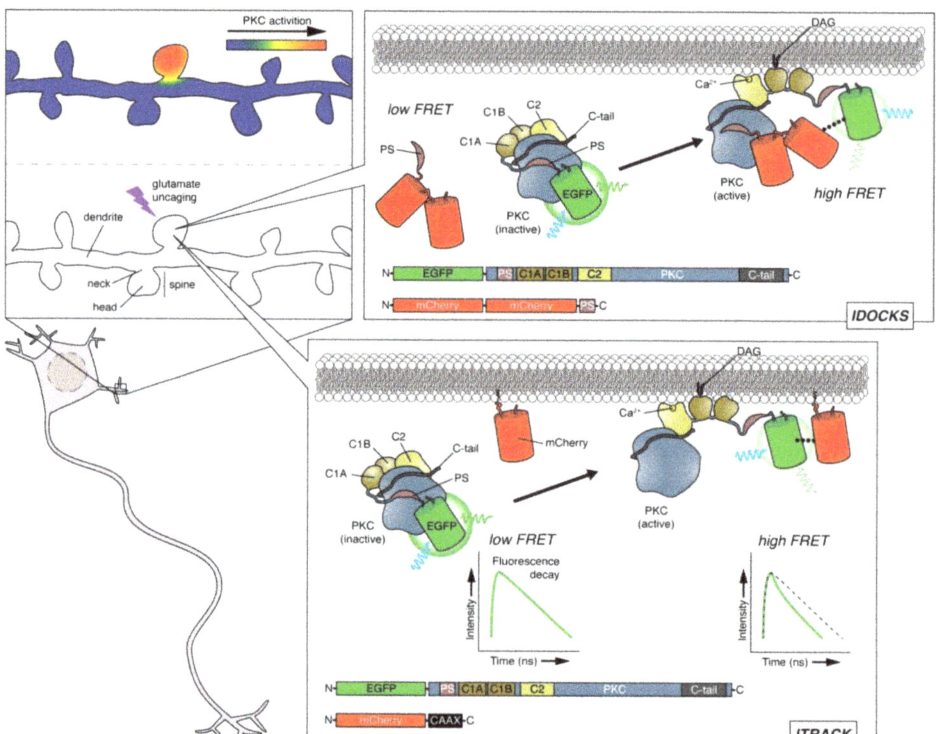

Figure 5.3. FLIM-FRET biosensors to study the role of isozyme-specific PKC signaling in dendritic spines. The bimolecular designs of IDOCKS (upper right) and ITRACK (lower right) both utilize a specific PKC isozyme directly fused with the EGFP FRET donor; however, in IDOCKS, a pair of mCherry FRET acceptors is fused to a PKC pseudosubstrate (PS) domain, while in ITRACK, a single mCherry acceptor is tethered directly to the plasma membrane. In both cases, PKC activation causes the donor part of the biosensor to translocate to the plasma membrane via binding to diacylglycerol (DAG), followed by PKC-activation-induced interaction between EGFP-PKC and PS-mCherry (IDOCKS) or simply increased proximity between EGFP-PKC and membrane-tethered mCherry (ITRACK) to yield a FRET increase, which is measured by a decrease in the donor (i.e., EGFP) fluorescence lifetime (illustrated for ITRACK). (upper left) Colgan *et al* [72] utilized these sensors to study isozyme-specific PKC signaling in individual dendritic spines responding to local glutamate uncaging.

monitor spatiotemporal PKC signaling in hippocampal neurons, the authors found that PKCα alone was required for synaptic plasticity in dendritic spines, which was due to a PDZ domain found only in PKCα.

The mechanisms of PKC signaling regulation in cancers are also being elucidated with the help of PKC biosensors. For a long time, PKC was thought to promote cancer progression, but the role of PKC as an oncoprotein is controversial since targeting PKC in cancers has been unsuccessful [73], raising the question whether loss of PKC rather than PKC activation actually promotes tumorigenesis. Antal *et al* characterized 46 cancer-associated mutations in PKC and evaluated how these mutations altered PKC activity, measured by CKAR, revealing that most cancer-associated in PKC mutations are loss-of-function. Furthermore, they found that a heterozygous PKC loss-of-function mutation found in colorectal cancer acts in a dominant-negative manner, and that correcting the mutation using CRISPR gene editing in patient-derived colon cancer cells reduced xenograft tumor size [74]. To further study this loss-of-function mechanism, Baffi *et al* used the PKC activity biosensor CKAR, the PKC activation biosensor Kinameleon, and a FRET-based translocation biosensor similar to ITRACK to study the critical conformational changes that PKC adopts for optimal response to second messengers [75]. By incorporating mutations and deletions into these different biosensors, they examined differential PKC activity, as well as translocation kinetics, during PKC activation. They discovered the autoinhibited state of PKC protects itself from dephosphorylation by PHLPP1 (a phosphatase that dephosphorylates PKC) and subsequent degradation, uncovering a quality control mechanism in which PHLPP1 regulates steady-state PKC levels by opposing the priming phosphorylation of newly synthesized PKC. Notably, high levels of PHLPP1 and low levels of PKC correlated with worsened survival in pancreatic cancer patients, highlighting a link between this quality control mechanism and PKC loss of function in cancer [75].

5.3 Urgent needs in the field

While we appreciate the unique advantages of using genetically encoded fluorescent biosensors to understand the spatiotemporal regulation of Ca^{2+}-triggered downstream signaling and its important functions in biological processes, there are still a number of challenges that need to be overcome in order to push this field forward.

- Ca^{2+} is involved in a very broad range of signaling pathways and influences nearly every facet of cellular life. Although we discussed only four major effectors downstream of Ca^{2+}, there are numerous other nodes in the Ca^{2+} signaling network, such as nitric oxide synthase (NOS), myosin light chain kinase (MLCK), troponin C, focal adhesion kinase 2 (FAK2), and phospholipase C δ (PLC δ), whose spatiotemporal dynamics remain largely unexplored. Biosensors for monitoring the activation/activity of these Ca^{2+} downstream effectors should provide new insights into Ca^{2+} signaling in physiological and pathological conditions such as cardiovascular disease.

- Biosensors designed to monitor Ca^{2+} effector activation/activity should also be optimized for easier distribution and application among the general research community. These efforts include improving the dynamic range, increasing sensitivity, and reducing phototoxicity. Improvements can be achieved through engineering of FPs with more desirable properties (lower phototoxicity, higher brightness, and better stability), or by screening biosensor variants via directed evolution.
- Increasing emphasis is being focused on *in vivo* imaging to visualize and correlate Ca^{2+} signaling dynamics with physiology and behavior in live animals [76]. However, the fluorescence signals of most current biosensors suffer from low tissue penetration depth due to light scattering and absorption, limiting the scope of *in vivo* imaging. A few notable examples have recently demonstrated the potential for improved *in vivo* Ca^{2+} imaging using biosensors with red-shifted fluorescence, including the use of infrared/near-infrared fluorescent proteins [77, 78]. Other non-fluorescent methods utilizing bioluminescence have also shown promising results for whole-animal imaging [79]. Further innovations along these lines will be needed to enable imaging of both Ca^{2+} and its effectors in animals to help us obtain more physiologically relevant data and translate basic research into benefits to human health.

5.4 Summary and outlook

Ca^{2+} regulates many aspects of cellular processes through its various downstream effector proteins. Thanks to the development and application of various biosensors in different cellular contexts, not only can we dissect the exquisite spatiotemporal mechanisms underlying physiological signaling, but we are also able to reveal the crucial role of effector proteins in pathological conditions, including neurodegenerative disease, cardiac disease and cancer. In recent decades, genetically encoded fluorescent biosensors have emerged as major tools for the study of cell signaling events downstream of Ca^{2+} due to their specificity, sensitivity and versatile designs. Given the importance of the fluorescent biosensor toolbox in probing the spatiotemporal regulation of cell signaling, many efforts are continually being dedicated to developing newer and better biosensors. With the advancement of chemical biology, synthetic biology, optical imaging, and gene editing techniques, the future of genetically encoded fluorescent biosensors to study Ca^{2+} effector signaling is bright.

Acknowledgements

This work was supported by National Institutes of Health Grants R35 CA197622, R01 DK073368, R01 DE030497, R01 HL162302, R01 CA262815 and RF1 MH126707 (to J Z) and Air Force Office of Scientific Research Grant FA9500-18-1-0051 (to J Z).

References

[1] Clapham D E 2007 Calcium signaling *Cell* **131** 1047–58

[2] Mehta S and Zhang J 2015 Dynamic visualization of calcium-dependent signaling in cellular microdomains *Cell Calcium* **58** 333–41

[3] Strehler E E and Treiman M 2004 Calcium pumps of plasma membrane and cell interior *Curr. Mol. Med.* **4** 323–35

[4] Hilgemann D W, Yaradanakul A, Wang Y and Fuster D 2006 Molecular control of cardiac sodium homeostasis in health and disease *J. Cardiovasc. Electrophysiol.* **17** S47–56

[5] Chad J E and Eckert R 1984 Calcium domains associated with individual channels can account for anomalous voltage relations of CA-dependent responses *Biophys. J.* **45** 993–9

[6] Berridge M J 2006 Calcium microdomains: organization and function *Cell Calcium* **40** 405–12

[7] Tsien R Y 1989 Fluorescent probes of cell signaling *Annu. Rev. Neurosci.* **12** 227–53

[8] Tsien R and Pozzan T 1989 Measurement of cytosolic free Ca^{2+} with quin2 *Meth. Enzymol.* **172** 230–62

[9] Minta A, Kao J P and Tsien R Y 1989 Fluorescent indicators for cytosolic calcium based on rhodamine and fluorescein chromophores *J. Biol. Chem.* **264** 8171–178

[10] Grynkiewicz G, Poenie M and Tsien R Y 1985 A new generation of Ca^{2+} indicators with greatly improved fluorescence properties *J. Biol. Chem.* **260** 3440–50

[11] Newman R H, Fosbrink M D and Zhang J 2011 Genetically encodable fluorescent biosensors for tracking signaling dynamics in living cells *Chem. Rev.* **111** 3614–66

[12] Zhou X, Clister T L, Lowry P R, Seldin M M, Wong G W and Zhang J 2015 Dynamic visualization of mTORC1 activity in living cells *Cell Rep.* **10** P1767–77

[13] Mehta S, Aye-Han N-N, Ganesan A, Oldach L, Gorshkov K and Zhang J 2014 Calmodulin-controlled spatial decoding of oscillatory Ca^{2+} signals by calcineurin *Elife* **3** e03765

[14] Clister T, Greenwald E C, Baillie G S and Zhang J 2019 AKAP95 organizes a nuclear microdomain to control local cAMP for regulating nuclear PKA *Cell Chem. Biol.* **26** 885–91

[15] Miyawaki A, Llopis J, Heim R, McCaffery J M, Adams J A, Ikura M and Tsien R Y 1997 Fluorescent indicators for Ca^{2+} based on green fluorescent proteins and calmodulin *Nature* **388** 882–7

[16] Greenwald E C, Mehta S and Zhang J 2018 Genetically encoded fluorescent biosensors illuminate the spatiotemporal regulation of signaling networks *Chem. Rev.* **118** 11707–94

[17] Campbell R E 2009 Fluorescent-protein-based biosensors: modulation of energy transfer as a design principle *Anal. Chem.* **81** 5972–9

[18] Jones G A and Bradshaw D S 2019 Resonance energy transfer: from fundamental theory to recent applications *Front. Phys.* **7** 100

[19] Periasamy A, Mazumder N, Sun Y, Christopher K G and Day R N 2015 FRET microscopy: basics, issues and advantages of FLIM-FRET imaging *Advanced Time-Correlated Single Photon Counting Applications Springer Series in Chemical Physics* ed W Becker (Cham: Springer Int. Publishing) pp 249–76

[20] Kwok S, Lee C, Sánchez S A, Hazlett T L, Gratton E and Hayashi Y 2008 Genetically encoded probe for fluorescence lifetime imaging of CaMKII activity *Biochem. Biophys. Res. Commun.* **369** 519–25

[21] Bajar B T, Wang E S, Zhang S, Lin M Z and Chu J 2016 A guide to fluorescent protein FRET pairs *Sensors (Basel)* **16** 1488

[22] Nakai J, Ohkura M and Imoto K 2001 A high signal-to-noise Ca^{2+} probe composed of a single green fluorescent protein *Nat. Biotechnol.* **19** 137–41

[23] Inoue M, Takeuchi A, Horigane S, Ohkura M, Gengyo-Ando K, Fujii H, Kamijo S, Takemoto-Kimura S, Kano M and Nakai J *et al* 2015 Rational design of a high-affinity, fast, red calcium indicator R-CaMP2 *Nat. Methods* **12** 64–70

[24] Tallini Y N, Ohkura M, Choi B-R, Ji G, Imoto K, Doran R, Lee J, Plan P, Wilson J and Xin H-B *et al* 2006 Imaging cellular signals in the heart *in vivo*: cardiac expression of the high-signal Ca^{2+} indicator GCaMP2 *Proc. Natl Acad. Sci. USA* **103** 4753–8

[25] Tian L, Hires S A, Mao T, Huber D, Chiappe M E, Chalasani S H, Petreanu L, Akerboom J, McKinney S A and Schreiter E R *et al* 2009 Imaging neural activity in worms, flies and mice with improved GCaMP calcium indicators *Nat. Methods* **6** 875–81

[26] Akerboom J, Chen T-W, Wardill T J, Tian L, Marvin J S, Mutlu S, Calderón N C, Esposti F, Borghuis B G and Sun X R *et al* 2012 Optimization of a GCaMP calcium indicator for neural activity imaging *J. Neurosci.* **32** 13819–40

[27] Mehta S, Zhang Y, Roth R H, Zhang J-F, Mo A, Tenner B, Huganir R L and Zhang J 2018 Single-fluorophore biosensors for sensitive and multiplexed detection of signalling activities *Nat. Cell Biol.* **20** 1215–25

[28] Means A R, VanBerkum M F, Bagchi I, Lu K P and Rasmussen C D 1991 Regulatory functions of calmodulin *Pharmacol. Ther.* **50** 255–70

[29] Romoser V A, Hinkle P M and Persechini A 1997 Detection in living cells of Ca^{2+}-dependent changes in the fluorescence emission of an indicator composed of two green fluorescent protein variants linked by a calmodulin-binding sequence. A new class of fluorescent indicators *J. Biol. Chem.* **272** 13270–4

[30] Heim R, Prasher D C and Tsien R Y 1994 Wavelength mutations and posttranslational autoxidation of green fluorescent protein *Proc. Natl Acad. Sci. USA* **91** 12501–4

[31] Delagrave S, Hawtin R E, Silva C M, Yang M M and Youvan D C 1995 Red-shifted excitation mutants of the green fluorescent protein *Biotechnology (NY)* **13** 151–4

[32] Chew T-L, Wolf W A, Gallagher P J, Matsumura F and Chisholm R L 2002 A fluorescent resonant energy transfer-based biosensor reveals transient and regional myosin light chain kinase activation in lamella and cleavage furrows *J. Cell Biol.* **156** 543–53

[33] Goeckeler Z M and Wysolmerski R B 1995 Myosin light chain kinase-regulated endothelial cell contraction: the relationship between isometric tension, actin polymerization, and myosin phosphorylation *J. Cell Biol.* **130** 613–27

[34] Poperechnaya A, Varlamova O, Lin P J, Stull J T and Bresnick A R 2000 Localization and activity of myosin light chain kinase isoforms during the cell cycle *J. Cell Biol.* **151** 697–708

[35] Chrzanowska-Wodnicka M and Burridge K 1996 Rho-stimulated contractility drives the formation of stress fibers and focal adhesions *J. Cell Biol.* **133** 1403–15

[36] Kishi H *et al* 2000 Stable transfectants of smooth muscle cell line lacking the expression of myosin light chain kinase and their characterization with respect to the actomyosin system *J. Biol. Chem.* **275** 1414–20

[37] Garcia J G, Lazar V and Gilbert L I 1997 Myosin light chain kinase in endothelium: molecular cloning and regulation *Am. J. Resp. Cell Mol. Biol.* **16** 489–94

[38] Huang L, Pike D, Sleat D E, Nanda V and Lobel P 2014 Potential pitfalls and solutions for use of fluorescent fusion proteins to study the lysosome *PLoS One* **9** e88893

[39] Bayer K U and Schulman H 2019 Cam kinase: still inspiring at 40 *Neuron* **103** 380–94

[40] Lisman J, Schulman H and Cline H 2002 The molecular basis of CaMKII function in synaptic and behavioural memory *Nat. Rev. Neurosci.* **3** 175–90
[41] Malinow R, Mainen Z F and Hayashi Y 2000 LTP mechanisms: from silence to four-lane traffic *Curr. Opin. Neurobiol.* **10** 352–7
[42] Takao K, Okamoto K-I, Nakagawa T, Neve R L, Nagai T, Miyawaki A, Hashikawa T, Kobayashi S and Hayashi Y 2005 Visualization of synaptic Ca^{2+}/calmodulin-dependent protein kinase II activity in living neurons *J. Neurosci.* **25** 3107–12
[43] Lee S-J R, Escobedo-Lozoya Y, Szatmari E M and Yasuda R 2009 Activation of CaMKII in single dendritic spines during long-term potentiation *Nature* **458** 299–304
[44] Yagishita S, Hayashi-Takagi A, Ellis-Davies G C R, Urakubo H, Ishii S and Kasai H 2014 A critical time window for dopamine actions on the structural plasticity of dendritic spines *Science* **345** 1616–20
[45] Fujii H, Inoue M, Okuno H, Sano Y, Takemoto-Kimura S, Kitamura K, Kano M and Bito H 2013 Nonlinear decoding and asymmetric representation of neuronal input information by CaMKIIα and calcineurin *Cell Rep.* **3** 978–87
[46] Erickson J R, Patel R, Ferguson A, Bossuyt J and Bers D M 2011 Fluorescence resonance energy transfer-based sensor Camui provides new insight into mechanisms of calcium/calmodulin-dependent protein kinase II activation in intact cardiomyocytes *Circ. Res.* **109** 729–38
[47] Lam A J, St-Pierre F, Gong Y, Marshall J D, Cranfill P J, Baird M A, McKeown M R, Wiedenmann J, Davidson M W and Schnitzer M J *et al* 2012 Improving FRET dynamic range with bright green and red fluorescent proteins *Nat. Methods* **9** 1005–12
[48] Lee S-J R, Escobedo-Lozoya Y, Szatmari E M and Yasuda R 2009 Activation of CaMKII in single dendritic spines during long-term potentiation *Nature* **458** 299–304
[49] Murakoshi H, Shibata A C E, Nakahata Y and Nabekura J 2015 A dark green fluorescent protein as an acceptor for measurement of Förster resonance energy transfer *Sci. Rep.* **5** 15334
[50] Nakahata Y, Nabekura J and Murakoshi H 2016 Dual observation of the ATP-evoked small GTPase activation and Ca^{2+} transient in astrocytes using a dark red fluorescent protein *Sci. Rep.* **6** 39564
[51] Ardestani G, West M C, Maresca T J, Fissore R A and Stratton M M 2019 FRET-based sensor for CaMKII activity (FRESCA): a useful tool for assessing CaMKII activity in response to Ca^{2+} oscillations in live cells *J. Biol. Chem.* **294** 11876–91
[52] Li H, Rao A and Hogan P G 2011 Interaction of calcineurin with substrates and targeting proteins *Trends Cell Biol.* **21** 91–103
[53] Rusnak F and Mertz P 2000 Calcineurin: form and function *Physiol. Rev.* **80** 1483–521
[54] Newman R H and Zhang J 2008 Visualization of phosphatase activity in living cells with a FRET-based calcineurin activity sensor *Mol. Biosyst.* **4** 496–501
[55] Persechini A and Cronk B 1999 The relationship between the free concentrations of Ca^{2+} and Ca^{2+}-calmodulin in intact cells *J. Biol. Chem.* **274** 6827–30
[56] Matsuzaki M, Honkura N and Ellis G C R 2004 Structural basis of long-term potentiation in single dendritic spines *Nature* **429** 761–6
[57] de la Pompa J L *et al* 1998 Role of the NF-ATc transcription factor in morphogenesis of cardiac valves and septum *Nature* **392** 182–6
[58] Kasahara A, Cipolat S, Chen Y, Dorn G W and Scorrano L 2013 Mitochondrial fusion directs cardiomyocyte differentiation via calcineurin and Notch signaling *Science* **342** 734–7

[59] Schaeffer P J *et al* 2009 Impaired contractile function and calcium handling in hearts of cardiac-specific calcineurin b1-deficient mice *Am. J. Physiol. Heart. Circ. Physiol.* **297** H1263–73
[60] Bazzazi H, Sang L, Dick I E, Joshi-Mukherjee R, Yang W and Yue D T 2015 Novel fluorescence resonance energy transfer-based reporter reveals differential calcineurin activation in neonatal and adult cardiomyocytes *J. Physiol. (Lond.)* **593** 3865–84
[61] Newton A C 2018 Protein kinase C: perfectly balanced *Crit. Rev. Biochem. Mol. Biol.* **53** 208–30
[62] Violin J D, Zhang J, Tsien R Y and Newton A C 2003 A genetically encoded fluorescent reporter reveals oscillatory phosphorylation by protein kinase C *J. Cell Biol.* **161** 899–909
[63] Antal C E, Violin J D, Kunkel M T, Skovsø S and Newton A C 2014 Intramolecular conformational changes optimize protein kinase C signaling *Chem. Biol.* **21** 459–69
[64] Braun D C, Garfield S H and Blumberg P M 2005 Analysis by fluorescence resonance energy transfer of the interaction between ligands and protein kinase Cdelta in the intact cell *J. Biol. Chem.* **280** 8164–71
[65] Schleifenbaum A, Stier G, Gasch A, Sattler M and Schultz C 2004 Genetically encoded FRET probe for PKC activity based on pleckstrin *J. Am. Chem. Soc.* **126** 11786–7
[66] Komatsu N, Aoki K, Yamada M, Yukinaga H, Fujita Y, Kamioka Y and Matsuda M 2011 Development of an optimized backbone of FRET biosensors for kinases and GTPases *Mol. Biol. Cell* **22** 4647–56
[67] Ross B L, Tenner B, Markwardt M L, Zviman A, Shi G, Kerr J P, Snell N E, McFarland J J, Mauban J R and Ward C W *et al* 2018 Single-color, ratiometric biosensors for detecting signaling activities in live cells *Elife* **7** e35458
[68] Sossin W S 2007 Isoform specificity of protein kinase Cs in synaptic plasticity *Learn. Mem.* **14** 236–46
[69] Callender J A and Newton A C 2017 Conventional protein kinase C in the brain: 40 years later *Neuronal Signal.* **1** NS20160005
[70] Kajimoto T, Caliman A D, Tobias I S, Okada T, Pilo C A, Van A-A N, Andrew McCammon J, Nakamura S-I and Newton A C 2019 Activation of atypical protein kinase C by sphingosine 1-phosphate revealed by an aPKC-specific activity reporter *Sci. Signal.* **12** eaat6662
[71] Kajimoto T, Sawamura S, Tohyama Y, Mori Y and Newton A C 2010 Protein kinase C δ-specific activity reporter reveals agonist-evoked nuclear activity controlled by Src family of kinases *J. Biol. Chem.* **285** 41896–910
[72] Colgan L A, Hu M, Misler J A, Parra-Bueno P, Moran C M, Leitges M and Yasuda R 2018 PKCα integrates spatiotemporally distinct Ca^{2+} and autocrine BDNF signaling to facilitate synaptic plasticity *Nat. Neurosci.* **21** 1027–37
[73] Mackay H J and Twelves C J 2007 Targeting the protein kinase C family: are we there yet? *Nat. Rev. Cancer* **7** 554–62
[74] Antal C E, Hudson A M, Kang E, Zanca C, Wirth C, Stephenson N L, Trotter E W, Gallegos L L, Miller C J and Furnari F B *et al* 2015 Cancer-associated protein kinase C mutations reveal kinase's role as tumor suppressor *Cell* **160** 489–502
[75] Baffi T R, Van A-A N, Zhao W, Mills G B and Newton A C 2019 Protein Kinase C quality control by phosphatase PHLPP1 unveils loss-of-function mechanism in cancer *Mol. Cell* **74** 378–92

[76] Russell J T 2011 Imaging calcium signals *in vivo*: a powerful tool in physiology and pharmacology *Br. J. Pharmacol.* **163** 1605–25

[77] Qian Y, Piatkevich K D, Mc Larney B, Abdelfattah A S, Mehta S, Murdock M H, Gottschalk S, Molina R S, Zhang W and Chen Y *et al* 2019 A genetically encoded near-infrared fluorescent calcium ion indicator *Nat. Methods* **16** 171–4

[78] Wu J, Abdelfattah A S, Miraucourt L S, Kutsarova E, Ruangkittisakul A, Zhou H, Ballanyi K, Wicks G, Drobizhev M and Rebane A *et al* 2014 A long stokes shift red fluorescent Ca^{2+} indicator protein for two-photon and ratiometric imaging *Nat. Commun.* **5** 5262

[79] Iwano S, Sugiyama M, Hama H, Watakabe A, Hasegawa N, Kuchimaru T, Tanaka K Z, Takahashi M, Ishida Y and Hata J *et al* 2018 Single-cell bioluminescence imaging of deep tissue in freely moving animals *Science* **359** 935–9

Chapter 6

Calcium-activated potassium channels

Alberto J Gonzalez-Hernandez, Aravind Kshatri and Teresa Giraldez

6.1 Introduction

Calcium-activated potassium (K_{Ca}) channels are unique members of the potassium ion channel family able to couple intracellular Ca^{2+} signals to membrane potential variations. K_{Ca} channels play fundamental roles ranging from regulating neuronal excitability to controlling muscle contraction. The K_{Ca} family of ion channels comprises three main subfamilies that are classified according to their single-channel conductance: SK (small conductance; ~4–14 pS), IK (intermediate conductance; ~32–39 pS) and BK (large conductance; ~200–300 pS). Although they present distinct biophysical and pharmacological characteristics, all members of the K_{Ca} family are expressed at the plasma membrane as tetramers of α subunits encoded by different genes. Phylogenetically, the genes encoding α subunits of SK/IK and those of BK belong to two separated groups (Wei *et al* 2005). This classification is paralleled by differences in their biophysical properties, Ca^{2+} sensitivity and regulatory mechanisms.

The following sections describe the function and physiological roles of K_{Ca} channel family members, as well as the underlying structural features. Additionally, different mechanisms regulating their function are discussed, as well as the pathologies that have been related to their malfunction.

6.2 BK channels

6.2.1 BK function and physiological roles

Large conductance voltage- and Ca^{2+}-activated K^+ channels (BK, MaxiK channels, Slo1 or KCa1.1 channels) were first described in the early 1980s as calcium and voltage-activated potassium currents (Marty 1981, Pallotta *et al* 1981). A distinctive feature was their sensitivity to the scorpion toxins charybdotoxin (Miller *et al* 1985) and iberiotoxin (Galvez *et al* 1990). In comparison to other K^+ channels,

doi:10.1088/978-0-7503-2009-2ch6

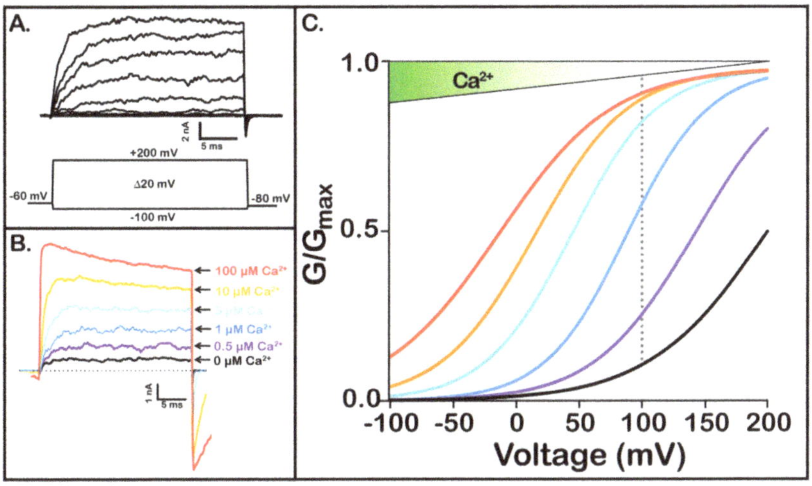

Figure 6.1. Macroscopic currents of BK channels. (A) Macroscopic currents of BK channels in response to a family of voltage pulses (schematized below the current traces) in the absence of Ca^{2+}. (B) Macroscopic currents of BK channels in response to increasing Ca^{2+} concentrations (color-coded in the legend) at a fixed voltage of +100 mV. (C) Increasing intracellular Ca^{2+} concentrations shift the voltage-dependent activation curves (*G–V* relations) towards more negative voltages. Ca^{2+} concentrations are color-coded as in (B).

BK channels exhibit an unusually large conductance ranging between 100 pS and 300 pS (Latorre *et al* 2017). The BKα subunit is encoded by the gene KCNMA1 (also known as *Slowpoke*), which was cloned from *Drosophila* (Adelman *et al* 1992, Atkinson *et al* 1991), mice (Butler *et al* 1993) and human (Dworetzky *et al* 1994, McCobb *et al* 1995). BK channels are synergistically activated by intracellular Ca^{2+} and membrane depolarization. The channel's conductance (*G*) increases with voltage, analogously to other voltage-gated potassium channels; in addition, this *G–V* relationship is shifted towards negative voltages as Ca^{2+} concentration increases (figure 6.1). This complex kinetic behavior was elegantly described by the work of the Magleby (Magleby 2003, McManus and Magleby 1988, Rothberg and Magleby 1999) and Aldrich (Cox *et al* 1997, Horrigan and Aldrich 1999, Horrigan *et al* 1999) laboratories, leading to the widely accepted allosteric model of Horrigan and Aldrich (Horrigan and Aldrich 2002). The physiological relevance of this activation mechanism is significant because of its wide dynamic range, covering voltage values from −200 mV to +300 mV and Ca^{2+} concentrations up to 10 mM (Magleby 2003). Not surprisingly, BK channels constitute an essential link between membrane excitability and intracellular [Ca^{2+}] signalling (Rothberg 2012).

Unlike other members of the K_{Ca} family, BK channels are widely expressed in a variety of excitable and non-excitable cells (Latorre *et al* 2017). In many cell types, BK activation relies on the local influx of Ca^{2+} to achieve micromolar levels (Fakler and Adelman 2008). In neurons and smooth muscle cells, membrane depolarization provided by an action potential triggers Ca^{2+} influx through voltage-dependent Ca^{2+} channels (VDCCs) (Griguoli *et al* 2016), coincidentally activating neighboring BK channels to repolarize the membrane and eventually closing VDCCs, terminating the Ca^{2+}signal (Marrion and Tavalin 1998). This negative feedback mechanism enables BK channels as important regulators of many physiological processes including smooth muscle contraction (Meredith *et al* 2004, Nelson *et al* 1995, Semenov *et al* 2011), insulin secretion (Houamed *et al* 2010), neurotransmitter release (Lingle *et al* 1996, Raffaelli *et al* 2004, Robitaille *et al* 1993), circadian rhythm (Meredith *et al* 2006, Whitt *et al* 2016), action potential termination (Montgomery and Meredith 2012, Muller *et al* 2007, Storm 1987) and heart regulation (Lai *et al* 2014).

Numerous studies have identified the VDCC subtypes mediating this coupling mechanism (Fakler and Adelman 2008). These include Cav2.1 (Berkefeld and Fakler 2008, Berkefeld *et al* 2006, Edgerton and Reinhart 2003, Womack *et al* 2004), Cav2.2 (Berkefeld *et al* 2006, Loane *et al* 2007, Prakriya and Lingle 1999), Cav1.2 (Berkefeld and Fakler 2008, Berkefeld *et al* 2006, Grunnet and Kaufmann 2004, Prakriya and Lingle 1999), Cav1.3 (Prakriya and Lingle 1999, Vivas *et al* 2017) and Cav3.2 (Gackiere *et al* 2013, Rehak *et al* 2013). Interestingly, although it was originally suggested that BK channels do not interact with Cav2.3 (Berkefeld *et al* 2006, Cerrada *et al* 2018), recent evidence suggests that functional association with these channels may occur in CA1 pyramidal neurons (Gutzmann *et al* 2019). Additionally, accessory subunits of Cav channels have been proposed to interact independently with BK channels. Zou *et al* (2008) demonstrated that interaction with Cavβ1 reduces Ca^{2+} sensitivity and slows down gating of the BK channel without affecting its relative membrane expression, although this finding remains to be shown in native tissues (Zou *et al* 2008). Another recent finding suggests that high affinity interactions of the Cavα2δ1 subunit with the BK channel N-terminus result in reduced membrane expression of Cav2 channels. This mechanism has been proposed to be of physiological relevance, since the administration of a BK N-terminus peptide leads to decreased inflammatory and neuropathic pain in mice (Zhang *et al* 2018a).

The functional association of BK to Ca^{2+} sources is quite versatile, not restricted to VDCC but including other Ca^{2+} sources, such as ryanodine (RyR) (Chavis *et al* 1998, Irie and Trussell 2017, Wang *et al* 2016, Whitt *et al* 2018, Yamamura *et al* 2012) and inositol 1,4,5-triphosphate (InsP3R) receptors (Zhao *et al* 2010). Additionally, Ca^{2+} influx through other non-selective cation-permeable channels, such as N-methyl-D-aspartate receptors (NMDAR) or transient receptor potential (TRP) channels have been shown to activate BK channels. NMDARs associate with BK channels in the neuronal postsynaptic membrane where they modulate excitability (Isaacson and Murphy 2001, Zhang *et al* 2018b) and synaptic plasticity (Gómez *et al* 2021). In the case of TRP channels, multiple subtypes of this large family of ion channels have been shown to form tight complexes with BK channels in a wide range of tissues. TRPV1 interacts with BK in dorsal root ganglion cells, where BK-mediated negative feedback has been proposed to modulate pain perception (Wu *et al* 2013). In cerebral artery smooth muscle cells, TRPV4 has been proposed to activate BK channels via RyR-mediated Ca^{2+} release to induce artery relaxation (Earley *et al* 2005, Liu *et al* 2020, Szarka *et al* 2018). In kidney podocytes, TRPC3 and TRPC6 physically interact with BK channels. These complexes have been proposed to be mechanically activated by glomeruli swelling, regulating glomerular filtration (Kim *et al* 2009). Lastly, Ca^{2+} influx through menthol activated TRPM8 channels has been also shown to activate BK channel currents in glioblastoma cells, linking these Ca^{2+} signaling complexes to tumor invasion (Wondergem and Bartley 2009).

The full functional picture of BK and its association to different calcium sources must take into account the regulation by auxiliary subunits and the characteristics of specific BK splice variants (see below). Although the dynamics of these complexes has not been fully explored in all physiological settings, it is tempting to hypothesize that the fine-tuning of this combinatorial complexity constitutes the basis for a large diversity of physiological outputs. An example of such complexity has been shown in the brain suprachiasmatic nucleus (SCN), where BK coupling to different calcium sources and auxiliary subunits during day and night leads to distinct excitability patterns (Whitt *et al* 2018).

6.2.2 BK regulation by auxiliary subunits

In contrast to voltage-gated potassium channels, where functional diversity arises from evolution of different genes, the variety of distinctive BK channel phenotypes in different cells and tissues is derived from modifications of the single KCNMA1 gene encoding the BKα subunit (Dworetzky *et al* 1996, McManus *et al* 1995). Such mechanisms include alternative splicing (Shipston 2001), post-translational modifications (Kyle and Braun 2014) and association with auxiliary subunits (Latorre *et al* 2017). Native BK channels are known to co-express with one of the two classes of structurally and functionally distinct modulatory subunits, β (1–4) and γ (1–4) (Brenner *et al* 2000, Knaus *et al* 1994, Wallner *et al* 1999, Yan and Aldrich 2012, Yan and Aldrich 2010). The association of these subunits modifies almost all aspects of BK channel gating such as its kinetics, voltage dependence, Ca^{2+} sensitivity as well as the pharmacological properties of BK channels (reviewed in Gonzalez-Perez and Lingle 2019).

The β subunits of the BK channel, coded by the genes KCNMB1-4, have a molecular size of ~20 KDa and share a common topology of two transmembrane homologous segments, TM1 and TM2, with a large extracellular loop and intracellular N and C-terminal regions (figure 6.2; reviewed in Latorre *et al* 2017). In the recently solved human BK- β4 cryogenic electron microscopy (cryo-EM) structure, the two TM helices of β subunits are sandwiched between the voltage sensor domains of the four BKα subunits and their large extracellular loop forms a crown over the central pore (Tao and MacKinnon 2019; see following section). Each of the four β subunits have different functional signatures that match their defined physiological roles. Despite sharing a common gating mechanism to modify the BK channel functions, the biophysical properties conferred to BK by each of these subunits are entirely unique (Yang and Cui 2015). The β1, β2, and β4 subunits modify the Ca^{2+} sensitivity of BK channel activation. β1 and β2 increase the apparent Ca^{2+} sensitivity and slow down channel gating kinetics (Latorre *et al* 2017, Gonzalez-Perez and Lingle 2019).

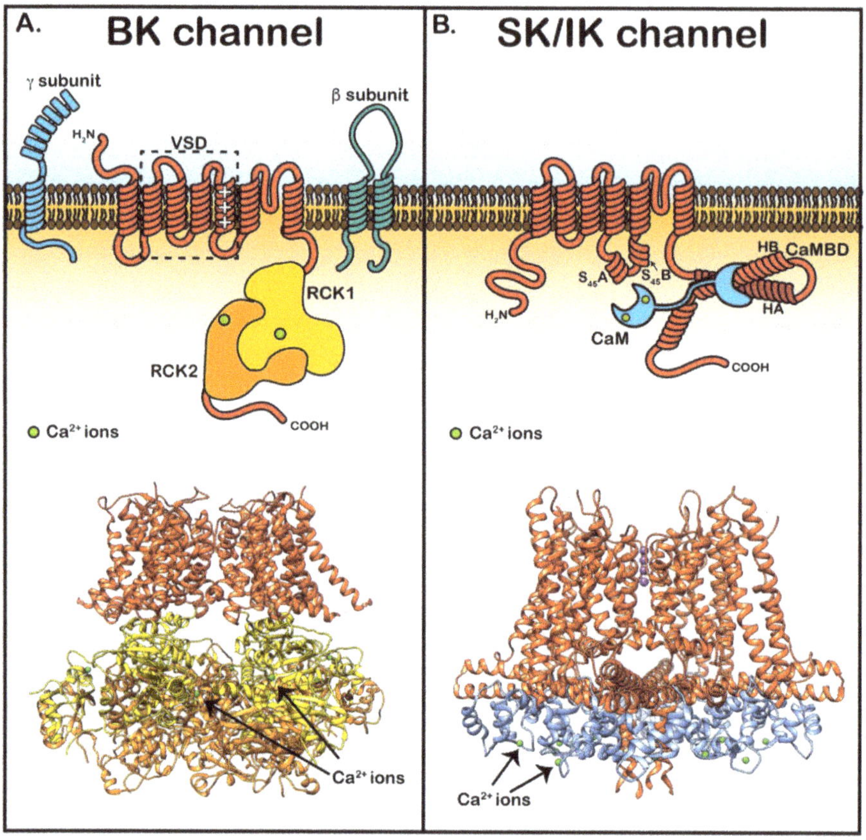

Figure 6.2. Topology and structures of BK and SK/IK channels. (A) *Top,* schematic topology of BKα (center) and auxiliary subunits β (right) and γ (left). In the BKα subunit, binding sites for divalent cations are located in the cytosolic C-terminal region of the channel. Each α subunit contains two high affinity Ca^{2+} binding sites (shown as green circles) and one low affinity Ca^{2+} and Mg^{2+} binding site, formed by residues from RCK1, S0–S1 and S2-S3 intracellular loops (not shown). *Bottom,* Ca^{2+}-bound BKα homotetramer full-length structure from human (PDB: 6V38). (B) *Top,* schematic protein topology of one SKα-subunit including CaM (in blue) bound to the CaMBD. Relevant regions involved in SK gating are indicated: helices $S_{45}A$, $S_{45}B$ and the CaMB formed by helices HA and HB. *Bottom,* full-length structure of the Ca^{2+}-CaM bound IK channel (PDB: 6CNN).

Although β4 exerts a more complex regulation in the presence of symmetrical K^+, it increases Ca^{2+} sensitivity under physiological conditions (Jaffe *et al* 2011, Latorre *et al* 2017). β1 and β4 subunits do not display either inactivation or instantaneous current rectification, which differentiates them from β2 and some splice variants of the β3 subunits (Latorre *et al* 2017, Gonzalez-Perez and Lingle 2019, Uebele *et al* 2000, Lingle *et al* 2001). With the exception of the β3 subunits, the β1, β2 and β4 subunits regulate BK channel gating by stabilizing the voltage sensor domain (VSD) in its active conformation (Contreras *et al* 2012). The association of BKα subunits with β subunits also confers sensitivity or resistance to various compounds of either pharmacological or physiological relevance (Latorre *et al* 2017). For instance, the regulation of BK by 17 β-estradiol requires the presence of the β1 or β4 subunits, which also modulate the sensitivity of BK channels to alcohol (Feinberg-Zadek *et al* 2008). The β4 subunit also confers BK channel resistance to iberiotoxin and charybdotoxin (Meera *et al* 2000).

All γ subunits (~35 KDa) belong to the superfamily of Leucine-Rich Repeat Containing (LRRC) proteins and are encoded by genes LRRC26 (γ1), LRRC52 (γ2), LRRC55 (γ3), and LRRC38 (γ4). They share a similar structure, consisting of an N-terminal cleavable signal peptide, a single transmembrane segment (TM), a large extracellular leucine-rich repeat domain (LRRD) and an intracellular C-terminus (figure 6.2); (Gonzalez-Perez and Lingle 2019, Yan and Aldrich 2012, Yan and Aldrich 2010). Unlike the complex effects of β subunits on the gating of BK channels, all γ subunits produce a hyperpolarizing shift in the voltage dependence of BK activation, to differing extents (γ1 > γ2 > γ3 > γ4). γ1 subunits produce the largest gating shifts (approximately −140 mV), resulting in BK opening at physiological voltages, even at resting intracellular Ca^{2+} concentrations (Yan and Aldrich 2010). Growing evidence points to γ1-containing BK channels as being relevant players in the physiology of epithelial cells (Gonzalez-Perez and Lingle 2019, Gonzalez-Perez *et al* 2021). Additionally, γ1 makes BK channels resistant to activators such as Mallotoxin and GoSlo (Almassy and Begenisich, 2012, Giraldez, 2017, Kshatri *et al* 2017). Another notable feature of these subunits is the 'all or none' gating mechanism, where their functional effects are independent of the number of subunits present in the BK channels (Gonzalez-Perez *et al* 2014). In other words, only one γ1 subunit is enough to produce the full gating effect on BK channels, whereas the functional effect of β subunits is incremental based on the number of β subunits associated with the BKα channels. It is also interesting to note that both γ1 and β2 subunits could assemble simultaneously with BK subunits resulting in an additive gating shift (Gonzalez-Perez *et al* 2015).

6.2.3 Structural basis of BK channel function

BK channels have unique features that distinguish them from other members of the calcium-gated and voltage-gated potassium channels families. BKα subunits show a modular structure (figure 6.2) consisting of an N-terminal domain embedded in the plasma membrane (MD) and a large C-terminal cytosolic domain that encompasses about one–third of the channel protein (Yang and Cui 2015). The MD core is reminiscent of voltage-gated potassium channels, consisting of six transmembrane (TM) segments S1–S6 which include a conserved region (S1–S4) known as the voltage sensor domain (VSD), and a pore domain (PD) region formed by segments S5-S6, which constitutes the ion conducting pathway (figure 6.2). One striking feature of BK channels is their large conductance, which stands out within the potassium channels superfamily. Rather than differences in the sequence of the selectivity filter, which is highly conserved (Neyton and Miller 1988), this characteristic is partly due to the presence of rings of negatively charged amino acids acting as an electrostatic bait to concentrate K^+ ions at the intracellular and extracellular cavities of the channel (Brelidze *et al* 2003, Carvacho *et al* 2008, Nimigean *et al* 2003).

Another relevant feature is the apparently large size of the inner cavity of around 8–10 Å, which results in inhibition by relatively large molecules (Tang *et al* 2009, Wilkens and Aldrich 2006) from the intracellular side (Li and Aldrich 2004) and provides accessibility to MTS reagents (Zhou *et al* 2011). In addition, BK has an extra TM segment S0, leading the NH_2 terminus of the protein to the extracellular side. The function of this segment has been related to VSD function, as well as to interactions with the auxiliary subunits (Castillo *et al* 2016, Koval *et al* 2007, Wallner *et al* 1996).

Full BK tetramers are embedded in the membrane with central PD regions forming a pore, surrounded by four VSDs and S0 segments at the periphery (figure 6.2). Similar to other members of the K_{Ca} family (see below) and bacterial Ca^{2+}-gated channels (MthK; Posson *et al* 2013), it has been proposed that the gate of the BK channel that controls ion permeation is located near the selectivity filter and not at the cytosolic end of S6, although this issue remains controversial and is not fully answered by available structural data (for thorough analyses of this topic, see Latorre *et al* (2017) and Zhou *et al* (2017)). Both movement of the VSD and Ca^{2+} binding to the C-terminal domain (see below) regulate the function of this gate.

Unlike other voltage-gated K^+ channels, the gating charges in BK's VSD are not confined to the S4 TM segment, but are distributed between the S2, S3 and S4 segments (Yang and Cui 2015). In fact, the voltage dependence of BK is relatively weak compared with K_V channels, (Horrigan and Aldrich 1999, Rothberg and Magleby 2000). It has been suggested that rather than undergoing a substantial movement of S4 as described in voltage-gated potassium channels, voltage-dependent gating in BK may involve small movements of S2, S3, and S4 (Ma *et al* 2006, Pantazis *et al* 2010). Additionally, interactions of the VSD with S0 (Koval *et al* 2007) and the C-terminal domain (see below) modulate the unique VSD movements associated with channel activation (Ma *et al* 2006, Pantazis *et al* 2010). Unexpectedly, the cryo-EM structures of BK showed that, contrary to voltage-gated K^+ channels, the VSD is not domain-swapped, i.e. is located next to the PD of the same subunit (Tao *et al* 2017).

The large cytosolic C-terminal domain of these channels contains two highly conserved domains known as Regulator of Conduction of K^+ (RCK1 and RCK2). Interestingly, RCK domains are also found in prokaryotic channels and other members of the Slo gene potassium channel family, playing a key role in gating and transporter activity (Giraldez and Rothberg 2017). In the BK tetramer, four RCK tandems form a characteristic structure known as the 'gating ring' (Hite *et al* 2017, Tao *et al* 2017, Wu *et al* 2010, Yuan *et al* 2011). This Ca^{2+}-sensing structural module contains distinct ion binding sites at the RCK1 and RCK2 domains (twelve in total in the full BK tetramer), which were identified in a large body of structure–function studies and account for the whole range of physiological Ca^{2+} concentrations activating the channel (Schreiber and Salkoff 1997, Shi *et al* 2002, Sweet and Cox 2008, Xia *et al* 2002, Zhang *et al* 2010). The high calcium sensitivity of BK channels has been attributed to a binding site containing a cluster of acidic residues within RCK2, known as the calcium bowl (Schreiber and Salkoff 1997, Schreiber *et al* 1999, Wei *et al* 1994). Two additional independent sites have been proposed to bind Ca^{2+}

within the RCK1 domain, a high affinity site within the more N-terminal part of RCK1 and a low affinity site at the C-terminal lobe of RCK1 (Xia *et al* 2002, Zhang *et al* 2010). The latter has been proposed to be normally occupied by Mg^{2+} under physiological conditions (Latorre *et al* 2017). Altogether, these three sites within the gating ring account for all of the Ca^{2+} sensitivity of the channel. In fact, substitution of the C-terminal domain by that of the pH-gated channel Slo3 (Xia *et al* 2004), or removal of the whole gating ring (Budelli *et al* 2013) abolishes all of the Ca^{2+} and Mg^{2+} sensitivity of the channel, ruling out the possibility that other Ca^{2+} binding sites exist outside of the C-terminal domain. An intriguing question in the field relates to the cooperativity between Ca^{2+} binding sites. Several studies addressing this question arrived at different conclusions, although in all cases the level of cooperativity found was low (Qian *et al* 2006, Sweet and Cox 2008). Intriguingly, the cryo-EM structures suggest that interactions between sites may occur, paving the way for new functional studies (Hite *et al* 2017, Zhou *et al* 2017). In addition to Ca^{2+} and Mg^{2+}, other divalent cations can also activate BK channels including Sr^{2+}, Cd^{2+}, Mn^{2+}, Co^{2+}, and Ni^{2+} (Zeng *et al* 2005). In addition, Ba^{2+} has been shown to activate the BK channel by binding to the calcium bowl (Zhou *et al* 2012). BK channels can also be activated by protons binding to the RCK1 site (Hou *et al* 2008).

Understanding how the binding of calcium to the gating ring is mechanically transduced into pore opening constitutes an important question regarding BK channel function. Based on the structure of the gating ring from a bacterial Ca^{2+}-activated channel (Jiang *et al* 2002), it was proposed that expansion of this region upon Ca^{2+} binding would be transmitted to the PD via a stretch of amino acids linking the RCK1 domain to the end of the S6 TM (known as the C-linker; (Niu *et al* 2004)). Structural data showing no significant conformational changes in the C-linker upon calcium binding together with further functional studies seem to draw a more complex picture (Zhou *et al* 2017). Thus, conformational changes of the gating ring could be partly transmitted to the PD partly via the C-linker, in addition to noncovalent interactions between the C-terminal domain and the MD (Lee and Cui 2010). In fact, cryo-EM structures of full BK channels unveiled close associations of the gating ring with the MD (Hite *et al* 2017, Tao *et al* 2017, Tao and MacKinnon 2019), which is consistent with previous findings showing alterations in the voltage dependence of gating charge movements upon Ca^{2+} binding (Horrigan and Aldrich 2002, Savalli *et al* 2012). Recent functional studies further support the idea that the interaction between the calcium sensing domain and VSDs constitutes an important mechanism modulating voltage-dependent gating and pore opening in BK channels (Geng *et al* 2020, Lorenzo-Ceballos *et al* 2019, Miranda *et al* 2018). Interestingly, some studies suggest that the two high-affinity sites (RCK1 and Ca^{2+} bowl) may use distinct pathways leading to channel activation (Yang and Cui 2015, Zhou *et al* 2017). Evidence supporting this idea includes the identification of a number of residues that seem to participate in BK activation by Ca^{2+} binding to the RCK1 site but not the Ca^{2+} bowl (Bao *et al* 2002, Yang *et al* 2010). In addition, fluorescence studies have shown that RCK1 and RCK2 domains move independently when their specific binding sites are occupied (Miranda *et al* 2016). Finally, both sites have shown different voltage sensitivities (Miranda *et al* 2013, Savalli *et al* 2012, Sweet and Cox 2008).

6.3 SK and IK channels

6.3.1 SK and IK function and physiological roles

The existence of calcium-activated and voltage-independent potassium channels was first proposed by Gárdos in erythrocytes, where he described a Ca^{2+}-mediated increase in potassium permeability induced by glycolysis inhibition (Gardos 1958, Maher and Kuchel 2003). Later work confirmed the role of intracellular Ca^{2+} ions in mediating rapid K^+-driven hyperpolarization of the neuronal membrane in invertebrates (Meech, 1972, 1974, Meech and Standen 1974, Thomas and Gorman 1977), amphibians (Barrett and Barret 1976) and mammals (Krnjevic *et al* 1975). SKα subunits are products of the KCNN1 (KCa2.1; SK1), KCNN2 (KCa2.2; SK2) and KCNN3 (KCa2.3; SK3) genes (Köhler *et al* 1996), whereas IKα (KCa3.1; IK1; SK4) are encoded by the KCNN4 gene (Ishii *et al* 1997b, Joiner *et al* 1997). In contrast to BK channels, the function of SK and IK is independent of membrane voltage (Hirschberg *et al* 1999). Consequently, their gating mechanisms can be attributed to a simpler model than BK, which was first proposed by the Marrion group twenty years ago (Hirschberg *et al* 1998). These channels are activated through a mechanism mediated by their intrinsic association with cytosolic calcium-binding protein calmodulin (CaM) (Adelman 2016). This tight association with CaM makes SK and IK channels highly sensitive to intracellular Ca^{2+} (~200–500 nM) (Köhler *et al* 1996, Xia *et al* 1998).

The SK and IK subfamilies have distinct tissue distribution and pharmacology. SK channels are mainly expressed in the nervous system (Stocker and Pedarzani 2000) and are selectively blocked by the bee venom peptide apamin at different concentrations depending on the particular isoform (Blatz and Magleby 1986, Grunnet *et al* 2001, Ishii *et al* 1997a). IK channels are predominantly distributed in blood cells, epithelial cells and in some peripheral neurons and are sensitive to charybdotoxin or iberiotoxin (de-Allie *et al* 1996) as well as by imidazoles such as clotrimazole (Brugnara *et al* 1993, Gardos 1958, Jensen *et al* 1998).

The physiological roles of SK and IK channels are very diverse, in tune with their well differentiated expression patterns. In non-excitable tissues, IK channels contribute to regulation of the membrane potential and Ca^{2+} signaling (Di *et al* 2010, Toyama *et al* 2008, Yu *et al* 2013). Expression of IK channels has also been shown by immunohistochemical staining in human enteric, sensory and sympathetic neurons (Bahia *et al* 2005, Furness *et al* 2004, Mongan *et al* 2005). Additionally, IK channels are localized to microglia and endothelial cells in the CNS (Pedarzani and Stocker 2008). Microglial IK channels have been shown to have diverse functional roles including in respiratory burst, cell proliferation, cell migration and lipopolysaccharide-mediated nitric oxide production (Kaushal *et al* 2007, Nguyen *et al* 2017). Finally, growing evidence points to a role of IK channels in astrogliosis, which is associated with most forms of CNS insult (Bouhy *et al* 2011,

Chen *et al* 2011, Yu *et al* 2014). In the CNS, SK channels play a major physiological role in the control of neuronal excitability, contributing to the action potential after hyperpolarization (Sah and McLachlan 1991, Zhang and McBain 1995) and modulating spike firing frequency (Stackman *et al* 2002). Additionally, they have been proposed to modulate synaptic plasticity (Faber *et al* 2005, Jones *et al* 2017, Ngo-Anh *et al* 2005) (reviewed in Faber 2009). In contrast to BK channels, close proximity to Ca^{2+} sources is not mandatory for SK and IK channels in most physiological contexts, due to their higher Ca^{2+} sensitivity. Therefore, SK channels show a wider spatial distribution from their Ca^{2+} sources (Fakler and Adelman 2008). In many (but not all) physiological contexts, SK channels provide negative feedback mechanisms on the associated Ca^{2+} sources similar to those discussed above for BK channels. For instance, SK channels have been shown to associate to L-type VGCCs in hippocampal neurons (Bowden *et al* 2001, Marrion and Tavalin 1998) and cardiac myocytes (Lu *et al* 2007). Additionally, they functionally associate to NMDAR and VGCC Cav2.3 in the postsynaptic terminals of hippocampal and cortical neurons, where they regulate neuronal plasticity by blunting NMDAR-driven Ca^{2+} influx (Jones *et al* 2017, Ngo-Anh *et al* 2005). In the postsynaptic density of Purkinje cells, SK channels have been shown to form unusually close associations with mGlu1α and Cav2.1 (<20 nm) in dendritic spines and shafts (Lujan *et al* 2018). Interestingly, these channels have also been proposed to interact with TRPV4 in paraventricular nuclei neurons (PVN), participating in the control of renal and cardiovascular function. In this context, Ca^{2+} entering through mechano-activated TRPV4 activates SK channels, which in turn lead to cell depolarization, decreasing the PVN firing and neurohormone secretion from these nuclei (Feetham *et al* 2015). Additionally, in auditory hair cells SK channels mediate inhibitory responses by coupling to Ca^{2+}-permeable nicotinic acetylcholine receptors (Oliver *et al* 2000).

6.3.2 Structural determinants of SK/IK channel function and regulation by CaM

The cloning of SK/IK-α channels revealed that they have a conserved topology reminiscent of the voltage-gated potassium channel family, containing six trans-membrane domains with intracellular N- and C-termini (Ishii *et al* 1997b, Joiner *et al* 1997, Köhler *et al* 1996). However, consistent with their voltage-independence, these channels lack VSDs in their structures (Hirschberg *et al* 1998, Hirschberg *et al* 1999, Köhler *et al* 1996). SK1-3 and IK channels share common features in their topology. All of them are constitutively associated with CaM, even in the absence of Ca^{2+}. Two conserved regions have been proposed to associate to CaM and mediate channel gating. The first consists of two α-helices in the intracellular S4–S5 linker, named $S_{45}A$ and $S_{45}B$ (figure 6.2). The second is made up of two long cytoplasmic helices in the C-terminal domain of the channel (HA and HB, known as the CaM binding-domain, CaMBD) (figure 6.2). CaM is a small acidic protein consisting of a central domain linking two globular regions (the lobes *N* and C); each lobe contains two EF hand motifs that bind Ca^{2+}

(Chattopadhyaya *et al* 1992). The main SK/IK-CaM interaction interface is formed by the CaMBD and the N- and C-terminal lobes of CaM. Upon Ca^{2+} binding, the $S_{45}A$ and $S_{45}B$ helices of one channel subunit interact with its bound CaM molecule and with the HA helix of the adjacent subunit in a domain-swapped conformation, allowing synergistic channel activation (see below; Lee and MacKinnon 2018).

Sequence identity between SK1, SK2 and SK3 is above 60%, whereas they show less than 40% homology with IK (Joiner *et al* 1997). However, the main residues underlying the gating machinery and interactions with CaM are notably conserved (Fanger *et al* 1999, Joiner *et al* 1997). This fact, taken together with similar functional profiles of all members of this subfamily, led to the proposal that they all share common mechanisms of Ca^{2+} gating. A combination of functional and structural studies have provided useful insights into these mechanisms. Partial crystal structures are available for SK2 CaMBD–Ca^{2+}–CaM (Schumacher *et al* 2001) and the apocalmodulin–SK complex (Schumacher *et al* 2004), whereas the full-length structures of the IK-apoCaM and IK–Ca^{2+}–CaM have been recently solved (Lee and MacKinnon 2018). Structural data confirmed, as had been largely demonstrated in functional studies (Adelman 2016), that SK and IK channels are pre-associated with CaM through the CaMBD, in a Ca^{2+}-independent manner, displaying 1:1 stoichiometry (Lee and MacKinnon 2018, Schumacher *et al* 2004, Schumacher *et al* 2001) (figure 6.2). Such CaM–SK conformational assembly prevents binding of Ca^{2+} at the CaM C-lobe. Therefore, the Ca^{2+}-free C-lobe holds SK–CaMBD while the N-lobe is able to move freely as an intrinsically disordered domain (Keen *et al* 1999, Lee and MacKinnon 2018, Schumacher *et al* 2001, 2004). In the presence of Ca^{2+}, Ca^{2+}-bound SK–CaM monomers are rearranged in pairs, forming a ‘dimer of dimers’. The Ca^{2+}–CaM N-lobe interacts with the $S_{45}A$ helix, pulling $S_{45}B$ and displacing S5 and S6 from the central cavity of the pore, resulting in channel opening (figure 6.2) (Lee and MacKinnon 2018).

Complementary structural studies have unveiled new insights into the mechanisms underlying the effect of endogenous and exogenous modulators of these channels. Phosphatidylinositol 4,5-bisphosphate (PIP_2), an endogenous positive modulator of SK function, binds to the CaM–SK interface and stabilizes their interaction (Zhang *et al* 2014). Similarly, binding of phenylurea (PHU) and its derivatives at the CaM–CaMBD interface enhances channel activation (Nam *et al* 2017, Zhang *et al* 2012). A proposed mechanism for this potentiation is the transition of an intrinsically disordered portion of the S6–CaMBD linker to a more stable ordered conformation, facilitating mechanical activation of the channel (Zhang *et al* 2013). Furthermore, a recent structural study demonstrates that hydrophobic interactions among the HA, $S_{45}A$ and $S_{45}B$ helices in SK2 and IK channels are crucial for their Ca^{2+} sensitivity and efficient coupling (Nam *et al* 2021a), which is consistent with previous IK functional data (Morales *et al* 2013). Structural data has led to the development of novel modulators of channel activity (Nam *et al* 2017, Zhang *et al* 2015a).

Understanding the characteristics of the PD and gate of SK and IK channels constitutes a relevant question related to the function of these channels. Using methanethiosulfonate compounds and substituted cysteine accesibility mutagenesis (SCAM), the groups of Maylie (SK channels; (Bruening-Wright *et al* 2002, 2007)) and Sauvé (IK channels; (Simoes *et al* 2002, Klein *et al* 2007)) suggested that, similarly to what has been proposed for BK channels (see above), the channel gate is close to the pore cavity, probably at the selectivity filter. The authors noted a narrow constriction in the permeation pathway generated by a valine in position 282 (V282). However, it did not block the entry of small molecules and, therefore, of K^+ ions. This constriction was also observed in the full-length structure of the Ca^{2+}-unbound IK channel (Lee and MacKinnon 2018).

A source of functional diversity of both SK and IK channels is their physiological regulation by phosphorylation/dephosphorylation. All of them include target residues for specific kinases in their sequences (Köhler *et al* 1996). In fact, in addition to their intrinsic association with CaM, SK channels have been shown to form constitutive complexes with the serine-threonine kinase CK2 and protein phosphatase 2A (PP2A) (Bildl *et al* 2004). Interestingly, the effects of CK2 and PP2A are state-dependent (Allen *et al* 2007). Thus, Ca^{2+} sensitivity is modulated by the relation between phosphorylation of closed channels by CK2 and dephosphorylation of open channels by PP2A (Adelman 2016, Allen *et al* 2007, Bildl *et al* 2004). Intriguingly, the target of phosphorylation by CK2 is located within the linker domain of CaM and not in the SK channel. Phosphorylation reduces the affinity of CaM for the CaMBD, resulting in a diminished Ca^{2+} sensitivity of SK and IK channels (Bildl *et al* 2004, Nam *et al* 2021b).

Reversible post-translational modifications and interaction with cofactors ultimately depend on the integration of the channel with upstream signalling pathways that may cooperate or antagonize each other to open or close the channels. For instance, activation of G-coupled protein receptors (GPCR) such as the receptor for the neurotransmitter norepinephrine mediate intracellular pathways activating CK2, resulting in sensitization of sensory neurons (Maingret *et al* 2008). Interestingly, CaM phosphorylation also inhibits PIP_2 modulation of SK channels, potentially increasing channel inhibition by G protein-mediated PIP2 hydrolysis (Zhang *et al* 2014). Other kinases that have been shown to actively modulate SK and IK channels in different physiological contexts include protein kinase A (PKA) and CaM kinase II (CaMKII). Phosphorylation by PKA of the CaMBD in IK channels leads to decreased CaM affinity for CaMBD, therefore reducing channel activity (Wong and Schlichter 2014). Finally, although CaMKII has been shown to interact with SK and IK channels, many aspects of this regulatory mechanism remain unclear (Ferreira *et al* 2015, Mizukami *et al* 2015, Shrestha *et al* 2019, Tenma *et al* 2018).

6.3.3 K_{Ca} channelopathies

Although the members of the K_{Ca} channels family (BK, IK and SK) display differences in their biophysical characteristics, molecular structures and pharmacological signatures, they all play an essential physiological role by coupling intracellular Ca^{2+} to changes in cell membrane potential. Consistent with the relevance of this process, inherited defects in the function of K_{Ca} channels lead to a wide array of human diseases. A list of mutations encoding K_{Ca} genes are presented in table 6.1. This section describes a selection of these mutations and their associated pathologies.

A frame shift mutation in exon 1 of SK3 channels has been reported in human schizophrenia patients (Bowen *et al* 2001). It was later demonstrated that this dominant-negative mutation suppresses SK-mediated currents by generating truncated channels (Miller *et al* 2001). Subsequently, the functional implication of this mutation in the pathogenesis of schizophrenia was established by Soden and collaborators (Soden *et al* 2013). In mice, expression of this mutation in dopamine neurons reduced the coupling between SK channels and NMDAR receptors leading to enhanced excitability and elevated dopamine release, eventually leading to impaired sensorimotor gating associated with psychotic behaviour in schizophrenia (Soden *et al* 2013). Koot and collaborators (Koot *et al* 2016) identified a de novo mutation (V450L) in SK3 channels associated with autosomal-dominant idiopathic non-cirrhotic portal hypertension. This amino acid substitution makes SK3 channels constitutively active even under basal conditions, generating a sustained K^+ conductance deleterious to liver cells, leading to portal hypertension (Koot *et al* 2016). A recent study reported a series of missense mutations (K269E, G350D, S436C) in SK3 channels causing Zimmermann–Laband syndrome (ZLS) (Bauer *et al* 2019). These gain-of-function (GOF) mutations were shown to enhance the Ca^{2+} sensitivity of SK3 channels and their activation rate. Although the exact pathophysiological mechanism of these mutations remains unclear, one plausible hypothesis is that the excessive K^+ conductance from SK3 channels results in arterial vasodilation, increasing capillary hydrostatic pressure. The ZLS phenotype could arise from vascular rupture or tissue damage during organogenesis (Bauer *et al* 2019).

IK channels are the major source for K^+ permeability in red blood cells (RBCs), where they are involved in maintaining water and solute homeostasis (Begenisich *et al* 2004). Many studies have identified *de novo* missense mutations in IK channels (R352H, V282M, V282E) that are linked to dehydrated hereditary stomatocytosis (DHSt) disorder (Andolfo *et al* 2015, Glogowska *et al* 2015, Rapetti-Mauss *et al* 2015). Importantly, the R352H mutation located in the CaMB domain was found to stabilize the activated state and increase the Ca^{2+} sensitivity of IK channels, leading to augmented K^+ efflux from RBCs and dehydration, followed by haemolytic anaemia (Rapetti-Mauss *et al* 2015). While it has not been tested functionally, V282M and V282E mutations would alter the ion conduction pathway due to their location in the S6 hydrophobic constriction (see previous section). In fact, V282G mutation produced constitutively open and 'leaky' channels (Garneau *et al* 2009).

Human BK channel channelopathies are primarily associated with neurological conditions such as seizures, movement disorders, developmental delay, and intellectual disability (reviewed in Bailey *et al* 2019). GOF mutations in BK channels have

Table 6.1. K_{Ca} channelopathies. This table summarizes mutations, pathophysiological phenotype, location within the protein, functional effects, and related references. Note that only mutations that have been functionally characterised are included. For extended information regarding human KCNMA1 mutations, see Miller *et al* (2021) and https://www.kciaf.org.

Gene/protein	Disease phenotype	Mutation/location in the protein	Functional effects	Reference
KCNMA1/BKα	A, CI, tremor and hypertelorism	S351Y/Pore	No current	Liang *et al* (2019)
KCNMA1/BKα	A, CA, CI, E + PD, ID	G354S/Pore	Reduced current and slower activation kinetics	Du *et al* (2020)
KCNMA1/BKα	A, CA, DD, H	G356R/Pore	No current	Liang *et al* (2019)
KCNMA1/BKα	CA, CM, DD, E, ID, H,	G375R/S6	No current	Liang *et al* (2019)
KCNMA1/BKα	A, CA, CM, DD, ID, H and strabismus	C413Y+N449fs/S6-RCK1 linker	Reduced current (*G–V* shift to depolarized potentials) (C413Y) + No current (N449fs)	Liang *et al* (2019)
KCNMA1/BKα	E + PD	D434G/RCK1 domain	Gain-of-function mutation leading to enhanced Ca^{2+} sensitivity	Du *et al* (2005), Yang *et al* (2010)
KCNMA1/BKα	Non-described	H444Q/RCK1	Loss-of-function mutation (*G–V* shift to depolarized potentials, slower activation and faster deactivation)	Moldenhauer *et al* (2020b)
KCNMA1/BKα	A, CA, PD	K457E/RCK1	No specified (putative loss-of-function mutation)	Buckley *et al* (2020)
KCNMA1/BKα	A, CA, E + PD, DD, ID	R458Ter/RCK1	Truncation	Yesil *et al* (2018)
KCNMA1/BKα	E, PD	I512V/RCK1	No difference	Tian *et al* (2018)
KCNMA1/BKα	E (not directly linked to this mutation)	K518N/RCK1	No difference	Li *et al* (2018)
KCNMA1/BKα	CA, PD, ID	N536H/RCK1	Gain-of-function mutation	Zhang *et al* (2020)
KCNMA1/BKα	E (not directly linked to this mutation)	E656A/RCK1-RCK2 linker	No difference	Li *et al* (2018)

KCNMA1/BKα	A, DD, ID, H and strabismus	I663V/RCK1-RCK2 linker	No current	Liang *et al* (2019)
KCNMA1/BKα	E, DD, ID	Y676Lfs*7/RCK1-RCK2 linker	Truncation	Tabarki *et al* (2016)
KCNMA1/BKα	CA, CI, DD, ID,	P805L/RCK2	Reduced current (*G–V* shift to depolarized potentials) and reduced expression	Liang *et al* (2019)
KCNMA1/BKα	DD, ID, PD	E884K/RCK2	Not available	Zhang *et al* (2015b)
KCNMA1/BKα	Non-described	D965V/RCK2	Loss-of-function mutation (*G–V* shift to depolarized potentials, slower activation)	Moldenhauer *et al* (2020b)
KCNMA1/BKα	CI, E, ID	D984N/RCK2	Not available	Liang *et al* (2019)
KCNMA1/BKα	E	N995S or N999S or N1053S and N999S+R1128W double mutant/RCK2 domain	De novo mutation Shifts the voltage dependence to more negative potentials without altering Ca^{2+} sensitivity	Li *et al* (2018), Plante *et al* (2019)
KCNMA1/BKα	Non-described	R1097H/RCK2 domain	Loss-of-function mutation (*G–V* shift to depolarized potential only at low $[Ca^{2+}]$)	Moldenhauer *et al* (2020a)
KCNMA1/BKα	E (not directly linked to this mutation)	N1159S/C-terminal	No difference	Li *et al* (2018)
KCNMB1/BKβ1	Diastolic hypertension	E65K/Extracellular loop connecting β1 two transmembrane segments.	Gain-of-function mutation rendering enhanced Ca^{2+} sensitivity	Fernandez-Fernandez *et al* (2004)
KCNMB3/BKβ3	Idiopathic generalized epilepsy	Del A750/C-terminal region (truncation of 21 amino acids)	BK inactivation	Hu *et al* (2003),Lorenz *et al* (2007)
KCNN3/SK3	Idiopathic non-cirrhotic portal hypertension	V450L/Cytoplasmic loop between the S4 and S5 transmembrane segments	Increased Ca^{2+} sensitivity and faster activation kinetics	Bauer *et al* (2019), Koot *et al* (2016)
KCNN3/SK3	Schizophrenia	hSK3Δ(L283fs287X mutation)/ N-terminal cytoplasmic region (deletion)	Suppresses SK currents/Disrupts SK-NMDAR coupling	Miller *et al* (2001), Soden *et al* (2013)

(Continued)

Table 6.1. (*Continued*)

Gene/protein	Disease phenotype	Mutation/location in the protein	Functional effects	Reference
KCNN3/SK3	Zimmermann–Laband syndrome	K269E/Cytoplasmic N-terminus	Increased Ca^{2+} sensitivity and faster activation kinetics	Bauer *et al* (2019)
KCNN3/SK3	Zimmermann–Laband syndrome	G350D/Cytoplasmic linker S2-S3	Increased Ca^{2+} sensitivity and faster activation kinetics	Bauer *et al* (2019)
KCNN3/SK3	Zimmermann–Laband syndrome	S436C/Cytoplasmic loop between the S4 and S5 transmembrane segments	Increased Ca^{2+} sensitivity and faster activation kinetics	Bauer *et al* (2019)
KCNN4/IK	Dehydrated hereditary stomatocytosis	R352H/Cytosolic CaMBD	Increased currents and Ca^{2+} sensitivity	Andolfo *et al* (2015), Rapetti-Mauss *et al* (2015)
KCNN4/IK	Dehydrated hereditary stomatocytosis	V282M/S6 cytosolic part		Andolfo *et al* (2015), Glogowska *et al* (2015)
KCNN4/IK	Dehydrated hereditary stomatocytosis	V282E/S6 cytosolic part	—	Glogowska *et al* (2015)

A: ataxia; CA: cerebral/cerebellar atrophy; CI: cognitive impairment; CM: congenital malformations; DD: developmental delay; E: epilepsy; H: hypotonia; ID: intelectual disability; PD: paroxysmal dyskinesia.

been implicated in the development of neuronal excitability disorders. A single GOF mutation—D434G—increases the Ca^{2+} sensitivity of BK channels and causes autosomal-dominant epilepsy with paroxysmal dyskinesias (Du *et al* 2005). Similarly, another GOF mutation—N995S—was found to enhance the voltage dependence of BK channel activation and cause epilepsy (Li *et al* 2018). Both of these mutations would enhance neuronal BK channel activity and therefore increase the fast after-hyperpolarization of the action potential, ultimately leading to increased neuronal excitability. In addition to the BKα subunits, mutations in its accessory subunits have also been reported to cause idiopathic generalized epilepsy (IGE) (Lorenz *et al* 2007). For instance, a deletion of base A450 in exon 4 of the β3 gene, results in a frame shift that alters three amino acids and truncates the protein by 18 amino acids (Hu *et al* 2003). Functionally, this alteration causes a rapid inactivation of BK channels and also shifts the activation curve rightwards, which has been associated with reduced synaptic inhibition, and therefore increased neuronal excitability and seizure susceptibility (Hu *et al* 2003). A recent study by Du *et al* (2020) identified a loss-of-function mutation in the selectivity filter of the BK channels (G354S) associated with progressive cerebellar degeneration, ataxia, and cognitive impairment. This mutation dramatically reduced BK single-channel conductance and ion selectivity, leading to depolarization and mitochondrial dysfunction. Subsequently, this was associated with a reduction in cellular viability and cerebellar ataxia (Du *et al* 2020).

6.4 Concluding remarks

In recent years, our understanding of the physiological roles, structural basis of function and pathophysiology of K_{Ca} channels has significantly evolved since they were first discovered almost forty years ago. In addition to a large body of functional and electrophysiological studies, we now have access to full-length structures of the BK channel with and without the β4 auxiliary subunit, as well as the human IK-CaM complex structures, allowing us to disentangle the mechanisms underlying gating and function with unprecedented precision. Significant advances have been made in our knowledge about the structure–function relations of the diverse structural architectures characterizing the members of the K_{Ca} family. A growing number of pathologies associated with K_{Ca} channel dysfunction has emerged in recent years, highlighting the importance of the physiological roles played by these channels, mainly related to the coupling of Ca^{2+} with membrane voltage signalling. Regardless of their different structural, biophysical and pharmacological features, members of the K_{Ca} family are commonly associated with other channels or regulatory proteins to form functional supercomplexes. Further studies of K_{Ca} channels within such functional complexes in their native environment will lead to a new level of fundamental discoveries. In the long run, this will pave the way to drug design studies or genetic strategies that recognize tissue-specific subunit combinations or K_{Ca}-protein supercomplexes and help enable more efficient treatments of K_{Ca} channelopathies.

References

Adelman J P 2016 SK channels and calmodulin *Channels (Austin)* **10** 1–6

Adelman J P *et al* 1992 Calcium-activated potassium channels expressed from cloned complementary DNAs *Neuron* **9** 209–16

Allen D *et al* 2007 Organization and regulation of small conductance Ca^{2+}-activated K^+ channel multiprotein complexes *J. Neurosci.* **27** 2369–76

Almassy J and Begenisich T 2012 The LRRC26 protein selectively alters the efficacy of BK channel activators *Mol. Pharmacol.* **81** 21–30

Andolfo I *et al* 2015 Novel Gardos channel mutations linked to dehydrated hereditary stomatocytosis (xerocytosis) *Am J. Hematol.* **90** 921–6

Atkinson N S *et al* 1991 A component of calcium-activated potassium channels encoded by the *Drosophila* slo locus *Science* **253** 551–5

Bahia P K *et al* 2005 A functional role for small-conductance calcium-activated potassium channels in sensory pathways including nociceptive processes *J. Neurosci.* **25** 3489–98

Bailey C S *et al* 2019 KCNMA1-linked channelopathy *J. Gen. Physiol.* **151** 1173–89

Bao L *et al* 2002 Elimination of the BK(Ca) channel's high-affinity Ca^{2+} sensitivity *J. Gen. Physiol.* **120** 173–89

Barrett E F and Barret J N 1976 Separation of two voltage-sensitive potassium currents, and demonstration of a tetrodotoxin-resistant calcium current in frog motoneurones *J. Physiol.* **255** 737–74

Bauer C K *et al* 2019 Gain-of-function mutations in KCNN3 encoding the small-conductance Ca^{2+}-activated K^+ channel SK3 cause zimmermann-laband syndrome *Am. J. Hum. Genet.* **104** 1139–57

Begenisich T *et al* 2004 Physiological roles of the intermediate conductance, Ca^{2+}-activated potassium channel Kcnn4 *J. Biol. Chem.* **279** 47681–7

Berkefeld H and Fakler B 2008 Repolarizing responses of BK_{Ca}–Cav complexes are distinctly shaped by their Cav subunits *J. Neurosci.* **28** 8238–45

Berkefeld H *et al* 2006 BKCa-Cav channel complexes mediate rapid and localized Ca^{2+}-activated K^+ signaling *Science* **314** 615–20

Bildl W *et al* 2004 Protein kinase CK2 is coassembled with small conductance Ca^{2+}-activated K^+ channels and regulates channel gating *Neuron* **43** 847–58

Blatz A L and Magleby K L 1986 Single apamin-blocked Ca-activated K^+ channels of small conductance in cultured rat skeletal muscle *Nature* **323** 718–20

Bouhy D *et al* 2011 Inhibition of the Ca^{2+}-dependent K^+ channel, KCNN4/KCa3.1, improves tissue protection and locomotor recovery after spinal cord injury *J. Neurosci.* **31** 16298–308

Bowden S E *et al* 2001 Somatic colocalization of rat SK1 and D class (Ca(v)1.2) L-type calcium channels in rat CA1 hippocampal pyramidal neurons *J. Neurosci.* **21** RC175

Bowen T *et al* 2001 Mutation screening of the KCNN3 gene reveals a rare frameshift mutation *Mol. Psychiatry* **6** 259–60

Brelidze T I *et al* 2003 A ring of eight conserved negatively charged amino acids doubles the conductance of BK channels and prevents inward rectification *Proc. Natl Acad. Sci. USA* **100** 9017–22

Brenner R *et al* 2000 Cloning and functional characterization of novel large conductance calcium-activated potassium channel beta subunits, hKCNMB3 and hKCNMB4 *J. Biol. Chem.* **275** 6453–61

Bruening-Wright A *et al* 2007 Evidence for a deep pore activation gate in small conductance Ca^{2+}-activated K^+ channels *J. Gen. Physiol.* **130** 601–10

Bruening-Wright A *et al* 2002 Localization of the activation gate for small conductance Ca^{2+}-activated K^+ channels *J. Neurosci.* **22** 6499–506

Brugnara C *et al* 1993 Inhibition of Ca^{2+}-dependent K^+ transport and cell dehydration in sickle erythrocytes by clotrimazole and other imidazole derivatives *J. Clin. Invest.* **92** 520–6

Buckley C *et al* 2020 Status dystonicus, oculogyric crisis and paroxysmal dyskinesia in a 25 year-old woman with a novel KCNMA1 variant, K457E *Tremor. Other Hyperkinet. Mov. (NY)* **10** 49

Budelli G *et al* 2013 Properties of Slo1 K^+ channels with and without the gating ring *Proc. Natl Acad. Sci. USA* **110** 16657–62

Butler A *et al* 1993 MSlo, a complex mouse gene encoding 'maxi' calcium-activated potassium channels *Science* **261** 221–4

Carvacho I *et al* 2008 Intrinsic electrostatic potential in the BK channel pore: role in determining single channel conductance and block *J. Gen. Physiol.* **131** 147–61

Castillo J P *et al* 2016 beta1-subunit-induced structural rearrangements of the Ca^{2+}- and voltage-activated K^+ (BK) channel *Proc. Natl Acad. Sci USA* **113** E3231–9

Cerrada A *et al* 2018 Quantitative analysis of subcellular nanodomains formed by BK and voltage-gated calcium channels *Biophys. J.* **114** 479a–80a

Chattopadhyaya R *et al* 1992 Calmodulin structure refined at 1.7 a resolution *J. Mol. Biol.* **228** 1177–92

Chavis P *et al* 1998 Modulation of big K^+ channel activity by ryanodine receptors and L-type Ca^{2+} channels in neurons *Eur J Neurosci.* **10** 2322–7

Chen Y J *et al* 2011 The KCa3.1 blocker TRAM-34 reduces infarction and neurological deficit in a rat model of ischemia/reperfusion stroke *J. Cereb. Blood Flow Metab.* **31** 2363–74

Contreras G F *et al* 2012 Modulation of BK channel voltage gating by different auxiliary beta subunits *Proc. Natl Acad. Sci. USA* **109** 18991–6

Cox D H *et al* 1997 Allosteric gating of a large conductance Ca-activated K^+ channel *J. Gen. Physiol.* **110** 257–81

de-Allie F A *et al* 1996 Characterization of Ca^{2+}-activated 86Rb+ fluxes in rat C6 glioma cells: a system for identifying novel IKCa-channel toxins *Br. J. Pharmacol.* **117** 479–87

Di L *et al* 2010 Inhibition of the K^+ channel KCa3.1 ameliorates T cell-mediated colitis *Proc. Natl Acad. Sci. USA* **107** 1541–6

Du W *et al* 2005 Calcium-sensitive potassium channelopathy in human epilepsy and paroxysmal movement disorder *Nat. Genet.* **37** 733–8

Du X *et al* 2020 Loss-of-function BK channel mutation causes impaired mitochondria and progressive cerebellar ataxia *Proc. Natl Acad. Sci. USA* **117** 6023–34

Dworetzky S I *et al* 1996 Phenotypic alteration of a human BK (hSlo) channel by hSlobeta subunit coexpression: changes in blocker sensitivity, activation/relaxation and inactivation kinetics, and protein kinase A modulation *J. Neurosci.* **16** 4543–50

Dworetzky S I *et al* 1994 Cloning and expression of a human large-conductance calcium-activated potassium channel *Brain Res. Mol. Brain Res.* **27** 189–93

Earley S *et al* 2005 TRPV4 forms a novel Ca^{2+} signaling complex with ryanodine receptors and BKCa channels *Circ. Res.* **97** 1270–9

Edgerton J R and Reinhart P H 2003 Distinct contributions of small and large conductance Ca^{2+}-activated K^+ channels to rat Purkinje neuron function *J. Physiol.* **548** 53–69

Faber E S 2009 Functions and modulation of neuronal SK channels *Cell Biochem. Biophys.* **55** 127–39
Faber E S *et al* 2005 SK channels regulate excitatory synaptic transmission and plasticity in the lateral amygdala *Nat. Neurosci.* **8** 635–41
Fakler B and Adelman J P 2008 Control of K(Ca) channels by calcium nano/microdomains *Neuron* **59** 873–81
Fanger C M *et al* 1999 Calmodulin mediates calcium-dependent activation of the intermediate conductance K_{Ca} channel, *IKCa1 J. Biol. Chem.* **274** 5746–54
Feetham C H *et al* 2015 TRPV4 and K_{Ca} ion channels functionally couple as osmosensors in the paraventricular nucleus *Br. J. Pharmacol.* **172** 1753–68
Feinberg-Zadek P L *et al* 2008 BK channel subunit composition modulates molecular tolerance to ethanol *Alcohol Clin. Exp. Res.* **32** 1207–16
Fernandez-Fernandez J M *et al* 2004 Gain-of-function mutation in the KCNMB1 potassium channel subunit is associated with low prevalence of diastolic hypertension *J. Clin. Invest.* **113** 1032–9
Ferreira R *et al* 2015 KCa3.1/IK1 channel regulation by cGMP-dependent protein kinase (PKG) via reactive oxygen species and CaMKII in microglia: an immune modulating feedback system? *Front. Immunol.* **6** 153
Furness J B *et al* 2004 Intermediate conductance potassium (IK) channels occur in human enteric neurons *Auton. Neurosci.* **112** 93–7
Gackiere F *et al* 2013 Functional coupling between large-conductance potassium channels and Cav3.2 voltage-dependent calcium channels participates in prostate cancer cell growth *Biol. Open* **2** 941–51
Galvez A *et al* 1990 Purification and characterization of a unique, potent, peptidyl probe for the high conductance calcium-activated potassium channel from venom of the scorpion *Buthus tamulus J. Biol. Chem.* **265** 11083–90
Gárdos G 1958 The function of calcium in the potassium permeability of human erythrocytes *Biochim. Biophys. Acta* **30** 653–4
Garneau L *et al* 2009 Hydrophobic interactions as key determinants to the KCa3.1 channel closed configuration. An analysis of KCa3.1 mutants constitutively active in zero Ca^{2+} *J. Biol. Chem.* **284** 389–403
Geng Y *et al* 2020 Coupling of Ca^{2+} and voltage activation in BK channels through the α B helix/voltage sensor interface *Proc. Natl Acad. Sci. USA* **117** 14512–21
Giraldez T 2017 The GoSlo family of BK channel activators: a no-go for gamma subunits? *Channels (Austin)* **11** 89–90
Giraldez T and Rothberg B S 2017 Understanding the conformational motions of RCK gating rings *J. Gen. Physiol.* **149** 431–41
Glogowska E *et al* 2015 Mutations in the Gardos channel (KCNN4) are associated with hereditary xerocytosis *Blood* **126** 1281–4
Gómez R *et al* 2021 NMDA receptor-BK channel coupling regulates synaptic plasticity in the barrel cortex PNAS**118** e2107026118
Gonzalez-Perez V and Lingle C J 2019 Regulation of BK channels by beta and gamma subunits *Annu. Rev. Physiol.* **81** 113–37
Gonzalez-Perez V *et al* 2021 Goblet cell LRRC26 regulates BK channel activation and protects against colitis in mice *Proc. Natl Acad. Sci. USA* **118** e2019149118

Gonzalez-Perez V *et al* 2014 Functional regulation of BK potassium channels by gamma1 auxiliary subunits *Proc. Natl Acad. Sci. USA* **111** 4868–73

Gonzalez-Perez V *et al* 2015 Two classes of regulatory subunits coassemble in the same BK channel and independently regulate gating *Nat. Commun.* **6** 8341

Griguoli M *et al* 2016 Presynaptic BK channels control transmitter release: physiological relevance and potential therapeutic implications *J. Physiol.* **594** 3489–500

Grunnet M *et al* 2001 Apamin interacts with all subtypes of cloned small-conductance Ca^{2+}-activated K^+ channels *Pflugers Arch.* **441** 544–50

Grunnet M and Kaufmann W A 2004 Coassembly of big conductance Ca^{2+}-activated K^+ channels and L-type voltage-gated Ca^{2+} channels in rat brain *J. Biol. Chem.* **279** 36445–53

Gutzmann J J *et al* 2019 Functional coupling of Cav2.3 and BK potassium channels regulates action potential repolarization and short-term plasticity in the mouse hippocampus *Front. Cell Neurosci.* **13** 27

Hirschberg B *et al* 1998 Gating of recombinant small-conductance Ca-activated K^+ channels by calcium *J. Gen. Physiol.* **111** 565–81

Hirschberg B *et al* 1999 Gating properties of single SK channels in hippocampal CA1 pyramidal neurons *Biophys. J.* **77** 1905–13

Hite R K *et al* 2017 Structural basis for gating the high-conductance Ca^{2+}-activated K^+ channel *Nature* **541** 52–7

Horrigan F T and Aldrich R W 1999 Allosteric voltage gating of potassium channels II. Mslo channel gating charge movement in the absence of Ca^{2+} *J. Gen. Physiol.* **114** 305–36

Horrigan F T and Aldrich R W 2002 Coupling between voltage sensor activation, Ca^{2+} binding and channel opening in large conductance (BK) potassium channels *J. Gen. Physiol.* **120** 267–305

Horrigan F T *et al* 1999 Allosteric voltage gating of potassium channels I. Mslo ionic currents in the absence of Ca^{2+} *J. Gen. Physiol.* **114** 277–304

Hou S *et al* 2008 Reciprocal regulation of the Ca^{2+} and H^+ sensitivity in the SLO1 BK channel conferred by the RCK1 domain *Nat. Struct. Mol. Biol.* **15** 403–10

Houamed K M *et al* 2010 BK channels mediate a novel ionic mechanism that regulates glucose-dependent electrical activity and insulin secretion in mouse pancreatic beta-cells *J. Physiol.* **588** 3511–23

Hu S *et al* 2003 Variants of the KCNMB3 regulatory subunit of maxi BK channels affect channel inactivation *Physiol. Genomics* **15** 191–8

Irie T and Trussell L O 2017 Double-nanodomain coupling of calcium channels, ryanodine receptors, and BK channels controls the generation of burst firing *Neuron* **96** 856–70

Isaacson J S and Murphy G J 2001 Glutamate-mediated extrasynaptic inhibition: direct coupling of NMDA receptors to Ca^{2+}-activated K^+ channels *Neuron* **31** 1027–34

Ishii T M *et al* 1997a Determinants of apamin and d-tubocurarine block in SK potassium channels *J. Biol. Chem.* **272** 23195–200

Ishii T M *et al* 1997b A human intermediate conductance calcium-activated potassium channel *Proc. Natl Acad. Sci. USA* **94** 11651–6

Jaffe D B *et al* 2011 Shaping of action potentials by type I and type II large-conductance Ca^{2+}-activated K^+ channels *Neuroscience* **192** 205–18

Jensen B S *et al* 1998 Characterization of the cloned human intermediate-conductance Ca^{2+}-activated K^+ channel *Am. J. Physiol.* **275** C848–56

Jiang Y *et al* 2002 Crystal structure and mechanism of a calcium-gated potassium channel *Nature* **417** 515–22

Joiner W J *et al* 1997 hSK4, a member of a novel subfamily of calcium-activated potassium channels *Proc. Natl Acad. Sci. USA* **94** 11013–8
Jones S L *et al* 2017 Dendritic small conductance calcium-activated potassium channels activated by action potentials suppress EPSPs and gate spike-timing dependent synaptic plasticity *Elife* **6** e30333
Kaushal V *et al* 2007 The Ca^{2+}-activated K^+ channel KCNN4/KCa3.1 contributes to microglia activation and nitric oxide-dependent neurodegeneration *J. Neurosci.* **27** 234–44
Keen J E *et al* 1999 Domains responsible for constitutive and Ca^{2+}-dependent interactions between calmodulin and small conductance Ca^{2+}-activated potassium channels *J. Neurosci.* **19** 8830–8
Kim E Y *et al* 2009 Canonical transient receptor potential channel (TRPC)3 and TRPC6 associate with large-conductance Ca^{2+}-activated K^+ (BKCa) channels: role in BKCa trafficking to the surface of cultured podocytes *Mol. Pharmacol.* **75** 466–77
Klein H, Garneau L, Banderali U, Simoes M, Parent L and Sauvé R *et al* 2007 Structural determinants of the closed KCa3.1 channel pore in relation to channel gating: results from a substituted cysteine accessibility analysis *J. Gen. Physiol.* **129** 299–315
Knaus H G *et al* 1994 Pharmacology and structure of high conductance calcium-activated potassium channels *Cell Signal.* **6** 861–70
Köhler M *et al* 1996 Small-conductance, calcium-activated potassium channels from mammalian brain *Science* **273** 1709–14
Koot B G *et al* 2016 A de novo mutation in KCNN3 associated with autosomal dominant idiopathic non-cirrhotic portal hypertension *J. Hepatol.* **64** 974–7
Koval O M *et al* 2007 A role for the S0 transmembrane segment in voltage-dependent gating of BK channels *J. Gen. Physiol.* **129** 209–20
Krnjevic K *et al* 1975 Evidence for Ca^{2+}-activated K^+ conductance in cat spinal motoneurons from intracellular EGTA injections *Can. J. Physiol. Pharmacol.* **53** 1214–8
Kshatri A S *et al* 2017 Differential efficacy of GoSlo-SR compounds on BKalpha and BKalphagamma1–4 channels *Channels (Austin)* **11** 66–78
Kyle B D and Braun A P 2014 The regulation of BK channel activity by pre- and post-translational modifications *Front. Physiol.* **5** 316
Lai M H *et al* 2014 BK channels regulate sinoatrial node firing rate and cardiac pacing *in vivo Am. J. Physiol. Heart. Circ. Physiol.* **307** H1327–38
Latorre R *et al* 2017 Molecular determinants of BK channel functional diversity and functioning *Physiol. Rev.* **97** 39–87
Lee C H and MacKinnon R 2018 Activation mechanism of a human SK-calmodulin channel complex elucidated by cryo-EM structures *Science* **360** 508–13
Lee U S and Cui J 2010 BK channel activation: structural and functional insights *Trends Neurosci.* **33** 415–23
Li W and Aldrich R W 2004 Unique inner pore properties of BK channels revealed by quaternary ammonium block *J. Gen. Physiol.* **124** 43–57
Li X *et al* 2018 De novo BK channel variant causes epilepsy by affecting voltage gating but not Ca^{2+} sensitivity *Eur. J. Hum. Genet.* **26** 220–9
Liang L *et al* 2019 De novo loss-of-function KCNMA1 variants are associated with a new multiple malformation syndrome and a broad spectrum of developmental and neurological phenotypes *Hum. Mol. Genet.* **28** 2937–51

Lingle C J *et al* 1996 Calcium-activated potassium channels in adrenal chromaffin cells *Ion Channels* **4** 261–301
Lingle C J, Zeng X-H, Ding J P and Xia X-M 2001 Inactivation of Bk channels mediated by the Nh2 terminus of the β3b auxiliary subunit involves a two-step mechanism: possible separation of binding and blockade *J. Gen. Physiol.* **117** 583–606
Liu N *et al* 2020 Channels that Cooperate with TRPV4 in the brain *J. Mol. Neurosci.* **70** 1812–20
Loane D J *et al* 2007 Co-assembly of N-type Ca^{2+} and BK channels underlies functional coupling in rat brain *J. Cell Sci.* **120** 985–95
Lorenz S *et al* 2007 Allelic association of a truncation mutation of the KCNMB3 gene with idiopathic generalized epilepsy *Am. J. Med. Genet. B Neuropsychiatr. Genet.* **144B** 10–3
Lorenzo-Ceballos Y *et al* 2019 Calcium-driven regulation of voltage-sensing domains in BK channels *Elife* **8** e44934
Lu L *et al* 2007 Molecular coupling of a Ca^{2+}-activated K^+ channel to L-type Ca^{2+} channels via alpha-actinin2 *Circ. Res.* **100** 112–20
Lujan R *et al* 2018 SK2 channels associate with $mGlu_{1\alpha}$ receptors and $Ca_V2.1$ channels in purkinje cells *Front. Cell Neurosci.* **12** 311
Ma Z *et al* 2006 Role of charged residues in the S1–S4 voltage sensor of BK channels *J. Gen. Physiol.* **127** 309–28
Magleby K L 2003 Gating mechanism of BK (Slo1) channels: so near, yet so far *J. Gen. Physiol.* **121** 81–96
Maher A D and Kuchel P W 2003 The Gárdos channel: a review of the Ca^{2+}-activated K^+ channel in human erythrocytes *Int. J. Biochem. Cell Biol.* **35** 1182–97
Maingret F *et al* 2008 Neurotransmitter modulation of small-conductance Ca^{2+}-activated K^+ channels by regulation of Ca^{2+} gating *Neuron* **59** 439–49
Marrion N V and Tavalin S J 1998 Selective activation of Ca^{2+}-activated K^+ channels by co-localized Ca^{2+} channels in hippocampal neurons *Nature* **395** 900–5
Marty A 1981 Ca-dependent K channels with large unitary conductance in chromaffin cell membranes *Nature* **291** 497–500
McCobb D P *et al* 1995 A human calcium-activated potassium channel gene expressed in vascular smooth muscle *Am. J. Physiol.* **269** H767–77
McManus O B *et al* 1995 Functional role of the beta subunit of high conductance calcium-activated potassium channels *Neuron* **14** 645–50
McManus O B and Magleby K L 1988 Kinetic states and modes of single large-conductance calcium-activated potassium channels in cultured rat skeletal muscle *J. Physiol.* **402** 79–120
Meech R W 1972 Intracellular calcium injection causes increased potassium conductance in aplysia nerve cells *Comp. Biochem. Physiol. A Comp. Physiol.* **42** 493–9
Meech R W 1974 The sensitivity of helix aspersa neurones to injected calcium ions *J. Physiol.* **237** 259–77
Meech R W and Standen N B 1974 Calcium-mediated potassium activation in helix neurones *J. Physiol.* **237** P 43–4
Meera P *et al* 2000 A neuronal beta subunit (KCNMB4) makes the large conductance, voltage- and Ca^{2+}-activated K^+ channel resistant to charybdotoxin and iberiotoxin *Proc. Natl Acad. Sci. USA* **97** 5562–7
Meredith A L *et al* 2004 Overactive bladder and incontinence in the absence of the BK large conductance Ca^{2+}-activated K^+ channel *J. Biol. Chem.* **279** 36746–52

Meredith A L *et al* 2006 BK calcium-activated potassium channels regulate circadian behavioral rhythms and pacemaker output *Nat. Neurosci.* **9** 1041–9

Miller C *et al* 1985 Charybdotoxin, a protein inhibitor of single Ca^{2+}-activated K^+ channels from mammalian skeletal muscle *Nature* **313** 316–8

Miller J P *et al* 2021 An emerging spectrum of variants and clinical features in KCNMA1-linked channelopathy *Channels (Austin).* **15** 447–64

Miller M J *et al* 2001 Nuclear localization and dominant-negative suppression by a mutant SKCa3 N-terminal channel fragment identified in a patient with schizophrenia *J. Biol. Chem.* **276** 27753–6

Miranda P *et al* 2013 State-dependent FRET reports calcium- and voltage-dependent gating-ring motions in BK channels *Proc. Natl Acad. Sci. USA* **110** 5217–22

Miranda P *et al* 2016 Interactions of divalent cations with calcium binding sites of BK channels reveal independent motions within the gating ring *Proc. Natl Acad. Sci. USA* **113** 14055–60

Miranda P *et al* 2018 Voltage-dependent dynamics of the BK channel cytosolic gating ring are coupled to the membrane-embedded voltage sensor *Elife* **7** e40664

Mizukami K *et al* 2015 Small-conductance Ca^{2+}-activated K^+ current is upregulated via the phosphorylation of CaMKII in cardiac hypertrophy from spontaneously hypertensive rats *Am. J. Physiol. Heart. Circ. Physiol.* **309** H1066–74

Moldenhauer H J *et al* 2020a Comparative gain-of-function effects of the KCNMA1-N999S mutation on human BK channel properties *J. Neurophysiol.* **123** 560–70

Moldenhauer H J *et al* 2020b Characterization of new human KCNMA1 loss-of-function mutations *Biophys. J.* **118** 114A

Mongan L C *et al* 2005 The distribution of small and intermediate conductance calcium-activated potassium channels in the rat sensory nervous system *Neuroscience* **131** 161–75

Montgomery J R and Meredith A L 2012 Genetic activation of BK currents *in vivo* generates bidirectional effects on neuronal excitability *Proc. Natl Acad. Sci. USA* **109** 18997–9002

Morales P *et al* 2013 Contribution of the KCa3.1 channel-calmodulin interactions to the regulation of the KCa3.1 gating process *J. Gen. Physiol.* **142** 37–60

Muller A *et al* 2007 Nanodomains of single Ca^{2+} channels contribute to action potential repolarization in cortical neurons *J. Neurosci.* **27** 483–95

Nam Y W *et al* 2021a Hydrophobic interactions between the HA helix and S4-S5 linker modulate apparent Ca^{2+} sensitivity of SK2 channels *Acta Physiol. (Oxf.)* **231** e13552

Nam Y W *et al* 2021b Differential modulation of SK channel subtypes by phosphorylation *Cell Calcium* **94** 102346

Nam Y W *et al* 2017 Structural insights into the potency of SK channel positive modulators *Sci. Rep.* **7** 17178

Nelson M T *et al* 1995 Relaxation of arterial smooth muscle by calcium sparks *Science* **270** 633–7

Neyton J and Miller C 1988 Discrete Ba^{2+} block as a probe of ion occupancy and pore structure in the high-conductance Ca^{2+}-activated K^+ channel *J. Gen. Physiol.* **92** 569–86

Ngo-Anh T J *et al* 2005 SK channels and NMDA receptors form a Ca^{2+}-mediated feedback loop in dendritic spines *Nat. Neurosci.* **8** 642–9

Nguyen H M *et al* 2017 Differential Kv1.3, KCa3.1, and Kir2.1 expression in 'classically' and 'alternatively' activated microglia *Glia* **65** 106–21

Nimigean C M *et al* 2003 Electrostatic tuning of ion conductance in potassium channels *Biochemistry* **42** 9263–8

Niu X *et al* 2004 Linker-gating ring complex as passive spring and Ca^{2+}-dependent machine for a voltage- and Ca^{2+}-activated potassium channel *Neuron* **42** 745–56
Oliver D *et al* 2000 Gating of Ca^{2+}-activated K^+ channels controls fast inhibitory synaptic transmission at auditory outer hair cells *Neuron* **26** 595–601
Pallotta B S *et al* 1981 Single channel recordings of Ca^{2+}-activated K^+ currents in rat muscle cell culture *Nature* **293** 471–4
Pantazis A *et al* 2010 Operation of the voltage sensor of a human voltage- and Ca^{2+}-activated K^+ channel *Proc. Natl Acad. Sci. USA* **107** 4459–64
Pedarzani P and Stocker M 2008 Molecular and cellular basis of small- and intermediate-conductance, calcium-activated potassium channel function in the brain *Cell Mol. Life Sci.* **65** 3196–217
Plante A E *et al* 2019 Effects of single nucleotide polymorphisms in human KCNMA1 on BK current properties *Front. Mol. Neurosci.* **12** 285
Posson D J *et al* 2013 The voltage-dependent gate in MthK potassium channels is located at the selectivity filter *Nat. Struct. Mol. Biol.* **20** 159–66
Prakriya M and Lingle C J 1999 BK channel activation by brief depolarizations requires Ca^{2+} influx through L- and Q-type Ca^{2+} channels in rat chromaffin cells *J. Neurophysiol.* **81** 2267–78
Qian X *et al* 2006 Intra- and intersubunit cooperativity in activation of BK channels by Ca^{2+} *J. Gen. Physiol.* **128** 389–404
Raffaelli G *et al* 2004 BK potassium channels control transmitter release at CA3–CA3 synapses in the rat hippocampus *J. Physiol.* **557** 147–57
Rapetti-Mauss R *et al* 2015 A mutation in the Gardos channel is associated with hereditary xerocytosis *Blood* **126** 1273–80
Rehak R *et al* 2013 Low voltage activation of KCa1.1 current by Cav3-KCa1.1 complexes *PLoS One* **8** e61844
Robitaille R *et al* 1993 Functional colocalization of calcium and calcium-gated potassium channels in control of transmitter release *Neuron* **11** 645–55
Rothberg B S 2012 The BK channel: a vital link between cellular calcium and electrical signaling *Protein Cell* **3** 883–92
Rothberg B S and Magleby K L 1999 Gating kinetics of single large-conductance Ca^{2+}-activated K^+ channels in high Ca^{2+} suggest a two-tiered allosteric gating mechanism *J. Gen. Physiol.* **114** 93–124
Rothberg B S and Magleby K L 2000 Voltage and Ca^{2+} activation of single large-conductance Ca^{2+}-activated K^+ channels described by a two-tiered allosteric gating mechanism *J. Gen. Physiol.* **116** 75–99
Sah P and McLachlan E M 1991 Ca^{2+}-activated K^+ currents underlying the afterhyperpolarization in guinea pig vagal neurons: a role for Ca^{2+}-activated Ca^{2+} release *Neuron* **7** 257–64
Savalli N *et al* 2012 The contribution of RCK domains to human BK channel allosteric activation *J. Biol. Chem.* **287** 21741–50
Schreiber M and Salkoff L 1997 A novel calcium-sensing domain in the BK channel *Biophys. J.* **73** 1355–63
Schreiber M *et al* 1999 Transplantable sites confer calcium sensitivity to BK channels *Nat. Neurosci.* **2** 416–21
Schumacher M A *et al* 2004 Crystal structures of apocalmodulin and an apocalmodulin/SK potassium channel gating domain complex *Structure* **12** 849–60

Schumacher M A *et al* 2001 Structure of the gating domain of a Ca^{2+}-activated K^+ channel complexed with Ca^{2+}/calmodulin *Nature* **410** 1120–4

Semenov I *et al* 2011 BK channel beta1 subunits regulate airway contraction secondary to M2 muscarinic acetylcholine receptor mediated depolarization *J. Physiol.* **589** 1803–17

Shi J *et al* 2002 Mechanism of magnesium activation of calcium-activated potassium channels *Nature* **418** 876–80

Shipston M J 2001 Alternative splicing of potassium channels: a dynamic switch of cellular excitability *Trends Cell Biol.* **11** 353–8

Shrestha A *et al* 2019 SK channel modulates synaptic plasticity by tuning CaMKIIalpha/beta dynamics *Front. Synaptic Neurosci.* **11** 18

Simoes M *et al* 2002 Cysteine mutagenesis and computer modeling of the S6 region of an intermediate conductance IKCa channel *J. Gen. Physiol.* **120** 99–116

Soden M E *et al* 2013 Disruption of dopamine neuron activity pattern regulation through selective expression of a human KCNN3 mutation *Neuron* **80** 997–1009

Stackman R W *et al* 2002 Small conductance Ca^{2+}-activated K^+ channels modulate synaptic plasticity and memory encoding *J. Neurosci.* **22** 10163–71

Stocker M and Pedarzani P 2000. Differential distribution of three Ca^{2+}-activated K^+ channel subunits, SK1, SK2, and SK3, in the adult rat central nervous system *Mol. Cell Neurosci.* **15** 476–93

Storm J F 1987 Action potential repolarization and a fast after-hyperpolarization in rat hippocampal pyramidal cells *J. Physiol.* **385** 733–59

Sweet T B and Cox D H 2008 Measurements of the BKCa channel's high-affinity Ca^{2+} binding constants: effects of membrane voltage *J. Gen. Physiol.* **132** 491–505

Szarka N *et al* 2018 Traumatic brain injury impairs myogenic constriction of cerebral arteries: role of mitochondria-derived H_2O_2 and TRPV4-dependent activation of BKca channels *J. Neurotrauma* **35** 930–9

Tabarki B *et al* 2016 Homozygous KCNMA1 mutation as a cause of cerebellar atrophy, developmental delay and seizures *Hum. Genet.* **135** 1295–8

Tang Q Y *et al* 2009 Closed-channel block of BK potassium channels by bbTBA requires partial activation *J. Gen. Physiol.* **134** 409–36

Tao X *et al* 2017 Cryo-EM structure of the open high-conductance Ca^{2+}-activated K^+ channel *Nature* **541** 46–51

Tao X and MacKinnon R 2019 Molecular structures of the human Slo1 K^+ channel in complex with β4 *Elife* **8** e51409

Tenma T *et al* 2018 Small-conductance Ca^{2+}-activated K^+ channel activation deteriorates hypoxic ventricular arrhythmias via CaMKII in cardiac hypertrophy *Am. J. Physiol. Heart. Circ. Physiol.* **315** H262–72

Thomas M V and Gorman A L 1977 Internal calcium changes in a bursting pacemaker neuron measured with arsenazo III *Science* **196** 531–3

Tian W T *et al* 2018 Proline-rich transmembrane protein 2-negative paroxysmal kinesigenic dyskinesia: Clinical and genetic analyses of 163 patients *Mov. Disord.* **33** 459–67

Toyama K *et al* 2008 The intermediate-conductance calcium-activated potassium channel KCa3.1 contributes to atherogenesis in mice and humans *J. Clin. Invest.* **118** 3025–37

Uebele V N, Lagrutta A, Wade T, Figueroa D J, Liu Y, McKenna E, Austin C P, Bennett P B and Swanson R 2000 Cloning and functional expression of two families of β-subunits of the large conductance calcium-activated K+ channel* *J. Biol. Chem.* **275** 23211–8

Vivas O *et al* 2017 Proximal clustering between BK and $Ca_V1.3$ channels promotes functional coupling and BK channel activation at low voltage *Elife* **6** e28029

Wallner M *et al* 1996 Determinant for beta-subunit regulation in high-conductance voltage-activated and Ca^{2+}-sensitive K^+ channels: an additional transmembrane region at the N terminus *Proc. Natl Acad. Sci. USA* **93** 14922–7

Wallner M *et al* 1999 Molecular basis of fast inactivation in voltage and Ca^{2+}-activated K^+ channels: a transmembrane beta-subunit homolog *Proc. Natl Acad. Sci. USA* **96** 4137–42

Wang B *et al* 2016 Knockout of the BK beta4-subunit promotes a functional coupling of BK channels and ryanodine receptors that mediate a fAHP-induced increase in excitability *J. Neurophysiol.* **116** 456–65

Wei A *et al* 1994 Calcium sensitivity of BK-type KCa channels determined by a separable domain *Neuron* **13** 671–81

Wei A D *et al* 2005 Int. Union of Pharmacology. LII. nomenclature and molecular relationships of calcium-activated potassium channels *Pharmacol. Rev.* **57** 463–72

Whitt J P *et al* 2018 Differential contribution of Ca^{2+} sources to day and night BK current activation in the circadian clock *J. Gen. Physiol.* **150** 259–75

Whitt J P *et al* 2016 BK channel inactivation gates daytime excitability in the circadian clock *Nat. Commun.* **7** 10837

Wilkens C M and Aldrich R W 2006 State-independent block of BK channels by an intracellular quaternary ammonium *J. Gen. Physiol.* **128** 347–64

Womack M D *et al* 2004 Calcium-activated potassium channels are selectively coupled to P/Q-type calcium channels in cerebellar Purkinje neurons *J. Neurosci.* **24** 8818–22

Wondergem R and Bartley J W 2009 Menthol increases human glioblastoma intracellular Ca^{2+}, BK channel activity and cell migration *J. Biomed. Sci.* **16** 90

Wong R and Schlichter L C 2014 PKA reduces the rat and human KCa3.1 current, CaM binding, and Ca^{2+} signaling, which requires Ser332/334 in the CaM-binding C terminus *J. Neurosci.* **34** 13371–83

Wu Y *et al* 2013 TRPV1 channels are functionally coupled with BK(mSlo1) channels in rat dorsal root ganglion (DRG) neurons *PLoS One* **8** e78203

Wu Y *et al* 2010 Structure of the gating ring from the human large-conductance Ca^{2+}-gated K^+ channel *Nature* **466** 393–7

Xia X M *et al* 1998 Mechanism of calcium gating in small-conductance calcium-activated potassium channels *Nature* **395** 503–7

Xia X M *et al* 2002 Multiple regulatory sites in large-conductance calcium-activated potassium channels *Nature* **418** 880–4

Xia X M *et al* 2004 Ligand-dependent activation of Slo family channels is defined by interchangeable cytosolic domains *J. Neurosci.* **24** 5585–91

Yamamura H *et al* 2012 Molecular assembly and dynamics of fluorescent protein-tagged single $K_{Ca}1.1$ channel in expression system and vascular smooth muscle cells *Am. J. Physiol. Cell Physiol.* **302** C1257–68

Yan J and Aldrich R 2012 BK potassium channel modulation by leucine-rich repeat-containing proteins *Proc. Natl Acad. Sci. USA* **109** 7917–22

Yan J and Aldrich R W 2010 LRRC26 auxiliary protein allows BK channel activation at resting voltage without calcium *Nature* **466** 513–6

Yang H and Cui J 2015 BK channels ed J Zheng and M C Trudeau *Handbook of Ion Channels* (Boca Raton, FL: CRC Press)

Yang J *et al* 2010 An epilepsy/dyskinesia-associated mutation enhances BK channel activation by potentiating Ca^{2+} sensing *Neuron* **66** 871–83

Yesil G *et al* 2018 Expanding the phenotype of homozygous *KCNMA1* mutations; dyskinesia, epilepsy, intellectual disability, cerebellar and corticospinal tract atrophy *Balkan Med. J.* **35** 336–9

Yu Z *et al* 2014 Targeted inhibition of KCa3.1 attenuates TGF-beta-induced reactive astrogliosis through the Smad2/3 signaling pathway *J. Neurochem.* **130** 41–9

Yu Z H *et al* 2013 Up-regulation of KCa3.1 promotes human airway smooth muscle cell phenotypic modulation *Pharmacol. Res.* **77** 30–8

Yuan P *et al* 2011 Open structure of the Ca^{2+} gating ring in the high-conductance Ca^{2+}-activated K^+ channel *Nature* **481** 94–7

Zeng X H *et al* 2005 Divalent cation sensitivity of BK channel activation supports the existence of three distinct binding sites *J. Gen. Physiol.* **125** 273–86

Zhang F X *et al* 2018a BK potassium channels suppress Cac$\alpha 2\delta$ subunit function to reduce inflammatory and neuropathic pain *Cell Rep.* **22** 1956–64

Zhang G *et al* 2020 A gain-of-function mutation in *KCNMA1* causes dystonia spells controlled with stimulant therapy *Mov. Disord.* **35** 1868–73

Zhang G *et al* 2010 Ion sensing in the RCK1 domain of BK channels *Proc. Natl Acad. Sci. USA* **107** 18700–5

Zhang J *et al* 2018b Glutamate-activated BK channel complexes formed with NMDA receptors *Proc. Natl Acad. Sci. USA* **115** E9006–14

Zhang L and McBain C J 1995 Potassium conductances underlying repolarization and after-hyperpolarization in rat CA1 hippocampal interneurones *J. Physiol.* **488** 661–72

Zhang M *et al* 2014 Selective phosphorylation modulates the PIP2 sensitivity of the CaM–SK channel complex *Nat. Chem. Biol.* **10** 753–9

Zhang M *et al* 2015a Molecular overlap in the regulation of SK channels by small molecules and phosphoinositides *Sci. Adv.* **1** e1500008

Zhang M *et al* 2012 Identification of the functional binding pocket for compounds targeting small-conductance Ca^{2+}-activated potassium channels *Nat. Commun.* **3** 1021

Zhang M *et al* 2013 Unstructured to structured transition of an intrisically disordered protein peptide in coupling Ca^{2+}-sensing and SK channel activation *Proc. Natl Acad. Sci. USA* **110** 4828–33

Zhang Z B *et al* 2015b De novo KCNMA1 mutations in children with early-onset paroxysmal dyskinesia and developmental delay *Mov Disord.* **30** 1290–2

Zhao G *et al* 2010 Type 1 IP3 receptors activate BKCa channels via local molecular coupling in arterial smooth muscle cells *J. Gen. Physiol.* **136** 283–91

Zhou Y *et al* 2011 Cysteine scanning and modification reveal major differences between BK channels and Kv channels in the inner pore region *Proc. Natl Acad. Sci. USA* **108** 12161–6

Zhou Y *et al* 2017 Threading the biophysics of mammalian Slo1 channels onto structures of an invertebrate Slo1 channel *J. Gen. Physiol.* **149** 985–1007

Zhou Y *et al* 2012 Barium ions selectively activate BK channels via the Ca^{2+}-bowl site *Proc. Natl Acad. Sci. USA* **109** 11413–8

Zou S *et al* 2008 The beta 1 subunit of L-type voltage-gated Ca^{2+} channels independently binds to and inhibits the gating of large-conductance Ca^{2+}-activated K^+ channels *Mol. Pharmacol.* **73** 369–78

Chapter 7

Calcium-activated chloride channels

Alec Kittredge, Aaron P Owji, Sara Ragi, Stephen H Tsang, Yu Zhang and Tingting Yang

7.1 Introduction

Calcium (Ca^{2+})-activated chloride (Cl^-) channels (CaCCs) mediate the transmembrane traffic of Cl^- and other monovalent anions in response to intracellular Ca^{2+} and play a crucial role in nearly all major tissue types. CaCC activity was first discovered in *Xenopus* oocytes and salamander photoreceptors [1, 2], and later found widely distributed throughout the body, primarily in secretive epithelium and excitable cells [3–6]. Essential physiological functions in which CaCCs participate include epithelial secretion, membrane excitability in cardiac muscle and neurons, olfactory and nociceptive transduction, regulation of vascular tone, and photoreception [6–9].

7.1.1 The role of Cl^-

As the most abundant physiological anion in many biological systems, Cl^- serves as the major compensatory ion for the movement of major cations, namely hydrogen (H^+), sodium (Na^+), potassium (K^+), and calcium (Ca^{2+}) [10]. Thus, Cl^- plays an important role in establishing membrane potentials by serving as a counterion to facilitate continued electrical signaling. Furthermore, the expression pattern of CaCCs overlaps with regions where the movement of Cl^- ions plays a role in physiology, such as the central nervous system (CNS) and the muscular system. The direction of the flow of Cl^- is dependent on three primary factors: (1) the plasma membrane potential, (2) the Cl^- concentration gradient, and (3) the intracellular concentration of Ca^{2+} [3]. The plasma membrane potential and the ion concentration gradient are collectively known as the 'electrochemical gradient,' which is the driving force behind ion (Cl^- in this case) flow through channels to cross the plasma membrane. While the extracellular concentration of Cl^- constantly remains around 100 mM, the intracellular Cl^- concentration varies greatly, dependent on the combined activities of Cl^- transporters and Cl^- channels, including CaCCs.

doi:10.1088/978-0-7503-2009-2ch7

7.1.2 Ca^{2+}-dependence

As Ca^{2+} is the major activating ligand for CaCCs, CaCCs are dependent on the intracellular concentration of Ca^{2+} ($[Ca^{2+}]_i$) for their activity. Changes in $[Ca^{2+}]_i$ are typically in response to one of two actions: (1) the activation of signaling pathways to release internal stores of Ca^{2+} (metabotropic signaling), or (2) the activation of calcium channels on the plasma membrane, such as voltage-gated Ca^{2+} channels, which allow Ca^{2+} to enter the cell across its electrochemical gradient (ionotropic signaling) [3]. Both pathways increase $[Ca^{2+}]_i$, but the former is much slower than the latter.

For example, in cardiac muscle, the resting potential of the membrane is more negative than the reversal potential of Cl^- (E_{Cl}), so when $[Ca^{2+}]_i$ rises, Cl^- exits the cell, depolarizing the membrane. This can also increase the open probability of voltage-gated Ca^{2+} channels, leading to more Ca^{2+} influx and further depolarization. Due to osmotic forces in response to Cl^- efflux, water and Na^+ move through their respective channels, generally following the direction of Cl^- movement. Conversely, if E_{Cl} is more positive than the membrane potential, the activation of CaCCs leads to hyperpolarization of the membrane. These changes in ion concentrations and membrane potentials are essential for proper function and signal propagation in cardiac muscle [3].

7.1.3 Ion selectivity and pharmacology

CaCCs are permeable to not only Cl^-, but also other anions (e.g. Br^-, I^-, NO_3^- and HCO_3^-, etc) and generally follow the 'Eisenman weak field strength' lyotrophic series in relation to permeability [11]. For this series, the relation between the ion and channel is weak, and thus the rate at which an ion enters a channel pore is inversely proportional to its dehydration energy: $I^- > Br^- > Cl^- > F^-$ [11–13]. As other types of anion channels, CaCCs have a low selectivity among anions. Pharmacologically, CaCCs are sensitive to non-selective Cl^- channel inhibitors such as niflumic acid (NFA) and 4,4'-Diisothiocyano-2,2'-stilbenedisulfonic acid (DIDS) [13, 14].

At least two families of CaCCs have been identified, namely the Bestrophin family and select members of the TMEM16 (also known as Anoctamin) family.

7.2 Bestrophins

7.2.1 Introduction

The bestrophin family was first defined with the discovery of the *BEST*1 (also known as *VMD2*) gene responsible for Best disease, a form of juvenile onset macular degeneration (figure 7.1) [15, 16, 42]. There are four members of the bestrophin family in humans, namely Best1-4. All these paralogs act as CaCCs when heterologously expressed in HEK293 cells [16, 17]. This family has 3–4 members in other mammals with distant relatives in fungi, plants, and prokaryotes [12, 18]. Eukaryotic bestrophins respond to $[Ca^{2+}]_i$, with EC_{50} in the low–mid nanomolar range in heterologous expression systems [17, 19–24].

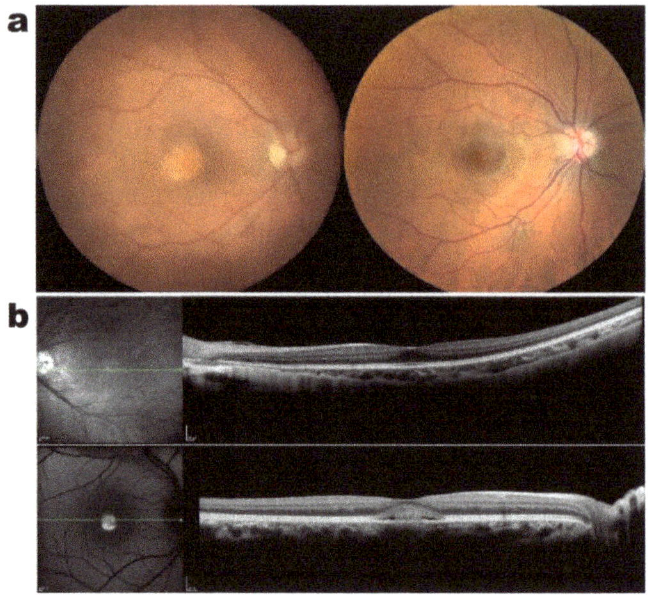

Figure 7.1. Clinical phenotypes associated with *BEST1* mutation. (a) Fundus infrared image of the maculae from a WT (left) and a patient (right) right eye, respectively. (b) Spectral Domain Optical Coherence Tomography (SDOCT) of the maculae from a WT (top) and a patient (bottom) right eye, respectively.

The bestrophin family varies greatly in regards to regulation and tissue distribution, while sharing functional properties such as anion selectivity and activity in response to $[Ca^{2+}]_i$ (figure 7.2). In addition to being permeable to halogen ions, they are permeable to HCO_3^-, and thus potentially play an important role in pH buffering systems [25, 26]. Moreover, Best1 can also pass the anionic amino acids gamma amino butyric acid (GABA) and glutamate in astrocytes [21].

The sequence identity among bestrophin paralogs is not particularly high, underscoring their evolutionarily conserved, yet physiologically divergent functions. In particular, the four human paralogs share between 56% and 66% sequence identity within the first 360 residues of the N-terminus, which contains the channel pore forming transmembrane domains, whereas the C-terminal regulatory domain is more varied within the family [19].

7.2.2 Structure

To date, five bestrophin structures have been reported, the human Best1 (hBest1) and Best2 (hBest2), the prokaryotic *Klebsiella pneumoniae* Best1 homolog KpBest, the *Gallus gallus* (chicken) Best1 homolog cBest1, and the *Bos taurus* (bovine) Best2 homolog bBest2 [28–33]. All five bestrophin structures exhibit the same overall architecture of a pentameric channel containing four transmembrane helices per protomer (20 per pentamer) with a flower-vase-shaped ion conduction pathway

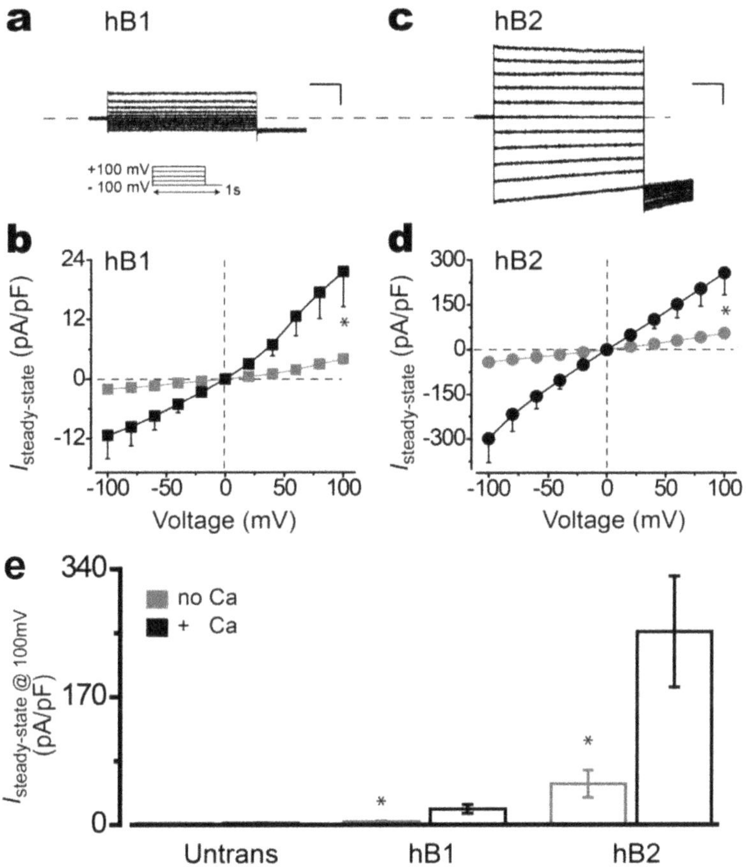

Figure 7.2. Ca^{2+}-dependent Cl^- current mediated by hBest1 and hBest2. (a-d) Representative current traces in the presence of 1.2 μM $[Ca^{2+}]_i$ (a and c), and population steady-state current density-voltage relationships (b and d) in the absence (gray) and presence (black) of 1.2 μM $[Ca^{2+}]_i$ from HEK293 cells expressing hBest1 (a and b) or hBest2 (c and d); n = 5–6 for each point. $*P < 0.05$ compared to currents without Ca^{2+}, using two-tailed unpaired Student t test. Scale bar, 300 pA (a), 1 nA (c), and 150 ms (a, c). Voltage protocol used to elicit currents is shown in *Inset*. (e) Bar chart showing the steady-state current densities from HEK293 cells expressing the indicated channels in the absence and presence of 1.2 μM $[Ca^{2+}]_i$, n = 5–12 for each bar. $*P < 0.05$ compared to currents without Ca^{2+} under the same condition, using two-tailed unpaired Student t test. All error bars in this figure represent s.e.m. Reprinted from [29], copyright 2020, The Authors, under exclusive licence to Springer Nature America, Inc. With permission of Springer.

spanning approximately 95 Å across the longest axis of the channel (figure 7.3). This underscores the overall highly conserved architecture of bestrophin channels. There are two major structural constrictions to the flow of ions down the channel axis, one (the 'neck') buried in the membrane, and the other (the 'aperture') at the end of the cytosolic vestibule (figure 7.3). Five highly conserved Ca^{2+} binding sites, one per protomer, are each located on the periphery of the cytosolic region, just below the transmembrane region (figure 7.3).

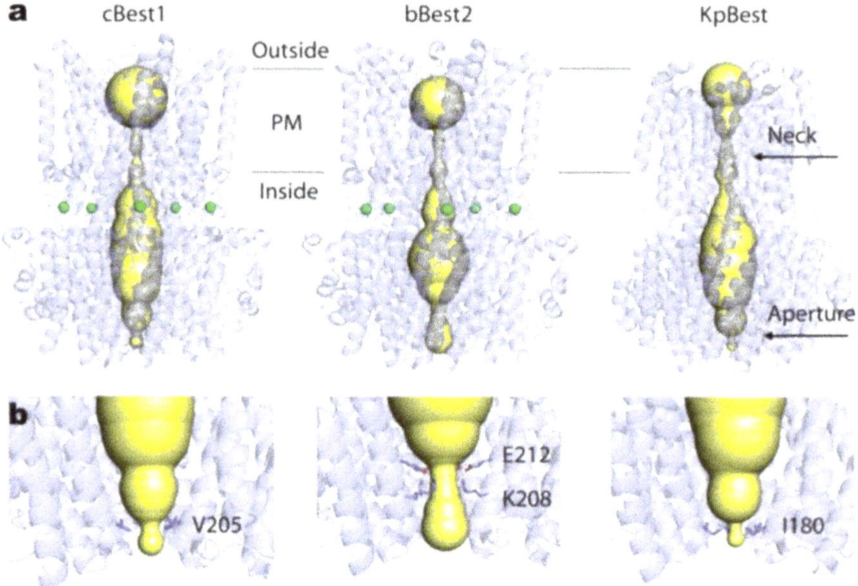

Figure 7.3. (a) Shown side-by-side are the Ca^{2+}-bound chicken Best1 (cBest1) crystal structure (4RDQ), the Ca^{2+}-bound bovine Best2 (bBest2) cryoEM structure (6VX7), and the prokaryotic Best1 homolog from *Klebsiella pneumoniae* (KpBest) crystal structure (4WD8). The three bestrophin channels are viewed from within the membrane (side view) with the ion conduction pathway shown as a yellow volume. The 'neck' sits at the inner membrane leaflet and is composed of three conserved hydrophobic residues. Just below the neck sits the five Ca^{2+} ions (green spheres) with one binding site per protomer, within the solvent-accessible acidic clasp. The cytosolic region extends the ion conduction pathway approximately 45 Å into the cytosol, where a second constriction, the aperture, contributes to gating and ion selectivity. (b) The chemical composition of the aperture differs for the known bestrophin structures. In cBest1, V205 forms the tightest constriction, while the conserved K208 and E212 form a salt bridge in bBest2, and in KpBest the constriction is formed by I180.

7.2.2.1 The neck

The first major constriction, termed the 'neck,' is composed of three conserved hydrophobic residues- I76, F80, and F84 in vertebrates, located at the level of the intracellular membrane leaflet of the transmembrane region (figure 7.3). The side chains of these three residues point directly towards the channel axis, forming a constriction with a radius of ~1.3 Å. This hydrophobic seal serves as a Ca^{2+}-dependent gate of the channel. Upon activation, a series of small movements, mostly consisting of side chain conformational changes, is propagated from the Ca^{2+}-binding site to the neck, allowing the neck helix (S2b) to rotate ~15° along its helical axis, resulting in the constrictive I76, F80, and F84 pointing away from the central axis and dilation of the neck to a radius of ~6.5 Å, which is large enough to allow passage of hydrated Cl^- ions, bulk solvent, and amino acids [6, 30, 33].

7.2.2.2 The aperture

The second major constriction, termed the 'aperture,' is located approximately 45 Å into the channel cavity, sitting at the farthest end of the cytosolic domain (figure 7.3).

The aperture is formed by a hydrophobic I or V in Best1, as exemplified by I180 and V205 in KpBest and cBest1, respectively, but changes to a charged K208-E212 duo in Best2 (figure 7.3(b)). As this cytosolic constriction controls both Ca^{2+}-dependent gating and anion selectivity, the structural divergence among paralogs underlies differences in channel activities, which are essential for their specific functions in different tissues. The radii of the closed apertures in KpBest, cBest1 and bBest2 are ~0.9, 1.3 and 2.2 Å, respectively, while the conformational changes involved in opening of the aperture are still unknown (figure 7.3) [6, 28–30, 33].

7.2.2.3 Ca^{2+} binding site

The Ca^{2+}-sensing apparatus is conserved, and consists of a cluster of acidic residues E300, D301, D302, D303, and D304 (figure 7.3(a)) [29, 31]. Of these, Ca^{2+} is bound directly by the side chain carboxyl groups of D301 and D304, while the other acidic side chains likely aid in Ca^{2+} recruitment. Ca^{2+} binding to this site recruits the carbonyl oxygen of A10 within the N-terminal extension, causing the N-terminal extension to become ordered. In the absence of Ca^{2+}, the C-terminal extension is disordered and in the presence of Ca^{2+} it becomes ordered, with implications for concentration and time dependence of the channel with regards to Ca^{2} [29, 30, 33].

7.3 Best1

7.3.1 Best1 function

Best1 is permeable to halogen ions, HCO_3^-, NO_3^-, SCN^-, $CH_3O_3^-$, GABA and glutamate [21, 25, 34]. Electrophysiological analysis and recent structural studies suggest that the neck and aperture collectively contribute to anion selectivity and Ca^{2+} dependent gating.

7.3.2 Ca^{2+}-dependent inactivation ('rundown')

While nanomolar to micromolar concentrations of intracellular Ca^{2+} activate Best1 opening, higher concentrations of Ca^{2+} or prolonged exposure result in channel 'rundown.' This Ca^{2+}-dependent channel rundown consists of a slow decline in Cl^- currents despite sufficient intracellular Ca^{2+} and depolarization, indicating that the channel is closing [20, 35, 36]. A conserved auto-inhibitory $_{358}$SF(M/G/L)GS$_{362}$ (numbering for mBest1) sequence in the C-terminus of the channel is responsible for this allosteric gating mechanism: in the presence of Ca^{2+}, the C-terminal extension becomes ordered and extends over the adjacent protomer to bind to its site on the next protomer over, forcing residues in the transmembrane core of the channel to swing inwards and close the hydrophobic neck [30, 31, 33, 35, 36]. In addition, Best1 rundown can be inhibited by the phosphorylation status of site S358 (within the auto-inhibitory $_{358}$SF(M/G/L)GS$_{362}$), which is regulated via protein kinase C (PKC) and phosphatase 2A (PP2A) [37].

7.3.3 ATP-dependent activity

ATP is a co-activator of Best1: ATP alone, in the absence of Ca^{2+}, does not evoke Cl^- currents, but enhances Cl^- currents by 3-fold in the presence of Ca^{2+} [38].

The ATP binding site was mapped to a cluster of conserved residues ($_{199}$GRIRD$_{203}$) adjacent to the aperture on the solvent-exposed surface of the cytoplasmic domain. As the aperture is a Ca^{2+}-dependent gate, ATP binding adjacent to this gate may provide a means of enhancing Best1's CaCC activity.

7.3.4 Best1 physiological roles

Best1 is predominantly expressed in the basolateral membrane of the retinal pigmented epithelium (RPE), and plays two critical roles: (1) regulating water transport essential for waste removal from the retina [9], and (2) generating light peak (LP), which is triggered by a substance, called 'light peak substance' (LPS), produced by photoreceptors in the presence of light [15, 39–44]. Both roles are apparently evident by the clinical phenotypes of retinal degenerative disorders caused by genetic mutation of the human *BEST1* gene, as the patients typically exhibit waste build-up in the macular and diminished LP in electrooculography (figure 7.1). At the molecular level, Cl^- current drives the transportation of water and waste, while Best1 is indispensable for Ca^{2+}-dependent Cl^- currents in human RPE, where LP is believed to represent a Ca^{2+}-dependent Cl^- current [44]. However, it should be noted that Best1 knockout mice do not show any retinal phenotype or impaired LP, underlining the different genetic requirements of CaCC among species [45–47].

Best1 is also detected in various regions of the CNS, such as hippocampus, cerebellum, olfactory bulb, dorsal spinal cord and dorsal root ganglion. The cell types in the CNS where Best1 expresses includes neurons, astrocytes, Purkinje cells and Bergmann glia, although the expression level is lower compared to that in RPE [13]. The exact role of Best1 in the CNS is still unknown, but likely related to its ability to permeate not only Cl^-, but also important neural transmitters GABA and glutamate. For instance, Best1 is permeable to glutamate in a Ca^{2+}- and GPCR-dependent manner in astrocytic microdomains near neuronal synapses, suggesting a role in the modulation of synaptic plasticity on targeted hippocampal pyramidal neurons [27].

In addition, Best1 is found at testes, kidneys, gastrointestinal tract, airways and cardiac tissue [13, 14, 48–50], where CaCC plays a critical role in physiology but the contribution of Best1 has not yet been established.

7.3.5 Best1 implications in diseases

The first disease characterized by dysfunction of Best1 was Best vitelliform macular dystrophy (also known as BVMD, or Best disease), a dominantly inherited disease caused by mutation of the *BEST1* gene [15]. In these patients, there is a buildup of waste materials, specifically around the photoreceptor outer segments which pulls water into the space behind the retina, eventually leading the retina to detach from the underlying connective tissue, and thus decreasing eyesight (figure 7.1) [3, 13, 51–54]. This also impacts the eye's ability to produce LP, resulting in abnormal electro-oculography recordings, a clinical feature of BVMD [55, 56].

Dysfunction of Best1 also causes ocular diseases in a mutation-specific dependency, including adult-onset foveomacular vitelliform dystrophy (AFVD), autosomal dominant vitreoretinochoroidopathy (ADVIRC), adult-onset microcornea, rod-cone dystrophy, early-onset cataract, and autosomal dominant microcornea rod -cone degeneration syndrome (ADMRCS) or posterior staphyloma syndrome, autosomal recessive bestrophinopathy (ARB), and a form of retinitis pigmentosa (RP) [57–62]. These diseases (including BVMD) resulting from *BEST1* mutation(s) are commonly termed bestrophinopathies, among which all except for ARB are autosomal dominant. In canines, dysfunction of the channel can result in canine multifocal retinopathy (CMR) [63]. So far, over 250 mutations in Best1 have been documented in humans.

Currently, there is no treatment for bestrophinopathies, but gene therapy is a promising solution because: (1) the disease is caused by a defined single gene mutation; (2) the eye is a self-inclusive organ with blood-ocular barriers. Two therapeutic strategies are available, namely gene augmentation and *in situ* gene correction. The former is accomplished by delivering a copy of functional (wild-type) *BEST1* gene into the mutant RPE cells, which has proven feasible in canines [64]; the latter requires correcting the mutated *BEST1* gene in the genome, which is potentiated by the recently developed clustered regularly interspaced short palindromic repeats (CRISPR) technique. Nevertheless, further optimizations, safety measurements and pathological studies are needed before the clinical applications of *BEST1* gene therapies.

As Best1 is located in the CNS and permeable to GABA, it is possible, but not yet proven, that Best1 can play a role in pathologies where GABA signaling is altered, such as elevated GABA release from astrocytes in Alzheimer's disease [65].

7.4 Best2

7.4.1 Best2 structure and function

Best2 CaCC activity was originally discovered in the early 21st century [16, 17]. Best1 and Best2 share roughly 63% sequence identity in their N-terminal domain (the first 364 amino acids), and 19% similarity in their C-terminal domain [19]. The relative ion permeabilities in mBest2 are $SCN^- > NO3^- = I^- > Br^- > Cl^-$, whereas the relative ion conductances are $NO3^- > I^- > Cl^{--} \geqslant Br^- > SCN^-$ [66]. In general, Best2 is less selective to ions compared to Best1.

The overall structure of Best2 is very similar to that of Best1, displaying a flower-vase-shaped ion conducting pathway with two major constrictions—the neck and the aperture (figure 7.3). However, despite conserved hydrophobic I76/F80/F84 at the neck, the aperture in Best2 is formed by a positively charged K208 connecting to a negatively charged E212 through a salt bridge, in sharp contrast to the hydrophobic I/V at the aperture in Best1 (figure 7.3). Importantly, the size of the closed aperture in Best2 is larger than that of a dehydrated Cl^-, unlike the smaller aperture in Best1. This structural difference underlies the distinct channel functions between these paralogs: Best2 has Ca^{2+}-independent conductivity for Cl^-, while Best1 conducts Cl^- strictly in a Ca^{2+}-dependent manner (figure 7.2). Moreover, two positively charged pore-facing residues in Best2- K265 and H91, form additional narrow points in the ion conducting pathway, which are not present in the Best1 structure.

7.4.2 Best2 physiological roles

Best2 is expressed in tissues throughout the body, including non-pigmented ciliary epithelium (NPE), colon, testes, olfactory sensory neurons, and sweat glands [26, 48, 66–69]. Best2 is localized at the basolateral plasma membrane of NPE, while Best2 knockout mice exhibit decreased intraocular pressure (IOP) [16, 66, 70], suggesting Best2 plays a critical role in IOP regulation. This idea is in accord with the major role of Cl^- current as a driving force for water transmembrane transport. However, as IOP is controlled by the balance between aqueous humor formation and drainage, whether Best2 participates in the former or the latter or both remains to be determined [66]. In the colon, Best2 is localized at the basolateral membrane and intracellular organelles of goblet cells, and primarily acts not as a Cl^- channel, but rather as a HCO_3^- channel to mediate vectorial transepithelial secretion [26]. The role of Best2 as a HCO_3^- channel is also seen in sweat glands, where Best2 is involved in sweat secretion [69]. In the ciliary layer of olfactory sensory neurons, mBest2 is co-localized with the major subunit of the cyclic nucleotide gated (CNG) channel responsible for the primary transduction current in those cells [68], and functions in axogenesis and ciliogenesis in developing olfactory neurons [67].

7.4.3 Best2 implications in disease

Although there is no known disease directly caused by the mutation of Best2, its involvement in regulating IOP provides a possible route of treatment for people inflicted with glaucoma, as high IOP is a risk factor for glaucoma, while lowering IOP slows the progression of the disease at both early and late stages [71–73]. Similarly, Best2's function as a HCO_3^- channel in the colon implies a potential therapeutic role in bowel inflammation diseases.

7.5 Best3 and Best4

7.5.1 Best3 and Best4 structure and function

Compared to the first two bestrophin paralogs, relatively little is known about Best3 and Best4 in regard to their structure and function, except for the general properties of the bestrophins, including the basic functionality as a CaCC channel, the overall architecture and the existence of a conserved auto-inhibitory sequence ($_{358}$SF(M/G/L)GS$_{362}$) at the C-terminus [35, 36, 74].

7.5.2 Best3 physiological roles

In cardiac smooth muscle cells, Best3 is expressed in both atrial and ventricular myocytes near the sarcolemma of the cells, essential for vasomotion of mesenteric arteries, but not required for tonic contractile responses [75]. In endothelial cells, Best3 is involved in mediating the NFκB inflammation pathway through negative interaction with the upstream regulators IKK and Iκβα [14, 76–79]. Best3 is also localized in the nucleus of renal proximal tubule cells, where it appears to have cell-protecting properties during endoplasmic reticulum stress by inhibiting induction of the pro-apoptotic transcription factor CHOP [78]. Interestingly, a C-terminally truncated (from amino acid 364) variant is expressed in mouse myoblasts and muscles, playing a role in lysosome and endoplasmic reticulum H^+ gradients, which in turn regulates cytosolic Ca^{2+} from internal stores [80].

Two main alternative splicing isoforms '−2–3+6' and '−2–3–6', corresponding to the variants with two exons (2 and 3) and three exons (2, 3 and 6) spliced out, respectively [81], are widely expressed in the brain, spinal cord, bone marrow, retina, kidney, lung, liver, testes, skeletal muscle and exocrine glands [82]. There is another splice variant, '−2–3–4' (lacking exons 2, 3 and 4), expressed solely in the brain [82]. These splicing variants are potentially involved in various physiological processes such as inflammation and fluid secretion, but they do not retain the CaCC activity.

7.5.3 Best3 implications in disease

No diseases have been definitively linked to Best3 dysfunction as of yet. However, Best3 is potentially involved in mandibular prognathism without maxillary hypoplasia, a form of facial deformity that can result in deficiencies in speech articulation and low efficiency of mastication [83]. Although globally observed, mandibular prognathism is more prominent in East Asian heritage, and occurs in almost 10% of the Japanese population [84, 85]. The implication of Best3 in mandibular prognathism is supported by the findings that Best3 is expressed in human skeletal muscle and cartilage, and a rare non-synonymous missense variant, L606I, is expressed in a Japanese pedigree [83, 86, 87].

In addition, Best3's various roles, from inflammation in the brain to apoptosis in proximal tubular cells in the kidneys and cardiac smooth muscle tissue, implicate it as a therapeutic target in a wide variety of chronic conditions [77, 78, 88].

7.5.4 Best4

Best4 mRNA is detected in the colon, brain, spinal cord, lungs, trachea, and testes, while the protein is primarily expressed in intestinal epithelial cells, where Best4 expression defines a subpopulation of absorptive-lineage cells and is induced during cell proliferation [86, 89]. To date, there is no known disease associated with Best4 dysfunction in humans, and in some species such as mice, Best4 is a pseudogene [90].

7.6 Introduction to the TMEM16/anoctamin family of CaCCs

There are 10 members of the Transmembrane 16 (TMEM16) family, also known as the Anoctamin (Ano) family. These proteins are named TMEM16A-K, excluding I, or Ano1-10. The *Anoct*amin nomenclature comes from the fact these anion-selective proteins were originally hypothesized to have 8 ('*oct*') transmembrane regions (this was later proven false, as they have 10). For consistency, we will use the TMEM16 nomenclature [7]. The TMEM16s are present in all eukaryotic phyla, but absent in prokaryotes, suggesting they appeared relatively later in evolution. Functionalities observed in the TMEM16 family proteins include CaCCs, lipid scramblases, and regulators of other channels [91–96]. Although all 10 members may act as CaCCs under the right experimental conditions, only a few, namely TMEM16A, TMEM16B, and TMEM16F have strong supporting evidence for being CaCCs under physiological conditions, with TMEM16F having dual CaCC and lipid scramblase functions [97, 98].

For these channels acting as CaCCs, their ion permeabilities, like the bestrophin family, follow the weak field strength lyotrophic series of $SCN^- > I^- > NO3^- > Br^- > Cl^- > F^-$ > gluconate [93, 99]. Channel activation is voltage-dependent at low $[Ca^{2+}]_i$, as the ions need to cross a majority of the transmembrane electric field to reach the binding site [100]. All members of the family have 10 transmembrane domains and cytosolic N- and C- termini, and form homodimers [91–93, 100, 101]. The overall architecture of TMEM16 paralogs are highly conserved, despite local differences which account for their divergent functions.

7.6.1 General structure of the TMEM16s

The TMEM16 proteins are homodimers with each monomer containing 10 TM helices, one membrane-spanning aqueous pore which may or may not be constricted, and at least two Ca^{2+}-binding sites. The main difference in these structures is in the diameter and composition of the pore, which confers the function of lipid scrambling or anion conduction [100, 102–107].

So far, there are three members of the TMEM16 family for which structures have been solved. The first structure of a TMEM16 homolog was from the fungus *Nectria haematococca*, nhTMEM16, which is a lipid scramblase. It is a rhombus-shaped homodimeric assembly with each monomer containing its own lipid scrambling path, which corresponds to the ion conduction pore in the TMEM16 CaCCs (figure 7.4) [108, 109]. When viewed from the extracellular side, the protomers are about 130×40 Å in dimension. 10 membrane-spanning α-helices preceded by two short α-helices (α0a and α0b) that peripherally interact with the inner membrane leaflet contribute to the transmembrane region (figure 7.4) [100]. Both the N- and C-termini are localized on the intracellular side of the membrane, and the termini contribute the largest interaction between the two dimer subunits (figure 7.4). In addition, there are two short α-helices, α5' located in the extracellular loop connecting α5 and α6, and α6' located in the intracellular loop between α6 and α7 (figure 7.4).

7.6.2 The ion conduction pathway

The nhTMEM16 lipid scramblase structure exhibits a hydrophilic membrane-exposed cavity with diameter of 8–11 Å that twists like a 'spiral staircase,' which forms the lipid translocation pathway (figure 7.4). The relative level of hydrophobicity of this cavity partially explains the mechanical difference in function between TMEM16 scramblases and ion channels. In the structure of the mouse TMEM16A (mTMEM16A), which is a CaCC, the helices forming this cavity have undergone structural rearrangements to seal the cavity off from the lipid bilayer (figure 7.4) [103, 108, 110, 111]. The channel pore, formed by helices a3 to a7 of the same subunit facing the membrane, appears as an 8–11 Å narrow crevice

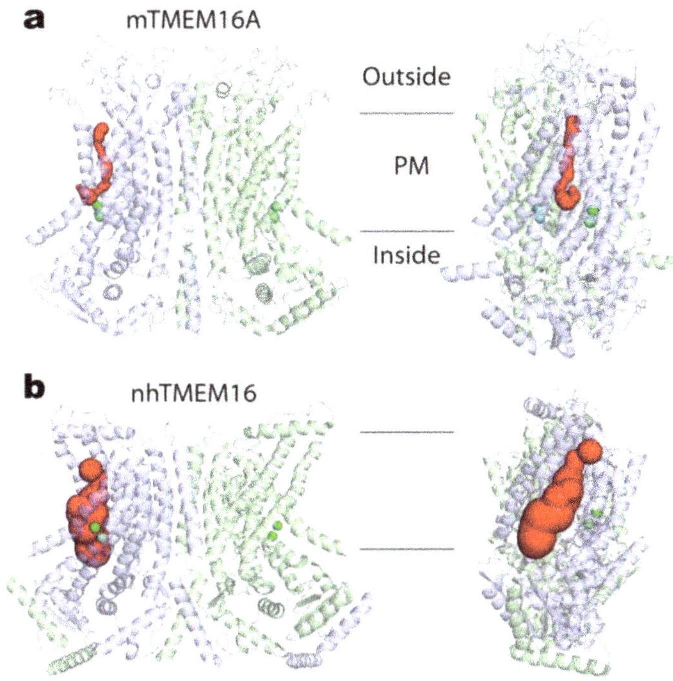

Figure 7.4. (a) cryoEM structure of the mouse TMEM16a Ca^{2+}-activated chloride channel (5OYB) in the Ca^{2+}-bound state, as seen from within the membrane (side view). The anion conduction pathway is shown as a red volume and four total Ca^{2+} ions, two per monomer, are shown as green spheres. One monomer is shown in green with the other shown in purple. PM, plasma membrane. (b) Crystal structure of the fungal nhTMEM16 (4WIS) lipid scramblase from *Nectria haematococca* as seen from within the membrane (side view). The lipid translocation pathway, shown in red, is larger than that of the Ca^{2+}-activated chloride channel mTMEM16a from the above panel.

(figure 7.4) [100]. Important residues for ion selectivity or ion conductance face the crevice, and are conserved between the TMEM16 CaCCs [99, 103, 108, 111].

7.6.3 The Ca^{2+} binding site and gating

The Ca^{2+} binding site in TMEM16 family proteins is comprised of five conserved acidic residues (E650, E698, E701, E730 and D734, mTMEM16A numbering) between TM helices 6–8 (figure 7.4) [100, 103, 109, 112]. This site is buried within the TM domain, approximately 15 Å away from the solvent on the intracellular side of the membrane. In the presence of Ca^{2+}, TM6 and TM8 are in close proximity to each other, with the negative charges of their conserved acidic residues neutralized by the binding of Ca^{2+}. In the absence of Ca^{2+}, the conserved acidic residues of TM6 and TM8 are farther from each other due to electrostatic repulsion and TM6 becomes more mobile, providing a solvent-exposed cavity that is accessible from the intracellular solution, facilitating Ca^{2+} entry to the site [104–107]. A conserved glycine residue (G644 in mTMEM16A) is required for Ca^{2+} -dependent activation in both TMEM16 lipid scramblases and anion channels [103, 106].

7.7 TMEM16A

7.7.1 TMEM16A function and structure

TMEM16A, identified as a CaCC in 2008, has been the most widely studied member of the TMEM16 family [91–93]. The gene encoding TMEM16A is located on the 11q13 amplicon, the same as Best1, and encodes a roughly 130 kDa monomer, or 260 kDa dimer corresponding to a 956 amino acid full length protein consisting of 25 exons [101, 113]. Although most TMEM16 channels form homodimers, TMEM16A and TMEM16B can form functional heterodimer channels in rat pineal glands to generate CaCCs and regulate melatonin secretion [114].

There are several structural differences between mTMEM16A CaCC and nhTMEM16 scramblase (figure 7.3). (1) mTMEM16A has a structured extracellular domain which is stabilized by disulfide bonds between three pairs of cysteine residues necessary for channel function, while nhTMEM16 does not [109, 110]. (2) the N- terminus of a mTMEM16A protomer loosely interacts with the C- terminus of the other protomer within the same dimeric assembly, whereas the interaction between nhTMEM16 protomers is more extensive, creating two more hydrophobic cavities in the scramblase [100, 103, 115]; (3) the extracellular portion of the ion conduction pore in mTMEM16A, composed of TM3 and TM4, is narrower than that of nhTMEM16, thus preventing lipid head groups from fitting into the pathway [102].

There is a Ca^{2+} binding site in each TMEM16A protomer. When a single Ca^{2+} is bound, the dimeric channel is 'primed' and subsequently gated by voltage, requiring membrane depolarization to fully open; when both protomers are bound by Ca^{2+}, the channel opens in a voltage-independent manner. Therefore, TMEM16A is voltage-dependent and -independent at low and high Ca^{2+}, respectively. Unlike the bestrophins, TMEM16A has a weak affinity for Cl^-, indicating there is unlikely a single strong anion binding site contributing to selectivity, but rather multiple sites that interact weakly with a passing anion resulting in general anion selectivity [103, 116].

There are four segments of the protein that may be alternatively spliced, termed a–d, and their alternative splicing leads to further differences in expression and functional specification of the gene product. All isoforms of the channel have CaCC activity; however, their properties, such as Ca^{2+} affinity and voltage dependence vary, and they are differentially distributed throughout various tissues. For example, TMEM16A(0), which lacks all four segments, is present in the testes, while the isoform without section *b* has a four-fold increase in calcium sensitivity [117].

7.7.2 TMEM16A physiological roles

TMEM16A is widely expressed throughout the body, including exocrine glands and organs where secretion is a major function (e.g. kidney, digestive tract, respiratory system, and reproductive tracts) [118–122], pulmonary arterial and cerebral artery smooth muscle cells, and sensory neurons [123–130].

In the gastrointestinal tract's secretory epithelia, TMEM16A functions in the salivary glands, exocrine pancreas hepatocytes, ileum, and large intestine, and may play a role in the motility of the interstitial cells of Cajal and connect the plasma membrane to the ER [123, 131, 132]. In the pancreas, TMEM16A mediates glucose-induced Cl^- efflux in beta cells and provides an exit pathway for HCO_3^- (thus regulating luminar pH) in acinar cells [133, 134]. In the airways, TMEM16A is expressed on the apical and lateral plasma membranes of polarized respiratory monolayers, where it acts as an apical conduit for Cl^- secretion and helps regulate lung liquid homeostasis and mucus production [119, 135, 136]. Moreover, the channel is expressed on the apical membranes of differentiating secretory epithelial cells and participates in ciliogenesis [137].

In vascular smooth muscle, the level of TMEM16A expression is as follows: portal vein > thoracic aorta > carotid artery > brain ventricles [138]. TMEM16A is involved in the regulation of coronary flow, and its expression in coronary arteries is higher than in mesenteric arteries, but lower than in pulmonary arteries [139].

In sensory neurons, TMEM16A plays important roles in olfactory signaling and nociception: (1) odorant molecules bind to odorant receptors in the cilia, leading to a neural transduction cascade which requires a TMEM16A regulated Cl^- electrochemical gradient between the mucus and intraciliary compartment for the activation of CNG channels and TMEM16B [140–143]; (2) TMEM16A enhances the excitability of dorsal root ganglion neurons during inflammation, augmenting formalin-mediated inflammatory pain, as well as neuropathic pain [129, 144]; (3) TMEM16A is activated by high temperatures above 44 °C and highly co-localized with the heat sensor TRPV1, resulting in the proliferation of nociceptive signals and pain sensation [129]. Moreover, TMEM16A expression is increased by neuropathic pain stimuli [145].

7.7.3 TMEM16A implications in disease

Overexpression of TMEM16A is a biomarker for numerous cancers, including breast cancer, squamous cell carcinomas and gastrointestinal stromal tumors [146–148]. Other human disorders associated with upregulation of TMEM16A include gastroparesis, acute pancreatitis, hypertension, and polycystric kidney disease [133, 136, 139, 149–151]. In addition, due to the channel's interaction with CFTR and overlapping function as a Cl^- channel, TMEM16A has also been proposed as a therapeutic target for neuropathic pain and cystic fibrosis [135]. Therefore, tissue-specific inhibition of TMEM16A function is a potential treatment for various human diseases.

7.8 TMEM16B

TMEM16B and TMEM16A are the most closely related members of the TMEM16 family, and the only two members that act strictly as CaCCs [152].

7.8.1 TMEM16B function and structure

Translation of the olfactory-specific open reading frame, which is the major isoform of TMEM16B, produces a 909 amino acid protein, corresponding to a ~104kDa predicted molecular weight [140]. This olfactory isoform lacks exon 13, encoding four amino acids ERSQ within the first intracellular loop which is important for channel gating [140]. The structure of TMEM16B has not been solved, but is predicted to resemble that of TMEM16A.

Like TMEM16A, TMEM16B is voltage-dependent at low $[Ca^{2+}]_i$, and -independent at high $[Ca^{2+}]_i$ [153, 154], but the overall affinity and sensitivity to Ca^{2+} are lower in TMEM16B compared to those in TMEM16A [153–156]. Glutamates E367 and $_{386}EEEEE_{390}$, located within the first intracellular putative loop, are important in the voltage dependence of the channel. The Hill coefficient of the protein is >1, indicating that more than one calcium ion is required for fully opening the channel.

7.8.2 TMEM16B physiological roles

TMEM16B protein is primarily expressed in the olfactory system. However, although TMEM16B is in olfactory sensory neurons (OSNs) and necessary for proper olfactory behaviors, it does not appear to be the main source of Cl^- currents in olfaction signaling. Instead, it plays important roles in interacting with other current-inducing channels [157, 158], spontaneous action potential firing, and glomerular targeting of OSNs [141]. TMEM16B is also involved in cerebellar motor learning, as it mediates inferior olivary repolarizing conductance in the medulla oblongata and is also expressed in the hippocampus [159].

In addition, TMEM16B is observed in other tissues/organs, such as retina, pancreas, respiratory system, esophagus and colon, where further investigation is needed to dissect its exact physiological roles [4, 138, 140, 159–167].

7.8.3 TMEM16B implications in disease

No human disease has yet been directly linked to TMEM16B mutations. TMEM16B is a potential target for body weight control, as CaCC current from TMEM16B in vagal afferents of nodose neurons is a major determinant of cholecystokinin (CCK)-induced satiety, while TMEM16B expression is downregulated in nodose ganglia of obese mice on a high fat diet [168].

7.9 TMEM16F

TMEM16F is a unique member in the family, because it has multi-functions as a Ca^{2+}-dependent lipid scramblase and even a Ca^{2+}-regulated non-selective cation channel, in addition to its activity as a CaCC [95, 108, 169]. In fact, the main role of TMEM16F at physiological Ca^{2+} concentrations is phospholipid scrambling, whereas the protein only acts as a Cl^- channel in more extreme conditions, such as at high Ca^{2+} or upon hypotonic stimuli [170, 171]. As a CaCC, TMEM16F has some distinct properties compared to TMEM16A and TMEM16B: (1) strong outward rectification, (2) poor cation/anion selectivity, (3) low Ca^{2+} sensitivity (>several μM $[Ca^{2+}]_i$,), and (4) slow activation (taking 3–5 min to reach peak levels after membrane break-in during patch clamp) [108, 169, 170, 172].

7.9.1 TMEM16F physiological roles

TMEM16F is widely expressed throughout the body, including platelets, macrophages, bone cells, T-lymphocytes, and podocytes of renal glomeruli. While the lipid scramblase activity of TMEM16F seems to dominate its physiological roles in blood coagulation, cell swelling, protein cleavage, cell migration, and immune defense [173–180], its CaCC activity is implicated in a cell death pathway, namely ferroptosis, and in bone mineralization [175, 181, 182, 183].

7.9.2 TMEM16F implications in disease

TMEM16F is mutated and loses functionality in patients with Scott Syndrome, a disorder of impaired blood coagulation due to dysfunction of TMEM16F's scramblase activity [95, 184, 185]. In addition, TMEM16F is a potential target for cancer due to its involvement in ferroptosis [183].

References

[1] Miledi R 1982 A calcium-dependent transient outward current in *Xenopus laevis* oocytes *Proc. R. Soc. Lond. B Biol. Sci.* **215** 491–7

[2] Bader C R, Bertrand D and Schwartz E A 1982 Voltage-activated and calcium-activated currents studied in solitary rod inner segments from the salamander retina *J. Physiol.* **331** 253–84

[3] Hartzell C, Putzier I and Arreola J 2005 Calcium-activated chloride channels *Annu. Rev. Physiol.* **67** 719–58

[4] Huang W C *et al* 2012 Calcium-activated chloride channels (CaCCs) regulate action potential and synaptic response in hippocampal neurons *Neuron* **74** 179–92

[5] Kim J A, Kang Y S and Lee Y S 2003 Role of Ca^{2+}-activated Cl^- channels in the mechanism of apoptosis induced by cyclosporin a in a human hepatoma cell line *Biochem. Biophys. Res. Commun.* **309** 291–7

[6] Owji A P, Kittredge A, Zhang Y and Yang T 2021 Structure and Function of the Bestrophin family of calcium-activated chloride channels *Channels (Austin)* **15** 604–23

[7] Hartzell H C *et al* 2009 Anoctamin/TMEM16 family members are Ca^{2+}-activated Cl^- channels *J. Physiol.* **587** 2127–39

[8] Duran C, Thompson C H, Xiao Q and Hartzell H C 2010 Chloride channels: often enigmatic, rarely predictable *Annu. Rev. Physiol.* **72** 95–121

[9] Xiao Q, Hartzell H C and Yu K 2010 Bestrophins and retinopathies *Pflugers Arch.* **460** 559–69

[10] Jentsch T J *et al* 2002 Molecular structure and physiological function of chloride channels *Physiol. Rev.* **82** 503–68

[11] Eisenman G and Horn R 1983 Ionic selectivity revisited: the role of kinetic and equilibrium processes in ion permeation through channels *J. Membr. Biol.* **76** 197–225

[12] Wright E M and Diamond J M 1977 Anion selectivity in biological systems *Physiol. Rev.* **57** 109–56

[13] Hartzell H C, Qu Z, Yu K, Xiao Q and Chein L-T 2008 Molecular physiology of bestrophins: multifunctional membrane proteins linked to best disease and other retinopathies *Physiol. Rev.* **88** 639–72

[14] O'Driscoll K E *et al* 2008 Expression, localization, and functional properties of Bestrophin 3 channel isolated from mouse heart *Am. J. Physiol. Cell Physiol.* **295** C1610–24
[15] Petrukhin K *et al* 1998 Identification of the gene responsible for best macular dystrophy *Nat. Genet.* **19** 241–7
[16] Sun H, Tsunenari T, Yau K-W and Nathans J 2002 The vitelliform macular dystrophy protein defines a new family of chloride channels *Proc. Natl Acad. Sci. USA* **99** 4008–13
[17] Qu Z, Fischmeister R and Hartzell C 2004 Mouse bestrophin-2 is a bona fide Cl(-) channel: identification of a residue important in anion binding and conduction *J. Gen. Physiol.* **123** 327–40
[18] Hagen A R, Barabote R D and Saier M H 2005 The bestrophin family of anion channels: identification of prokaryotic homologues *Mol. Membr. Biol.* **22** 291–302
[19] Tsunenari T *et al* 2003 Structure-function analysis of the bestrophin family of anion channels *J. Biol. Chem.* **278** 41114–25
[20] Xiao Q *et al* 2008 Regulation of bestrophin Cl channels by calcium: role of the C terminus *J. Gen. Physiol.* **132** 681–92
[21] Lee S, Yoon B-E, Berglund K, Oh S-J, Park H, Shin H-S, Augustine G J and Lee C J 2010 Channel-mediated tonic GABA release from glia *Science.* **330** 790–6
[22] Qu Z *et al* 2003 Two bestrophins cloned from *Xenopus laevis* oocytes express Ca^{2+}-activated Cl^{-} currents *J. Biol. Chem.* **278** 49563–72
[23] Tsunenari T, Nathans J and Yau K W 2006 Ca^{2+}-activated Cl^{-} current from human bestrophin-4 in excised membrane patches *J. Gen. Physiol.* **127** 749–54
[24] Vaisey G, Miller A N and Long S B 2016 Distinct regions that control ion selectivity and calcium-dependent activation in the bestrophin ion channel *Proc. Natl Acad. Sci. USA* **113** E7399–408
[25] Qu Z and Hartzell H C 2008 Bestrophin Cl- channels are highly permeable to HCO_3^{-} *Am. J. Physiol. Cell Physiol.* **294** C1371–7
[26] Yu K *et al* 2010 Bestrophin-2 mediates bicarbonate transport by goblet cells in mouse colon *J. Clin. Invest.* **120** 1722–35
[27] Oh S J and Lee C J 2017 Distribution and function of the bestrophin-1 (best1) channel in the brain *Exp. Neurobiol.* **26** 113–21
[28] Yang T *et al* 2014 Structure and selectivity in bestrophin ion channels *Science* **346** 355–9
[29] Owji A P *et al* 2020 Structural and functional characterization of the bestrophin-2 anion channel *Nat. Struct. Mol. Biol.* **27** 382–91
[30] Miller A N, Vaisey G and Long S B 2019 Molecular mechanisms of gating in the calcium-activated chloride channel bestrophin *Elife* **8** e43231
[31] Vaisey G and Long S B 2018 An allosteric mechanism of inactivation in the calcium-dependent chloride channel BEST1 *J. Gen. Physiol.* **150** 1484–97
[32] Dickson V K, Pedi L and Long S B 2014 Structure and insights into the function of a Ca(2+)-activated Cl(-) channel *Nature* **516** 213–8
[33] Owji A P, Wang J, Kittredge A, Clark Z, Zhang Y, Hendrickson W A and Yang T 2022 Structures and gating mechanisms of human bestrophin anion channels *Nat. Commun.* **13** 3836
[34] Woo D H, Han K-S, Shim J W, Yoon B-E, Kim E, Bae J Y, Oh S-J, Hwang E M, Marmorstein A D, Bae Y C, Park J-Y and Lee C J 2012 TREK-1 and Best1 channels mediate fast and slow glutamate release in astrocytes upon GPCR activation *Cell* **151** 25–40

[35] Qu Z, Cui Y and Hartzell C 2006 A short motif in the C-terminus of mouse bestrophin 3 [corrected] inhibits its activation as a Cl channel *FEBS Lett.* **580** 2141–6
[36] Qu Z Q *et al* 2007 Activation of bestrophin Cl^- channels is regulated by C-terminal domains *J. Biol. Chem.* **282** 17460–7
[37] Xiao Q *et al* 2009 Dysregulation of human bestrophin-1 by ceramide-induced dephosphorylation *J. Physiol.* **587** 4379–91
[38] Zhang Y *et al* 2018 ATP activates bestrophin ion channels through direct interaction *Nat. Commun.* **9** 3126
[39] Gallemore R P and Steinberg R H 1993 Light-evoked modulation of basolateral membrane Cl^- conductance in chick retinal pigment epithelium: the light peak and fast oscillation *J. Neurophysiol.* **70** 1669–80
[40] Gouras P *et al* 2009 Bestrophin detected in the basal membrane of the retinal epithelium and drusen of monkeys with drusenoid maculopathy *Graefe's Arch. Clin. Exp. Ophthalmol.* **247** 1051–6
[41] Marmorstein A D *et al* 2000 Bestrophin, the product of the best vitelliform macular dystrophy gene (VMD2), localizes to the basolateral plasma membrane of the retinal pigment epithelium *Proc. Natl Acad. Sci. USA* **97** 12758–63
[42] Marquardt A *et al* 1998 Mutations in a novel gene, VMD2, encoding a protein of unknown properties cause juvenile-onset vitelliform macular dystrophy (Best's disease) *Hum. Mol. Genet.* **7** 1517–25
[43] Bakall B *et al* 2003 Expression and localization of bestrophin during normal mouse development *Invest. Ophthalmol. Vis. Sci.* **44** 3622–8
[44] Li Y *et al* 2017 Patient-specific mutations impair BESTROPHIN1's essential role in mediating Ca *Elife* **6** e29914
[45] Carter-Dawson L D and LaVail M M 1979 Rods and cones in the mouse retina. I. Structural analysis using light and electron microscopy *J. Comp Neurol.* **188** 245–62
[46] Szel A *et al* 1996 Distribution of cone photoreceptors in the mammalian retina *Microsc. Res. Tech.* **35** 445–62
[47] Marmorstein L Y *et al* 2006 The light peak of the electroretinogram is dependent on voltage-gated calcium channels and antagonized by bestrophin (best-1) *J. Gen. Physiol.* **127** 577–89
[48] Braun J *et al* 2010 Quantitative expression analyses of candidates for alternative anion conductance in cystic fibrosis mouse models *J. Cyst Fibros.* **9** 351–64
[49] Park H *et al* 2009 Bestrophin-1 encodes for the Ca^{2+}-activated anion channel in hippocampal astrocytes *J. Neurosci.* **29** 13063–73
[50] Park H *et al* 2013 High glutamate permeability and distal localization of best1 channel in CA1 hippocampal astrocyte *Mol. Brain.* **6** 54
[51] Eldred G E 1995 Lipofuscin fluorophore inhibits lysosomal protein degradation and may cause early stages of macular degeneration *Gerontology* **41** 15–28
[52] Besharse J 1982 The daily light-dark cycle and rhythmic metabolism in the photoreceptor—Pigment epithelial complex *Progress in Retinal Research.* (Amsterdam: Elsevier) ch 2 pp 82–118
[53] LaVail M M 1983 Outer segment disc shedding and phagocytosis in the outer retina *Trans. Ophthalmol. Soc. UK* **103** 397–404
[54] Nguyen-Legros J and Hicks D 2000 Renewal of photoreceptor outer segments and their phagocytosis by the retinal pigment epithelium *Int Rev Cytol.* **196** 245–313

[55] Arden G B 1962 Alterations in the standing potential of the eye associated with retinal disease *Trans. Ophthalmol. Soc. UK* **82** 63–72

[56] Cross H E and Bard L 1974 Electro-oculography in best's macular dystrophy *Am. J. Ophthalmol.* **77** 46–50

[57] Reddy M A *et al* 2003 A clinical and molecular genetic study of a rare dominantly inherited syndrome (MRCS) comprising of microcornea, rod-cone dystrophy, cataract, and posterior staphyloma *Br. J. Ophthalmol.* **87** 197–202

[58] Yardley J *et al* 2004 Mutations of VMD2 splicing regulators cause nanophthalmos and autosomal dominant vitreoretinochoroidopathy (ADVIRC) *Invest. Ophthalmol. Vis. Sci.* **45** 3683–9

[59] Michaelides M *et al* 2006 Evidence of genetic heterogeneity in MRCS (microcornea, rod-cone dystrophy, cataract, and posterior staphyloma) syndrome *Am. J. Ophthalmol.* **141** 418–20

[60] Burgess R *et al* 2008 Biallelic mutation of BEST1 causes a distinct retinopathy in humans *Am. J. Hum. Genet.* **82** 19–31

[61] Burgess R *et al* 2009 ADVIRC is caused by distinct mutations in BEST1 that alter pre-mRNA splicing *J. Med. Genet.* **46** 620–5

[62] Davidson A E *et al* 2009 Missense mutations in a retinal pigment epithelium protein, bestrophin-1, cause retinitis pigmentosa *Am. J. Hum. Genet.* **85** 581–92

[63] Guziewicz K E *et al* 2007 Bestrophin gene mutations cause canine multifocal retinopathy: a novel animal model for best disease *Invest. Ophthalmol. Vis. Sci.* **48** 1959–67

[64] Guziewicz K E *et al* 2018 Gene therapy corrects a diffuse retina-wide microdetachment modulated by light exposure *Proc. Natl Acad. Sci. USA* **115** E2839–48

[65] Elorza-Vidal X, Gaitán-Peñas H and Estévez R 2019 Chloride channels in astrocytes: structure, roles in brain homeostasis and implications in disease *Int. J. Mol. Sci.* **20** 1034

[66] Zhang Y, Patil R V and Marmorstein A D 2010 Bestrophin 2 is expressed in human non-pigmented ciliary epithelium but not retinal pigment epithelium *Mol. Vis.* **16** 200–6

[67] Klimmeck D *et al* 2009 Bestrophin 2: an anion channel associated with neurogenesis in chemosensory systems *J. Comp. Neurol.* **515** 585–99

[68] Pifferi S *et al* 2006 Bestrophin-2 is a candidate calcium-activated chloride channel involved in olfactory transduction *Proc. Natl Acad. Sci. USA* **103** 12929–34

[69] Cui C Y *et al* 2012 Forkhead transcription factor foxA1 regulates sweat secretion through bestrophin 2 anion channel and Na–K–Cl cotransporter 1 *Proc. Natl Acad. Sci. USA* **109** 1199–203

[70] Bakall B *et al* 2008 Bestrophin-2 is involved in the generation of intraocular pressure *Invest. Ophthalmol. Vis. Sci.* **49** 1563–70

[71] Heijl A *et al* 2002 Reduction of intraocular pressure and glaucoma progression: results from the early manifest glaucoma trial *Arch. Ophthalmol.* **120** 1268–79

[72] Kass M A *et al* 2002 The ocular hypertension treatment study: a randomized trial determines that topical ocular hypotensive medication delays or prevents the onset of primary open-angle glaucoma *Arch. Ophthalmol.* **120** 701–13 discussion 829–30

[73] Lichter P R *et al* 2001 Interim clinical outcomes in the collaborative initial glaucoma treatment study comparing initial treatment randomized to medications or surgery *Ophthalmology* **108** 1943–53

[74] Qu Z *et al* 2010 A PI3 kinase inhibitor found to activate bestrophin 3 *J. Cardiovasc. Pharmacol.* **55** 110–5

[75] Broegger T *et al* 2011 Bestrophin is important for the rhythmic but not the tonic contraction in rat mesenteric small arteries *Cardiovasc. Res.* **91** 685–93
[76] Song W, Yang Z and He B 2014 Bestrophin 3 ameliorates TNFα-induced inflammation by inhibiting NF-κB activation in endothelial cells *PLoS One* **9** e111093
[77] Jiang L *et al* 2013 Mitochondria dependent pathway is involved in the protective effect of bestrophin-3 on hydrogen peroxide-induced apoptosis in basilar artery smooth muscle cells *Apoptosis* **18** 556–65
[78] Lee W K *et al* 2012 ERK1/2-dependent bestrophin-3 expression prevents ER-stress-induced cell death in renal epithelial cells by reducing CHOP *Biochim. Biophys. Acta.* **1823** 1864–76
[79] Matchkov V V *et al* 2008 Bestrophin-3 (vitelliform macular dystrophy 2-like 3 protein) is essential for the cGMP-dependent calcium-activated chloride conductance in vascular smooth muscle cells *Circ. Res.* **103** 864–72
[80] Wu L *et al* 2016 A C-terminally truncated mouse best3 splice variant targets and alters the ion balance in lysosome-endosome hybrids and the endoplasmic reticulum *Sci. Rep.* **6** 27332
[81] Golubinskaya V *et al* 2015 Bestrophin-3 is differently expressed in normal and injured mouse glomerular podocytes *Acta Physiol. (Oxf).* **214** 481–96
[82] Srivastava A *et al* 2008 A variant of the Ca^{2+}-activated Cl channel best3 is expressed in mouse exocrine glands *J. Membr. Biol.* **222** 43–54
[83] Kajii T S *et al* 2019 Whole-exome sequencing in a Japanese pedigree implicates a rare non-synonymous single-nucleotide variant in BEST3 as a candidate for mandibular prognathism *Bone* **122** 193–8
[84] Singh G D 1999 Morphologic determinants in the etiology of class III malocclusions: a review *Clin. Anat.* **12** 382–405
[85] Nakasima A, Ichinose M and Nakata S 1986 Genetic and environmental factors in the development of so-called pseudo- and true mesiocclusions *Am. J. Orthod. Dentofacial Orthop.* **90** 106–16
[86] Stöhr H *et al* 2002 Three novel human VMD2-like genes are members of the evolutionary highly conserved RFP-TM family *Eur. J. Hum. Genet.* **10** 281–4
[87] Kumagai K *et al* 2016 Activation of a chondrocyte volume-sensitive Cl^- conductance prior to macroscopic cartilage lesion formation in the rabbit knee anterior cruciate ligament transection osteoarthritis model *Osteoarth. Cartil.* **24** 1786–94
[88] Golubinskaya V *et al* 2019 Bestrophin-3 expression in a subpopulation of astrocytes in the neonatal brain after hypoxic-ischemic injury *Front. Physiol.* **10** 23
[89] Ito G *et al* 2013 Lineage-specific expression of bestrophin-2 and bestrophin-4 in human intestinal epithelial cells *PLoS One* **8** e79693
[90] Krämer F, Stöhr H and Weber B H 2004 Cloning and characterization of the murine Vmd2 RFP-TM gene family *Cytogenet. Genome Res.* **105** 107–14
[91] Caputo A *et al* 2008 TMEM16A, a membrane protein associated with calcium-dependent chloride channel activity *Science* **322** 590–4
[92] Schroeder B C *et al* 2008 Expression cloning of TMEM16A as a calcium-activated chloride channel subunit *Cell* **134** 1019–29
[93] Yang Y D *et al* 2008 TMEM16A confers receptor-activated calcium-dependent chloride conductance *Nature* **455** 1210–5
[94] Okada Y 1997 Volume expansion-sensing outward-rectifier Cl^- channel: fresh start to the molecular identity and volume sensor *Am. J. Physiol.* **273** C755–89

[95] Suzuki J *et al* 2010 Calcium-dependent phospholipid scrambling by TMEM16F *Nature* **468** 834–8
[96] Huang F *et al* 2013 TMEM16C facilitates Na^+-activated K^+ currents in rat sensory neurons and regulates pain processing *Nat. Neurosci.* **16** 1284–90
[97] Malvezzi M *et al* 2013 Ca^{2+}-dependent phospholipid scrambling by a reconstituted TMEM16 ion channel *Nat. Commun.* **4** 2367
[98] Tian Y, Schreiber R and Kunzelmann K 2012 Anoctamins are a family of Ca^{2+}-activated Cl^- channels *J. Cell Sci.* **125** 4991–8
[99] Reyes J P *et al* 2014 Anion permeation in calcium-activated chloride channels formed by TMEM16A from *Xenopus tropicalis Pflugers Arch.* **466** 1769–77
[100] Brunner J D *et al* 2014 X-ray structure of a calcium-activated TMEM16 lipid scramblase *Nature* **516** 207–12
[101] Sheridan J T *et al* 2011 Characterization of the oligomeric structure of the Ca^{2+}-activated Cl^- channel Ano1/TMEM16A *J. Biol. Chem.* **286** 1381–8
[102] Paulino C *et al* 2017 Structural basis for anion conduction in the calcium-activated chloride channel TMEM16A *Elife* **6** e26232
[103] Dang S *et al* 2017 Cryo-EM structures of the TMEM16A calcium-activated chloride channel *Nature* **552** 426–9
[104] Bushell S R *et al* 2019 The structural basis of lipid scrambling and inactivation in the endoplasmic reticulum scramblase TMEM16K *Nat. Commun.* **10** 3956
[105] Falzone M E *et al* 2019 Structural basis of Ca^{2+}-dependent activation and lipid transport by a TMEM16 scramblase *Elife* **8** e43229
[106] Alvadia C *et al* 2019 Cryo-EM structures and functional characterization of the murine lipid scramblase TMEM16F *Elife* **8** e44365
[107] Kalienkova V *et al* 2019 Stepwise activation mechanism of the scramblase nhTMEM16 revealed by cryo-EM *Elife* **8** e44364
[108] Yang H *et al* 2012 TMEM16F forms a Ca^{2+}-activated cation channel required for lipid scrambling in platelets during blood coagulation *Cell* **151** 111–22
[109] Yu K *et al* 2012 Explaining calcium-dependent gating of anoctamin-1 chloride channels requires a revised topology *Circ. Res.* **110** 990–9
[110] Paulino C, Kalienkova V, Lam A K M, Neldner Y and Dutzler R 2017 Activation mechanism of the calcium-activated chloride channel TMEM16A revealed by cryo-EM *Nature* **552** 421–5
[111] Peters C J *et al* 2015 Four basic residues critical for the ion selectivity and pore blocker sensitivity of TMEM16A calcium-activated chloride channels *Proc. Natl Acad. Sci. USA* **112** 3547–52
[112] Tien J *et al* 2014 A comprehensive search for calcium binding sites critical for TMEM16A calcium-activated chloride channel activity *Elife* **3** e02772
[113] Fallah G *et al* 2011 TMEM16A(a)/anoctamin-1 shares a homodimeric architecture with CLC chloride channels *Mol. Cell Proteomics* **10** M110.004697
[114] Yamamura H *et al* 2018 TMEM16A and TMEM16B channel proteins generate Ca *J. Biol. Chem.* **293** 995–1006
[115] De Jesús-Pérez J J *et al* 2018 Phosphatidylinositol 4,5-bisphosphate, cholesterol, and fatty acids modulate the calcium-activated chloride channel TMEM16A (ANO1) *Biochim. Biophys. Acta, Mol. Cell. Biol. Lipids* **1863** 299–312

[116] Qu Z and Hartzell H C 2000 Anion permeation in Ca^{2+}-activated Cl^- channels *J. Gen. Physiol.* **116** 825–44
[117] Ferrera L *et al* 2009 Regulation of TMEM16A chloride channel properties by alternative splicing *J. Biol. Chem.* **284** 33360–8
[118] Henriques T *et al* 2019 TMEM16A calcium-activated chloride currents in supporting cells of the mouse olfactory epithelium *J. Gen. Physiol.* **151** 954–66
[119] Rock J R *et al* 2009 Transmembrane protein 16A (TMEM16A) is a Ca^{2+}-regulated Cl^- secretory channel in mouse airways *J. Biol. Chem.* **284** 14875–80
[120] Hwang S J *et al* 2009 Expression of anoctamin 1/TMEM16A by interstitial cells of Cajal is fundamental for slow wave activity in gastrointestinal muscles *J. Physiol.* **587** 4887–904
[121] Faria D *et al* 2014 The calcium-activated chloride channel Anoctamin 1 contributes to the regulation of renal function *Kidney Int.* **85** 1369–81
[122] Ousingsawat J *et al* 2009 Loss of TMEM16A causes a defect in epithelial Ca^{2+}-dependent chloride transport *J. Biol. Chem.* **284** 28698–703
[123] Huang F *et al* 2009 Studies on expression and function of the TMEM16A calcium-activated chloride channel *Proc. Natl Acad. Sci. USA* **106** 21413–8
[124] Huang F *et al* 2012 Calcium-activated chloride channel TMEM16A modulates mucin secretion and airway smooth muscle contraction *Proc. Natl Acad. Sci. USA* **109** 16354–9
[125] Manoury B, Tamuleviciute A and Tammaro P 2010 TMEM16A/anoctamin 1 protein mediates calcium-activated chloride currents in pulmonary arterial smooth muscle cells *J. Physiol.* **588** 2305–14
[126] Thomas-Gatewood C *et al* 2011 TMEM16A channels generate Ca^2+-activated Cl^- currents in cerebral artery smooth muscle cells *Am. J. Physiol. Heart. Circ. Physiol.* **301** H1819–27
[127] Romanenko V G *et al* 2010 Tmem16A encodes the Ca^{2+}-activated Cl^- channel in mouse submandibular salivary gland acinar cells *J. Biol. Chem.* **285** 12990–3001
[128] Gomez-Pinilla P J *et al* 2009 Ano1 is a selective marker of interstitial cells of Cajal in the human and mouse gastrointestinal tract *Am. J. Physiol. Gastrointest. Liver Physiol.* **296** G1370–81
[129] Cho H *et al* 2012 The calcium-activated chloride channel anoctamin 1 acts as a heat sensor in nociceptive neurons *Nat. Neurosci.* **15** 1015–21
[130] Maurya D K and Menini A 2014 Developmental expression of the calcium-activated chloride channels TMEM16A and TMEM16B in the mouse olfactory epithelium *Dev. Neurobiol.* **74** 657–75
[131] Zhu M H *et al* 2009 A Ca^{2+}-activated Cl^- conductance in interstitial cells of Cajal linked to slow wave currents and pacemaker activity *J. Physiol.* **587** 4905–18
[132] Zhu M H *et al* 2015 Intracellular Ca^{2+} release from endoplasmic reticulum regulates slow wave currents and pacemaker activity of interstitial cells of Cajal *Am. J. Physiol. Cell Physiol.* **308** C608–20
[133] Han Y, Shewan A M and Thorn P 2016 HCO_3^- transport through anoctamin/transmembrane protein ANO1/TMEM16A in pancreatic acinar cells regulates luminal pH *J. Biol. Chem.* **291** 20345–52
[134] Jung J *et al* 2013 Dynamic modulation of ANO1/TMEM16A HCO_3^- permeability by Ca^{2+}/calmodulin *Proc. Natl Acad. Sci. USA* **110** 360–5

[135] Veit G *et al* 2012 Proinflammatory cytokine secretion is suppressed by TMEM16A or CFTR channel activity in human cystic fibrosis bronchial epithelia *Mol. Biol. Cell.* **23** 4188–202
[136] Benedetto R *et al* 2019 TMEM16A is indispensable for basal mucus secretion in airways and intestine *FASEB J.* **33** 4502–12
[137] Ruppersburg C C and Hartzell H C 2014 The Ca^{2+}-activated Cl^- channel ANO1/ TMEM16A regulates primary ciliogenesis *Mol. Biol. Cell* **25** 1793–807
[138] Davis A J *et al* 2010 Expression profile and protein translation of TMEM16A in murine smooth muscle *Am. J. Physiol. Cell Physiol.* **299** C948–59
[139] Askew Page H R *et al* 2019 TMEM16A is implicated in the regulation of coronary flow and is altered in hypertension *Br. J. Pharmacol.* **176** 1635–48
[140] Stephan A B *et al* 2009 ANO2 is the cilial calcium-activated chloride channel that may mediate olfactory amplification *Proc. Natl Acad. Sci. USA* **106** 11776–81
[141] Pietra G *et al* 2016 The Ca^{2+}-activated Cl^- channel TMEM16B regulates action potential firing and axonal targeting in olfactory sensory neurons *J. Gen. Physiol.* **148** 293–311
[142] Dibattista M *et al* 2017 The long tale of the calcium activated Cl *Channels (Austin)* **11** 399–414
[143] Zak J D *et al* 2018 Calcium-activated chloride channels clamp odor-evoked spike activity in olfactory receptor neurons *Sci. Rep.* **8** 10600
[144] García G *et al* 2014 *Evidence for the participation of Ca^{2+}-activated chloride channels in formalin-induced acute and chronic nociception Brain Res.* **1579** 35–44
[145] Chen Q Y *et al* 2019 Mechanism of persistent hyperalgesia in neuropathic pain caused by chronic constriction injury *Neural Regen. Res.* **14** 1091–8
[146] Kashyap M K *et al* 2009 Genomewide mRNA profiling of esophageal squamous cell carcinoma for identification of cancer biomarkers *Cancer Biol. Ther.* **8** 36–46
[147] West R B *et al* 2004 The novel marker, DOG1, is expressed ubiquitously in gastrointestinal stromal tumors irrespective of KIT or PDGFRA mutation status *Am. J. Pathol.* **165** 107–13
[148] Wilkerson P M and Reis-Filho J S 2013 The 11q13-q14 amplicon: clinicopathological correlations and potential drivers *Genes Chromo. Cancer* **52** 333–55
[149] Mazzone A *et al* 2011 Altered expression of Ano1 variants in human diabetic gastroparesis *J. Biol. Chem.* **286** 13393–403
[150] Papp R *et al* 2019 Targeting TMEM16A to reverse vasoconstriction and remodelling in idiopathic pulmonary arterial hypertension *Eur. Respir. J.* **53** 1800965
[151] Schreiber R *et al* 2019 Lipid peroxidation drives renal cyst growth *J. Am. Soc. Nephrol.* **30** 228–42
[152] Milenkovic V M *et al* 2010 Evolution and functional divergence of the anoctamin family of membrane proteins *BMC Evol. Biol.* **10** 319
[153] Cenedese V *et al* 2012 The voltage dependence of the TMEM16B/anoctamin2 calcium-activated chloride channel is modified by mutations in the first putative intracellular loop *J. Gen. Physiol.* **139** 285–94
[154] Pifferi S, Dibattista M and Menini A 2009 TMEM16B induces chloride currents activated by calcium in mammalian cells *Pflugers Arch.* **458** 1023–38
[155] Ponissery Saidu S *et al* 2013 Channel properties of the splicing isoforms of the olfactory calcium-activated chloride channel anoctamin 2 *J. Gen. Physiol.* **141** 691–703

[156] Adomaviciene A *et al* 2013 Putative pore-loops of TMEM16/anoctamin channels affect channel density in cell membranes *J. Physiol.* **591** 3487–505

[157] Cenedese V *et al* 2015 Assessment of the olfactory function in Italian patients with type 3 von willebrand disease caused by a homozygous 253 Kb deletion involving VWF and TMEM16B/ANO2 *PLoS One.* **10** e0116483

[158] Billig G M *et al* 2011 Ca^{2+}-activated Cl^- currents are dispensable for olfaction *Nat. Neurosci.* **14** 763–9

[159] Zhang Y *et al* 2017 Inferior Olivary TMEM16B mediates cerebellar motor learning *Neuron* **95** 1103–11

[160] Yu T T *et al* 2005 Differentially expressed transcripts from phenotypically identified olfactory sensory neurons *J. Comp. Neurol.* **483** 251–62

[161] Stöhr H *et al* 2009 TMEM16B, a novel protein with calcium-dependent chloride channel activity, associates with a presynaptic protein complex in photoreceptor terminals *J. Neurosci.* **29** 6809–18

[162] Hengl T *et al* 2010 Molecular components of signal amplification in olfactory sensory cilia *Proc. Natl Acad. Sci. USA* **107** 6052–7

[163] Rasche S *et al* 2010 Tmem16b is specifically expressed in the cilia of olfactory sensory neurons *Chem. Senses* **35** 239–45

[164] Sagheddu C *et al* 2010 Calcium concentration jumps reveal dynamic ion selectivity of calcium-activated chloride currents in mouse olfactory sensory neurons and TMEM16b-transfected HEK 293T cells *J. Physiol.* **588** 4189–204

[165] Mayer U *et al* 2009 The proteome of rat olfactory sensory cilia *Proteomics* **9** 322–34

[166] Pifferi S, Cenedese V and Menini A 2012 Anoctamin 2/TMEM16B: a calcium-activated chloride channel in olfactory transduction *Exp. Physiol.* **97** 193–9

[167] Keckeis S *et al* 2017 Anoctamin2 (TMEM16B) forms the Ca *Exp. Eye Res.* **154** 139–50

[168] Wang R *et al* 2019 TMEM16B determines cholecystokinin sensitivity of intestinal vagal afferents of nodose neurons *JCI Insight* **4** e122058

[169] Grubb S *et al* 2013 TMEM16F (Anoctamin 6), an anion channel of delayed Ca^{2+} activation *J. Gen. Physiol.* **141** 585–600

[170] Scudieri P *et al* 2015 Ion channel and lipid scramblase activity associated with expression of TMEM16F/ANO6 isoforms *J. Physiol.* **593** 3829–48

[171] Kunzelmann K *et al* 2014 Molecular functions of anoctamin 6 (TMEM16F): a chloride channel, cation channel, or phospholipid scramblase? *Pflugers Arch.* **466** 407–14

[172] Shimizu T *et al* 2013 TMEM16F is a component of a Ca^{2+}-activated Cl^- channel but not a volume-sensitive outwardly rectifying Cl^- channel *Am. J. Physiol. Cell Physiol.* **304** C748–59

[173] Schenk L K *et al* 2018 Regulation and function of TMEM16F in renal podocytes *Int. J. Mol. Sci.* **19** 1798

[174] Ousingsawat J *et al* 2015 Anoctamin 6 mediates effects essential for innate immunity downstream of P2X7 receptors in macrophages *Nat. Commun.* **6** 6245

[175] Ousingsawat J *et al* 2015 Anoctamin-6 controls bone mineralization by activating the calcium transporter NCX1 *J. Biol. Chem.* **290** 6270–80

[176] Baig A A *et al* 2016 TMEM16F-mediated platelet membrane phospholipid scrambling Is critical for hemostasis and thrombosis but not thromboinflammation in mice-brief report *Arterioscl. Thromb. Vasc. Biol.* **36** 2152–7

[177] Sommer A *et al* 2016 Phosphatidylserine exposure is required for ADAM17 sheddase function *Nat. Commun.* **7** 11523

[178] Mattheij N J *et al* 2016 Survival protein anoctamin-6 controls multiple platelet responses including phospholipid scrambling, swelling, and protein cleavage *FASEB J.* **30** 727–37

[179] Ehlen H W *et al* 2013 Inactivation of anoctamin-6/Tmem16f, a regulator of phosphatidylserine scrambling in osteoblasts, leads to decreased mineral deposition in skeletal tissues *J. Bone Miner. Res.* **28** 246–59

[180] Jacobsen K S *et al* 2013 The role of TMEM16A (ANO1) and TMEM16F (ANO6) in cell migration *Pflugers Arch.* **465** 1753–62

[181] Martins J R *et al* 2011 Anoctamin 6 is an essential component of the outwardly rectifying chloride channel *Proc. Natl Acad. Sci. USA* **108** 18168–72

[182] Juul C A *et al* 2014 Anoctamin 6 differs from VRAC and VSOAC but is involved in apoptosis and supports volume regulation in the presence of Ca^{2+} *Pflugers Arch.* **466** 1899–910

[183] Mou Y, Wang J, Wu J, He D, Zhang C, Duan C and Li B 2019 Ferroptosis, a new form of cell death: opportunities and challenges in cancer *J. Hematol. Oncol* **12** 34

[184] Castoldi E *et al* 2011 Compound heterozygosity for 2 novel TMEM16F mutations in a patient with scott syndrome *Blood* **117** 4399–400

[185] Zwaal R F A, Comfurius P and Bevers E M 2004 Scott syndrome, a bleeding disorder caused by defective scrambling of membrane phospholipids *Biochim. Biophys. Acta* **1636** 119–28

Part III

Physiological roles of calcium

Chapter 8

Excitation–contraction coupling in skeletal muscle: fast Ca^{2+} signaling for muscle activation

Martin F Schneider and Erick O Hernández-Ochoa

Adapted in part from chapter 57 of Schneider and Hernández-Ochoa (2012) with permission from Elsevier.

8.1 Introduction to Ca^{2+} signaling in skeletal muscle

Calcium is one of the most abundant elements in the Universe, planet earth and oceans, and the most abundant metal found in the human body. Evolution seized advantage of Ca^{2+} for use as a universal signaling element along with Na^+ and other monovalent ions for fast cellular communication. In contrast with monovalent ions, that can be used to shape electrical signals and generate electrochemical gradients, Ca^{2+} ion can also act as a signaling factor itself by interacting with molecules so as to modify their function upon Ca^{2+} binding (Berridge *et al* 2000). Despite its relative abundance, the concentration of free ionized Ca^{2+} in the serum and extracellular fluids is maintained relatively low (1.3 mM in humans) when compared to other ions. Under resting conditions, the Ca^{2+} concentration is maintained at vastly lower levels in the intracellular milieu when compared to Na^+ or K^+ ions (Pozzan *et al* 1994). In skeletal muscle cells, when Ca^{2+} is called into action, following an action potential (AP), intracellular Ca^{2+} levels rapidly rise to induce conformational changes of contractile proteins, a process known as excitation–contraction coupling (ECC) (Sandow 1952). The elevation in myoplasmic $[Ca^{2+}]$ induced by APs is also important for metabolism and transcriptional activity regulation (Gundersen 2011, Diaz-Vegas *et al* 2019).

The story of Ca^{2+} signaling began in the muscle physiology field, with the seminal experiments of Ringer (1887) using striated muscle as model of study (Rudolf *et al* 2003). He was the first one to serendipitously demonstrate the impact of Ca^{2+} on muscle function. In his experiments using a bath solution lacking Ca^{2+}, he noticed

doi:10.1088/978-0-7503-2009-2ch8

that the sartorius muscle of the frog, a skeletal muscle, was able to twitch in response to stimulation, while the isolated heart maintained in the same solution was unable to sustain spontaneous contractions. The addition of a buffer containing Ca^{2+} (which was accidently made with tap water) to the recording chamber had little effect on skeletal muscle, but was able to prolong the spontaneous contraction of the isolated heart. In an unanticipated way, these experiments illustrated one of the major distinctions between skeletal and cardiac muscle in terms of requirements of extracellular Ca^{2+} and Ca^{2+} influx for ECC in cardiac but not skeletal muscle (Ringer 1887). Followed Ringer's experiments, the observation that intracellular Ca^{2+} is required for muscle contraction (Kamada and Kinoshita 1943), the discovery of Ca^{2+}-ATPase of the sarcoplasmic reticulum (Hasselbach 1964, Ebashi 1974), the regulation of actomyosin interaction (Weber 1959) via Ca^{2+}-tropomyosin regulation (Ebashi 1974) demonstrated the role of Ca^{2+} as a critical regulator of muscle contraction. In the intervening years the identification of voltage-dependent activation of Ca^{2+} release in skeletal muscle (Schneider and Chandler 1973, Rios and Brum 1987), the identification of the critical substrates for ECC, voltage-gated Ca^{2+} channels $Ca_v1.1$ and Ca^{2+} release channel RyR1 (Franzini-Armstrong and Nunzi 1983, Beam *et al* 1986; Meissner *et al* 1986, Imagawa *et al* 1987, Inui *et al* 1987, Takahashi *et al* 1987, Tanabe *et al* 1987, 1988, 1990a, Takeshima *et al* 1989, Nakai *et al* 1998b) through the most recent characterization of their high-resolution molecular structure (Yan *et al* 2015, Zalk *et al* 2015, Wu *et al* 2016), has broadened our understanding of Ca^{2+} signaling in skeletal ECC.

8.2 Structural organization of Ca^{2+} channels in skeletal muscle fibers

The skeletal muscle fiber AP transforms the muscle from the resting to the activated state (Sandow 1952). The AP is initiated at the neuro-muscular junction (NMJ) in response to the local depolarizing end-plate potential (EPP) caused by acetylcholine release from the motor nerve endings innervating the muscle fiber at the NMJ located near the mid-point of the fiber (Andersson-Cedergren 1959, Hubbard 1973, Katz 1996) (figure 8.1(a)). The initiated muscle fiber AP then propagates away from the NMJ (Adrian and Marshall 1977), both longitudinally along the length of the muscle fiber and radially into the muscle fiber along tubular invaginations of the surface membrane known as transverse tubules (TTs) (Huxley and Taylor 1958, Adrian *et al* 1969) (figure 8.1(a)). In mammalian skeletal muscle fibers, two sets of transverse tubules encircle each myofibril at the level of the junction of the sarcomeric A (anisotropic; thick filament myosin-containing) and I (isotropic; thin filament actin-containing) bands (figure 8.1(a)) (Huxley and Straub 1958, Franzini-Armstrong and Porter 1964, Franzini-Armstrong and Jorgensen 1994). Thus, at the typical resting sarcomere length of about 2 um per sarcomere, the TTs are about 1 um apart. Importantly for muscle Ca^{2+} signaling, the voltage-dependent Ca^{2+} channels ($Ca_V1.1$ channels, aka dihydropyridine receptors, DHPRs) are located within the TT membrane, directly apposed to the sarcoplasmic reticulum ryanodine receptor (RyR1) Ca^{2+} release channels (figure 8.1(b) and (c))

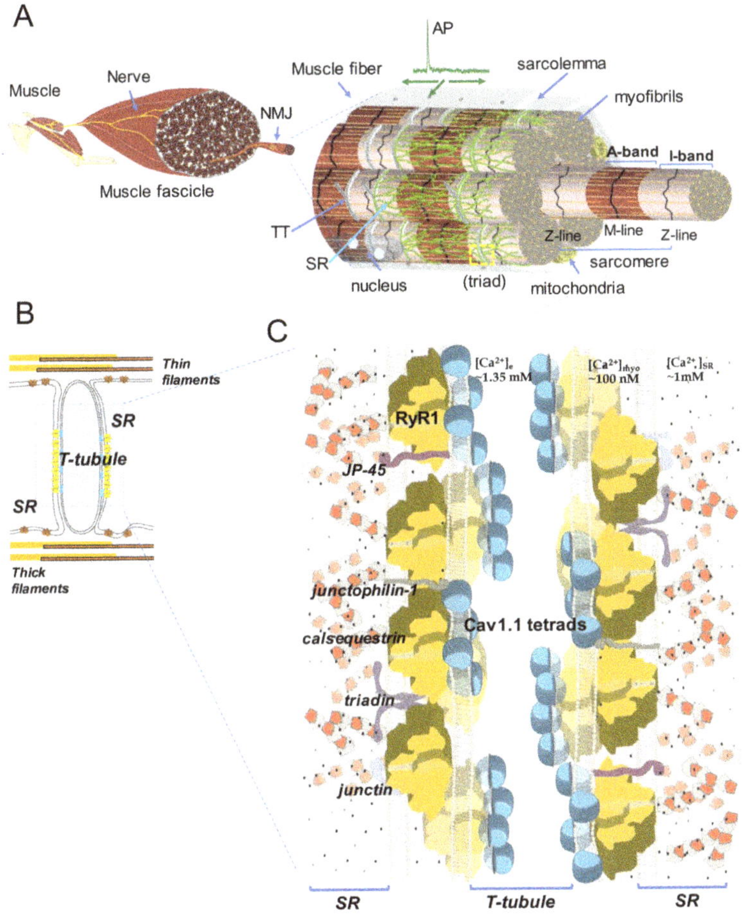

Figure 8.1. Skeletal muscle structure and subcellular organization. (a) Cartoon of a skeletal muscle visualized at different magnifications, from whole muscle and muscle fiber down to the sarcomere, the functional contractile unit of muscle. The muscle fiber is activated by an AP conducted via the nerve of a spinal motor neuron. The motor nerve AP causes acetylcholine (Ach) release at the end-plate of the NMJ. This is followed by the activation of Ach receptors located in the muscle fiber which in turn trigger a skeletal muscle AP that travels longitudinally along the sarcolemma and radially into the transverse tubule (TT) system. (b) Transverse zoom-in of the yellow boxed region shown in A at the level triad. $Ca_V1.1$s are located in the TT membrane, while the RyR1 are juxtaposed in the SR membrane. Each TT is flanked by two terminal cysternae of the SR forming a triad. An AP activates voltage sensors ($Ca_V1.1$ channels) located in the TT system which are coupled to and activate Ca^{2+} release channels (RyR1) located in the intracellular sarcoplasmic reticulum (SR). The $Ca_V1.1$-dependent activation of RyR1 elicits a massive release of Ca^{2+} from the SR Ca^{2+} stores that results in the activation of the contractile filaments to ultimately produce fiber contraction and movement in a process known as excitation–contraction coupling (ECC). (c), Zoom-in of the blue boxed region shown in B depicting $Ca_V1.1$ tetrads-RyR1 interactions and auxiliary proteins. Junctophilins (JPH1 and JPH2) maintain the triad geometry via TT and SR ultrastructural interactions and bind to and regulate both RyR1 and $Ca_V1.1$ (Ito *et al* 2001, Landstrom *et al* 2014, Perni *et al* 2017). Triadin and junctin bind to RyR1 and connect it to intraluminal SR calsequestrin, a SR Ca^{2+}-binding protein that regulates SR Ca^{2+} release (Zhang *et al* 1997). JP-45 connects $Ca_V1.1$ directly to calsequestrin (Anderson *et al* 2003). Reprinted from Klein *et al* 1988, copyright (1988) with permission from Elsevier.

(Andersson-Cedergren 1959, Peachey 1965a, 1965b, Franzini-Armstrong and Nunzi 1983, Rios and Brum 1987, Franzini-Armstrong and Kish 1995), which puts the release sites in close proximity (within about ½ um) of the contractile filaments where the released Ca^{2+} binds to regulatory proteins to remove the inhibition of thick and thin filament interaction (Ebashi 1974, 1976). This allows thick filament myosin heads to now interact with thin filament actin sites for force production and muscle fiber shortening (Huxley 1971, 1974, 1975). In contrast to cardiac muscle, skeletal muscle Ca^{2+} entry into the cytoplasm via TT $Ca_V1.1$ channels is not required for muscle activation (Armstrong *et al* 1972, Dayal *et al* 2017). Instead, $Ca_V1.1$ molecular movement in response to TT membrane potential changes serves as voltage (V) sensor (Schneider and Chandler 1973, Rios and Brum 1987) for regulation of RyR1 Ca^{2+} release channels in the sarcoplasmic reticulum the subcellular Ca^{2+} storage location in skeletal muscle (Porter and Palade 1957, Hasselbach 1964, Peachey 1965a).

8.3 Muscle fiber Ca^{2+} transients

The measurement of myoplasmic free Ca^{2+} concentration represents a rich tradition of biophysical measurements and analyses (Rudolf *et al* 2003). Skeletal muscle requires the transfer of a large number of Ca^{2+} ions from their intracellular storage location within the SR to the muscle fiber sarcoplasm for binding to thin filament regulatory binding sites on troponin C. The accompanying relatively large sarcoplasmic Ca^{2+} transient has provided an experimental system for developing techniques for monitoring intracellular free Ca^{2+} concentration ($[Ca^{2+}]$) using indicator dyes or photo-proteins that alter their optical properties when they bind Ca^{2+} (Ridgway and Ashley 1967, Kovacs *et al* 1979, Baylor *et al* 1983, Melzer *et al* 1984). Use of Ca^{2+} indicator dyes for monitoring the time course of free myoplasmic Ca^{2+} ($[Ca^{2+}]$) required: (a) developing dye molecules that alter their optical properties when they bind Ca^{2+}; (b) introducing the dye into muscle fibers; (c) recording the change in the dye signal in response to fiber depolarization; and (d) calculating the $[Ca^{2+}]$ time course from the measured dye signal. In addition, it was of paramount interest for understanding membrane potential control of SR Ca^{2+} release to: (e) determine the time course of Ca^{2+} bound to myoplasmic Ca^{2+} binding sites. Calculation of the time derivative of the time course of total myoplasmic Ca^{2+}, both free and bound to all myoplasmic Ca^{2+} binding sites; and (f) provide a means of calculating the rate of release of Ca^{2+} from the SR that underlies a Ca^{2+} transient.

8.3.1 Ca^{2+} indicator dyes used in muscle fibers

Early studies utilized the indicator dyes murexide (Eusebi *et al* 1980, Ogawa *et al* 1980), arsenazo III (Miledi *et al* 1977, Baylor *et al* 1983) or antipyrylazo III (Kovacs *et al* 1979, Ogawa *et al* 1980, Rios and Schneider 1981, Palade and Vergara 1982) that gave a very small change in absorbance spectrum in response to Ca^{2+} binding (figure 8.2(a)). Detection of such small signals required highly stabilized light sources and very low levels of mechanical vibration in the illumination and recording system. These dyes had low affinity for Ca^{2+} due to a relatively high off rate of bound Ca^{2+}, which contributed to the small signal amplitude, but provided high

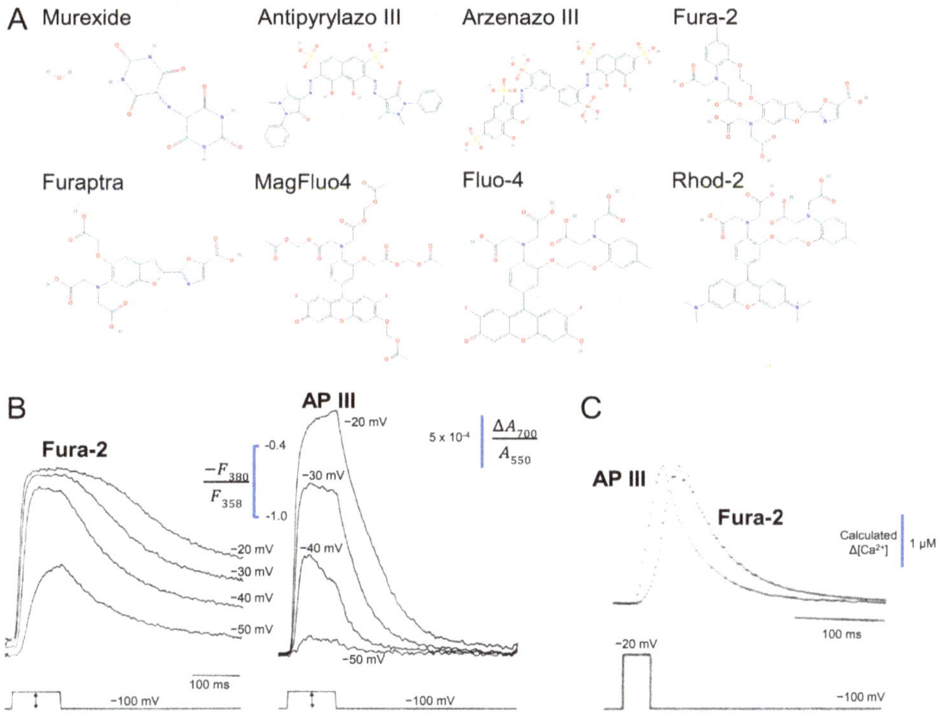

Figure 8.2. Ca^{2+} indicators and Ca^{2+} signals in skeletal muscle ECC. (a) 2D structure depiction of low- and high-affinity Ca^{2+} indicators used in studies of ECC and Ca^{2+} signaling in skeletal muscle (chemical structures were obtained from: https://pubchem.ncbi.nlm.nih.gov/). (b) Comparison of simultaneously recorded time course of Δ-ratiometric signal from high-affinity Ca^{2+} indicator fura-2 (left) and Δ-absorbance signal from low-affinity Ca^{2+} indicator antipyrylazo III (AP-III, right). Note that for depolarizations of increasing amplitude (between −40 and −20) fura-2 signals displayed saturation whereas AP-III absorbance was far from saturation. (c) Comparison of the time course of [Ca^{2+}] transient calculated from simultaneous records of AP-III absorbance and fura-2 fluorescence elicited by a 30 ms voltage step to −20 mV. Note that the fura-2 [Ca^{2+}] transient was normalized to match the peak of the AP-III signal to allow for a comparison of their kinetics. Data derived from a frog skeletal muscle fiber cut at both ends and voltage clamped using a double vaseline gap system. Reprinted from Klein *et al* 1988, copyright 1988, with permission from Elsevier.

temporal fidelity, with the absorbance signal matching the time course of sarcoplasmic [Ca^{2+}] (see below).

Subsequently, Dr Roger Tsien introduced a class of ratiometric fluorescent Ca^{2+} indicator dyes based on conjugating a fluorophore to the Ca^{2+} buffer BAPTA (Grynkiewicz *et al* 1985), which he had developed earlier (Tsien 1980). His indicator fura-2 is a relatively high-affinity Ca^{2+} binding molecule that provided a relatively large fluorescence change signal for small elevations of [Ca^{2+}] (figure 8.2(b), left). Thus, fura-2 could be used for monitoring resting [Ca^{2+}] and small relatively slow elevations of [Ca^{2+}], but with the disadvantage that the signal saturates at higher [Ca^{2+}] as achieved in muscle fibers during depolarization (figure 8.2(b), left). In addition, fura-2 provides a relatively slow response in comparison to a muscle

[Ca^{2+}] transient (figure 8.2(c)). Subsequentl,y Dr Tsien developed a lower affinity, and more rapidly equilibrating Ca^{2+} indicator dye furaptra, which much more faithfully tracked the time course of myoplasmic [Ca^{2+}] in muscle (Konishi *et al* 1991).

8.3.2 Introducing Ca^{2+} indicator dyes into muscle fibers

The typical Ca^{2+} indicator dyes are multiply charged and not membrane permeant. The salt form of the dye could be introduced into a single muscle fiber by pressure injection from an intracellular 'sharp' electrode pipette inserted into a muscle fiber (Konishi and Baylor 1991). In an alternative approach a single muscle fiber could be mounted in a vaseline gap chamber, and one or both ends of the fiber could be cut or permeabilized beyond the gap(s) allowing the dye included in the end pool solution to enter the fiber in the recording pool by diffusion (Kovacs *et al* 1979). These procedures for dye introduction involved variable extent of fiber disruption. As an alternative approach, charged carboxyl groups on the Ca^{2+} indicator molecules were esterified forming uncharged acetoxy methyl ester or so called 'AM' form of the dyes (Tsien 1981). These Ca^{2+} insensitive AM esters cross the plasma membrane, and are de-esterified by endogenous esterases present in the muscle fiber myoplasm, thereby regaining their Ca^{2+} sensitivity. Use of the AM form of the dye provides a fully non-invasive procedure for dye introduction into a muscle fiber. One caveat regarding this approach is that some AM dye may also cross intracellular membranes and enter intracellular compartments before de-esterification, thereby contaminating the cytoplasmic signal with a signal from another compartment (Tsien 1981). However, this may also be taken advantage of in terms of monitoring the Ca^{2+} concentration within intracellular organelles, in particular the muscle fiber SR (Kabbara and Allen 2001, Launikonis *et al* 2005), from which Ca^{2+} ions are released and in which luminal [Ca^{2+}] could consequently decrease during Ca^{2+} release.

8.3.3 The [Ca^{2+}–dye] signal during fiber depolarization

Ca^{2+} binding to a Ca^{2+} indicator dye causes a change in one or more optical properties of the dye that can be used for calculating the time course of the Ca^{2+}–dye complex, and from that the [Ca^{2+}] time course as discussed below (Tsien 1980, Rudolf *et al* 2003). However, fiber force generation can cause mechanical movement and corresponding movement artifacts in the optical signal unless appropriate precautions are taken. The major initial approach for eliminating optical artifacts of fiber movement was to stretch a muscle fiber to unphysiologically long sarcomere lengths where force production is no longer generated despite elevated myoplasmic [Ca^{2+}] due to lack of overlap of thick and thin myofilaments. Optical signals for illumination over 10s of μm of the fiber then provided average myoplasmic [Ca^{2+}–dye] signals induced by action potentials, trains of action potentials or voltage-clamp depolarizations of frog muscle fibers (Baylor *et al* 1983, Kovacs *et al* 1983).

An alternative approach to avoiding optical artifacts of fiber movement, which our lab has more recently utilized with mouse intact muscle fibers at physiological sarcomere length, is to monitor fluorescence signals along a line scan across a muscle fiber. Using a high-speed confocal line scan system (Zeiss LSM 5 LIVE) on an inverted microscope we acquire time courses at multiple points along the line, both within and outside a single muscle fiber, which is attached non-uniformly along its length to the coated glass coverslip bottom of the chamber, but can still undergo variable extents of predominantly lateral movement (Prosser *et al* 2008, 2009, 2010, Olojo *et al* 2011a, 2011b, Robison *et al* 2011, Hernández-Ochoa and Schneider 2012, Hernández-Ochoa *et al* 2014, 2016). By carefully choosing signals, only from a sub-segment of the scan line that remains clearly within the fiber at all times during fiber depolarization and repolarization, we are able to obtain a signal of average myoplasmic Ca^{2+} indicator fluorescence despite the fiber movement accompanying fiber activation and relaxation. This approach gives an average cytoplasmic signal, and like the large spot approach used earlier with stretched fibers, records an average myoplasmic [Ca^{2+}–dye] (Rios and Schneider 1981, Kovacs *et al* 1983, Melzer *et al* 1986, 1987, Klein *et al* 1988). If the selected line sub-segment does not remain within the fiber during contraction, then movement artifacts obscure the later part of the [Ca^{2+}–dye] signal.

The development and evolution of the procedures described above and below involved contributions from multiple laboratories over several decades. For convenience we will here utilize records obtained in our laboratory with the high-speed line scan method (preceding paragraph) to illustrate the general approach and conclusions. For a single muscle fiber action potential AP, the Ca^{2+}–dye signal for a rapidly equilibrating Ca^{2+} indicator consists of a few ms rise and fall that follows the AP depolarization by a few ms, and considerably precedes force generation (figures 8.3(a)–(c), left column). For repetitive trains of APs, Ca^{2+} dye signals, illustrated here for the low-affinity indicator MagFluo4, first remain discrete with increasing frequency (here 20 Hz, figure 8.3(b) middle) and then begin to partially fuse with further increased frequency (here 100 Hz; figure 8.3(b), right). Thus, Ca^{2+} remains relatively high but with oscillation around the mean level, during the 100 Hz train (figure 8.3(b), right). However, the peak [Ca^{2+}] does not exhibit major 'creep up', as might be expected for repeated additions of a constant amount of Ca^{2+} to the myoplasm with each AP. This foreshadows the finding that SR Ca^{2+} release is relatively rapidly inactivated, with considerable inactivation produced even by a single AP (see below). For the same patterns of repetitive APs, force measured in parallel experiments (Iyer *et al* 2016) partially fuses at 20 Hz (figure 8.3(c) middle) and exhibits a smooth tetanic rise during 100 Hz stimulation (figure 8.3(c) right).

8.3.4 Calculating the free [Ca^{2+}] time course from the measured [Ca^{2+}–dye] signals

For a given wavelength excitation and emission, dye fluorescence (F) changes linearly in proportion to the fractional occupancy of the indicator dye by Ca^{2+}, going from fluorescence F_{min} at zero Ca^{2+} bound dye to F_{max} at fully Ca^{2+} bound dye (Tsien 1980, Rudolf *et al* 2003). F, F_{max} and F_{min} are all proportional to dye

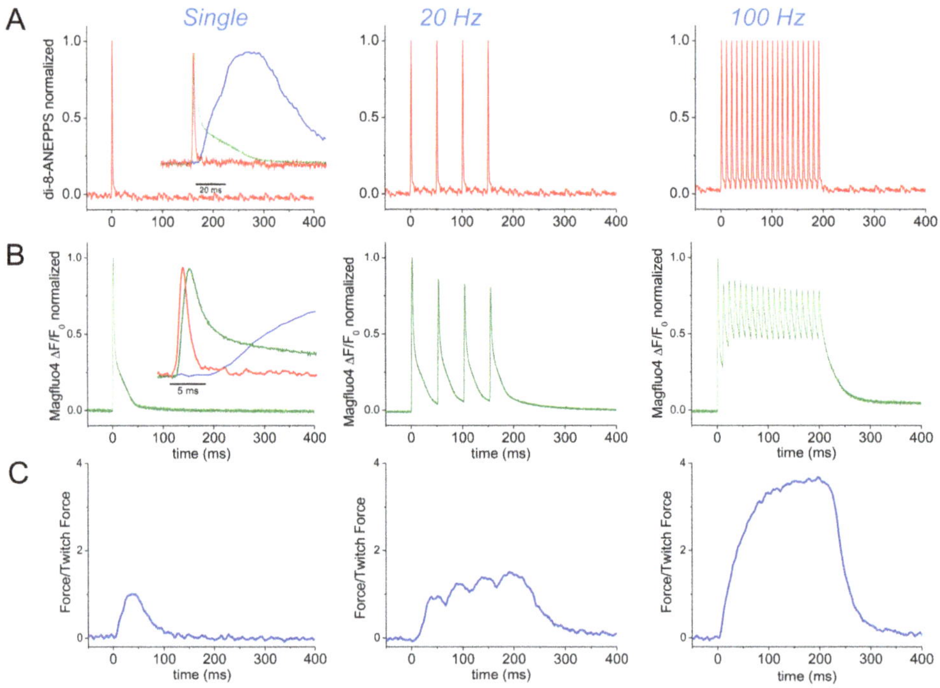

Figure 8.3. Comparison of the time course of the action potential (AP) (a), Ca^{2+} transients (b) and muscle force (c) for a single external stimulus and for 200 ms trains of external stimuli at 20 Hz (center) and 100 Hz (right) measured independently. Insets in A and B show AP, Ca^{2+} transient and force traces for the single external stimulus on different expanded time scales (data from Schneider and Hernandez-Ochoa 2012).

concentration, path length, illumination area and other instrumental factors. However, if the dye changes either its absorbance or emission spectra when Ca^{2+} is bound, then that dye is considered a 'ratiometric' dye and the ratio R of signals recorded at two different wavelengths can be used. R changes linearly with dye saturation by Ca^{2+}, from R_{min} at zero Ca^{2+} bound dye to R_{max} at fully Ca^{2+} saturated dye. It provides the important advantage that the values of R, R_{min} and R_{max} are independent of dye concentration, path length or illuminated area, which cancel out in taking the ratios. The $F(t)$, F_{min} and F_{max} values, or the $R(t)$, R_{min} and R_{max} values provide direct experimental measurement of the time course of the fraction of dye bound to Ca^{2+}, since the optical change is essentially instantaneous with formation of the Ca^{2+}–dye complex on the time scale of the muscle $[Ca^{2+}]$ transients. The fluorescence expression for fraction of the dye bound to Ca^{2+} must be set equal to a kinetic expression for the fraction of dye bound by Ca^{2+} for the Ca^{2+} dye binding reaction. If the change in $[Ca^{2+}]$ is sufficiently slow that free $[Ca^{2+}]$ is in equilibrium with the Ca^{2+} indicator dye, the equilibrium equation for $[Ca^{2+}\text{–dye}]$/[dye total] would be used for calculating myoplasmic free$[Ca^{2+}]$, giving

$$[Ca^{2+}] = K_D(F - F_{min})/(F_{max} - F) \tag{8.1}$$

for single-wavelength recording (i.e., using a non-ratiometric dye), where F_{min} and F_{max} have to be experimentally determined or estimated in each fiber (Grynkiewicz *et al* 1985). Using a ratiometric dye, the analogous equilibrium equation for ratiometric recording is

$$[Ca^{2+}] = K_D\beta(R - R_{min})/(R_{max} - R) \tag{8.2}$$

where the constant β is the value of F_{max}/F_{min} at the reference wavelength and R_{min} and R_{max} are constants (Grynkiewicz *et al* 1985). If the reference wavelength is the isosbestic wavelength, then the value of β is 1 (Grynkiewicz *et al* 1985, Klein *et al* 1988).

Alternatively, if the rate of Ca^{2+} binding/unbinding to the dye is not sufficiently rapid to reach instantaneous equilibration with the free $[Ca^{2+}]$ time course, as is the case for high-affinity Ca^{2+} indicator dyes and the skeletal muscle fiber AP induced Ca^{2+} transient, then a kinetic correction is required. Here the differential equation for Ca^{2+} binding to dye is used to express the time course of $[Ca^{2+}\text{–dye}]$, giving

$$[Ca^{2+}] = K_D((1/k_{off})(dF/dt) + (F - F_{min}))/(F_{max} - F) \tag{8.3}$$

for non-ratiometric recording (Klein *et al* 1988). This and the analogous equation for ratiometric recording can be solved numerically by calculating dF/dt from the recorded fluorescence time course (Klein *et al* 1988). No such kinetic correction term is needed for slowly changing or resting signals, where the $d\mathrm{F}/dt$ term is negligible, or at the time of peak F, since at that time dF/dt is zero. However, rising and falling phases of fluorescence transients recorded with relatively slowly equilibrating dyes such as fura-2 do require correction for the kinetic delay in order to accurately track muscle fiber $[Ca^{2+}]$ transients (Klein *et al* 1988). Note that K_D is just a scale factor in equations (8.1)–(8.3), so its value does not affect the relative time course of the calculated $[Ca^{2+}]$ transient.

8.3.5 Ca^{2+} binding to TnC

In active muscle, force and shortening is produced as a result of the ATP dependent interaction of cross bridges (myosin heads) on the muscle thick filaments with their binding sites on actin molecules in the adjacent thin filaments (Huxley 1974, 1975, 1995). In contrast, in resting muscle, the myosin cross bridges on the thick contractile filaments are blocked from interacting with their binding sites on the actin molecules in the adjacent thin filaments, so the filaments cannot interact, and the muscle is maintained in a mechanically relaxed state (Huxley 1995). Prevention of mechanical activation in resting muscle is achieved by thin filament tropomyosin molecules blocking access of thick filament myosin cross bridges to their thin filament actin binding sites (Ebashi 1974, 1976, Huxley 1995). A rise in myoplasmic free $[Ca^{2+}]$ serves as the signal to turn off the tropomyosin inhibition of myosin/actin interaction and allow mechanical activity (Ebashi 1984, Moss 1992). The Ca^{2+} sensitive switch is provided by troponin C (TnC) molecules residing on the thin filament (Grabarek *et al* 1992, Gergely *et al* 1993).

Skeletal muscle troponin C (TnC) is an approximately 20 KD protein having four 'EF hand' type Ca^{2+} binding sites (Parmacek and Leiden 1991, Grabarek *et al* 1992). Two sites are 'low-affinity' sites that bind only Ca^{2+} at the [Ca^{2+}] and [Mg^{2+}] levels encountered in muscle, whereas the other two sites have higher affinity, and bind Mg^{2+} at resting [Ca^{2+}] and [Mg^{2+}] levels in skeletal muscle (Baylor *et al* 1983, Grabarek *et al* 1992). During elevated [Ca^{2+}], as in response to an AP, the Mg^{2+} bound to the high-affinity sites exchanges for Ca^{2+}, but is not believed to provide a regulatory change since in either case the sites are occupied by a divalent cation (Grabarek *et al* 1992, Moss 1992, Gergely *et al* 1993). In contrast, the transition from Ca^{2+} free to Ca^{2+} bound status at the regulatory sites when [Ca^{2+}] rises causes a conformational change in TnC, which is transmitted to tropomyosin, resulting in a rearrangement of tropomyosin on the thin filament, resulting in the availability of the thin filament actin site for myosin binding, force and shortening (Ebashi 1984, Grabarek *et al* 1992, Moss 1992, Gordon and Ridgway 1993, Gergely *et al* 1993). In parallel, and by a slower enzymatic reaction, the speed and efficiency of the myosin ATPase, and thus the cross bridge mechanochemical cycle kinetics, is modulated by the Ca^{2+}-calmodulin dependent activation of myosin light chain kinase and the resulting phosphorylation of myosin light chains (Manning and Stull 1979).

8.3.6 Time course of Ca^{2+} binding to thin filament TnC in response to a muscle action potential

Based on the preceding paragraph, an important question in ECC is what is the time course of Ca^{2+}-binding to thin filament TnC regulatory sites during the [Ca^{2+}] transient initiated by a single action potential? Reversible binding of Ca^{2+} to each of the regulatory Ca^{2+} binding sites on thin filament TnC follows the reaction scheme,

$$Ca^{2+} + \mathrm{TnC} \leftrightarrow Ca^{2+}-\mathrm{TnC} \tag{8.4}$$

Thus the rate of change of Ca^{2+} bound to the TnC regulatory sites (Ca^{2+}–TnC) follows the differential equation

$$d[Ca^{2+}-\mathrm{TnC}]/dt = (k_{on}\mathrm{TnC})[Ca^{2+}][\mathrm{TnC}] - (k_{off}\mathrm{TnC})[Ca^{2+}-\mathrm{TnC}] \tag{8.5}$$

Using the measured time course of [Ca^{2+}] (above), together with biochemically determined values for the *on* and *off* rate constants k_{on}TnC and k_{off}TnC, respectively, for Ca^{2+} binding and unbinding to TnC regulatory sites, and the total TnC concentration in a muscle fiber, numerical integration of equation (8.5) can be carried out to give the time courses of the concentrations of free and Ca-bound TnC regulatory sites (Baylor *et al* 1983). Interestingly, the Ca^{2+} transient for a single action potential results in a peak value of the fraction (f) of TnC regulatory sites occupied by Ca^{2+}, approaching 1 (i.e., full occupancy of the regulatory sites by Ca^{2+}) (Baylor *et al* 1983, Baylor and Hollingworth 2012). If the two regulatory sites must both be occupied for fiber activation, factional activation would be equal to f^2, which was only slightly smaller than f after a single AP.

8.3.7 Calculating the rate of release of Ca^{2+} from the SR during fiber depolarization

Comparison of the values of [Ca^{2+}] and [Ca^{2+}–TnC] induced by a single action potential indicates that the amount of Ca^{2+} in the free [Ca^{2+}] transient represents only a small fraction of the total amount of Ca^{2+} released from the SR in response to the action potential (Baylor and Hollingworth 2003). A much larger amount of released Ca^{2+} is not free in the muscle fiber water but is instead bound to numerous Ca^{2+} binding sites, including those on TnC, within the muscle fiber volume. In addition to the TnC regulatory sites, the SR Ca^{2+} pump and ATP provide the major rapidly equilibrating Ca^{2+} binding sites for released Ca^{2+} in the muscle fiber (Timmer *et al* 1998, Baylor and Hollingworth 2003).

Binding to each of these Ca^{2+} specific sites can be calculated using a similar approach as illustrated above for the TnC regulatory sites (Timmer *et al* 1998, Baylor and Hollingworth 2003). In addition, the Ca^{2+}/Mg^{2+} sites on TnC, as well as the sites on Parvalbumin (Parv) are predominantly Mg^{2+} bound in the resting fiber, but bind Ca^{2+} in exchange for Mg^{2+} during periods of elevated [Ca^{2+}]. Using the *on* and *off* rate constants for Ca^{2+} (and for Mg^{2+}, if applicable) binding to each of these sites together with the experimentally measured time course of [Ca^{2+}], the time course of total Ca^{2+} ($[Ca^{2+}]_T$), free and bound to all these sites outside the SR can be calculated (Timmer *et al* 1998, Baylor and Hollingworth 2003). The rate of Ca^{2+} release from the SR is then calculated as $d[Ca^{2+}]_T/dt$. As an extension of this procedure, additional exogenous Ca^{2+} buffers can be introduced into the myoplasm (Shirokova *et al* 1996, Struk *et al* 1998). Ca^{2+} binding to these additional sites can be calculated from the *on* and *off* rate constants for Ca^{2+} binding to the added sites, and their contribution can be added to the calculated Ca^{2+} bound to intrinsic sites and free in the myoplasm (Shirokova *et al* 1996, Struk *et al* 1998, Royer *et al* 2008, Prosser *et al* 2009).

8.3.8 Ca^{2+} release activation and inactivation during fiber depolarization

The procedures for monitoring [Ca^{2+}] transients and calculating their underlying Ca^{2+} release waveform described in the preceding sections were developed and used by several groups studying ECC, first in frog skeletal muscle fibers and subsequently in rat and mouse fibers (Miledi *et al* 1977, Kovacs *et al* 1979, Rios and Schneider 1981, Baylor *et al* 1983, Kovacs *et al* 1983, Melzer *et al* 1984, 1986, 1987, Schneider *et al* 1985, 1987, Klein *et al* 1988, Simon and Schneider 1988, Lamb and Stephenson 1991, Simon *et al* 1991, Simon and Hill 1992, Garcia and Schneider 1993, Shirokova *et al* 1996, Royer *et al* 2008, Prosser *et al* 2009, 2010, Olojo *et al* 2011a, 2011b). Here we again use fluo-4 records from our lab for illustrative purposes. A single AP causes a few ms burst of Ca^{2+} release (figure 8.4(a), bottom), which generates a similarly brief [Ca^{2+}] transient (figure 8.4(a), top), see also (Baylor and Hollingworth 2003). Repeating the APs at 10 ms intervals (100 Hz) causes the Ca^{2+} release to decline markedly from the first to second AP, even though the resulting [Ca^{2+}] fluctuations slightly increment with successive APs in this fiber (figure 8.4(b), top). In other fibers and using prolonged trains of APs, [Ca^{2+}] exhibits relatively constant fluctuations around the

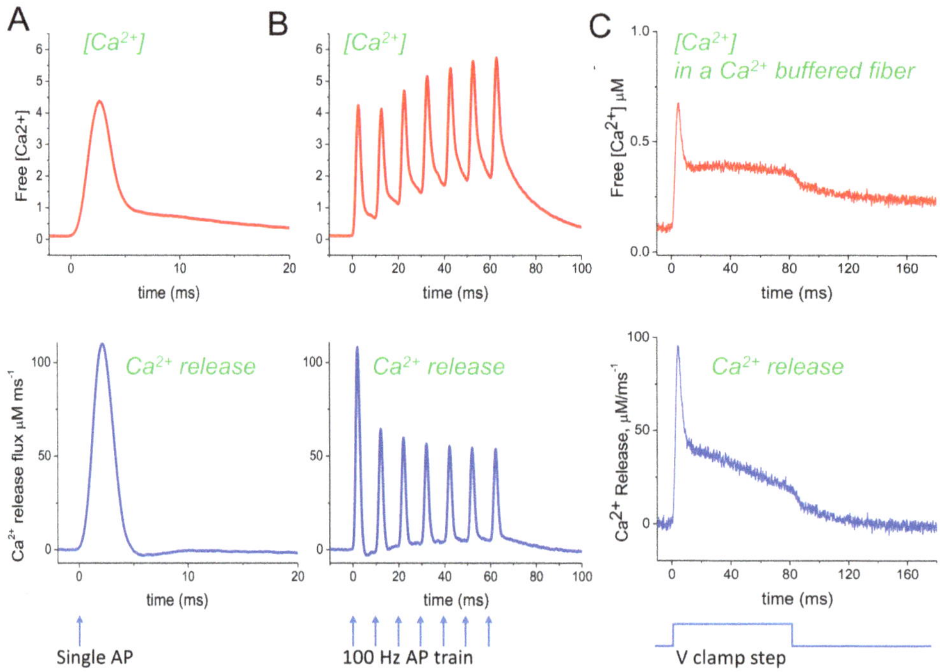

Figure 8.4. Time course of $\Delta[Ca^{2+}]$ and Ca^{2+} release flux evoked by APs or a voltage-clamp depolarization. (a and b) Time course of the $\Delta[Ca^{2+}]$ (top) and Ca^{2+} release flux (bottom) derived from an intact skeletal muscle fiber loaded with fluo-4 AM in response to external stimulation that elicited either a single AP (a) or a 60 ms 100 Hz APs train (b) as described by Timmer *et al* (1998) and Prosser *et al* (2010). Arrows indicate the application of external stimuli. (c) Time course of $\Delta[Ca^{2+}]$ and Ca^{2+} release flux elicited by a 80 ms voltage step depolarization to +20 mV. Records obtained in the whole-cell configuration and derived from a fiber loaded with rhod-2 and EGTA/Ca^{2+} via the patch pipette. Ca^{2+} release waveforms were calculated as described by Olojo *et al* (2011a).

mean level, which maintains the level of contractile activation (figure 8.3(b) right). The decline in release with successive pulses in a 100 Hz train (figure 8.4(b)) is a consistent finding in frog (Baylor *et al* 1983) and rodent muscle (Baylor and Hollingworth 2003, Prosser *et al* 2010) that implies an inactivation of the Ca^{2+} release process with time during the train of APs. Alternatively, using a voltage-clamp step depolarization and the Ca^{2+} indicator rhod-2 shows that Ca^{2+} release declines markedly during constant depolarization (figure 8.4(c), bottom; (Melzer *et al* 1984, Prosser *et al* 2009)). Note that figure 8.4(c) was carried out using exogenous Ca^{2+} buffer introduced into the myoplasm via a whole-cell patch pipette, so the $[Ca^{2+}]$ time course is distorted by the presence of Ca^{2+} buffer (figure 8.4(c), top), but the Ca^{2+} release (figure 8.4(c), bottom) is calculated taking account of the exogenous myoplasmic $[Ca^{2+}]$ buffer, and is thus well approximated.

The Ca^{2+} release waveform declines in two distinct phases during a V clamp depolarization (figure 8.4(c), bottom; see also figure 8.5(a), P1). The faster phase,

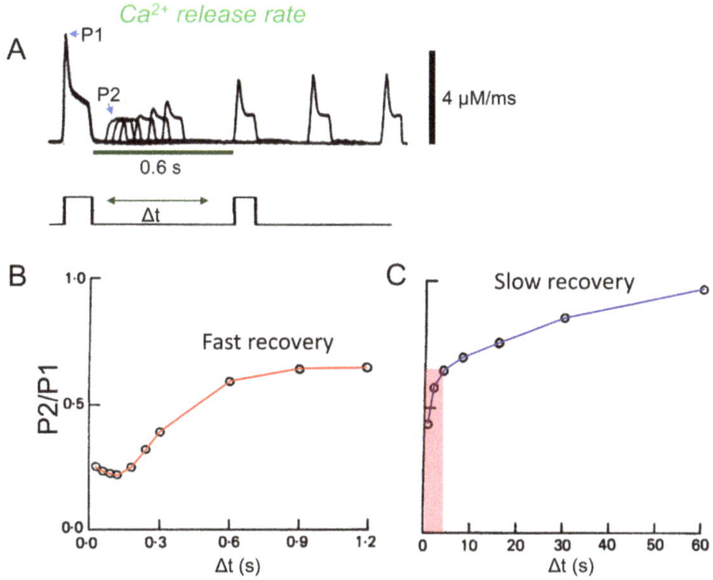

Figure 8.5. Inactivation and recovery of SR Ca^{2+} release in amphibian skeletal muscle. (a) Ca^{2+} release records from a double-pulse protocol studying the recovery of the Ca^{2+} release following a conditioning prepulse. A 120 ms conditioning prepulse to −20 mV was followed by a test pulse to the same amplitude after a variable interval (*t*). P1 is the peak of release in the conditioning prepulse and P2 is the peak of release in the test pulse. Note the absence of the fast early peak in the rate of Ca^{2+} release for short recovery intervals (data from Schneider and Simon 1988). Fast recovery (b) and slow phase recovery (c) of the peak rate of Ca^{2+} release following a conditioning depolarization. The ratio of the peak of Ca^{2+} release for the test pulse (P2) to peak rate of release for the conditioning pulse (P1) is plotted as function of time. Note the compressed time scale in (c) compared with (b) (data from Schneider *et al* 1987). The red box in panel C cover the fast recovery from panel (b). Reprinted from Schneider et al (1988) John Wiley & Sons, copyright 1988 The Physiological Society.

which is completed within the first 10s of ms during a large step depolarization appears to be due to Ca^{2+} dependent inactivation of the RyR1 Ca^{2+} release channel, which occurs relatively early during the depolarization-induced elevated [Ca^{2+}] (Schneider *et al* 1987, Schneider and Simon 1988, Simon *et al* 1991, Ma *et al* 2003). The slower phase of decline of Ca^{2+} release during maintained depolarization has been attributed to Ca^{2+} depletion from the SR during continued release (Schneider *et al* 1987) but other possibilities have also been suggested (Rios *et al* 2006, Royer *et al* 2008, Sztretye *et al* 2011). Double-pulse recovery experiments show two phases of recovery from suppressed release after a conditioning depolarization (Schneider *et al* 1987). The fast phase of decline of Ca^{2+} release during depolarization recovers relatively rapidly with recovery time after repolarization (figure 8.5(b)) and is complete within less than 1 s in frog fibers at reduced temperature. Recovery of the overall release takes 10s of s in the same fibers (figure 8.5(c)) and represents a recovery from the slow phase of decline of release that occurred during the conditioning pulse.

8.4 Sub-sarcomeric Ca^{2+} gradients within the macroscopic signals?

Thus far we have assumed spatially uniform [Ca^{2+}] transients. However, the Ca^{2+} release units are positioned at discrete regular positions along the fiber (figure 8.1(c)), which could lead to [Ca^{2+}] gradients during release activation (Baylor and Hollingworth 2011, 2012, Hollingworth *et al* 2012, Hollingworth and Baylor 2013). The repeating sarcomeric pattern along an amphibian or mammalian muscle fiber has a repeat length of about 2 μm in a fiber at rest length, corresponding to the Z disk-to-Z disk or M band-to-M band repeats (figure 8.1(a)). The triads, which are formed by each TT together with its two neighboring SR terminal cisternae (figure 8.1(b) and (c)), are the sarcomeric sources for SR Ca^{2+} release. Triads are regularly positioned as a 'belt' wound around the myofibril in register with the sarcomeric repeat (figure 8.1(a)) and provide Ca^{2+} for TnC binding by both longitudinal and radial diffusion away from the triad (figure 8.1(a)–(c)). In frog muscle the TTs are present at each Z line, and thus at 2 μm from each other along the fiber length (Block *et al* 1988, Franzini-Armstrong and Jorgensen 1994) . In rodent muscle the triads are positioned at the boundary of the A and I bands, and thus at a longitudinal separation of only about 1 μm in rest length muscle (Block *et al* 1988, Franzini-Armstrong and Jorgensen 1994). Since each triad only needs to supply Ca^{2+} to half the distance to its neighboring triads, the frog triad supplies 1 μm in each longitudinal direction, whereas the rodent triad supplies only 0.5 μm in each direction at resting fiber length. The question as to whether [Ca^{2+}] gradients exist between neighboring triads during periods of high Ca^{2+} release has been examined both experimentally (Zoghbi *et al* 2000, Gomez *et al* 2006) and theoretically (Zoghbi *et al* 2000, Baylor and Hollingworth 2007, Holash and MacIntosh 2019).

8.5 Unitary Ca^{2+} release events

Local, brief and stochastic elevations of myoplasmic Ca^{2+}, termed 'Ca^{2+} sparks', are readily measurable in mammalian cardiomyocytes as well as in amphibian skeletal muscle fibers using fluo-4, a Ca^{2+} indicator dye having very low resting fluorescence at low [Ca^{2+}]. These signals represent underlying unitary calcium release events (Cheng *et al* 1993, Tsugorka *et al* 1995, Klein *et al* 1996, Klein and Schneider 2006). In both cardiac myocytes and frog skeletal muscle fibers such elementary events occur at much higher frequency during small depolarizations than in resting fibers. Under conditions of low likelihood of release activation, Ca^{2+} sparks have been shown to summate during larger depolarizations to form the global Ca^{2+} transient, with the timing of occurrence of sparks during a relatively large depolarization giving rise to the trajectory of the macroscopic Ca^{2+} transient (Klein *et al* 1996, Schneider and Klein 1996, Jiang *et al* 1999, Klein *et al* 1999, Schneider 1999). Unitary Ca^{2+} release events are believed to occur in mammalian skeletal muscle fibers, yet they have not been clearly resolved under physiological conditions (Baylor *et al* 2002), possibly because of being too small and/or too brief for detection with available imaging techniques (Klein and Schneider 2006). However, Ca^{2+} sparks are detected in developing mammalian skeletal muscle, disappear during the postnatal period and reappear in adult fibers undergoing dedifferentiation in culture, changes that parallel the radial organization and disorganization of the TT system, with corresponding formation and dissociation of TT–SR junctions in development and dedifferentiation, respectively (Conklin *et al* 1999, Shirokova *et al* 1999, Chun *et al* 2003, Zhou *et al* 2006, Brown *et al* 2007). In addition, Ca^{2+} sparks are detected in mammalian skeletal muscle fibers under conditions of extreme osmotic or mechanical stress (Wang *et al* 2005), which may disrupt the TT–SR coupling. These observations suggest that the occurrence of cardiac/frog skeletal-like Ca^{2+} sparks requires local groups of RyRs that are uncoupled from $Ca_V1.1$, which is the case for all cardiac RyRs and for the 'parajunctional' RyRs present in frog skeletal muscle (Felder and Franzini-Armstrong 2002, Pouvreau *et al* 2007). In contrast, when alternate RyRs are coupled to $Ca_V1.1$ (figure 8.1(c)), as is the case in adult mammalian skeletal muscle fibers under physiological conditions, the Ca^{2+} sparks may be smaller and/or short-term, and thus less readily detected. The type of Ca^{2+} sparks that appear in mammalian fibers when RyRs are uncoupled from $Ca_V1.1$ may also appear in muscle disease or degenerative states due to disruption of $Ca_V1.1$–RyR coupling (Wang *et al* 2005, Cheng and Lederer 2008).

8.6 local non-sarcomeric Ca^{2+} signals: $Ca_V1.1$ and perinuclear Ca^{2+} signals

Local Ca^{2+} signals in skeletal muscle fibers have been predominantly investigated as elementary Ca^{2+} release events from the sarcoplasmic reticulum. Interestingly, in skeletal muscle the nuclei are located at the periphery of the cell, with part of the surface area of each nucleus apposed to the sarcolemma, while the other part of the nuclear surface sees the myoplasmic space (Georgiev *et al* 2015). The perinuclear regions have shown the occurrence of nifedipine (L-type channel blocker)-sensitive and dantrolene (RyR1 inhibitor)-sensitive local Ca^{2+} signals in the perinuclear domains under hypertonic conditions. It has been suggested that these local perinuclear Ca^{2+} signals can target specific subnuclear regions and be isolated from the global RyR1-dependent Ca^{2+} signals and regulate subnuclear processes (Echevarria *et al* 2003, Georgiev *et al* 2015). These perinuclear local Ca^{2+} signals could be important in excitation–transcription coupling—the regulation of gene expression controlled by muscle electrical activity—in nuclear subdomains and contribute to adaptive responses. However, the specific contribution of calcium influx via $Ca_V1.1$ or RyR1 to local Ca^{2+} perinuclear domains as well as its role in skeletal muscle during physiological or pathological conditions requires further investigation.

8.7 $Ca_V1.1$: voltage sensors for ECC

Membrane depolarization of the TT system, during an AP (or via voltage-clamp depolarization), is detected by voltage sensors located within $Ca_V1.1$ channels and measured as charge movement (macroscopic gating arising predominantly from $Ca_V1.1$) (figure 8.6(b)). Skeletal muscle $Ca_V1.1$ voltage sensors were initially identified using electrophysiological approaches (Schneider and Chandler 1973, Adrian and Almers 1976, Chandler *et al* 1976, Stanfield 1977, Sanchez and Stefani 1978),

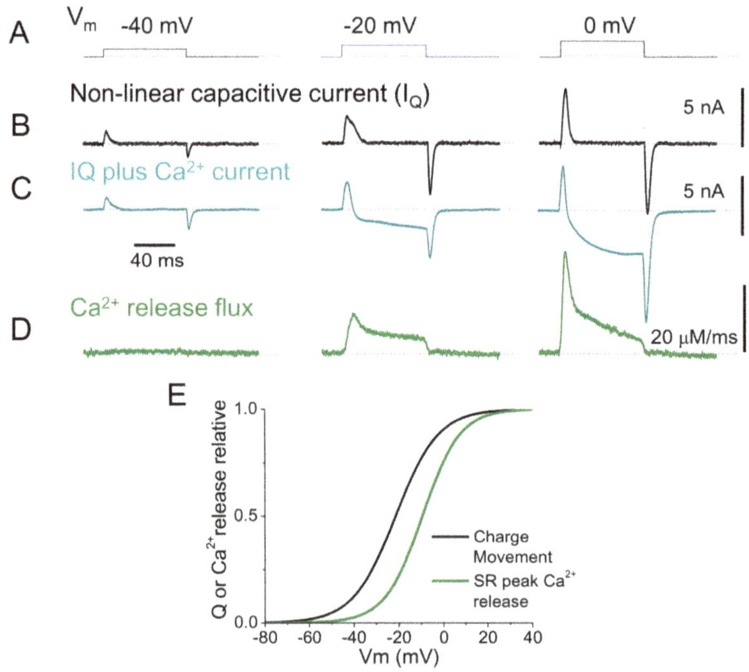

Figure 8.6. Voltage dependence of charge movement, Ca^{2+} currents and Ca^{2+} release flux in murine skeletal muscle. (a) voltage step depolarizations. (b) TT voltage sensor charge movement currents. (c) L-type Ca^{2+} currents together with charge movement currents. (d) SR Ca^{2+} release flux. Records obtained in whole-cell voltage clamped mouse fast skeletal muscle fiber. Non-linear capacitive currents and Ca^{2+} currents were recorded with the 'P/4' pulse protocol (Armstrong and Bezanilla 1974) to eliminate linear currents. Fibers were exposed to solutions containing an appropriate 'cocktail' of channel blockers to suppress ionic conductances (b) or in a recording solution with 2 mM Ca^{2+} and no block of Ca^{2+} channels (c). Ca^{2+} release flux calculated from the same fiber under the same conditions as in B. (e) Voltage dependence of charge moved by the TT voltage sensor (black curve) and peak Ca^{2+} release flux (green curve). Unpublished results from whole-cell voltage clamped FDB fiber with 20 mM EGTA, 11.5 mM Ca^{2+}, 6 mM Mg^{2+}, 10 mM ATP and 100 μM rhod-2 in the fiber interior to clamp intracellular $[Ca^{2+}]$ around 110 nM, suppress fiber contraction and calculate Ca^{2+} release flux as described by Olojo *et al* (2011a) and Hernández-Ochoa *et al* (2014). Reprinted from Schneider *et al* (1988) John Wiley & Sons, copyright 1988 The Physiological Society.

which was followed by biochemical, and molecular characterization of $Ca_V1.1$ (Beam *et al* 1986, Takahashi *et al* 1987, Tanabe *et al* 1987, 1988, 1990a, Takeshima *et al* 1989, Catterall 1991, Nakai *et al* 1998b).

$Ca_V1.1$ channels are principally expressed in the membrane of the TT system of adult skeletal muscle fibers and are members of a diverse family of voltage-dependent Ca^{2+} channels. While TTs of skeletal muscle exhibits the highest concentration of L-type Ca^{2+} channels, the L-type Ca^{2+} current is small (figure 8.6(c)), activates more slowly than Ca^{2+} release (figure 8.6(d)), and is not needed for muscle activation (Dayal *et al* 2017). In stark contrast, the presence of the $Ca_V1.1$ working as a voltage sensor is essential for ECC (Beam *et al* 1986, Knudson *et al* 1989).

Molecular details of the voltage-gated Ca^{2+} channels from skeletal muscle were first identified by binding, purification, and reconstitution assays (Takahashi and Catterall 1987). Using molecular biology techniques, their amino acid sequences were characterized by cDNA cloning and sequencing (Tanabe *et al* 1987). Contrasting with their prominent functional status, TT voltage sensor ($Ca_V1.1$ channels) were left behind from the saga of structure–function studies for other types of voltage-gated Ca^{2+} channels (Yue 2004) due to their difficulty to express and their potential modulation by interaction with RyR1 channels in the abutting Sr.

8.7.1 $Ca_V1.1$ has pseudo 4 fold symmetry of V sensor domains, but full L-type Ca^{2+} channel is highly asymmetric

The skeletal muscle Ca^{2+} channel complex is comprised of a voltage-sensing and pore-forming $\alpha 1$ subunit, that interacts with ancillary β, $\alpha 2\delta$, and γ subunits (Takahashi *et al* 1987, Catterall 1991, 2000) (figure 8.7(a) and (b)) and other critical ligands (i.e., STAC3 and calmodulin) (Nelson *et al* 2013). The $Ca_V1.1$ $\alpha 1$ subunit also confers most of the pharmacological sensitivity to channel blockers and modulators (Takahashi *et al* 1987, Tanabe *et al* 1987). The $Ca_V1.1$ $\alpha 1$ subunit consists of a single polypeptide of ~2000 residues, with four comparable but non-identical intramembrane domains (I–IV), each containing six transmembrane (TM) alpha helixes (S1–S6), shown in a linear representation in figure 8.7(c) (Takahashi *et al* 1987, Tanabe *et al* 1987), as well as cytoplasmic amino and carboxyl terminals. While the organization of the TM domains of the $Ca_V1.1$ $\alpha 1$ subunit has a strong pseudo four-fold symmetry in the plane of the TT membrane (figure 8.7(b)), the intracellular regions of $Ca_V1.1$, including the single β subunit is asymmetrical (figure 8.7(a)), which could be relevant for $Ca_V1.1$-RyR1 interactions. The helixes S1–S4 of each domain of the $\alpha 1$ subunit form a voltage-sensing domain (VSD) (Bezanilla 2000, Catterall 2011), whereas S5 and S6 from all four intramembrane domains contribute to the Ca^{2+}-conductive pore (figure 8.7(c)). The fourth TM helical segment (S4) harbors several positively charged residues (Arg and Lys), separated by two hydrophobic residues (figure 8.7(c)) (Tanabe *et al* 1987, Catterall 2011). S4 helixes are believed to move outward across the plane of the TT in response to depolarization, establishing the determinants for voltage-sensitivity (Rios and Pizarro 1991, Schneider 1994). Historically, there was a considerable time gap between recognition of the biological importance of Ca^{2+} channels and their structural examination (Yue 2004). Recent cryo-EM studies (Wu *et al* 2016), have unveiled details about the molecular architecture of the $Ca_V1.1$ channel of skeletal muscle with its complete set of auxiliary subunits (figure 8.7(a)). The central $\alpha 1$-subunit of $Ca_V1.1$ has a core structure and is associated with an extracellular $\alpha 2\delta$-subunit, an intracellular β-subunit, and a 4-TM γ-subunit.

8.7.2 SR Ca^{2+} release channels in skeletal muscle

The skeletal muscle SR Ca^{2+} release channel (aka ryanodine receptor type-1, RyR1) provides the Ca^{2+} needed for muscle contraction. The RyR1 is a very large protein of approximately 2.3 MDa formed by four identical subunits (Imagawa *et al* 1987,

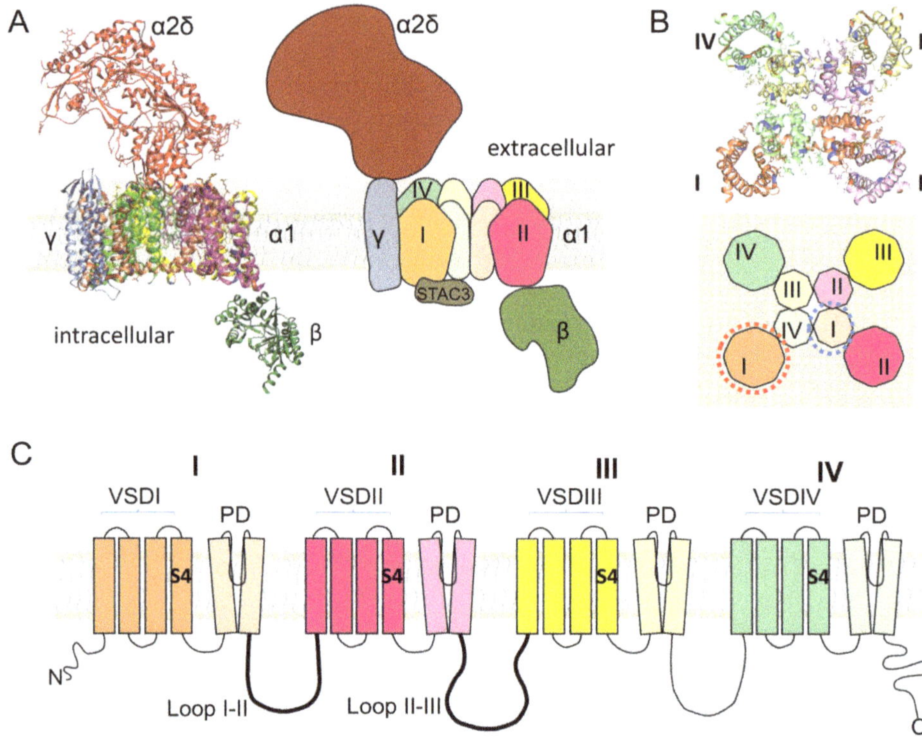

Figure 8.7. Architecture and membrane topology of the $Ca_V1.1$. (a) Side view of a cryo-EM reconstruction (left) and cartoon (right) of the $Ca_V1.1$ and its auxiliary subunits (Protein Data Bank (PDB) 5GJW) (Wu *et al* 2016). Domains of the $Ca_V1.1$ α1 subunit are color-coded. Auxiliary β-, α2δ-, and γ-subunits are colored in green, red, and dark blue, respectively. STAC3 is also shown in the cartoon and is considered an essential auxiliary component for ECC (Nelson *et al* 2013). Structure–function studies of STAC3 indicate that it binds to both the II–III loop and C-terminal of $Ca_V1.1$ (see Polster *et al* 2015, 2018, Niu *et al* 2018). (b) Extracellular view (looking from the TT lumen) of a cryo-EM reconstruction (top) and cartoon (bottom) of the $Ca_V1.1$ α1 subunit. The red dashed circle indicates the location of S1–S4 from domain I and the blue dashed circle shows its corresponding pore domain (S5–S6). Note that each voltage sensor domain (S1–S4) is not adjacent to its corresponding pore domain (S5–S6). These models were created in Chimera (Pettersen *et al* 2004). In panel (a), the cryo-EM resolved intracellular segments of $Ca_V1.1$ α1 are arbitrary not shown for a clearer identification of β subunit. (c) Membrane topology of the $Ca_V1.1$ α1 subunit. The α1 subunit is composed by an interconnect array of four homologous (but not identical) domains (I–IV), consisting of six transmembrane alpha helices, S1–S6. S1–S4 from each domain form a voltage sensor domain (VSD), whereas S5 and S6 from all four domains form the pore domain (PD). Intracellular loops connect the domains, the loops II–III and I–II are important for ECC. Reprinted from Schneider *et al* (1988) John Wiley & Sons, copyright 1988 The Physiological Society.

Inui *et al* 1987). Each subunit contains an intramembrane region, located within the C-terminal region and representing only a fraction of the total protein, plus a cytoplasmic region that represents most of the total protein, a segment known as the 'foot region' (figure 8.8) (Radermacher *et al* 1992, Serysheva *et al* 1995, Samso *et al* 2005, Yan *et al* 2015, Zalk *et al* 2015). The cytoplasmic region of the RyR1 channel (280 Å × 280 Å × 120 Å) is continuous with the transmembrane region

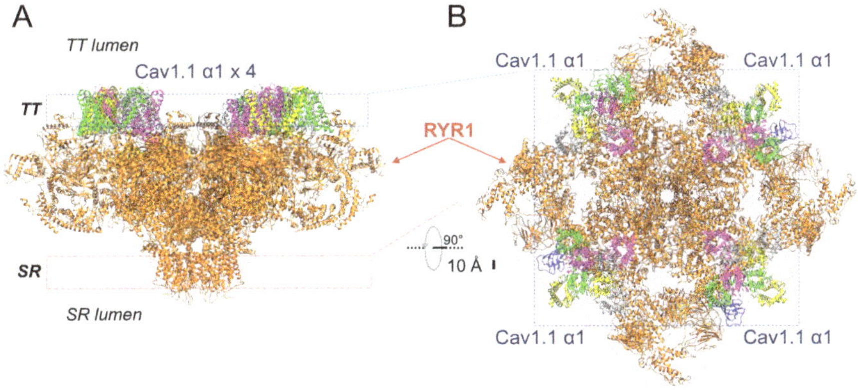

Figure 8.8. Hypothetical array of four $Ca_V1.1$ α1 subunits (domains colored as in figure 8.7) with one RyR1 homotetramer (tan color). Side view (a) and TT luminal views (b) of cryo-EM reconstructions of RyR1 (PDB, 5TAL) (Des Georges *et al* 2016) and four superimposed $Ca_V1.1$ α1 subunits (PDB 5GJW) (Wu *et al* 2016), forming a tetrad. This model was created in Chimera (Pettersen *et al* 2004) using the cryo-EM maps with relative location of the $Ca_V1.1$ subunits as suggested by Samso (2017). Reprinted with permission from Hernández-Ochoa and Schneider (2018), copyright 2018 The Authors.

(120 Å × 120 Å × 60 Å; figure 8.8(a)) (Lanner *et al* 2010). The RyR1 SR transmembrane region forms the Ca^{2+} release channel (Ludtke *et al* 2005, Samso *et al* 2009, Yan *et al* 2015, Zalk *et al* 2015).

RyR1s are arranged in a regular array within the terminal cisternae of the junctional SR (figures 8.1(b) and (c)) (Franzini-Armstrong and Nunzi 1983, Franzini-Armstrong and Jorgensen 1994, Franzini-Armstrong *et al* 1998). Similarly, $Ca_V1.1$ channels are clustered in groups of four (or tetrads) in the TT membrane that is adjacent to the junctional SR (Block *et al* 1988, Franzini-Armstrong and Jorgensen 1994), with a $Ca_V1.1$ tetrad facing every other RyR1 in the junctional SR RyR1 array (figures 8.1(c) and 8.8(b)). The arrays of tetrads mechanically interact with RyR1 juxtaposed in the SR membrane, and ancillary proteins are also known as 'couplons' (Stern *et al* 1997). It has been proposed that each $Ca_V1.1$ molecule composing a tetrad is oriented in the same coordinated position relative to the subunits of its juxtaposed RyR1 (figure 8.8(a)) (Block *et al* 1988, Franzini-Armstrong and Jorgensen 1994). Yet, the actual orientation of each $Ca_V1.1$ within the tetrad and relative to the RyR1 is unknown. Since these interfaces take place at alternate RyR1s, half of the RyR1s are 'uncoupled' with $Ca_V1.1$s, and half are coupled to $Ca_V1.1$s (figure 8.1(c)) (Franzini-Armstrong and Kish 1995).

The location of RyR1 (coupled and uncoupled) determines the organization of the $Ca_V1.1$ channels in the juxtaposed TT membrane, creating a 'chessboard' display of coupled and uncoupled RyR1s that produces the ordered $Ca_V1.1$ distribution typical of skeletal muscle (figure 8.1(c)) (Yin and Lai 2000). Action potential-evoked activation of RyR1 is believed to be mediated via direct or indirect communication with TT voltage sensors (Rios and Brum 1987, Nakai *et al* 1998a, 1998b, Protasi *et al* 2002). However, despite extensive biophysical and ultrastructural studies, the molecular basis for TT voltage sensor function and the chemical mechanisms that support TT voltage regulated RyR1 SR Ca^{2+} release have remained unclear.

8.7.3 The $Ca_V1.1$ is the voltage sensor and the dominant regulator of RyR1

The $Ca_V1.1$ working as a voltage sensor for RyR1 can be considered as the master ligand for controlling SR Ca^{2+} release in fully differentiated skeletal muscle fibers and numerous studies validate this concept. For, example, no other cellular event or component (e.g., ATP concentration), except possibly Ca^{2+} influx into the TT/SR gap, would be anticipated to change drastically during the action potential (1–2 ms duration) which increases myoplasmic Ca^{2+} sufficiently to cause a twitch contraction. Yet, Ca^{2+} influx is not needed for skeletal muscle activation (Armstrong *et al* 1972, Gonzalez-Serratos *et al* 1982, Brum *et al* 1987, Dayal *et al* 2017).

Experiments monitoring elementary Ca^{2+} release events (Tsugorka *et al* 1995, Klein *et al* 1996, Shirokova *et al* 1998, Jiang *et al* 1999, Schneider 1999, Baylor *et al* 2002, Rios and Brum 2002) in amphibian muscle fibers, indicate that in functioning muscle fibers, the RyR1 SR Ca^{2+} release channels are in low open probability (Po) states when the voltage sensors are in the resting condition, but switch to high Po during the AP or voltage-clamp depolarization that activates the voltage sensors (Schneider 1999, Rios and Brum 2002, Zhou *et al* 2004). Under resting conditions in mature functioning muscle fibers, RyR1s coupled to TT voltage sensors are locked in the 'off' configuration due to an inhibitory influence of the voltage sensor in its resting configuration (Zhou *et al* 2006). Movement of the voltage sensor into the active configuration during depolarization removes this 'lock' on RyR1, facilitating Po of RyR1 channels (Melzer *et al* 1984). When the voltage sensor returns to the resting configuration at the end of the depolarization the RyR1 returns to its resting low Po state.

One crucial discovery for the establishment of $Ca_V1.1$ as a critical component of ECC was the characterization of a naturally occurring 'knock out' of the $Ca_V1.1$ $\alpha 1$ subunit ('dysgenic' mouse; (Klaus *et al* 1983)). This finding demonstrated that myotubes derived from the dysgenic mice lacked ECC and intramembrane charge movement; the transfection of $\alpha 1$s subunit (skeletal muscle isoform) in these cells restored 'skeletal' type of ECC, which does not requires Ca^{2+} influx (Tanabe *et al* 1988). The expression of the cardiac isoform ($\alpha 1$c subunit) of the Ca_V1 channels did not restore skeletal ECC (Tanabe *et al* 1990b). The inability of $Ca_V1.2$, and of other Ca^{2+} channel subtypes (Flucher *et al* 2000, Wilkens and Beam 2003) to rescue the ECC allowed the investigation of essential elements for skeletal muscle ECC via chimeric channels and identified $Ca_V1.1$ α subunit as a critical component for muscle function.

Another important model that advanced the characterization of the molecular players of the ECC and their interactions was the RyR1-knockout mouse (the dyspedic mouse (Takeshima *et al* 1995)), which allowed for the expression of RyR mutants and/or different $Ca_V1.1$ $\alpha 1$/RyR1 combinations (Nakai *et al* 1996, 1997). These approach also identified that the skeletal RyR1 as an essential component for the skeletal muscle function.

8.7.4 Regions of $Ca_V1.1$ that are important for activating RyR1

Chimeric channels made with α1 subunits of $Ca_V1.1$ and $Ca_V1.2$ demonstrated that a region in the intracellular loop between the second and third domains (II–III loop), specifically, the region spanning residues 720–764/5, was important for this function (figure 8.7(c)) (Tanabe *et al* 1990a, Kugler *et al* 2004). Interestingly, in the cryo-EM structure of $Ca_V1.1$ the structure of the II–III loop is undefined (Wu *et al* 2016), whereas the I–II and III–IV/C-terminal regions are defined. While the identification of the $Ca_V1.1$ regions that are critical for ECC has been more active, perhaps due to the smaller size of the $Ca_V1.1$ channel, the identification of binding domains in the RyR1 for the II–III loop has been less fruitful. Only a few reports, where deletions of large segments of the RyR1 successfully altered ECC, concluded that several regions of the RyR1 are involved in the interaction with $Ca_V1.1$ (Leong and MacLennan 1998, Casarotto *et al* 2000, Proenza *et al* 2002). Recent approaches are revisiting the role of the loop II–III and its association with the adaptor protein STAC3 on ECC (Polster *et al* 2018). These new results revitalize the notion that the II–III linker plays an important role in ECC.

The $Ca_V1.1$ I–II loop is the site for interaction with the β1a subunit (figure 8.7(c)) (Catterall 2011). The β1a subunit, is important for several aspects of ECC. The β1a subunit is needed for the functional expression of $Ca_V1.1$ α1 subunit (Gregg *et al* 1996), is crucial for enhancement of $Ca_V1.1$ α1 triad expression (Strube *et al* 1996), assembly of $Ca_V1.1$ α1 in tetrads (Schredelseker *et al* 2005, 2009), and elicitation of $Ca_V1.1$ α1 charge movement (Dayal *et al* 2013). Skeletal-type ECC is reduced in muscle cells lacking the expression of β1a (Strube *et al* 1996) and is rescued by expression of β1a (Beurg *et al* 1997). The use of chimeric constructs of β1a (Beurg *et al* 1999) with other β subunits, as well as the use of synthetic peptides (Hernandez-Ochoa *et al* 2014), allowed the identification of the C-terminal region of β1a as an important domain for possible interaction with RyR1 during TT voltage-dependent SR Ca^{2+} release.

Thus, several sites in the $Ca_V1.1$ α1 subunit, in addition to the II–III loop, contribute to the overall $Ca_V1.1$/RyR interaction. These include the loop I–II and β-subunit (Chen *et al* 2004, Van Petegem *et al* 2004), and more indirectly, the III–IV loop (Bannister *et al* 2008) and the C-terminal domain of the $Ca_V1.1$ α1 (Flucher *et al* 2000) (figure 8.7(c)).

As mentioned above, the interaction of the TT $Ca_V1.1$ voltage sensor with the RyR appears to require the skeletal (but not the cardiac) II–III loop of the α 1 subunit (Tanabe *et al* 1990a, Nakai *et al* 1994). TT depolarization causes the reorientation of intramembrane charged helices in the $Ca_V1.1$ toward the TT lumen, i.e., away from the RyR (figure 8.7(c)), which would occur during the action potential depolarization of the TT. A straightforward hypothesis is that movement of the $Ca_V1.1$ S4 helices away from the RyR removes an inhibitory influence on Ca^{2+} release, and thereby activates release (Chandler *et al* 1976), and it has been proposed that the conformational change of the $Ca_V1.1$ may result in decreased Mg^{2+} affinity of the RyR, and thereby remove the Mg^{2+} suppression of

RyR channel opening (Lamb and Stephenson 1991). In addition, the Cav β1a subunit also seems to participate in the interaction (Gregg *et al* 1996), together with other accessory cytoplasmic (e.g., calmodulin, STAC3, S100A1) and junctional membrane and SR luminal proteins (e.g., triadin, junctin, calsequestrin, JP-45 and juntophilin, figure 8.1(c)) (Brillantes *et al* 1994, Chen and MacLennan 1994, Tripathy *et al* 1995, Prosser *et al* 2008, Beard *et al* 2009, Treves *et al* 2009) that may modulateCa_V1.1-RyR activation.

8.7.5 Ca_V1.1 channel gating modified in Ca_V1.1s coupled to RyR1 and Ca_V1.1 uncoupled RYR1

In addition to the 'orthograde' coupling from Ca_V1.1 to RyR (Rios and Brum 1987, Tanabe *et al* 1990a, Nakai *et al* 1998b), there is also a 'retrograde' coupling whereby activation of the RyR promote and modify the calcium current via the L-type Ca^{2+} channel (Nakai *et al* 1996, 1998a, Protasi *et al* 2002, Sheridan *et al* 2006).

It is also noteworthy that RyR Ca^{2+} release channels in the mammalian TT–SR junction are either structurally coupled or uncoupled to TT voltage sensors (Block *et al* 1988). One possibility for activation of such 'uncoupled' RyR channels is secondary activation by calcium-induced calcium release mediated by calcium released by calcium released via immediately adjacent RyR channels that are coupled to TT voltage sensors. Another possibility is lateral 'coupled gating' between neighboring RyRs (Marx *et al* 1998).

8.7.6 Ca_V1.1 provides voltage sensor for both intrinsic (in Ca_V1.1) and extrinsic (in RyR1) Ca^{2+} channels

In voltage-gated sodium channels, a channel structurally and evolutionary similar to Ca_V1.1 (Catterall 2000), voltage-clamp fluorometry studies revealed that the four VSDs, each linked to a partial pore-forming region, may be distinctly and allosterically coupled to the pore to various degrees of involvement in the control of channel gating (Chanda and Bezanilla 2002). VSDs I–III activate in parallel and fast to mediate Na^+ channel opening, whereas VSD IV activates more slowly and its time course parallels that of Nav current fast inactivation (Chanda and Bezanilla 2002). In cardiac isoform Ca_V1.2 individual VSDs showed differential function for each domain; VSDs II and III exhibited voltage-dependent and kinetic characteristics compatible with channel activation (Pantazis *et al* 2014). However the cardiac isoform (α1c subunit) is unable to support RyR1 activation in myotubes lacking α1s (Tanabe *et al* 1990a, 1990b).

Similarly, the voltage dependence and timing of Ca^{2+} entry via Ca_V1.1, as well as the voltage dependence and timing of TT voltage-dependent RyR1 Ca^{2+} release, are expected to be functions of the α1-subunit of Ca_V1.1, which also contains four highly similar but non-identical VSDs, I–IV, figures 8.7(b) and 8.9(a) (Takahashi *et al* 1987, Tanabe *et al* 1987, Catterall 2011, Wu *et al* 2016).

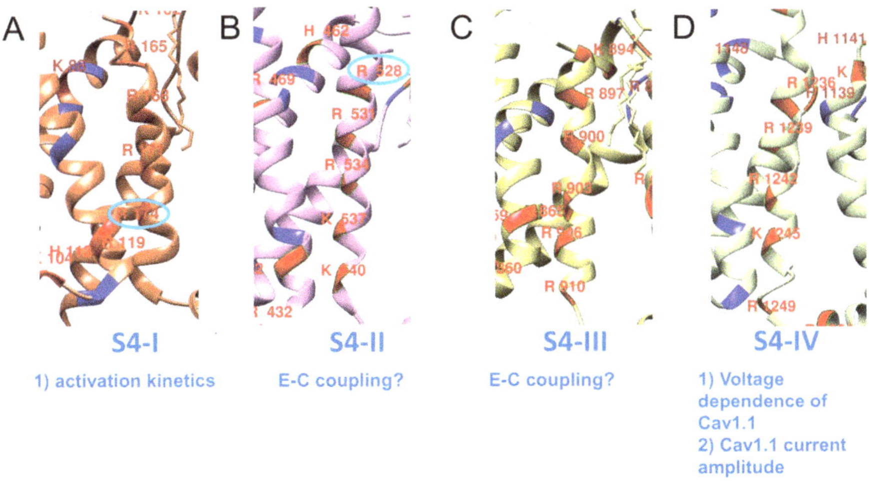

Figure 8.9. Voltage-sensing domains of $Ca_V1.1$. Side view at the level of the S4s of the cryo-EM reconstruction of $Ca_V1.1$ (PDB 5GJW) (Wu *et al* 2016). (a) S4-I, (b) S4-II, (c) S4-III and (d) S4-IV. The positively charged residues of the S4s are red-colored. Negatively charged amino acids are shown in blue. These models were created using Chimera (Pettersen *et al* 2004). References for S4-I (Eltit *et al* 2012, Tuluc *et al* 2016a), S4-II (Wu *et al* 2012, Tuluc *et al* 2016a), S4-III (Tuluc *et al* 2016a, 2016b) and S4-IV (Tuluc *et al* 2009, 2016a, 2016b). Reprinted with permission from Hernández-Ochoa and Schneider (2018), copyright 2018 The Authors.

8.7.7 All four V sensors may not be active in each coupled $Ca_V1.1$

Evidence for a differential role of each VSD in $Ca_V1.1$ channel operation using chimeric studies (interchange of VD SI region, $Ca_V1.1 \leftrightarrow Ca_V1.2$) and alternative splicing of VSD IV of $Ca_V1.1$ suggests that VSDs I and IV control current activation and its voltage-dependence, respectively (Tuluc *et al* 2009, 2016a, 2016b). Because VSD I and VSD IV appear to be linked to the slow activating $Ca_V1.1$ ionic current, it was hypothesized that VSDs I and IV do not contribute to the more rapid $Ca_V1.1$-dependent SR Ca^{2+} release (Tuluc and Flucher 2011, Flucher 2016) (see figure 8.9(a) and (d)).

In support of this hypothesis, a functional study of a mutation causing malignant hyperthermia susceptibility (R174W) in S4 of VSD I of $Ca_V1.1$, revealed that this mutation reduces $Ca_V1.1$ ionic current, but does not affect $Ca_V1.1$-dependent SR Ca^{2+} release (Eltit *et al* 2012) (figure 8.9(a)). Wu and colleagues (Wu *et al* 2012) used a mouse model for hypokalemic periodic paralysis with a targeted $Ca_V1.1$ R528H mutation in S4 of VSD II. Muscle fibers from the $Ca_V1.1$ R528H homozygous mouse exhibited impaired depolarization-induced Ca^{2+} release, insinuating that VSD II could participate in $Ca_V1.1$-dependent SR Ca^{2+} release (Wu *et al* 2012) (figure 8.9(b)). Altogether, these results represent partial and indirect evidence of the role of each VSD in $Ca_V1.1$. Currently, the role of each individual VSDs to the voltage dependence of $Ca_V1.1$ pore opening and activation of RyR1 Ca^{2+} release in functioning skeletal muscle remains unexplored. Thus, one (of the four) VSDs of each $Ca_V1.1$ could couple to one RyR1 monomer in the homo tetrameric channel.

This VSD in each of the 4 $Ca_V1.1$s linked to each RyR1 homotetramer would have to be active to open the RyR1 channel.

8.8 Relaxation and SR Ca^{2+} uptake

In addition to the RyR Ca^{2+} release channels in the junctional SR, the entire SR membrane contains a high concentration of Ca^{2+} ATPase molecules, the SR/ER (SERCA) Ca^{2+} pump (Hasselbach 1964, 1998). This transport system transfers Ca^{2+} ions against their concentration gradient and establishes the resting 4-order of magnitude concentration gradient of roughly 0.1 μM cytosolic [Ca^{2+}] and 1 mM or higher SR luminal [Ca^{2+}]. In the resting fiber the cytosolic [Ca^{2+}] is low and the SERCA pump cycles correspondingly slowly. When cytosolic Ca^{2+} rises, Ca^{2+} ions bind to and are transported back into the SR by the SERCA pump. Considering that the concentration of pump sites in the myoplasmic volume is similar to the concentration of troponin C Ca^{2+} regulatory sites also referred to the myoplasmic volume, simple binding of Ca^{2+} to the pump could relax the muscle after a twitch, but cycling is required for successive activation (Timmer *et al* 1998, Baylor and Hollingworth 2003). In fast twitch fibers, the soluble Ca^{2+} binding protein parvalbumin provides a way station for binding Ca^{2+} during relaxation prior to its ultimate return to the SR (Gerday and Gillis 1976, Pechere *et al* 1977). The final uptake of released Ca^{2+} back into the SR, mediated by the ATP dependent Ca^{2+} pump in the SR membrane, reprimes the initial state of the resting fiber, with essentially all of the released Ca^{2+} returned to the SR Ca^{2+}, and terminates the ECC process.

8.9 $Ca_V1.1$ modulation

$Ca_V1.1$ belong to the Ca_V1 family of voltage-gated Ca^{2+} channels, which includes $Ca_V1.2$ (expressed in cardiomyocytes), and $Ca_V1.3$ and $Ca_V1.4$ (expressed predominantly in neurons) (Catterall 2011, Dolphin 2016). Robust regulation of the L-type Ca^{2+} current by calmodulin (CaM), manifested as Ca^{2+} dependent inactivation (CDI) and/or Ca^{2+} dependent facilitation (CDF), has been demonstrated in expression systems and native systems for $Ca_V1.2/1.3/1.4$ (Peterson *et al* 2000, Liang *et al* 2003, Dick *et al* 2008, Tadross *et al* 2008, Ben-Johny and Yue 2014), whereas Ca^{2+} current-dependent and CaM-mediated regulation for $Ca_V1.1$ has remained controversial (Ohrtman *et al* 2008, Halling *et al* 2009, Bannister and Beam 2013). Studies in skeletal muscle fibers have shown that $Ca_V1.1$ current exhibit CDI (Almers *et al* 1981, Francini and Stefani 1989), but this form of inactivation is insensitive to Ba^{2+} substitution in the recording solution (Cota and Stefani 1989, Hernández-Ochoa and Schneider 2012), suggesting that CDI in $Ca_V1.1$ could derive from SR-Ca^{2+} release and not via Ca^{2+} influx (Bannister and Beam 2013). Studies in an homologous expression system (myotubes) have shown $Ca_V1.1$ CaM-dependent CDI and ECC deficits (Stroffekova 2008, 2011), leading to the idea that CaM could be important for ECC, however, it is important to consider that ECC in myotubes exhibit a developmental form of ECC which may substantially differ from that of adult skeletal muscle (Olsson *et al* 2015, Bannister 2016).

Interestingly, studies of $Ca_V1.1$ in heterologous expression system have shown that CaM (and STAC3) could be important in regulating $Ca_V1.1$ membrane trafficking and gating (Niu *et al* 2018). As mentioned above, it seems that the $Ca_V1.1$. Ca^{2+} current is not essential for ECC (Dayal *et al* 2017), however, it could be important for SR Ca^{2+} store refilling (Robin and Allard 2015), muscle performance (Lee *et al* 2015) and excitation-metabolism coupling (Georgiou *et al* 2015). Whether CaM is involved in the regulation of $Ca_V1.1$ trafficking, gating, CDI or CDF in adult skeletal muscle remains to be determined.

8.10 RyR1 modulation

The TT voltage sensor ($Ca_V1.1$) is believed to be the principal partner that uniquely enables RyR1 opening in functioning skeletal muscle fibers, however, additional RyR1 interacting proteins (i.e., CaM, S100A1, calsequestrin, triadin, junctin, junctophilin, JP-45), metabolites, ions and drugs (ATP, Ca^{2+}, Mg^{2+}, caffeine), as well as post-translational modifications allow the RyR1's function to be fine-tuned (MacLennan and Wong 1971, Meissner *et al* 1986, Fuentes *et al* 1994, Ikemoto *et al* 1995, Tripathy *et al* 1995, Treves *et al* 1997, 2009, Marengo *et al* 1998, Eu *et al* 2000, Aracena *et al* 2003, Serysheva *et al* 2005, Samso *et al* 2009, Zalk *et al* 2015). RyRs, including RyR1, exhibit *U*-shape Ca^{2+} dependence in its function; they are activated by Ca^{2+} levels in the μM range and inactivated by sub mM Ca^{2+} levels (Meissner *et al* 1986). ATP is a stabilizer of the Ca^{2+}-induced opening of RyR1 (Meissner *et al* 1986). Caffeine, is a RyR1 pharmacological activator (Rousseau *et al* 1988). Structural details of binding sites for Ca^{2+}, ATP and caffeine, but not for CaM, have been identified for RyR1. For more in-depth reviews about RyR1 function-regulation see: (Meissner 2017, Ogawa *et al* 2020).

8.11 RYR1 and disease states

Malignant hyperthermia susceptibility (MHS) (Rosenberg *et al* 1993) is a pharmacogenetic disease that is seeing after anesthesia during surgical procedures and could be fatal if not identified and managed appropriately (Rosenberg *et al* 2007, 2015). It is s triggered by exposure to inhaled anesthetics or the muscle relaxant succinylcholine. It is characterized by extreme hypermetabolic state with severe fever, muscle contractures, myoglobinuria and kidney damage. About 60%–70% of MHS cases are caused by mutations in RyR1 (Robinson *et al* 2006, Maclennan and Zvaritch 2011, Rosenberg *et al* 2015). The main defect in MH is an increased RyR1 Po. Estimates of MH clinical incidence ranges from 1:10 000 to 1:250 000 (Ording 1985, Rosenberg *et al* 2015).

RyR1-related myopathies (central core disease, multi-minicore disease, core-rod myopathy, centronuclear myopathy and congenital fiber-type disproportion) are the most common class of congenital myopathies (Jungbluth *et al* 2002, Jungbluth 2007, Jungbluth and Gautel 2014), with new emerging phenotypes (Dowling *et al* 2011, 2014, Colombo *et al* 2015, Matthews *et al* 2018, Witting *et al* 2018). These conditions are characterized by childhood onset of skeletal muscle hypotonia, progressive muscle weakness and delayed motor development

(Amburgey *et al* 2011, Witting *et al* 2017). RyR1-related myopathies is an umbrella term that includes histopathologically and clinically distinct syndromes of variable severity. Mutations responsible of RyR1-related myopathies have been identified along the entire sequence of the RyR1 (Shuaib *et al* 1987, Quane *et al* 1993, Zhang *et al* 1993, Avila and Dirksen 2001, Avila *et al* 2001, Robinson *et al* 2006, Maclennan and Zvaritch 2011, Dowling *et al* 2014). For deeper insights into RyR1 physiology and mechanisms of MH and RyR1-related myopathies see Rosenberg *et al* (1993), Dowling *et al* (2014), Jungbluth *et al* (2018).

8.12 Ca_V1.1 channelopathies

Mutations in *CACNA1S*, the gene encoding the Ca_V1.1 α subunit, has been linked to hypokalemic periodic paralysis (hypoPP) (Cannon 2015), a rare autosomic dominant myopathy that is characterized by muscle weakness and paralysis attacks that are coincident with reductions in serum K^+ (Weber and Lehmann-Horn 1993). HypoPP is triggered by periods of rest after bouts of intense exercise, carbohydrate-rich evening meals, stress, cold, alcohol or salt intake and anesthesia (Statland *et al* 2018). Interestingly, most of these mutations occur in positively charge residues located at the S4 helixes of the VSDs (Jurkat-Rott *et al* 1994, Cannon 2015, 2017, Fuster *et al* 2017a, 2017b). One of the most common mutations seen in affected individuals is the R528H (Cannon 2015). Using genetically engineered mice carrying this R528H mutation, a murine model that recapitulates many of the clinical features seen in humans, recent studies have shown Ca_V1.1 channels with this mutation exhibit a 'gating pore current', a cationic conductive crevice formed in the vicinity of the VSD (Wu *et al* 2012). This gating pore current is believed to be active at resting potential and cause anomalous membrane depolarization enough to promote Na_V1.4 inactivation, leading to the loss of excitability during hypoPP attacks (Wu *et al* 2012). For authoritative reviews and updates in this topic see Weber and Lehmann-Horn (1993) and Cannon (2015).

MHS is caused primarily by mutations in the RyR1 release channel (see above), however, several reports have shown that MH cases (~1%) are associated to mutations in the Ca_V1.1 α1 subunit (Monnier *et al* 1997, Eltit *et al* 2012). Two of these Ca_V1.1 mutations, R174W and R1086H seems to promote RYR1 Ca^{2+} leak at resting conditions and increase the RyR1 sensitivity to inhaled anesthetics.

A more recent study identified several Ca_V1.1 mutations that are associated with muscle weakness and hypotonia (Schartner *et al* 2017). In these compound mutations, characterized by frameshifts, deletions, non-sense mutations and missense mutations, common observed features were SR dilations, myofibrilar disarrangements and ECC deficits due to reduced functional expression of Ca_V1.1 (Schartner *et al* 2017, 2019).

8.13 Conclusion

Intense and productive research on the field of Ca^{2+} signaling in skeletal muscle has led to finding answers to many interrogations related to ECC, the critical players (Ca_V1.1 and RyR1) and their biophysical and molecular structural properties, yet, leading to more new questions in need of response. The quintessential question of how $Ca_V1.1$ interacts with RyR1 to trigger Ca^{2+} release remains unanswered. The Schneider and Chandler hypothesis that the excitatory signal passes from the TT to the SR membrane via voltage sensors reorienting in the TT membrane connecting with the junctional feet of the SR Ca^{2+} release channel started the track to answer this question. We still do not know how the movement of transmembrane helixes of the VSD of $Ca_V1.1$ conveys an intramolecular signal initially across $Ca_V1.1$ critical regions (II–III loop and possibly I–II loop-β1a interactions) and then via intermolecular communication with the RyR1 to ultimately trigger Ca^{2+} release.

The cryo-EM available structures for RyR1 and $Ca_V1.1$ were unveiled at near atomic resolution and provided substantial structural details. However, important regions, such has II–III loop of $Ca_V1.1$, C-terminal residues of β1a and residues 3495–3502 of helical domain 2 of RyR1, considered important for β1a binding to RyR1, are still unresolved. A new generation of biophysicists armed with high-resolution molecular imaging, along with electrophysiological and molecular manipulations may provide a further understanding of the intricacies between $Ca_V1.1$ and RyR1.

Funding

This publication was supported by the National Institute of Arthritis and Musculoskeletal and Skin Diseases of the National Institutes of Health under Award Number R37-AR055099 and R01AR075726 (to M F S). The content is solely the responsibility of the authors and does not necessarily represent the official views of the National Institutes of Health.

Author contributions

MFS and EOH-O wrote, edited and approved the manuscript.

Acknowledgments

We thank all our past and present collaborators, the contributors of the cited work, and apologize to the authors in the field of skeletal muscle ECC that unintentionally were not referenced in our manuscript.

References

Adrian R H and Almers W 1976 Charge movement in the membrane of striated muscle *J. Physiol.* **254** 339–60

Adrian R H, Costantin L L and Peachey L D 1969 Radial spread of contraction in frog muscle fibres *J. Physiol.* **204** 231–57

Adrian R H and Marshall M W 1977 Sodium currents in mammalian muscle *J. Physiol.* **268** 223–50

Almers W, Fink R and Palade P T 1981 Calcium depletion in frog muscle tubules: the decline of calcium current under maintained depolarization *J. Physiol.* **312** 177–207

Amburgey K *et al* 2011 Prevalence of congenital myopathies in a representative pediatric united states population *Ann. Neurol.* **70** 662–65

Anderson A A *et al* 2003 The novel skeletal muscle sarcoplasmic reticulum JP-45 protein. Molecular cloning, tissue distribution, developmental expression, and interaction with alpha 1.1 subunit of the voltage-gated calcium channel *J. Biol. Chem.* **278** 39987–92

Andersson-Cedergren E 1959 Ultrastructure of motor end-plate and sarcoplasmic reticulum components of mouse skeletal muscle fiber as revealed by three-dimensional reconstructions from serial sections *J. Ultrastruct. Res.* **1** 1–191

Aracena P, Sanchez G, Donoso P, Hamilton S L and Hidalgo C 2003 S-glutathionylation decreases Mg^{2+} inhibition and S-nitrosylation enhances Ca^{2+} activation of RyR1 channels *J. Biol. Chem.* **278** 42927–35

Armstrong C M, Bezanilla F and Horowicz P 1974 Charge movement associated with the opening and closing of the activation gates of the Na channels *J. Gen. Physiol.* **63** 533–52

Armstrong C M, Bezanilla F M and Horowicz P 1972 Twitches in the presence of ethylene glycol bis(β-aminoethyl ether)-*N*,*N′*-tetracetic acid *Biochim. Biophys. Acta* **267** 605–8

Avila G and Dirksen R T 2001 Functional effects of central core disease mutations in the cytoplasmic region of the skeletal muscle ryanodine receptor *J. Gen. Physiol.* **118** 277–90

Avila G, O'Brien J J and Dirksen R T 2001 Excitation–contraction uncoupling by a human central core disease mutation in the ryanodine receptor *Proc. Natl. Acad. Sci. USA* **98** 4215–20

Bannister R A 2016 Bridging the myoplasmic gap II: more recent advances in skeletal muscle excitation-contraction coupling *J. Exp. Biol.* **219** 175–82

Bannister R A and Beam K G 2013 $Ca_V1.1$: the atypical prototypical voltage-gated Ca^{2+} channel *Biochim. Biophys. Acta* **1828** 1587–97

Bannister R A, Grabner M and Beam K G 2008 The alpha(1S) III–IV loop influences 1,4-dihydropyridine receptor gating but is not directly involved in excitation-contraction coupling interactions with the type 1 ryanodine receptor *J. Biol. Chem.* **283** 23217–23

Baylor S M, Chandler W K and Marshall M W 1983 Sarcoplasmic reticulum calcium release in frog skeletal muscle fibres estimated from Arsenazo III calcium transients *J. Physiol.* **344** 625–66

Baylor S M and Hollingworth S 2003 Sarcoplasmic reticulum calcium release compared in slow-twitch and fast-twitch fibres of mouse muscle *J. Physiol.* **551** 125–38

Baylor S M and Hollingworth S 2007 Simulation of Ca^{2+} movements within the sarcomere of fast-twitch mouse fibers stimulated by action potentials *J. Gen. Physiol.* **130** 283–302

Baylor S M and Hollingworth S 2011 Calcium indicators and calcium signalling in skeletal muscle fibres during excitation-contraction coupling *Prog. Biophys. Mol. Biol.* **105** 162–79

Baylor S M and Hollingworth S 2012 Intracellular calcium movements during excitation-contraction coupling in mammalian slow-twitch and fast-twitch muscle fibers *J. Gen. Physiol.* **139** 261–72

Baylor S M, Hollingworth S and Chandler W K 2002 Comparison of simulated and measured calcium sparks in intact skeletal muscle fibers of the frog *J. Gen. Physiol.* **120** 349–68

Beam K G, Knudson C M and Pwell J A 1986 A lethal mutation in mice eliminates the slow calcium current in skeletal muscle cells *Nature* **320** 168–70

Beard N A, Wei L and Dulhunty A F 2009 Ca^{2+} signaling in striated muscle: the elusive roles of triadin, junctin, and calsequestrin *Eur. Biophys. J.* **39** 27–36

Ben-Johny M and Yue D T 2014 Calmodulin regulation (calmodulation) of voltage-gated calcium channels *J. Gen. Physiol.* **143** 679–92

Berridge M J, Lipp P and Bootman M D 2000 The versatility and universality of calcium signalling *Nat. Rev. Mol. Cell Biol.* **1** 11–21

Beurg M *et al* 1999 Involvement of the carboxy-terminus region of the dihydropyridine receptor beta1a subunit in excitation-contraction coupling of skeletal muscle *Biophys. J.* **77** 2953–67

Beurg M, Sukhareva M, Strube C, Powers P A, Gregg R G and Coronado R 1997 Recovery of Ca^{2+} current, charge movements, and Ca^{2+} transients in myotubes deficient in dihydropyridine receptor beta 1 subunit transfected with beta 1 cDNA *Biophys. J.* **73** 807–18

Bezanilla F 2000 The voltage sensor in voltage-dependent ion channels *Physiol. Rev.* **80** 555–92

Block B A, Imagawa T, Campbell K P and Franzini-Armstrong C 1988 Structural evidence for direct interaction between the molecular components of the transverse tubule/sarcoplasmic reticulum junction in skeletal muscle *J. Cell Biol.* **107** 2587–600

Brillantes A B *et al* 1994 Stabilization of calcium release channel (ryanodine receptor) function by FK506-binding protein *Cell* **77** 513–23

Brown L D, Rodney G G, Hernández-Ochoa E, Ward C W and Schneider M F 2007 Ca^{2+} sparks and T tubule reorganization in dedifferentiating adult mouse skeletal muscle fibers *Am. J. Physiol. Cell Physiol.* **292** C1156–66

Brum G, Stefani E and Rios 1987 Simultaneous measurements of Ca^{2+} currents and intracellular Ca^{2+} concentrations in single skeletal muscle fibers of the frog *Can. J. Physiol. Pharmacol.* **65** 681–5

Cannon S C 2015 Channelopathies of skeletal muscle excitability *Comprehens. Physiol.* **5** 761–90

Cannon S C 2017 An atypical $Ca_V1.1$ mutation reveals a common mechanism for hypokalemic periodic paralysis *J. Gen. Physiol.* **149** 1061–64

Casarotto M G, Gibson F, Pace S M, Curtis S M and Mulcair M 2000 A structural requirement for activation of skeletal ryanodine receptors by peptides of the dihydropyridine receptor II–III loop *J. Biol. Chem.* **275** 11631–7

Catterall W A 1991 Excitation-contraction coupling in vertebrate skeletal muscle: a tale of two calcium channels *Cell* **64** 871–4

Catterall W A 2000 Structure and regulation of voltage-gated Ca^{2+} channels *Annu. Rev. Cell Dev. Biol.* **16** 521–55

Catterall W A 2011 Voltage-gated calcium channels *Cold Spring Harb. Perspect. Biol.* **3** a003947

Chanda B and Bezanilla F 2002 Tracking voltage-dependent conformational changes in skeletal muscle sodium channel during activation *J. Gen. Physiol.* **120** 629–45

Chandler W K, Rakowski R F and Schneider M F 1976 A non-linear voltage dependent charge movement in frog skeletal muscle *J. Physiol.* **254** 245–83

Chen S R and MacLennan D H 1994 Identification of calmodulin-, Ca^{2+}-, and ruthenium red-binding domains in the Ca^{2+} release channel (ryanodine receptor) of rabbit skeletal muscle sarcoplasmic reticulum *J. Biol. Chem.* **269** 22698–704

Chen Y H, Li M H, Zhang Y, He L-l, Yamada Y, Fitzmaurice A, Shen Y, Zhang H, Tong L and Yang J 2004 Structural basis of the alpha1-beta subunit interaction of voltage-gated Ca^{2+} channels *Nature* **429** 675–80

Cheng H and Lederer W J 2008 Calcium sparks *Physiol. Rev.* **88** 1491–545

Cheng H, Lederer W J and Cannell M B 1993 Calcium sparks: elementary events underlying excitation-contraction coupling in heart muscle *Science* **262** 740–4

Chun L G, Ward C W and Schneider M F 2003 Ca^{2+} sparks are initiated by Ca^{2+} entry in embryonic mouse skeletal muscle and decrease in frequency postnatally *Am. J. Physiol. Cell Physiol.* **285** C686–97

Colombo I *et al* 2015 Congenital myopathies: natural history of a large pediatric cohort *Neurology* **84** 28–35

Conklin M W, Barone V, Sorrentino V and Coronado R 1999 Contribution of ryanodine receptor type 3 to Ca^{2+} sparks in embryonic mouse skeletal muscle *Biophys. J.* **77** 1394–403

Cota G and Stefani E 1989 Voltage-dependent inactivation of slow calcium channels in intact twitch muscle fibers of the frog *J. Gen. Physiol.* **94** 937–51

Dayal A, Bhat V, Franzini-Armstrong C and Grabner M 2013 Domain cooperativity in the beta1a subunit is essential for dihydropyridine receptor voltage sensing in skeletal muscle *Proc. Natl Acad. Sci. USA* **110** 7488–93

Dayal A, Schrötter K, Pan Y, Föhr K, Melzer W and Grabner M 2017 The Ca^{2+} influx through the mammalian skeletal muscle dihydropyridine receptor is irrelevant for muscle performance *Nat. Commun.* **8** 475

des Georges A *et al* 2016 Structural basis for gating and activation of RyR1 *Cell* **167** 145–57

Diaz-Vegas A, Eisner V and Jamovich E 2019 Skeletal muscle excitation-metabolism coupling *Arch. Biochem. Biophys.* **664** 89–94

Dick I E, Tadross M R, Liang H, Tay L H, Yang W and Yue D T 2008 A modular switch for spatial Ca^{2+} selectivity in the calmodulin regulation of Ca_V channels *Nature* **451** 830–4

Dolphin A C 2016 Voltage-gated calcium channels and their auxiliary subunits: physiology and pathophysiology and pharmacology *J. Physiol.* **594** 5369–90

Dowling J J, Lawlor M W and Dirksen R T 2014 Triadopathies: an emerging class of skeletal muscle diseases *Neurotherapeutics* **11** 773–85

Dowling J J *et al* 2011 King-Denborough syndrome with and without mutations in the skeletal muscle ryanodine receptor (RYR1) gene *Neuromusc. Disord.* **21** 420–7

Ebashi S 1974 Regulatory mechanism of muscle contraction with special reference to the Ca-troponin-tropomyosin system *Essays Biochem.* **10** 1–36

Ebashi S 1976 Excitation-contraction coupling *Annu. Rev. Physiol.* **38** 293–313

Ebashi S 1984 Ca^{2+} and the contractile proteins *J. Mol. Cell. Cardiol.* **16** 129–36

Echevarria W, Leite M F, Guerra M T, Zipfel W R and Nathanson M H 2003 Regulation of calcium signals in the nucleus by a nucleoplasmic reticulum *Nat. Cell Biol.* **5** 440–6

Eltit J M *et al* 2012 Malignant hyperthermia susceptibility arising from altered resting coupling between the skeletal muscle L-type Ca^{2+} channel and the type 1 ryanodine receptor *Proc. Natl Acad. Sci. USA* **109** 7923–8

Eu J P, Sun J, Xu L, Stamler J S and Meissner G 2000 The skeletal muscle calcium release channel: coupled O_2 sensor and NO signaling functions *Cell* **102** 499–509

Eusebi F, Miledi R and Takahashi T 1980 Calcium transients in mammalian muscles *Nature* **284** 560–61

Felder E and Franzini-Armstrong C 2002 Type 3 ryanodine receptors of skeletal muscle are segregated in a parajunctional position *Proc. Natl. Acad. Sci. USA* **99** 1695–700

Flucher B E 2016 Specific contributions of the four voltage-sensing domains in L-type calcium channels to gating and modulation *J. Gen. Physiol.* **148** 91–5

Flucher B E, Kasielke N and Grabner M 2000 The triad targeting signal of the skeletal muscle calcium channel is localized in the COOH terminus of the alpha(1S) subunit *J. Cell Biol.* **151** 467–78

Francini F and Stefani E 1989 Decay of the slow calcium current in twitch muscle fibers of the frog is influenced by intracellular EGTA *J. Gen. Physiol.* **94** 953–69

Franzini-Armstrong C and Jorgensen A O 1994 Structure and development of E-C coupling units in skeletal muscle *Annu. Rev. Physiol.* **56** 509–34

Franzini-Armstrong C and Kish J W 1995 Alternate disposition of tetrads in peripheral couplings of skeletal muscle *J. Muscle Res. Cell Motil.* **16** 319–24

Franzini-Armstrong C and Nunzi G 1983 Junctional feet and particles in the triads of a fast-twitch muscle fibre *J. Muscle Res. Cell Motil.* **4** 233–52

Franzini-Armstrong C and Porter K R 1964 Sarcolemmal Invaginations constituting the T system in fish muscle fibers *J. Cell Biol.* **22** 675–96

Franzini-Armstrong C, Protasi F and Ramesh V 1998 Comparative ultrastructure of Ca^{2+} release units in skeletal and cardiac muscle *Ann. NY Acad. Sci.* **853** 20–30

Fuentes O, Valdivia C, Vaughan D, Coronado R and Valdivia H H 1994 Calcium-dependent block of ryanodine receptor channel of swine skeletal muscle by direct binding of calmodulin *Cell Calcium* **15** 305–16

Fuster C, Perrot J, Berthier C, Jacquemond V and Allard B 2017a Elevated resting H^+ current in the R1239H type 1 hypokalaemic periodic paralysis mutated Ca^{2+} channel *J. Physiol.* **595** 6417–28

Fuster C, Perrot J, Berthier C, Jacquemond V, Charnet P and Allard B 2017b Na leak with gating pore properties in hypokalemic periodic paralysis V876E mutant muscle Ca channel *J. Gen. Physiol.* **149** 1139–48

Garcia J and Schneider M F 1993 Calcium transients and calcium release in rat fast-twitch skeletal muscle fibres *J. Physiol.* **463** 709–28

Georgiev T, Svirin M, Jaimovich E and Fink R H A 2015 Localized nuclear and perinuclear Ca^{2+} signals in intact mouse skeletal muscle fibers *Front. Physiol.* **6** 263

Georgiou D K *et al* 2015 Ca^{2+} binding/permeation via calcium channel, $Ca_V1.1$, regulates the intracellular distribution of the fatty acid transport protein, CD36, and fatty acid metabolism *J. Biol. Chem* **290** P23751–65

Gerday C and Gillis J M 1976 Proceedings: the possible role of parvalbumins in the control of contraction *J. Physiol.* **258** P96–7P

Gergely J, Grabarek Z and Tao T 1993 The molecular switch in troponin C *Adv. Exp. Med. Biol.* **332** 117–23

Gómez J, Ñeco P, DiFranco M and Vergara J L 2006 Calcium release domains in mammalian skeletal muscle studied with two-photon imaging and spot detection techniques *J. Gen. Physiol.* **127** 623–37

Gonzalez-Serratos H, Valle-Aguilera R, Lathrop D A and del Carmen Garcia M 1982 Slow inward calcium currents have no obvious role in muscle excitation-contraction coupling *Nature* **298** 292–4

Gordon A M and Ridgway E B 1993 Cross-bridges affect both TnC structure and calcium affinity in muscle fibers *Adv. Exp. Med. Biol.* **332** 183–92 (discussion 192–94)

Grabarek Z, Tao T and Gergely J 1992 Molecular mechanism of troponin-C function *J. Muscle Res. Cell Motil.* **13** 383–93

Gregg R G, Messing, Strube C and Powers P A 1996 Absence of the beta subunit (cchb1) of the skeletal muscle dihydropyridine receptor alters expression of the alpha 1 subunit and eliminates excitation-contraction coupling *Proc. Natl. Acad. Sci. USA* **93** 13961–6

Grynkiewicz G, Poenie M and Tsien R Y 1985 A new generation of Ca^{2+} indicators with greatly improved fluorescence properties *J. Biol. Chem.* **260** 3440–50

Gundersen K 2011 Excitation-transcription coupling in skeletal muscle: the molecular pathways of exercise *Biol. Rev. Camb. Philos. Soc.* **86** 564–600

Halling D B *et al* 2009 Determinants in Ca_V1 channels that regulate the Ca^{2+} sensitivity of bound calmodulin *J. Biol. Chem.* **284** 20041–51

Hasselbach W 1964 ATP-driven active transport of calcium in the membranes of the sarcoplasmic reticulum *Proc. R. Soc Lond. B Biol. Sci.* **160** 501–4

Hasselbach W 1998 The Ca^{2+}-ATPase of the sarcoplasmic reticulum in skeletal and cardiac muscle. an overview from the very beginning to more recent prospects *Ann. NY Acad. Sci.* **853** 1–8

Hernández-Ochoa E O, Olojo R O, Rebbeck R T and Dulhunty A F 2014 β1a490-508, a 19-residue peptide from C-terminal tail of Cav1.1 β1a subunit, potentiates voltage-dependent calcium release in adult skeletal muscle fibers *Biophys. J.* **106** 535–47

Hernández-Ochoa E O and Schneider M F 2012 Voltage clamp methods for the study of membrane currents and SR Ca^{2+} release in adult skeletal muscle fibres *Prog. Biophys. Mol. Biol.* **108** 98–118

Hernández-Ochoa E O and Schneider M F 2012 Calcium dependent inactivation of $Ca_V1.1$ channels in adult skeletal muscle: a possible role of RyR1 channels *Biophys. J.* **102** 125a

Hernández-Ochoa E O, Vanegas C, Iyer S R, Lovering R M and Schneider M F 2016 Alternating bipolar field stimulation identifies muscle fibers with defective excitability but maintained local Ca^{2+} signals and contraction *Skelet. Muscle* **6** 6

Hernández-Ochoa E O and Schneider M F 2018 Voltage sensing mechanism in skeletal muscle excitation-contraction coupling: coming of age or midlife crisis? *Skelet. Muscle* **8** 22

Holash R J and MacIntosh B R 2019 A stochastic simulation of skeletal muscle calcium transients in a structurally realistic sarcomere model using MCell *PLoS Comput. Biol.* **15** e1006712

Hollingworth S and Baylor S M 2013 Comparison of myoplasmic calcium movements during excitation-contraction coupling in frog twitch and mouse fast-twitch muscle fibers *J. Gen. Physiol.* **141** 567–83

Hollingworth S, Kim M M and Baylor S M 2012 Measurement and simulation of myoplasmic calcium transients in mouse slow-twitch muscle fibres *J. Physiol.* **590** 575–94

Hubbard J I 1973 Microphysiology of vertebrate neuromuscular transmission *Physiol. Rev.* **53** 674–723

Huxley A F 1971 The activation of striated muscle and its mechanical response *Proc. R. Soc. Lond. B Biol. Sci.* **178** 1–27

Huxley A F 1974 Muscular contraction *J. Physiol.* **243** 1–43

Huxley A F 1975 The origin of force in skeletal muscle *Ciba Found. Symp.* **31** 271–90

Huxley A F 1995 Muscle contraction. Crossbridge tilting confirmed *Nature* **375** 631–2

Huxley A F and Straub R W 1958 Local activation and interfibrillar structures in striated muscle *J. Physiol.* **143 Suppl.** 40P

Huxley A F and Taylor R E 1958 Local activation of striated muscle fibres *J. Physiol.* **144** 426–41

Ikemoto T, Iino M and Endo M 1995 Enhancing effect of calmodulin on Ca^{2+}-induced Ca^{2+} release in the sarcoplasmic reticulum of rabbit skeletal muscle fibres *J. Physiol.* **487** 573–82

Imagawa T, Smith J S, Coronado R and Campbell K P 1987 Purified ryanodine receptor from skeletal muscle sarcoplasmic reticulum is the Ca^{2+}-permeable pore of the calcium release channel *J. Biol. Chem.* **262** 16636–43

Inui M, Saito A and Fleischer S 1987 Purification of the ryanodine receptor and identity with feet structures of junctional terminal cisternae of sarcoplasmic reticulum from fast skeletal muscle *J. Biol. Chem.* **262** 1740–7

Ito K, Komazaki S, Yoshida M, Nishi M, Kitamura K and Takeshima H 2001 Deficiency of triad junction and contraction in mutant skeletal muscle lacking junctophilin type 1 *J. Cell Biol.* **154** 1059–67

Iyer S R, Valencia A P, Hernández-Ochoa E O and Lovering R M 2016 *In vivo* assessment of muscle contractility in animal studies *Methods Mol. Biol.* **1460** 293–307

Jiang Y H, Klein M G and Schneider M F 1999 Numerical simulation of Ca^{2+} 'sparks' in skeletal muscle *Biophys. J.* **77** 2333–57

Jungbluth H 2007 Central core disease *Orphanet J. Rare Dis.* **2** 25

Jungbluth H and Gautel M 2014 Pathogenic mechanisms in centronuclear myopathies *Front. Aging Neurosci.* **6** 339

Jungbluth H *et al* 2002 Autosomal recessive inheritance of RYR1 mutations in a congenital myopathy with cores *Neurology* **59** 284–7

Jungbluth H, Treves S, Zorzato F, Sarkozy A, Ochala J, Sewry C, Phadke R, Gautel M and Muntoni F 2018 Congenital myopathies: disorders of excitation-contraction coupling and muscle contraction *Nat. Rev. Neurol.* **14** 151–67

Jurkat-Rott K *et al* 1994 A calcium channel mutation causing hypokalemic periodic paralysis *Hum. Mol. Genet.* **3** 1415–9

Kabbara A A and Allen D G 2001 The use of the indicator fluo-5N to measure sarcoplasmic reticulum calcium in single muscle fibres of the cane toad *J. Physiol.* **534** 87–97

Kamada T and Kinoshita H 1943 Disturbances initiated from the naked surface of muscle protoplasm *Japan. J. Zool.* **10** 469–93

Katz B 1996 Neural transmitter release: from quantal secretion to exocytosis and beyond. The Fenn Lecture *J. Neurocytol.* **25** 677–86

Klaus M M, Scordilis S P, Rapalus J M, Briggs R T and Powell J A 1983 Evidence for dysfunction in the regulation of cytosolic Ca^{2+} in excitation-contraction uncoupled dysgenic muscle *Dev. Biol.* **99** 152–65

Klein M G, Cheng H, Santana L F, Jiang Y-H, Lederer W J and Schneider M F 1996 Two mechanisms of quantized calcium release in skeletal muscle *Nature* **379** 455–8

Klein M G, Lacampagne A and Schneider M F 1999 A repetitive mode of activation of discrete Ca^{2+} release events (Ca^{2+} sparks) in frog skeletal muscle fibres *J. Physiol.* **515** 391–411

Klein M G and Schneider M F 2006 Ca^{2+} sparks in skeletal muscle *Prog. Biophys. Mol. Biol.* **92** 308–32

Klein M G, Simon B J, Szucs G and Schneider M F 1988 Simultaneous recording of calcium transients in skeletal muscle using high- and low-affinity calcium indicators *Biophys. J.* **53** 971–88

Knudson C M, Chaudhari N, Sharp A H, Powell J A, Beam G G and Campbell K P 1989 Specific absence of the alpha 1 subunit of the dihydropyridine receptor in mice with muscular dysgenesis *J. Biol. Chem.* **264** 1345–8

Konishi M and Baylor S M 1991 Myoplasmic calcium transients monitored with purpurate indicator dyes injected into intact frog skeletal muscle fibers *J. Gen. Physiol.* **97** 245–70

Konishi M, Hollingworth S, Arkins A B and Bayor S M 1991 Myoplasmic calcium transients in intact frog skeletal muscle fibers monitored with the fluorescent indicator furaptra *J. Gen. Physiol.* **97** 271–301

Kovacs L, Ríos E and Schneider M F 1979 Calcium transients and intramembrane charge movement in skeletal muscle fibres *Nature* **279** 391–6

Kovacs L, Rios E and Schneider M F 1983 Measurement and modification of free calcium transients in frog skeletal muscle fibres by a metallochromic indicator dye *J. Physiol.* **343** 161–96

Kugler G, Weiss R G, Flucher B E and Grabner M 2004 Structural requirements of the dihydropyridine receptor alpha1S II–III loop for skeletal-type excitation-contraction coupling *J. Biol. Chem.* **279** 4721–8

Lamb G D and Stephenson D G 1991 Effect of Mg^{2+} on the control of Ca^{2+} release in skeletal muscle fibres of the toad *J. Physiol.* **434** 507–28

Landstrom A P, Beavers D L and Wehrens X H T 2014 The junctophilin family of proteins: from bench to bedside *Trends Mol. Med.* **20** 353–62

Lanner J T, Georgiou D K, Joshi A D and Hamilton S L 2010 Ryanodine receptors: structure, expression, molecular details, and function in calcium release *Cold Spring Harb. Perspect. Biol.* **2** a003996

Launikonis B S, Zhou J, Royer L, Shannon T R, Brum G and Ríos E 2005 Confocal imaging of $[Ca^{2+}]$ in cellular organelles by SEER, shifted excitation and emission ratioing of fluorescence *J. Physiol.* **567** 523–43

Lee C S *et al* 2015 Ca^{2+} permeation and/or binding to $Ca_V1.1$ fine-tunes skeletal muscle Ca^{2+} signaling to sustain muscle function *Skelet. Muscle* **5** 4

Leong P and MacLennan D H 1998 A 37-amino acid sequence in the skeletal muscle ryanodine receptor interacts with the cytoplasmic loop between domains II and III in the skeletal muscle dihydropyridine receptor *J. Biol. Chem.* **273** 7791–4

Liang H, DeMaria C D, Erickson M G, Mori M X, Alseikhan B A and Yue D T 2003 Unified mechanisms of Ca^{2+} regulation across the Ca^{2+} channel family *Neuron* **39** 951–60

Ludtke S J, Serysheva I I, Hamilton S L and Chiu W 2005 The pore structure of the closed RyR1 channel *Structure* **13** 1203–11

Ma K, Mallidis C, Bhasin S, Mahabadi V, Araza J, Gonzalez-Cadavid, Arias J and Salehian B 2003 Glucocorticoid-induced skeletal muscle atrophy is associated with upregulation of myostatin gene expression *Am. J. Physiol. Endocrinol. Metab.* **285** E363–71

MacLennan D H and Wong P T 1971 Isolation of a calcium-sequestering protein from sarcoplasmic reticulum *Proc. Natl. Acad. Sci. USA* **68** 1231–35

Maclennan D H and Zvaritch E 2011 Mechanistic models for muscle diseases and disorders originating in the sarcoplasmic reticulum *Biochim. Biophys. Acta* **1813** 948–64

Manning D R and Stull J T 1979 Myosin light chain phosphorylation and phosphorylase A activity in rat extensor digitorum longus muscle *Biochem. Biophys. Res. Commun.* **90** 164–70

Marengo J J, Hidalgo C and Bull R 1998 Sulfhydryl oxidation modifies the calcium dependence of ryanodine-sensitive calcium channels of excitable cells *Biophys. J.* **74** 1263–77

Marx S O, Ondrias K and Marx A R 1998 Coupled gating between individual skeletal muscle Ca^{2+} release channels (ryanodine receptors) *Science* **281** 818–21

Matthews E *et al* 2018 Atypical periodic paralysis and myalgia: a novel RYR1 phenotype *Neurology* **90** e412–8

Meissner G 2017 The structural basis of ryanodine receptor ion channel function *J. Gen. Physiol.* **149** 1065–89

Meissner G, Darling E and Eveleth J 1986 Kinetics of rapid Ca^{2+} release by sarcoplasmic reticulum. Effects of Ca^{2+}, Mg^{2+}, and adenine nucleotides *Biochemistry* **25** 236–44

Melzer W, Ríos E and Schneider M F 1984 Time course of calcium release and removal in skeletal muscle fibers *Biophys. J.* **45** 637–41

Melzer W, Ríos E and Schneider M F 1987 A general procedure for determining the rate of calcium release from the sarcoplasmic reticulum in skeletal muscle fibers *Biophys. J.* **51** 849–63

Melzer W, Schneider M F, Simon B J and Szucs G 1986 Intramembrane charge movement and calcium release in frog skeletal muscle *J. Physiol.* **373** 481–511

Miledi R, Parker I and Schalow G 1977 Measurement of calcium transients in frog muscle by the use of arsenazo III *Proc. R. Soc. Lond. B Biol. Sci* **198** 201–10

Monnier N, Procaccio V, Stieglitz P and Lunardi J 1997 Malignant-hyperthermia susceptibility is associated with a mutation of the alpha 1-subunit of the human dihydropyridine-sensitive L-type voltage-dependent calcium-channel receptor in skeletal muscle *Am. J. Hum. Genet.* **60** 1316–25

Moss R L 1992 Ca^{2+} regulation of mechanical properties of striated muscle. Mechanistic studies using extraction and replacement of regulatory proteins *Circ. Res.* **70** 865–84

Nakai J, Adams B A, Imoto K and Beam K G 1994 Critical roles of the S3 segment and S3-S4 linker of repeat I in activation of L-type calcium channels *Proc. Natl Acad. Sci. USA* **91** 1014–18

Nakai J, Dirksen R T, Nguyen H T, Pessah I N, Beam K G and Allen P 1996 Enhanced dihydropyridine receptor channel activity in the presence of ryanodine receptor *Nature* **380** 72–5

Nakai J, Ogura T, Protasi F, Franzini-Armstrong C, Allen P D and Beam K G 1997 Functional nonequality of the cardiac and skeletal ryanodine receptors *Proc. Natl. Acad. Sci. USA* **94** 1019–22

Nakai J, Sekiguchi N, Rando T A, Allen P D and Beam K G 1998a Two regions of the ryanodine receptor involved in coupling with L-type Ca^{2+} channels *J. Biol. Chem.* **273** 13403–6

Nakai J, Tanabe T, Konno T, Adams B and Beam K G 1998b Localization in the II–III loop of the dihydropyridine receptor of a sequence critical for excitation-contraction coupling *J. Biol. Chem.* **273** 24983–6

Nelson B R, Wu F, Liu Y, Anderson D M, McNally J, Lin W, Cannon S C, Bassel-Duby R and Olson E N 2013 Skeletal muscle-specific T-tubule protein STAC3 mediates voltage-induced Ca^{2+} release and contractility *Proc. Natl Acad. Sci. USA* **110** 11881–6

Niu J, Yang W, Yue D T, Inoue T and Ben-Johny M 2018 Duplex signaling by CaM and Stac3 enhances $Ca_V1.1$ function and provides insights into congenital myopathy *J. Gen. Physiol.* **150** 1145–61

Ogawa H, Kurebayashi N, Yamazawa T and Murayama T 2020 Regulatory mechanisms of ryanodine receptor/Ca^{2+} release channel revealed by recent advancements in structural studies *J. Muscle Res. Cell Motil.* **42** 291–304

Ogawa Y, Harafuji H and Kurebayashi N 1980 Comparison of the characteristics of four metallochromic dyes as potential calcium indicators for biological experiments *J. Biochem.* **87** 1293–303

Ohrtman J, Ritter B, Polster A, Beam K G and Papadopoulos S 2008 Sequence differences in the IQ motifs of $Ca_V1.1$ and $Ca_V1.2$ strongly impact calmodulin binding and calcium-dependent inactivation *J. Biol. Chem.* **283** 29301–11

Olojo R O, Hernández-Ochoa E O, Ikemoto N and Schneider M F 2011a Effects of conformational peptide probe DP4 on bidirectional signaling between DHPR and RyR1 calcium channels in voltage-clamped skeletal muscle fibers *Biophys. J.* **100** 2367–77

Olojo R O, Ziman A P, Hernández-Ochoa E O, Allen P D, Schneider M F and Ward C W 2011b Mice null for calsequestrin 1 exhibit deficits in functional performance and sarcoplasmic reticulum calcium handling *PLoS One* **6** e27036

Olsson K, Cheng A J, Seher A, Al-Ameri M, Rullman E, Westerblad H, Lanner J T, Bruton J D and Gustafsson T 2015 Intracellular Ca^{2+}-handling differs markedly between intact human muscle fibers and myotubes *Skelet. Muscle* **5** 26

Ording H 1985 Incidence of malignant hyperthermia in Denmark *Anesth. Analg.* **64** 700–4

Palade P and Vergara J 1982 Arsenazo III and antipyrylazo III calcium transients in single skeletal muscle fibers *J. Gen. Physiol.* **79** 679–707

Pantazis A, Savalli N, Sigg D, Neely A and Olcese R 2014 Functional heterogeneity of the four voltage sensors of a human L-type calcium channel *Proc. Natl. Acad. Sci. USA* **111** 18381–6

Parmacek M S and Leiden J M 1991 Structure, function, and regulation of troponin C *Circulation* **84** 991–1003

Peachey L D 1965a The sarcoplasmic reticulum and transverse tubules of the frog's sartorius *J. Cell Biol.* **25** Suppl: 209–31

Peachey L D 1965b Transverse tubules in excitation-contraction coupling *Fed. Proc.* **24** 1124–34

Pechere J F, Derancourt J and Haiech J 1977 The participation of parvalbumins in the activation-relaxation cycle of vertebrate fast skeletal-muscle *FEBS Lett.* **75** 111–4

Perni S, Lavorato M and Beam K G 2017 De novo reconstitution reveals the proteins required for skeletal muscle voltage-induced Ca^{2+} release *Proc. Natl Acad. Sci. USA* **114** 13822–27

Peterson B Z, Lee J S, Mulle J G, Wang Y, de Leon M and Yue D T 2000 Critical determinants of Ca^{2+}-dependent inactivation within an EF-hand motif of L-type Ca^{2+} channels *Biophys. J.* **78** 1906–20

Pettersen E F, Goddard T D, Huang C C, Couch G S, Greenblatt D M, Meng E C and Ferrin T E 2004 UCSF Chimera – a visualization system for exploratory research and analysis *J. Comput. Chem.* **25** 1605–12

Polster A, Nelson B R, Papadopoulos S, Olson E N and Beam K G 2018 Stac proteins associate with the critical domain for excitation-contraction coupling in the II–III loop of $Ca_V1.1$ *J. Gen. Physiol.* **150** 613–24

Polster A, Perni S, Bichraoui H and Beam K G 2015 Stac adaptor proteins regulate trafficking and function of muscle and neuronal L-type Ca^{2+} channels *Proc. Natl Acad. Sci. USA* **112** 602–6

Porter K R and Palade G E 1957 Studies on the endoplasmic reticulum. III. Its form and distribution in striated muscle cells *J. Biophys. Biochem. Cytol.* **3** 269–300

Pouvreau S, Royer L, Yi J, Brum G, Meissner G, Ríos E and Zhou J 2007 Ca^{2+} sparks operated by membrane depolarization require isoform 3 ryanodine receptor channels in skeletal muscle *Proc. Natl Acad. Sci. USA* **104** 5235–40

Pozzan T, Rizzuto R, Volpe P and Melolesi J 1994 Molecular and cellular physiology of intracellular calcium stores *Physiol. Rev.* **74** 595–636

Proenza C, O'Brien J, Nakai J, Mukherjee S and Allen P D 2002 Identification of a region of RyR1 that participates in allosteric coupling with the alpha(1S) ($Ca_V1.1$) II–III loop *J. Biol. Chem.* **277** 6530–35

Prosser B L, Hernández-Ochoa E O, Lovering R M, Andronache Z, Zimmer D B, Melzer W and Schneider M F 2010 S100A1 promotes action potential-initiated calcium release flux and force production in skeletal muscle *Am. J. Physiol. Cell Physiol.* **299** C891–902

Prosser B L, Hernández-Ochoa E O, Zimmer D B and Schneider M F 2009 Simultaneous recording of intramembrane charge movement components and calcium release in wild-type and S100A1−/− muscle fibres *J. Physiol.* **587** 4543–59

Prosser B L *et al* 2008 S100A1 binds to the calmodulin-binding site of ryanodine receptor and modulates skeletal muscle excitation-contraction coupling *J. Biol. Chem.* **283** 5046–57

Protasi F, Paolini C, Nakai J, Beam K G, Franzini-Armstrong C and Allen P D 2002 Multiple regions of RyR1 mediate functional and structural interactions with alpha(1S)-dihydropyridine receptors in skeletal muscle *Biophys. J.* **83** 3230–44

Quane K A *et al* 1993 Mutations in the ryanodine receptor gene in central core disease and malignant hyperthermia *Nat. Genet.* **5** 51–5

Radermacher M, Wagenknecht T, Grassucci R, Frank J, Inui M, Chadwick C and Fleischer S 1992 Cryo-EM of the native structure of the calcium release channel/ryanodine receptor from sarcoplasmic reticulum *Biophys. J.* **61** 936–40

Ridgway E B and Ashley C C 1967 Calcium transients in single muscle fibers *Biochem. Biophys. Res. Commun.* **29** 229–34

Ringer S 1887 Regarding the action of lime, potassium and sodium salts on skeletal muscle *J. Physiol.* **8** 20–4

Ríos E and Brum G 1987 Involvement of dihydropyridine receptors in excitation-contraction coupling in skeletal muscle *Nature* **325** 717–20

Ríos E and Brum G 2002 Ca^{2+} release flux underlying Ca^{2+} transients and Ca^{2+} sparks in skeletal muscle *Front. Biosci.* **7** d1195–1211

Ríos E, Launikonis B S, Royer L, Brum G and Zhou J 2006 The elusive role of store depletion in the control of intracellular calcium release *J. Muscle Res. Cell Motil.* **27** 337–50

Ríos E and Pizarro G 1991 Voltage sensor of excitation-contraction coupling in skeletal muscle *Physiol. Rev.* **71** 849–908

Ríos E and Schneider M F 1981 Stoichiometry of the reactions of calcium with the metal-lochromic indicator dyes antipyrylazo III and arsenazo III *Biophys. J.* **36** 607–21

Robin G and Allard B 2015 Voltage-gated Ca^{2+} influx through L-type channels contributes to sarcoplasmic reticulum Ca^{2+} loading in skeletal muscle *J. Physiol.* **593** 4781–97

Robinson R, Carpenter D, Shaw M-A, Halsall J and Hopkins P 2006 Mutations in RYR1 in malignant hyperthermia and central core disease *Hum. Mutat.* **27** 977–89

Robison P, Hernández-Ochoa E O and Schneider M F *et al* 2011 Adherent primary cultures of mouse intercostal muscle fibers for isolated fiber studies *J. Biomed. Biotechnol.* **2011** 393740

Rosenberg H, Davis M, James D, Pollock N and Stowell K 2007 Malignant hyperthermia *Orphanet J. Rare Dis.* **2** 21

Rosenberg H, Pollock N, Schiemann A, Bulger T and Stowell K 2015 Malignant hyperthermia: a review *Orphanet J. Rare Dis.* **10** 93

Rosenberg H, Sambuughin N and Riazi S 1993 Malignant hyperthermia susceptibility. *GeneReviews* ed M P Adam *et al* (Seattle, WA: University of Washington) https://www.ncbi.nlm.nih.gov/books/NBK1146/

Rousseau E, Ladine J, Liu Q-Y and Meissner G 1988 Activation of the Ca^{2+} release channel of skeletal muscle sarcoplasmic reticulum by caffeine and related compounds *Arch. Biochem. Biophys.* **267** 75–86

Royer L, Pouvreau S and Ríos *et al* 2008 Evolution and modulation of intracellular calcium release during long-lasting, depleting depolarization in mouse muscle *J. Physiol.* **586** 4609–29

Rudolf R, Mongillo M, Rizzuto R and Pozzan T 2003 Looking forward to seeing calcium *Nat. Rev. Mol. Cell Biol.* **4** 579–86

Samso M 2017 A guide to the 3D structure of the ryanodine receptor type 1 by cryoEM *Protein Sci.* **26** 52–68

Samsó M, Feng W, Pessah I N and Allen P D 2009 Coordinated movement of cytoplasmic and transmembrane domains of RyR1 upon gating *PLoS Biol.* **7** e85

Samso M, Wagenknecht T and Allen P D 2005 Internal structure and visualization of trans-membrane domains of the RyR1 calcium release channel by cryo-EM *Nat. Struct. Mol. Biol.* **12** 539–44

Sanchez J A and Stefani E 1978 Inward calcium current in twitch muscle fibres of the frog *J. Physiol.* **283** 197–209

Sandow A 1952 Excitation-contraction coupling in muscular response *Yale J. Biol. Med.* **25** 176–201
Schartner V, Laporte J and Böhm J 2019 Abnormal excitation-contraction coupling and calcium homeostasis in myopathies and cardiomyopathies *J. Neuromusc. Dis.* **6** 289–305
Schartner V *et al* 2017 Dihydropyridine receptor (DHPR, CACNA1S) congenital myopathy *Acta Neuropathol.* **133** 517–33
Schneider M F 1994 Control of calcium release in functioning skeletal muscle fibers *Annu. Rev. Physiol.* **56** 463–84
Schneider M F 1999 Ca^{2+} sparks in frog skeletal muscle: generation by one, some, or many SR Ca^{2+} release channels? *J. Gen. Physiol.* **113** 365–72
Schneider M F and Chandler W K 1973 Voltage dependent charge movement of skeletal muscle: a possible step in excitation-contraction coupling *Nature* **242** 244–6
Schneider M F and Hernandez-Ochoa E O 2012 Skeletal muscle excitation-contraction coupling *Muscle: Fundamental Biology and Mechanisms of Disease* vol 2 . ed J A Hill and E Olson (London: Academic) pp 811–22
Schneider M F and Klein M G 1996 Sarcomeric calcium sparks activated by fiber depolarization and by cytosolic Ca^{2+} in skeletal muscle *Cell Calcium* **20** 123–8
Schneider M F, Rios E and Melzer W 1985 Use of a metallochromic indicator to study intracellular calcium movements in skeletal muscle *Cell Calcium* **6** 109–18
Schneider M F and Simon B J 1988 Inactivation of calcium release from the sarcoplasmic reticulum in frog skeletal muscle *J. Physiol.* **405** 727–45
Schneider M, Simon B J and Szucs G 1987 Depletion of calcium from the sarcoplasmic reticulum during calcium release in frog skeletal muscle *J. Physiol.* **392** 167–92
Schredelseker J, Dayal A, Schwerte T and Franzini-Armstrong C 2009 Proper restoration of excitation-contraction coupling in the dihydropyridine receptor beta1-null zebrafish relaxed is an exclusive function of the beta1a subunit *J. Biol. Chem.* **284** 1242–51
Schredelseker J, Di Biase V, Obermair G J, Tatiana-Felder E, Flucher B E, Fanzini-Armstrong C and Grabner M 2005 The beta 1a subunit is essential for the assembly of dihydropyridine-receptor arrays in skeletal muscle *Proc. Natl. Acad. Sci. USA* **102** 17219–24
Serysheva I I, Hamilton S L, Chiu W and Ludtke S J 2005 Structure of Ca^{2+} release channel at 14 Å resolution *J. Mol. Biol.* **345** 427–31
Serysheva I I, Orlova E V, Chiu W, Sherman M B, Hamilton S L and van Heel M 1995 Electron cryomicroscopy and angular reconstitution used to visualize the skeletal muscle calcium release channel *Nat. Struct. Biol.* **2** 18–24
Sheridan D C, Takekura H, Franzini-Armstrong C, Beam K G, Allen P D and Perez C F 2006 Bidirectional signaling between calcium channels of skeletal muscle requires multiple direct and indirect interactions *Proc. Natl Acad. Sci. USA* **103** 19760–5
Shirokova N, Garcia J and Ríos E 1996 Ca^{2+} release from the sarcoplasmic reticulum compared in amphibian and mammalian skeletal muscle *J. Gen. Physiol.* **107** 1–18
Shirokova N, Garcia J and Ríos E 1998 Local calcium release in mammalian skeletal muscle *J. Physiol.* **512** 377–84
Shirokova N, Shirokov R, Rossi D, González A, Kirsche W G, Gárcia J, Sorrentino V and Ríos E 1999 Spatially segregated control of Ca^{2+} release in developing skeletal muscle of mice *J. Physiol.* **521** 483–95
Shuaib A, Paasuke R T, Brownell W and Keith A 1987 Central core disease. Clinical features in 13 patients *Medicine (Baltimore)* **66** 389–96

Simon B J and Hill D A 1992 Charge movement and SR calcium release in frog skeletal muscle can be related by a Hodgkin–Huxley model with four gating particles *Biophys. J.* **61** 1109–16

Simon B J, Klein M G and Schneider M F 1991 Calcium dependence of inactivation of calcium release from the sarcoplasmic reticulum in skeletal muscle fibers *J. Gen. Physiol.* **97** 437–71

Simon B J and Schneider M F 1988 Time course of activation of calcium release from sarcoplasmic reticulum in skeletal muscle *Biophys. J.* **54** 1159–63

Stanfield P R 1977 A calcium dependent inward current in frog skeletal muscle fibres *Pflugers Arch.* **368** 267–70

Statland J M *et al* 2018 Review of the diagnosis and treatment of periodic paralysis *Muscle Nerve* **57** 522–30

Stern M D, Pizarro G and Rios E 1997 Local control model of excitation-contraction coupling in skeletal muscle *J. Gen. Physiol.* **110** 415–40

Stroffekova K 2008 Ca^{2+}/CaM-dependent inactivation of the skeletal muscle L-type Ca^{2+} channel ($Ca_V1.1$) *Pflugers Arch.* **455** 873–84

Stroffekova K 2011 The IQ motif is crucial for $Ca_V1.1$ function *J. Biomed. Biotechnol.* **2011** 504649

Strube C, Beurg M, Powers P A, Gregg R G and Coronado R 1996 Reduced Ca^{2+} current, charge movement, and absence of Ca^{2+} transients in skeletal muscle deficient in dihydropyridine receptor beta 1 subunit *Biophys. J.* **71** 2531–43

Struk A, Szucs G, Kemmer H and Melzer W 1998 Fura-2 calcium signals in skeletal muscle fibres loaded with high concentrations of EGTA *Cell Calcium* **23** 23–32

Sztretye M, Yi J, Figueroa L, Zhou J, Royer L, Allen P, Brum G and Ríos E 2011 Measurement of RyR permeability reveals a role of calsequestrin in termination of SR Ca^{2+} release in skeletal muscle *J. Gen. Physiol.* **138** 231–47

Tadross M R, Dick I E and Yue D T 2008 Mechanism of local and global Ca^{2+} sensing by calmodulin in complex with a Ca^{2+} channel *Cell* **133** 1228–40

Takahashi M and Catterall W A 1987 Identification of an alpha subunit of dihydropyridine-sensitive brain calcium channels *Science* **236** 88–91

Takahashi M, Seagar M J, Jones J F, Reber B F and Catterall W A 1987 Subunit structure of dihydropyridine-sensitive calcium channels from skeletal muscle *Proc. Natl Acad. Sci. USA* **84** 5478–82

Takeshima H *et al* 1989 Primary structure and expression from complementary DNA of skeletal muscle ryanodine receptor *Nature* **339** 439–45

Takeshima H, Yamazawa T, Ikemoto T, Takehura H, Nishi M, Noda T and Lino M 1995 Ca^{2+}-induced Ca^{2+} release in myocytes from dyspedic mice lacking the type-1 ryanodine receptor *EMBO J.* **14** 2999–3006

Tanabe T, Beam K G, Adams B A, Niidome T and Numa S 1990a Regions of the skeletal muscle dihydropyridine receptor critical for excitation-contraction coupling *Nature* **346** 567–69

Tanabe T, Beam K G, Powell J A and Numa S 1988 Restoration of excitation-contraction coupling and slow calcium current in dysgenic muscle by dihydropyridine receptor complementary DNA *Nature* **336** 134–9

Tanabe T, Mikami A, Numa S and Beam K G 1990b Cardiac-type excitation-contraction coupling in dysgenic skeletal muscle injected with cardiac dihydropyridine receptor cDNA *Nature* **344** 451–3

Tanabe T, Takeshima H, Mikami A, Flockerzi V, Takahashi H, Kangawa K,, Kojima M, Matsuo H, Hirose T and Numa S 1987 Primary structure of the receptor for calcium channel blockers from skeletal muscle *Nature* **328** 313–8

Timmer J, Müller T and Melzer W 1998 Numerical methods to determine calcium release flux from calcium transients in muscle cells *Biophys. J.* **74** 1694–707

Treves S, Scutari E, Robert M, Groh S, Ottolia M, Prestipino G, Ronjat M and Zorzato F 1997 Interaction of S100A1 with the Ca^{2+} release channel (ryanodine receptor) of skeletal muscle *Biochemistry* **36** 11496–503

Treves S, Vukcevic M, Maj M, Thurnheer R, Mosca B and Zorzato F 2009 Minor sarcoplasmic reticulum membrane components that modulate excitation-contraction coupling in striated muscles *J. Physiol.* **587** 3071–9

Tripathy A, Xu L, Mann G and Meissner G 1995 Calmodulin activation and inhibition of skeletal muscle Ca^{2+} release channel (ryanodine receptor) *Biophys. J.* **69** 106–19

Tsien R Y 1980 New calcium indicators and buffers with high selectivity against magnesium and protons: design, synthesis, and properties of prototype structures *Biochemistry* **19** 2396–404

Tsien R Y 1981 A non-disruptive technique for loading calcium buffers and indicators into cells *Nature* **290** 527–8

Tsugorka A, Rios E and Blatter L A 1995 Imaging elementary events of calcium release in skeletal muscle cells *Science* **269** 1723–6

Tuluc P, Benedetti B, Coste de Bagneaux P, Grabner M and flucher B E 2016a Two distinct voltage-sensing domains control voltage sensitivity and kinetics of current activation in $Ca_V1.1$ calcium channels *J.Gen. Physiol.* **147** 437–49

Tuluc P and Flucher B E 2011 Divergent biophysical properties, gating mechanisms, and possible functions of the two skeletal muscle $Ca_V1.1$ calcium channel splice variants *J. Muscle Res. Cell Motil.* **32** 249–56

Tuluc P, Molenda N, Schlick B, Obermair G J and Flucher B E 2009 A $Ca_V1.1$ Ca^{2+} channel splice variant with high conductance and voltage-sensitivity alters EC coupling in developing skeletal muscle *Biophys. J.* **96** 35–44

Tuluc P, Yarov-Yarovoy V, Benedetti B and Flucher B E 2016b Molecular interactions in the voltage sensor controlling gating properties of Ca_V calcium channels *Structure* **24** 261–71

Van Petegem F, Clark K A, Chatelain F C and Minor D L Jr 2004 Structure of a complex between a voltage-gated calcium channel beta-subunit and an alpha-subunit domain *Nature* **429** 671–5

Wang X *et al* 2005 Uncontrolled calcium sparks act as a dystrophic signal for mammalian skeletal muscle *Nat. Cell Biol.* **7** 525–30

Weber A 1959 On the role of calcium in the activity of adenosine 5′-triphosphate hydrolysis by actomyosin *J. Biol. Chem.* **234** 2764–9

Weber F and Lehmann-Horn F 1993 Hypokalemic periodic paralysis *GeneReviews* ed M P Adam *et al* (Seattle, WA: University of Washington)

Wilkens C M and Beam K G 2003 Insertion of alpha1S II–III loop and C terminal sequences into alpha1H fails to restore excitation-contraction coupling in dysgenic myotubes *J. Muscle Res. Cell Motil.* **24** 99–109

Witting N *et al* 2018 Phenotype and genotype of muscle ryanodine receptor rhabdomyolysis-myalgia syndrome *Acta Neurol. Scand.* **137** 452–61

Witting N, Werlauff U, Duno M and Vissing J 2017 Phenotypes, genotypes, and prevalence of congenital myopathies older than 5 years in Denmark *Neurol. Genet.* **3** e140

Wu F, Mi W, Hernández E O, Burns D K, Fu Y, Gray H F, Struyk A F, Schneider M F and Cannon S C 2012 A calcium channel mutant mouse model of hypokalemic periodic paralysis *J. Clin. Invest.* **122** 4580–91

Wu J, Yan Z, Li Z, Qian X, Lu S, Dong M, Zhou Q and Yan N 2016 Structure of the voltage-gated calcium channel Ca(v)1.1 at 3.6 A resolution *Nature* **537** 191–6

Yan Z *et al* 2015 Structure of the rabbit ryanodine receptor RyR1 at near-atomic resolution *Nature* **517** 50–5

Yin C C and Lai F A 2000 Intrinsic lattice formation by the ryanodine receptor calcium-release channel *Nat. Cell Biol.* **2** 669–71

Yue D T 2004 The dawn of high-resolution structure for the queen of ion channels *Neuron* **42** 357–9

Zalk R and Clarke O B *et al* 2015 Structure of a mammalian ryanodine receptor *Nature* **517** 44–9

Zhang L, Kelley J, Schmeisser G, Kobayashi Y M and Jones L R 1997 Complex formation between junctin, triadin, calsequestrin, and the ryanodine receptor. Proteins of the cardiac junctional sarcoplasmic reticulum membrane *J. Biol. Chem.* **272** 23389–97

Zhang Y, Chen H S, Khanna V K, De Leon S, Phillips M S and Schappert K, Britt B A, Brownell K W and MacLennan D H 1993 A mutation in the human ryanodine receptor gene associated with central core disease *Nat. Genet.* **5** 46–50

Zhou J, Stern M D, Brum G and Rios E 2004 The elementary events of Ca^{2+} release elicited by membrane depolarization in mammalian muscle *J. Physiol.* **557** 43–58

Zhou J, Yi J, Royer L, Launikonis B S, Gonzázez A, Gárcia J and Ríos E 2006 A probable role of dihydropyridine receptors in repression of Ca^{2+} sparks demonstrated in cultured mammalian muscle *Am. J. Physiol. Cell Physiol.* **290** C539–53

Zoghbi M E, Bolanos P, Villaba-Galea C, Marcano A, Hernández, Fill A and Escobar A L 2000 Spatial Ca^{2+} distribution in contracting skeletal and cardiac muscle cells *Biophys. J.* **78** 164–73

Chapter 9

Voltage-gated Ca^{2+} channels and excitation–secretion coupling at the synapse

Brittany Williams, Jessica R Thomas and Amy Lee

At chemical synapses, voltage-gated (Ca_V) Ca^{2+} channels mediate the influx of Ca^{2+} ions that initiate the exocytotic release of neurotransmitters. In coupling electrical activity to neurosecretion, Ca_V channels are powerfully situated to shape the flow of information through the nervous system. This chapter describes the factors that regulate the contributions of presynaptic Ca_V channels to neurotransmitter release, with a particular emphasis on Ca^{2+}-dependent modulators of Ca_V channels at distinct types of synapses.

9.1 Introduction

Chemical synapses are specialized sites of information transfer between neurons where the presynaptic release of neurotransmitter modifies the activity of post-synaptic partners. Voltage-gated Ca^{2+} (Ca_V) channels are essential players in this process. Tightly clustered near synaptic vesicles, Ca_V channels open in response to presynaptic depolarization and the resulting Ca^{2+} influx triggers a series of biochemical events that culminates in vesicle fusion and exocytosis of neuroactive chemicals into the synaptic cleft. Because small changes in the presynaptic Ca^{2+} concentration can have dramatic effects on the properties of neurotransmitter release, factors that modulate the activity of Ca_V channels can profoundly influence how information is encoded in the nervous system. This chapter focuses on the roles of Ca_V channels in regulating various aspects of synaptic transmission, and how modulation of these channels by Ca^{2+} sensors influences these roles.

9.2 Ca_V subtypes modulate release properties at different types of synapses

Ca_V channels are multi-protein complexes consisting mainly of a pore-forming α_1 subunit, an intracellular β subunit, and an extracellular $\alpha_2\delta$ subunit (Simms and

doi:10.1088/978-0-7503-2009-2ch9

Table 9.1. Voltage-gated Ca^{2+} (Ca_V) channel genetic nomenclature, protein and physiology classification, pharmacological modulators.

Gene	Protein	α_1 Subunit name	Physiology	Pharmacological modulators
CACNA1S	$Ca_V1.1$	α1S	L-type	Dihydropyridines
CACNA1C	$Ca_V1.2$	α1C		Phenylalkylamines
CACNA1D	$Ca_V1.3$	α1D		Benzothiazepines
CACNA1F	$Ca_V1.4$	α1F		
CACNA1A	$Ca_V2.1$	α1A	P/Q type	ω-agatoxin
CACNA1B	$Ca_V2.2$	α1B	N-type	ω-conotoxin
CACNA1E	$Ca_V2.3$	α1E	R-type	SNX-482
CACNA1G	$Ca_V3.1$	α1G	T-type	MibefradilTTA-P2
CACNA1H	$Ca_V3.2$	α1H		
CACNA1I	$Ca_V3.3$	α1I		

Zamponi 2014). Ca_V channels are named for their α_1 subunit, such that $Ca_V2.1$ refers to channels containing the α_{1A} subunit (table 9.1, Ertel *et al* (2000)).

9.2.1 Ca_V2 subtypes mediate fast neurotransmitter release at most conventional synapses

Based initially on pharmacological evidence, most synapses have been found to rely on $Ca_V2.x$ channels. At excitatory synapses utilizing glutamate as a neurotransmitter, release is blocked generally by antagonists of $Ca_V2.1$ and/or $Ca_V2.2$ (Takahashi and Momiyama 1993, Wheeler *et al* 1994). At inhibitory synapses, the Ca_V2 subtype involved in the release of GABA differs according to cell-type. In the cerebellum, GABA-ergic synapses utilize primarily $Ca_V2.1$ (Iwasaki *et al* 2000, Stephens *et al* 2001), whereas in the hippocampus, presynaptic $Ca_V2.2$ or $Ca_V2.1$ mediate GABA release from different classes of interneurons (Hefft and Jonas 2005).

Some synapses are characterized by multiple Ca_V2 subtypes with distinct properties, which may help tailor synaptic responses to certain patterns of activity. For example, all the Ca_V2 subtypes are found in the mossy fiber boutons of the dentate granule cells that form synapses with CA3 pyramidal cells in the hippocampus. Here, $Ca_V2.1$ is activated by standard action potential waveforms, contributing to 'nanodomain' Ca^{2+} that supports fast glutamate release. Compared to the other Ca_V2 subtypes, $Ca_V2.3$ activates at more negative voltages and thus can promote release during subthreshold depolarizations (Dietrich *et al* 2003). In addition, repetitive (i.e., tetanic) stimulation facilitates the opening of $Ca_V2.3$ channels, which augments a spatially discrete microdomain of Ca^{2+} that triggers long-term and post-tetanic potentiation (Breustedt *et al* 2003, Dietrich *et al* 2003). Functional specialization of Ca_V2 subtypes is an efficient strategy by which neuromodulators could selectively trigger synaptic plasticity without affecting basal release properties.

At a number of synapses, including the Calyx of Held synapse in the auditory brainstem, there is a developmental shift from a mixed population of Ca_V2 subtypes to one which is dominated by $Ca_V2.1$ (Iwasaki and Takahashi 1998, Ishikawa *et al* 2005, Nakamura *et al* 2015). $Ca_V2.1$ is generally associated with 'tight' coupling to synaptic vesicles, where Ca^{2+} entry through relatively few channels triggers release (Eggermann *et al* 2011). This may be due to selective clustering of $Ca_V2.1$ by active zone proteins. For example, at parallel fiber-Purkinje cell synapses, $Ca_V2.1$ and $Ca_V2.2$ are localized within and surrounding the active zone, respectively. Deletion of one of the active zone proteins, Munc13–3, disrupts this localization as well as the maturation of the synaptic release properties from loose, microdomain coupling mediated by both $Ca_V2.1$ and $Ca_V2.2$ to tight, nanodomain coupling mediated by $Ca_V2.1$ (Kusch *et al* 2018).

9.2.2 Ca_V1 channels mediate transmission at sensory ribbon synapses

The primary cell-types that mediate our sense of vision, hearing, and balance utilize synapses that are characterized by a specialized organelle called a ribbon. In retinal photoreceptors and bipolar cells, as well as in cochlear and vestibular hair cells, the ribbon helps prime vesicles for exocytosis (Snellman *et al* 2011), and is important for maintaining tonic glutamate release during prolonged stimulation (Matthews and Fuchs 2010). Unlike the phasic action potential-driven release at most synapses, ribbon synapses mediate analog transmission that is modulated by graded changes in the membrane potential (Matthews and Fuchs 2010).

In addition to the ribbon, a second trait of these synapses is a reliance on Ca_V1 channels. Pharmacological and genetic evidence suggests that the primary Ca_V1 subtypes expressed in cochlear inner hair cells (IHCs) and retinal photoreceptors are $Ca_V1.3$ and $Ca_V1.4$, respectively (Barnes and Hille 1989, Platzer *et al* 2000, Schnee and Ricci 2003, Mansergh *et al* 2005, Baig *et al* 2011, Waldner *et al* 2018). A critical feature of these Ca_V1 subtypes is that they inactivate relatively slowly—an adaptation that supports the sustained release properties of ribbon synapses. Like other Ca_V channels, Ca_V1 channels undergo inactivation that is regulated by voltage (voltage-dependent inactivation, VDI) or by Ca^{2+} (Ca^{2+}-dependent inactivation, CDI). The co-assembly of the Ca_V β_2 subunit with $Ca_V1.3$ and $Ca_V1.4$ contributes to the limited VDI exhibited by these channels in IHCs and photoreceptors, respectively (reviewed in Pangrsic *et al* 2018). As will be discussed later, the mechanisms suppressing CDI appear to be distinct for $Ca_V1.3$ and $Ca_V1.4$ channels in these cell-types.

9.3 Ca^{2+}-dependent modulation of Ca_V channels and neurotransmitter release

During trains of depolarizations, Ca^{2+} currents mediated by $Ca_V2.1$ inactivate faster than Ba^{2+} currents (i.e., exhibit CDI), but undergo an initial increase in amplitude (Ca^{2+}-dependent facilitation, CDF). CDI but not CDF is blunted by high intracellular concentrations of EGTA (10 mM), indicating a difference in the Ca^{2+}

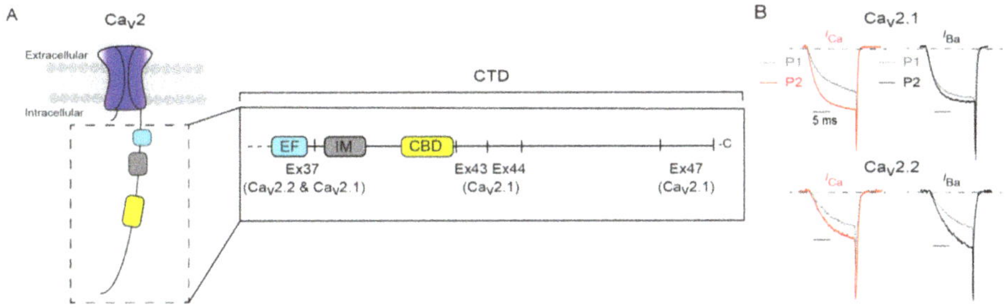

Figure 9.1. Molecular determinants that mediate CaM regulation of Ca_V2 channels. (a) Schematic represents C-terminal structures essential for CDI and CDF of Ca_V2 channels. These include two CaM interaction sites known as the IQ-like domain (IM, gray) and CaM-binding domain (CBD, yellow). In addition, a single EF-hand (EF, light blue) lies upstream of the IQ domain and plays a key role in the modulation of CDI and CDF. For $Ca_V2.1$, alternative splicing of exons 43 and 44 modulate current density, while splicing of exons 37 and 47 modulates CDF. (b) Representative Ca^{2+} (I_{Ca}, left) and Ba^{2+} (I_{Ba}, right) currents evoked before (P1, gray trace) and after (P2, red for I_{Ca} and black for I_{Ba}) a prepulse to 20 mV for $Ca_V2.1$ (top) and $Ca_V2.2$ (bottom). Representative traces were modified from Thomas *et al* (2018), copyright 2018 orignally published in *Journal of General Physiology*.

sensitivities of the two processes (Lee *et al* 1999, 2000). This difference involves the EF-hand Ca^{2+} binding domains of CaM, which are present in pairs in the N-terminal and C-terminal lobe of the protein (Kawasaki and Kretsinger 2017). There is a reliance of CDI on the CaM N-lobe, and of CDF on the CaM C-lobe (DeMaria *et al* 2001, Lee *et al* 2003). CaM binds to a consensus IQ domain within the C-terminal domain (CTD) of $Ca_V2.1$ that is required for CDF but can also interact with a downstream CaM-binding domain (CBD) which modulates CDI (figure 9.1(a)) (Lee *et al* 1999, 2000, DeMaria *et al* 2001). With low concentrations of intracellular EGTA (i.e., 0.5 mM), CDI is also evident for $Ca_V2.2$ (N-type) and $Ca_V2.3$ (R-type) channels (Liang *et al* 2003).

9.3.1 Alternative splicing modulates CDI of $Ca_V2.1$

While CDI of $Ca_V1.2$ and $Ca_V1.3$ channels depends on local Ca^{2+} signals emanating from individual channels, CDI for Ca_V2 channels relies on global Ca^{2+} signals supported by the Ca^{2+} influx through multiple channels (Ben-Johny and Yue 2014). Thus, factors that increase the whole-cell current density should lead to greater CDI of Ca_V2 channels. This point is well-illustrated by alternative splice variants containing exons 43 and/or 44 in the distal CTD of $Ca_V2.1$ (P/Q-type) channels (figure 9.1(a)). Inclusion of either of these exons causes a reduction in current density and corresponding decrease in CDI, with the reduction in CDI being maximal for variants containing both exons (Soong *et al* 2002). Exactly how exons 43 and 44 might inhibit $Ca_V2.1$ current density is unknown. The mechanism could involve conformational changes in the CTD when these exons are included, which destabilize the channel protein and/or its trafficking to the cell surface. Since the CTD of $Ca_V2.1$ is a hotspot for multiple protein interactions

(Maximov *et al* 1999, Kaeser *et al* 2011), it is also possible that inclusion of exons 43 or 44 could disrupt a binding site for a protein needed to stabilize $Ca_V2.1$ channels within the plasma membrane.

9.3.2 Molecular determinants for CDF in $Ca_V2.1$ and regulation by alternative splicing

CDF of $Ca_V2.1$ can be measured in a triple pulse voltage protocol in which test current amplitudes are compared before (P1) and after (P2) a conditioning prepulse. At prepulse voltages evoking significant inward Ca^{2+} current, CDF is evident as an increase in the amplitude of the P2 compared to the P1 current for $Ca_V2.1$ but not $Ca_V2.2$ (figure 9.1(b)) (Thomas and Lee 2016). The unique ability of $Ca_V2.1$ to undergo CDF is illustrated at the Calyx of Held synapse, where $Ca_V2.1$ mediates prominent presynaptic Ca^{2+} currents that support neurotransmitter release, and which undergo CDF (Cuttle *et al* 1998, Forsythe *et al* 1998). In $Ca_V2.1$ knockout mice, presynaptic Ca_V2 currents, largely $Ca_V2.2$, do not exhibit CDF (Inchauspe *et al* 2007).

We recently used a chimeric approach to determine why CDF is not observed for $Ca_V2.2$ (Thomas *et al* 2018). Consistent with the importance of the IQ domain for CDF of $Ca_V2.1$, transfer of the proximal CTD containing the EF-hand through the IQ domain from $Ca_V2.1$ to $Ca_V2.2$ produces strong CDF of the chimeric channels. $Ca_V2.2$ channels containing just the EF-hand, IQ domain, or the intervening sequence (PreIQ) from $Ca_V2.1$ do not exhibit CDF, indicating the importance of the entire EF-IQ module for this process. Although peptides containing the PreIQ-IQ domain of $Ca_V2.2$ bind CaM weakly compared to those of $Ca_V2.1$ (Peterson *et al* 1999, Liang *et al* 2003), inclusion of the EF-hand along with the PreIQ-IQ sequence nullifies this difference (Thomas *et al* 2018). Thus, the absence of CDF in $Ca_V2.2$ does not arise from weaker binding of CaM but through an inability of the EF-PreIQ-IQ domain to transduce CaM binding into an ability of $Ca_V2.2$ to support CDF.

The importance of the EF-hand for CDF is further illustrated by $Ca_V2.1$ splice variants containing exons 37a or 37b ($Ca_V2.1e \times 37a$, $Ca_V2.1e \times 37b$, respectively; figure 9.1(a)). These variants differ by 10 amino acids within the EF-hand domain (Soong *et al* 2002). $Ca_V2.1e \times 37b$ exhibits significantly reduced CDF as compared $Ca_V2.1e \times 37a$. However, when exon 47 is deleted from $Ca_V2.1e \times 37b$, large-amplitude Ca^{2+} currents exhibit robust CDF (Chaudhuri *et al* 2004). Apparently, exclusion of exon 47 in the context of $Ca_V2.1e \times 37b$ transforms CDF from a process driven by local Ca^{2+} signals to one that depends on global Ca^{2+} elevations. In rat cerebellar Purkinje neurons, an upregulation of $Ca_V2.1e \times 37a$ expression between postnatal days 6 and 11 correlates with increased CDF (Chaudhuri *et al* 2005), and may play a role in developmental processes such as those involved in cerebellar associative learning (Freeman and Nicholson 2004).

9.3.3 Ca^{2+}-dependent modulation of Ca_V2 channels and presynaptic plasticity

CDI and CDF of $Ca_V2.1$ both contribute to short-term synaptic plasticity at a variety of synapses, which can be modified by CaBPs and related CaS proteins (reviewed in Nanou and Catterall 2018). However, many of these results were obtained in electrophysiological recordings at room temperature with relatively high concentrations of extracellular Ca^{2+}. Under more physiological conditions, CDF is relatively nominal and does not contribute significantly to short-term facilitation at synapses in the hippocampus, cerebellum, and brainstem (Weyrer *et al* 2019). CaBP1, binding to the CaS site, prevents CDF and enhances CDI of $Ca_V2.1$ (Lee *et al* 2002), and causes short-term depression of PV-expressing interneuron-CA1 pyramidal neuron synapses in the hippocampus (Nanou *et al* 2018). Thus, CaS proteins could oppose CDF of $Ca_V2.1$ under physiological conditions in nerve terminals, helping to limit the potential for vesicle depletion at some synapses.

The modest level of $Ca_V2.1$ CDF is expected to offset the effect of CDI in promoting synaptic depression (Weyrer *et al* 2019). This could be prominent at CA3–CA1 hippocampal synapses, which rely on both $Ca_V2.1$ and $Ca_V2.2$, since $Ca_V2.2$ channels undergo CDI but not CDF. The inability of $Ca_V2.2$ to undergo CDF may be beneficial in nociceptive neurons and sympathetic neurons where, if present, CDF would oppose the inhibition of $Ca_V2.2$ by neurotransmitters acting on G-protein coupled receptors in the control of pain transmission and sympathetic outflow.

9.4 Ca^{2+}-dependent modulation of Ca_V1 channels at ribbon synapses

As has been described for Ca_V2 channels, $Ca_V1.3$ and some $Ca_V1.4$ splice variants can undergo CDI which also depends on CaM binding to an IQ domain in the CTD of these channels (figure 9.2(a)). However in both IHCs and photoreceptors, CDI of Ca_V1 channels is extremely limited (Corey *et al* 1984, Barnes and Hille 1989, Schnee and Ricci 2003, Grant and Fuchs 2008). This section will describe some of the proposed mechanisms that oppose CDI in these cell-types.

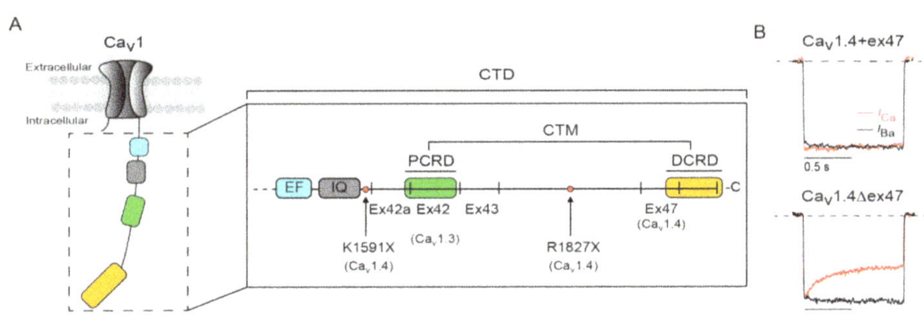

Figure 9.2. Molecular determinants that mediate CaM regulation of Ca_V1 channels. (a) Schematic represents key structural elements important for CaM regulation of Ca_V1 channels. These include a single EF-hand (light blue), a CaM interaction site (IQ, gray), and two regulatory domains referred to as the distal and proximal C-terminal regulatory domain, DCRD (orange) and PCRD (green), respectively. Exons that undergo alternative splicing as well as disease-causing mutations (red dots) that modulate CDI are indicated. (b) Representative I_{Ca} (red) and I_{Ba} (black) for $Ca_V1.4$ channels containing (top) or excluding exon 47 (bottom). $Ca_V1.4$ channels lacking exon 47 ($Ca_V1.4\Delta e \times 47$) undergo CDI in that I_{Ca} decays during a 1 s depolarizing test pulse more readily compared to I_{Ba}; however, this decay is absent for I_{Ca} mediated by $Ca_V1.4$ channels contain exon 47 ($Ca_V1.4 + e \times 47$) indicating a lack of CDI. Representative traces were modified from Williams *et al* (2018), originally published in Journal of General Physiology.

9.4.1 CaBPs as modulators of Ca_V1 CDI

CaBPs are a family of CaM-like Ca^{2+} binding proteins which suppress CDI by competing with CaM for binding to Ca_V1 channels (reviewed in Hardie and Lee 2016). CaBPs are highly expressed in IHCs and antagonize CDI of $Ca_V1.3$ in transfected cells (Yang *et al* 2006, Cui *et al* 2007). Mutations in the *CaBP2* gene that cause autosomal recessive hearing impairment in humans disrupted the ability of CaBP2 to suppress CDI of $Ca_V1.3$ in transfected cells (Schrauwen *et al* 2012). These results raised the possibility that CaBP2 was a prominent inhibitor of CDI in IHCs. However, VDI but not CDI of $Ca_V1.3$ was increased in IHCs of CaBP2 knockout (KO) mice (Picher *et al* 2017). The lack of changes in CDI in CaBP2 KO IHCs could be due to the compensatory action of CaBP1, which is also expressed in IHCs and strongly suppresses CDI of $Ca_V1.3$ in transfected cells (Cui *et al* 2007, Yang *et al* 2016).

9.4.2 A C-terminal modulator suppresses CDI and modulates activation of Ca_V1 subtypes

In the synaptic terminals of retinal photoreceptors, $Ca_V1.4$ plays an analogous role to $Ca_V1.3$ in inner hair cells in mediating sustained Ca^{2+} influx that is coupled to tonic exocytosis (Pangrsic *et al* 2018). Consistent with this function, molecular cloning of the gene encoding $Ca_V1.4$ (*CACNA1F*) revealed that these channels undergo almost no CDI, despite the conservation of the IQ domain (figure 9.2(b)) (Koschak *et al* 2003, Baumann *et al* 2004). The weak CDI of $Ca_V1.4$ is due to a C-terminal automodulatory domain (CTM) in the distal CTD, which weakens the ability of CaM to bind to the IQ domain (figure 9.2(a)) (Singh *et al* 2006, Liu *et al* 2010) (but see Wahl-Schott *et al* 2006, Griessmeier *et al* 2009). The CTM consists of a distal and proximal C-terminal regulatory domain (DCRD and PCRD, respectively; figure 9.2(a)) which physically interact in order to suppress CDI. A CTM is also present in $Ca_V1.3$ where it plays a role similar to that in $Ca_V1.4$ (Singh *et al* 2008).

The function of the CTM was discovered upon characterization of a *CACNA1F* mutation (K1591X) linked to congenital stationary night blindness type 2 (Singh *et al* 2006). Due to the insertion of a premature stop codon just downstream of the IQ domain (figure 9.2(a)), K1591X causes the removal of the entire CTM in $Ca_V1.4$ thus enabling CaM binding and CDI (Singh *et al* 2006). The mutation also causes the channel to activate at more negative voltages compared to wild-type channels (Singh *et al* 2006). Similar alterations in $Ca_V1.4$ function were reported for another mutation in *CACNA1F* (R1827X; figure 9.2(a)) which deleted the DCRD but not the PCRD (Burtscher *et al* 2014).

9.4.3 Alternative splicing of the CTD modulates Ca_V1 CDI

Like Ca_V2 channels, Ca_V1 channels are subject to extensive alternative splicing, which can dramatically modify their properties (Lipscombe *et al* 2013). For $Ca_V1.3$, splice variants completely lacking the IQ domain and CaM-dependent

CDI are expressed in the inner ear as well as in the brain (Erickson *et al* 2003, Bock *et al* 2011). Other $Ca_V1.3$ variants in the brain have functional alterations similar to those caused by the K1591X and R1827X mutations in *CACNA1F*. Short $Ca_V1.3$ variants lacking the entire CTM ($Ca_V1.3$ex42A) or just the DCRD ($Ca_V1.3e \times 43s$) exhibit stronger CDI and more negative voltage-dependence of activation than long variants containing an intact CTM ($Ca_V1.3e \times 42$) (figure 9.2(a)) (Singh *et al* 2008, Bock *et al* 2011, Tan *et al* 2011). The physiological relevance of the intramolecular interaction of the CTM in $Ca_V1.3$ was demonstrated in a knock-in mouse in which the insertion of a hemagglutinin tag (HA) within the DCRD prevented its interaction with the PCRD (Scharinger *et al* 2015). While this manipulation would be expected to increase CDI and cause a positive shift in voltage-dependent activation of $Ca_V1.3e \times 42$, these alterations were not observed for the $Ca_V1.3$ channels expressed endogenously in inner hair cells of the knock-in mice. Increased CDI but no change in voltage-dependent activation was found for $Ca_V1.3$-mediated currents in adrenal chromaffin cells from these mice (Scharinger *et al* 2015). Thus, the significance of CTM modulation of CDI varies according to cellular context.

Our group has characterized a $Ca_V1.4$ splice variant that deletes a portion of the DCRD and is highly expressed in primate but not rodent retina (Haeseleer *et al* 2016). The partial deletion of the DCRD results from skipping of exon 47 ($Ca_V1.4\Delta e \times 47$; figure 9.2(a)). Like K1591X that causes the complete deletion of the CTM, $Ca_V1.4\Delta e \times 47$ displays robust CDI (figure. 9.2(b)) and activation at more negative voltages compared to $Ca_V1.4$ channels containing exon 47. However, Ca^{2+} currents mediated by $Ca_V1.4\Delta e \times 47$ decay more slowly within the first 200 ms compared to those mediated by K1591X mutant channels (Williams *et al* 2018). Notably, $Ca_V1.3e \times 43s$ variants show similar slower initial decay as compared to $Ca_V1.3e \times 42a$ (Bock *et al* 2011), suggesting that the PCRD may decelerate inactivation in the context of both $Ca_V1.3$ and $Ca_V1.4$. The slow initial phase of CDI in $Ca_V1.4\Delta e \times 47$ requires Ca^{2+} binding to the Ca^{2+} binding sites (i.e., EF-hand motifs) in the N-terminal lobe (N-lobe) of CaM (Williams *et al* 2018), which normally senses moderate global cytosolic elevations fueled by multiple channels in contrast to the CaM C-lobe, which senses strong local Ca^{2+} signals near the pore of individual Ca_V channels (Liang *et al* 2003). Ca^{2+} binding to CaM N-lobe also contributes to the enhanced voltage-dependent activation of $Ca_V1.4\Delta e \times 47$, but not K1591X mutants (Williams *et al* 2018). These findings suggest that the residual CTM present in $Ca_V1.4\Delta e \times 47$ enables a unique modulation by CaM N-lobe, which may diversify Ca^{2+} signals required for photoreceptor signaling in the retina.

9.5 Conclusions

Research focusing on the expression profiles, intrinsic properties, alternative splicing, and modulation of Ca_V channels has enabled a better understanding of how these channels influence synaptic transmission. However, there are a number of lingering questions. First, the physiological impact of some forms of Ca_V channel regulation remain unclear. For instance, findings that CDF does not contribute to short-term plasticity at a number of synapses (Weyrer *et al* 2019) raise questions regarding when and where CDF may be important. CDF and CDI are generally less dramatic for $Ca_V2.1$ channels when endogenously expressed in neurons than when exogenously expressed in transfected HEK293 cells (Lee *et al* 2000, DeMaria *et al* 2001, Benton and Raman 2009, Kreiner *et al* 2010). Thus, factors not present in HEK293 cells could modulate CDF and CDI of $Ca_V2.1$ *in vivo*. The identification of these factors could yield insights into the neurophysiological contributions of CDF and CDI in neurons.

Second, despite evidence for an array of Ca_V splice variants, their roles with respect to synaptic transmission are largely unknown. Splice variation affecting the CTM is expected to lead to $Ca_V1.3$ and $Ca_V1.4$ channels with vastly different levels of CDI and voltage-dependence of activation (Tan *et al* 2011, 2012, Haeseleer *et al* 2016, Williams *et al* 2018). With respect to $Ca_V1.3$, such differences in activation could contribute to heterogeneity among active zones in IHCs, which is thought to be important for sound encoding (Ohn *et al* 2016). $Ca_V1.4$ variants exhibiting robust CDI (i.e., lacking exon 47 (Williams *et al* 2018)) may undergo developmental regulation as has been shown for Ca_V2 channels (Chaudhuri *et al* 2005). Alternatively, such $Ca_V1.4$ variants may be differentially expressed in distinct cell-types to allow for the efficient transmission of visual information in the parallel ON and OFF pathways of the retina.

Finally, it is now well-established that CaBPs and CaM can differentially modulate Ca_V channels (Hardie and Lee 2016), but the interplay of these proteins in altering synaptic transmission remains to be elucidated. From *in vitro* studies, it seems that CaBPs can compete with CaM for binding to Ca_V channels under Ca^{2+}-free conditions (Findeisen *et al* 2013, Oz *et al* 2013). Since presynaptic Ca^{2+} levels are regulated in an activity-dependent manner, reversible modulation of Ca_V channels by CaM and CaBPs could alter the propensity of some synapses to undergo short-term synaptic plasticity (Zucker 1999).

In summary, the contributions of CaM, CaBPs, and other forms of Ca_V regulation to excitation–secretion coupling are likely to vary between synapses and neuronal cell-types. State-of-the-art techniques in optical imaging and electrophysiology at well-characterized synapses are likely to yield further rich insights into how such forms of Ca_V regulation affect synaptic output and information encoding by neural circuits.

Acknowledgments

This work was supported by NIH grants EY026817 and NS115653 (AL), and F31-EY026477 (BW).

References

Baig S M *et al* 2011 Loss of $Ca_V1.3$ (*CACNA1D*) function in a human channelopathy with bradycardia and congenital deafness *Nat. Neurosci.* **14** 77–84

Barnes S and Hille B 1989 Ionic channels of the inner segment of tiger salamander cone photoreceptors *J. Gen. Physiol.* **94** 719–43

Baumann L, Gerstner A, Zong X, Biel M and Wahl-Schott C 2004 Functional characterization of the L-type Ca^{2+} channel $Ca_V1.4\alpha1$ from mouse retina *Invest. Ophthalmol. Vis. Sci.* **45** 708–13

Ben-Johny M and Yue D T 2014 Calmodulin regulation (calmodulation) of voltage-gated calcium channels *J. Gen. Physiol.* **143** 679–92

Benton M D and Raman I M 2009 Stabilization of Ca current in Purkinje neurons during high-frequency firing by a balance of Ca-dependent facilitation and inactivation *Channels* **3** 393–401

Bock G *et al* 2011 Functional properties of a newly identified C-terminal splice variant of $Ca_V1.3$ L-type Ca^{2+} channels *J. Biol. Chem.* **286** 42736–48

Breustedt J, Vogt K E, Miller R J, Nicoll R A and Schmitz D 2003 α_{1E}-containing Ca^{2+} channels are involved in synaptic plasticity *Proc. Natl Acad. Sci. USA* **100** 12450–55

Burtscher V *et al* 2014 Spectrum of Cav1.4 dysfunction in congenital stationary night blindness type 2 *Biochim. Biophys. Acta* **1838** 2053–65

Chaudhuri D, Alseikhan B A, Chang S Y, Soong T W and Yue D T 2005 Developmental activation of calmodulin-dependent facilitation of cerebellar P-type Ca^{2+} current *J. Neurosci.* **25** 8282–94

Chaudhuri D, Chang S Y, DeMaria C D, Alvania R S, Soong T W and Yue D T 2004 Alternative splicing as a molecular switch for Ca^{2+}/calmodulin-dependent facilitation of P/Q-type Ca^{2+} channels *J. Neurosci.* **24** 6334–42

Corey D P, Dubinsky J M and Schwartz E A 1984 The calcium current in inner segments of rods from the salamander (Ambystoma tigrinum) retina *J. Physiol.* **354** 557–75

Cui G, Meyer A C, Calin-Jageman I, Neef J, Haeseleer F, Moser T and Lee A 2007 Ca^{2+}-binding proteins tune Ca^{2+}-feedback to $Ca_V1.3$ channels in auditory hair cells *J. Physiol.* **585** 791–803

Cuttle M F, Tsujimoto T, Forsythe I D and Takahashi T 1998 Facilitation of the presynaptic calcium current at an auditory synapse in rat brainstem *J. Physiol.* **512** 723–29

DeMaria C D, Soong T, Alseikhan B A, Alvania R S and Yue D T 2001 Calmodulin bifurcates the local Ca^{2+} signal that modulates P/Q-type Ca^{2+} channels *Nature* **411** 484–9

Dietrich D, Kirschstein T, Kukley M, Pereverzev A, von der Brelie C, Schneider T and Beck H 2003 Functional specialization of presynaptic $Ca_V2.3$ Ca^{2+} channels *Neuron* **39** 483–96

Eggermann E, Bucurenciu I, Goswami S P and Jonas P 2011 Nanodomain coupling between Ca^{2+} channels and sensors of exocytosis at fast mammalian synapses *Nat. Rev. Neurosci.* **13** 7–21

Erickson M G, Liang H, Mori M X and Yue D T 2003 FRET two-hybrid mapping reveals function and location of L-type Ca^{2+} channel CaM preassociation *Neuron* **39** 97–107

Ertel E A *et al* 2000 Nomenclature of voltage-gated calcium channels *Neuron* **25** 533–35

Findeisen F, Rumpf C H and Minor D L Jr. 2013 Apo states of calmodulin and CaBP1 control Ca_V1 voltage-gated calcium channel function through direct competition for the IQ domain *J. Mol. Biol.* **425** 3217–34

Forsythe I D, Tsujimoto T, Barnes-Davies M, Cuttle M F and Takahashi T 1998 Inactivation of presynaptic calcium current contributes to synaptic depression at a fast central synapse *Neuron* **20** 797–807

Freeman J H Jr. and Nicholson D A 2004 Developmental changes in the neural mechanisms of eyeblink conditioning *Behav. Cogn. Neurosci. Rev.* **3** 3–13

Grant L and Fuchs P 2008 Calcium- and calmodulin-dependent inactivation of calcium channels in inner hair cells of the rat cochlea *J. Neurophysiol.* **99** 2183–93

Griessmeier K, Cuny H, Roetzer K, Griesbeck O, Harz H, Biel M and Wahl-Schott C 2009 Calmodulin is a functional regulator of $Ca_V1.4$ L-type Ca^{2+} channels *J. Biol. Chem.* **284** P29809–16

Haeseleer F, Williams B and Lee A 2016 Characterization of C-terminal splice variants of $Ca_V1.4$ Ca^{2+} channels in human retina *J. Biol. Chem.* **291** 15663–73

Hardie J and Lee A 2016 Decalmodulation of Cav1 channels by CaBPs *Channels* **10** 33–7

Hefft S and Jonas P 2005 Asynchronous GABA release generates long-lasting inhibition at a hippocampal interneuron-principal neuron synapse *Nat. Neurosci.* **8** 1319–28

Inchauspe C G, Forsythe I D and Uchitel O D 2007 Changes in synaptic transmission properties due to the expression of N-type calcium channels at the calyx of Held synapse of mice lacking P/Q-type calcium channels *J. Physiol.* **584** 835–51

Ishikawa T, Kaneko M, Shin H S and Takahashi T 2005 Presynaptic N-type and P/Q-type Ca^{2+} channels mediating synaptic transmission at the calyx of Held of mice *J. Physiol.* **568** 199–209

Iwasaki S, Momiyama A, Uchitel O D and Takahashi T 2000 Developmental changes in calcium channel types mediating central synaptic transmission *J. Neurosci.* **20** 59–65

Iwasaki S and Takahashi T 1998 Developmental changes in calcium channel types mediating synaptic transmission in rat auditory brainstem *J. Physiol.* **509** 419–23

Kaeser P S, Deng L, Wang Y, Dulubova I, Liu X, Rizo J and Sudhof T C 2011 RIM proteins tether Ca^{2+} channels to presynaptic active zones via a direct PDZ-domain interaction *Cell* **144** 282–95

Kawasaki H and Kretsinger R H 2017 Structural and functional diversity of EF-hand proteins: evolutionary perspectives *Protein Sci.* **26** 1898–920

Koschak A, Reimer D, Walter D, Hoda J C, Heinzle T, Grabner M and Striessnig J 2003 $Ca_v1.4\alpha1$ subunits can form slowly inactivating dihydropyridine-sensitive L-type Ca^{2+} channels lacking Ca^{2+}-dependent inactivation *J. Neurosci.* **23** 6041–49

Kreiner L, Christel C J, Benveniste M, Schwaller B and Lee A 2010 Compensatory regulation of $Ca_V2.1$ Ca^{2+} channels in cerebellar Purkinje neurons lacking parvalbumin and calbindin D-28k *J. Neurophysiol.* **103** 371–81

Kusch V *et al* 2018 Munc13-3 Is required for the developmental localization of Ca^{2+} channels to active zones and the nanopositioning of $Ca_V2.1$ near release sensors *Cell Rep.* **22** 1965–73

Lee A, Scheuer T and Catterall W A 2000 Ca^{2+}/calmodulin-dependent facilitation and inactivation of P/Q-type Ca^{2+} channels *J. Neurosci.* **20** 6830–8

Lee A, Westenbroek R E, Haeseleer F, Palczewski K, Scheuer T and Catterall W A 2002 Differential modulation of $Ca_V2.1$ channels by calmodulin and Ca^{2+}-binding protein 1 *Nat. Neurosci.* **5** 210–17

Lee A, Wong S T, Gallagher D, Li B, Storm D R, Scheuer T and Catterall W A 1999 Ca^{2+}/calmodulin binds to and modulates P/Q-type calcium channels *Nature* **399** 155–9

Lee A, Zhou H, Scheuer T and Catterall W A 2003 Molecular determinants of Ca^{2+}/calmodulin-dependent regulation of $Ca_V2.1$ channels *Proc. Natl Acad. Sci. USA* **100** 16059–64

Liang H, DeMaria C D, Erickson M G, Mori M X, Alseikhan B and Yue D T 2003 Unified mechanisms of Ca^{2+} regulation across the Ca^{2+} channel family *Neuron* **39** 951–60

Lipscombe D, Allen S E and Toro C P 2013 Control of neuronal voltage-gated calcium ion channels from RNA to protein *Trends Neurosci.* **36** 598–609

Liu X, Yang P S, Yang W and Yue D T 2010 Enzyme-inhibitor-like tuning of Ca^{2+} channel connectivity with calmodulin *Nature* **463** 968–72

Mansergh F, Orton N C, Vessey J P, Lalonde M R, Stell W K, Tremblay F, Barnes S, Rancourt D E and Bech-Hansen N T 2005 Mutation of the calcium channel gene *Cacna1f* disrupts calcium signaling, synaptic transmission and cellular organization in mouse retina *Hum. Mol. Genet.* **14** 3035–46

Matthews G and Fuchs P 2010 The diverse roles of ribbon synapses in sensory neurotransmission *Nat. Rev. Neurosci.* **11** 812–22

Maximov A, Sudhof T C and Bezprozvanny I 1999 Association of neuronal calcium channels with modular adaptor proteins *J. Biol. Chem.* **274** 24453–6

Nakamura Y, Harada H, Kamasawa N, Matsui K, Rothman J S, Shigemoto R, Silver R A, DiGregorio D A and Takahashi T 2015 Nanoscale distribution of presynaptic Ca^{2+} channels and its impact on vesicular release during development *Neuron* **85** 145–58

Nanou E and Catterall W A 2018 Calcium channels, synaptic plasticity, and neuropsychiatric disease *Neuron* **98** 466–81

Nanou E, Lee A and Catterall W A 2018 Control of excitation/inhibition balance in a hippocampal circuit by calcium sensor protein regulation of presynaptic calcium channels *J. Neurosci.* **38** 4430–40

Ohn T L, Rutherford M A, Jing Z, Jung S, Duque-Afonso C J, Hoch G, Picher M M, Scharinger A, Strenzke N and Moser T 2016 Hair cells use active zones with different voltage dependence of Ca^{2+} influx to decompose sounds into complementary neural codes *Proc. Natl Acad. Sci. USA* **113** E4716–25

Oz S, Benmocha A, Sasson Y, Sachyani D, Almagor L, Lee A, Hirsch J A and Dascal N 2013 Competitive and non-competitive regulation of calcium-dependent inactivation in $Ca_V1.2$ L-type Ca^{2+} channels by calmodulin and Ca^{2+}-binding protein 1 *J. Biol. Chem.* **288** 12680–91

Pangrsic T, Singer J H and Koschak A 2018 Voltage-gated calcium channels: key players in sensory coding in the retina and the inner ear *Physiol. Rev.* **98** 2063–96

Peterson B Z, DeMaria C D, Adelman J P and Yue D T 1999 Calmodulin is the Ca^{2+} sensor for Ca^{2+}-dependent inactivation of L-type calcium channels *Neuron* **22** 549–58

Picher M M *et al* 2017 Ca^{2+}-binding protein 2 inhibits Ca^{2+}-channel inactivation in mouse inner hair cells *Proc. Natl Acad. Sci. USA* **114** E1717–26

Platzer J, Engel J, Schrott-Fischer A, Stephan K, Bova S, Chen H, Zheng H and Striessnig J 2000 Congenital deafness and sinoatrial node dysfunction in mice lacking class D L-type Ca^{2+} channels *Cell* **102** 89–97

Scharinger A *et al* 2015 Cell-type-specific tuning of $Ca_V1.3$ Ca^{2+}-channels by a C-terminal automodulatory domain *Front. Cell. Neurosci.* **9** 309

Schnee M E and Ricci A J 2003 Biophysical and pharmacological characterization of voltage-gated calcium currents in turtle auditory hair cells *J. Physiol.* **549** 697–717

Schrauwen I *et al* 2012 A mutation in CABP2, expressed in cochlear hair cells, causes autosomal-recessive hearing impairment *Am. J. Hum. Genet.* **91** 636–45

Simms B A and Zamponi G W 2014 Neuronal voltage-gated calcium channels: structure, function, and dysfunction *Neuron* **82** 24–45

Singh A, Gebhart M, Fritsch R, Sinnegger-Brauns M J, Poggiani C, Hoda J C, Engel J, Romanin C, Striessnig J and Koschak A 2008 Modulation of voltage- and Ca^{2+}-dependent gating of $Ca_V1.3$ L-type calcium channels by alternative splicing of a C-terminal regulatory domain *J. Biol. Chem.* **283** 20733–44

Singh A, Hamedinger D, Hoda J C, Gebhart M, Koschak A, Romanin C and Striessnig J 2006 C-terminal modulator controls Ca^{2+}-dependent gating of $Ca_V1.4$ L-type Ca^{2+} channels *Nat. Neurosci.* **9** 1108–16

Snellman J, Mehta B, Babai N, Bartoletti T M, Akmentin W, Francis A, Matthews G, Thoreson W and Zenisek D 2011 Acute destruction of the synaptic ribbon reveals a role for the ribbon in vesicle priming *Nat. Neurosci.* **14** 1135–41

Soong T W, DeMaria C D, Alvania R S, Zweifel L S, Liang M C, Mittman S, Agnew W S and Yue D T 2002 Systematic identification of splice variants in human P/Q-type channel $\alpha_12.1$ subunits: implications for current density and Ca^{2+}-dependent inactivation *J. Neurosci.* **22** 10142–52

Stephens G J, Morris N P, Fyffe R E and Robertson B 2001 The $Ca_V2.1/\alpha1A$ (P/Q-type) voltage-dependent calcium channel mediates inhibitory neurotransmission onto mouse cerebellar Purkinje cells *Eur. J. Neurosci.* **13** 1902–12

Takahashi T and Momiyama A 1993 Different types of calcium channels mediate central synaptic transmission *Nature* **366** 156–8

Tan B Z, Jiang F, Tan M Y, Yu D, Huang H, Shen Y and Soong T W 2011 Functional characterization of alternative splicing in the C terminus of L-type $Ca_V1.3$ channels *J. Biol. Chem.* **286** 42725–35

Tan G M, Yu D, Wang J and Soong T W 2012 Alternative splicing at C terminus of $Ca_V1.4$ calcium channel modulates calcium-dependent inactivation, activation potential, and current density *J. Biol. Chem.* **287** 832–47

Thomas J R, Hagen J, Soh D and Lee A 2018 Molecular moieties masking Ca^{2+}-dependent facilitation of voltage-gated $Ca_V2.2$ Ca^{2+} channels *J. Gen. Physiol.* **150** 83–94

Thomas J R and Lee A 2016 Measuring Ca^{2+}-dependent modulation of voltage-gated Ca^{2+} channels in HEK-293T cells *Cold Spring Harb. Protoc.* **2016** prot087213

Wahl-Schott C, Baumann L, Cuny H, Eckert C, Griessmeier K and Biel M 2006 Switching off calcium-dependent inactivation in L-type calcium channels by an autoinhibitory domain *Proc. Natl Acad. Sci. USA* **103** 15657–62

Waldner D M, Bech-Hansen N T and Stell W K 2018 Channeling vision: $Ca_V1.4$-A critical link in retinal signal transmission *BioMed. Res. Int.* **2018** 7272630

Weyrer C, Turecek J, Niday Z, Liu P W, Nanou E, Catterall W A, Bean B P and Regehr W G 2019 The role of $Ca_V2.1$ channel facilitation in synaptic facilitation *Cell Rep.* **26** 2289–97

Wheeler D B, Randall A and Tsien R W 1994 Roles of N-type and Q-type Ca^{2+} channels in supporting hippocampal synaptic transmission *Science* **264** 107–11

Williams B, Haeseleer F and Lee A 2018 Splicing of an automodulatory domain in $Ca_V1.4$ Ca^{2+} channels confers distinct regulation by calmodulin *J. Gen. Physiol.* **150** 1676–87

Yang P S, Alseikhan B A, Hiel H, Grant L, Mori M X, Yang W, Fuchs P A and Yue D T 2006 Switching of Ca^{2+}-dependent inactivation of $Ca_V1.3$ channels by calcium binding proteins of auditory hair cells *J. Neurosci.* **26** 10677–89

Yang T, Scholl E S, Pan N, Fritzsch B, Haeseleer F and Lee A 2016 Expression and localization of CaBP Ca^{2+} binding proteins in the mouse cochlea *PLoS One* **11** e0147495

Zucker R S 1999 Calcium- and activity-dependent synaptic plasticity *Curr. Opin. Neurobiol.* **9** 305–13

IOP Publishing

Calcium Signals
From single molecules to physiology
Leslie S Satin, Manu Ben-Johny and Ivy E Dick

Chapter 10

Calcium signaling in context: case studies in endocrine cells

Patrick A Fletcher and Arthur S Sherman

10.1 Introduction

In this chapter we consider calcium signaling from a particular point of view. We aim to show that in addition to the presence or absence of particular calcium and calcium-regulated channels and calcium pumps, the degree of expression is critical. Beyond that, each of these proteins acts in a context set by the other elements present in the cell. Thus, it is the constellation of calcium-involved mechanisms that determines how input signals are transduced into calcium signals. We will show how some typical output patterns, such as amplitude coding and frequency coding, can arise and how they depend on both spatial and temporal aspects of cell organization.

We will illustrate these general principles by cases studies taken from two tissue systems that have a long history of mathematical modeling, the endocrine (anterior) pituitary and the beta cells of the pancreatic islets of Langerhans. The cells found in these structures come from similar developmental processes that equip them with similar complements of channels, pumps and organelles but exhibit a wide range of calcium signaling motifs. We will employ simplified models that highlight similarities and differences in mechanisms and behaviors among cell types rather than striving for accurate representation in detail of particular cell types.

These cell types are also interesting in their own right because of their important roles in whole-body physiology and because they contribute to disease when their function is impaired. Although the tissue mass of each cell type is small, they are often the sole sources of the hormones they secrete, and inadequate or excessive secretion can have dramatic effects [1]. Some examples: inadequate growth hormone secretion from pituitary somatotrophs results in short stature (dwarfism), while excessive secretion leads to gigantism. Corticotrophs are implicated in Addison's disease if under-active and Cushing's disease if over-active. Improper temporal

doi:10.1088/978-0-7503-2009-2ch10

patterning of gonadotropins may lead to infertility. Absence of insulin secretion due to autoimmune destruction of the beta cells causes type 1 diabetes, while a relative shortfall of insulin secretion in the face of insulin resistance is a major contributor to type 2 diabetes [2]. Insulin secretion that is inappropriately high when blood glucose is not elevated leads to congenital hyperinsulinism and dangerous hypoglycemia [3]. Here we restrict consideration to the calcium dynamics and associated electrical dynamics of the cells and refer readers to the broader literature for the biomedical implications.

10.2 Calcium signal encoding via membrane potential oscillations in an endocrine pituitary cell model

Endocrine pituitary cells produce and secrete hormones in response to physiological input signals, which they encode as calcium signals to drive calcium-dependent exocytosis. They are relatively small cells (~10–20 μm diameter) that rely on global calcium increases to drive exocytosis. Like many types of neurons and muscle cells, endocrine cells express a complement of ion channels and pumps that allow for electrical excitability and repetitive firing of action potentials [4]. There are five hormone producing cell types in the mammalian anterior pituitary gland, each specializing in the production of different hormones: growth hormone in somatotrophs, prolactin in lactotrophs, adrenocorticotropic hormone in corticotrophs, thyroid-stimulating hormone in thyrotrophs, and luteinizing hormone and follicle stimulating hormone in gonadotrophs. Despite their differing physiological roles, they share a common set of components that underlie their excitability and calcium signaling [5]. We will use a simple mathematical model based on these common features to demonstrate how the interplay between ionic currents and spatial compartmentalization of calcium signals allows endocrine cells to encode the strength of an input stimulus into both localized and whole-cell calcium signals.

Our model endocrine cell is spherical, with a radius of 10 μm and ion channels and pumps uniformly distributed on the plasma membrane (figure 10.1(a)). As in other excitable cells, action potentials are generated from the interplay between a rapidly activating voltage-dependent inward current, in this case a voltage-activated calcium current (I_{CaV}), and a more slowly activating voltage-dependent outward potassium current (I_{K}). These are responsible for regenerative depolarization and delayed repolarization of the membrane potential during spikes, respectively. Pituitary cells are known to express several types of sodium channels and voltage-gated calcium channels (VGCCs) [5], but we consider only a simple model with L-type calcium channels. Similarly, several additional types of voltage-gated potassium channels have also been reported, but we omit these for simplicity. We also consider a voltage-insensitive background sodium current (I_{Nab}) and an inwardly rectifying potassium current (I_{Kir}), which control the resting membrane potential. These pacemaker currents can provide a small net depolarization, supporting spontaneous repetitive firing of action potentials in several pituitary cell types. The input stimulus is modeled by increasing a voltage-insensitive inward current (I_{in}), as might result, for example,

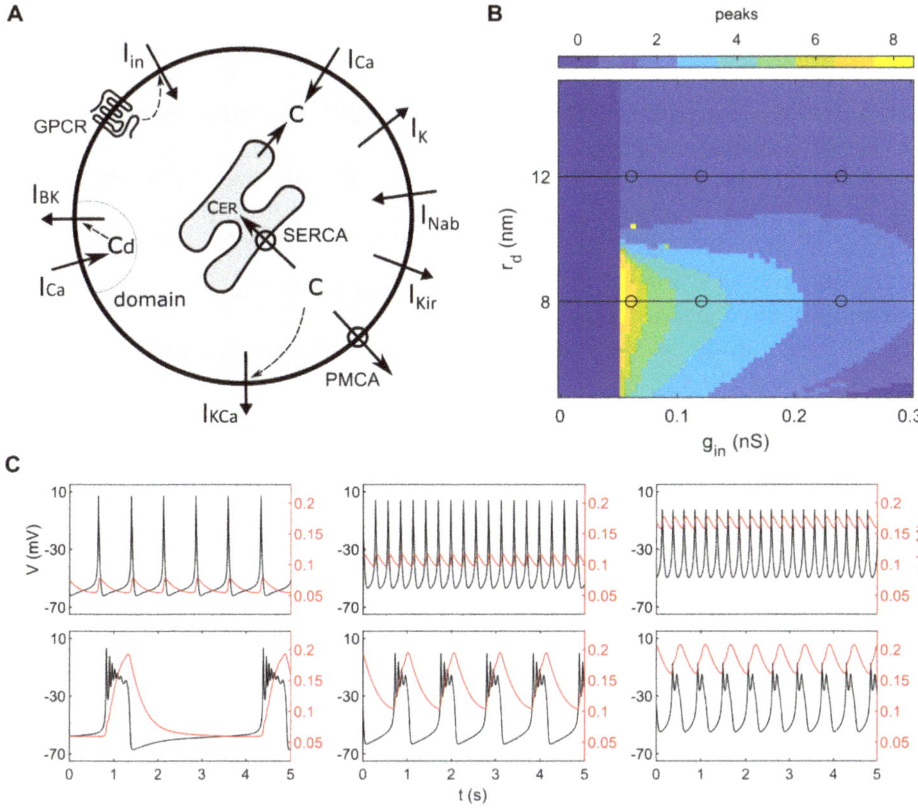

Figure 10.1. Calcium signal encoding in a simple excitable cell. (a) Diagram of the cell components considered in the pituitary endocrine cell model. GPCR, G-protein coupled receptor; I_{in}, GPCR-activated inward current; I_{Ca}, voltage-gated calcium current; I_K, delayed-rectifier potassium current; I_{Nab}, background sodium current; I_{Kir}, inwardly rectifying potassium current; I_{KCa}, calcium-activated potassium current; I_{BK}, large conductance calcium-activated potassium channel; c, cytosolic calcium concentration; c_d, domain calcium concentration; c_{ER}, ER calcium concentration; PMCA, plasma membrane calcium ATPase; SERCA, sarcoplasmic-endoplasmic reticulum calcium ATPase. (b) Variation in the number of peaks per action potential as a function of the conductance of the GPCR-activated inward current (g_{in}) and the microdomain size (r_d), the latter representing the distance between the BK channel calcium sensor and the calcium channel. Simulations were done for every point on a grid of 64 by 64 combinations of r_d and g_{in}. (c) Membrane potential (V, black) and whole-cell calcium (c, red) trajectories corresponding to the black circles in (b), with r_d = 12 nm (top row) or 8 nm (bottom row) and g_{in} increasing from left to right.

from activation of the G_S-coupled growth hormone-releasing hormone receptor in somatotrophs or corticotropin-releasing hormone receptor in corticotrophs.

To understand how this input stimulus is encoded into a calcium signal, we also model calcium concentration and fluxes. The calcium concentration in the cell rises in response to opening of VGCCs, directly coupling calcium signals to membrane potential fluctuations. It follows that voltage and calcium oscillations will share the same oscillation frequency. However, as we shall see, whether the calcium oscillations are of sufficient amplitude to relay frequency-coded information depends on the spatial and temporal organization of the cell.

In our model, cytosolic calcium concentration is reduced by the plasma membrane calcium ATPase (PMCA) and by sequestration into the endoplasmic reticulum (ER) via the sarco/endoplasmic reticulum calcium ATPase (SERCA). We also include a passive leak of calcium between ER and cytosol, the mechanism of which is not modeled in detail. Importantly, calcium diffusion in the cytosol is restricted by calcium buffering proteins; this fact is accounted for implicitly in the model by assuming that only a small fraction (~1%) of Ca^{2+} ions are free to participate in the dynamics. Because of this buffering, the whole-cell calcium concentration changes slowly relative to membrane potential.

Calcium-sensitive potassium currents are key players in shaping pituitary cell electrical activity, and here we focus on two: a voltage-insensitive SK-like potassium current (I_{KCa}) and the voltage- and large conductance calcium-activated potassium (BK) current (I_{BK}). These currents provide feedback of intracellular calcium levels onto the membrane potential. I_{KCa} provides an important negative feedback message that follows the time scale of intracellular calcium dynamics, contributing to spike and burst termination and spike frequency adaptation.

Depending on context, I_{BK} can have a more interesting and perhaps counter-intuitive effect. When sufficiently expressed and localized near VGCCs, the rapid activation of these potassium channels can have the stimulatory effect of converting spiking activity to bursting. We illustrate this by modeling calcium concentration in a microdomain (c_d) surrounding the calcium channels. BK channels, which have relatively low affinity for calcium (16.6 μM, [6]) are modeled as sensing the microdomain calcium rather than the bulk cytosolic calcium. If the microdomain is too large (or equivalently, if the BK channels are too far from VGCCs), c_d cannot reach sufficient levels to activate BK channels appreciably during electrical activity, and the cell exhibits spiking activity (figure 10.1(b), upper horizontal line; figure 10.1(c) top panels). However, if the microdomain is sufficiently small, BK channels are robustly activated and the cell exhibits bursting activity (figure 10.1(b), lower horizontal line; figure 10.1(c), bottom panels). The mechanism for this type of bursting relies on the fact that BK channel activation is fast, which in turn follows from the rapid changes in calcium concentration in the small microdomain. The rapidly developing I_{BK} pre-empts I_K and prevents its activation, as first appreciated by van Goor *et al* [7]. This leads to a depolarized plateau with small oscillations due to I_{Ca} and I_{BK} and results in a large calcium influx. The burst finally terminates once enough calcium enters the cell to activate the SK-like I_{KCa}. BK channel density must be sufficiently high for this form of bursting to occur; gonadotrophs, which lack BK do not show it. Experimentally, dynamic clamp has been used to artificially mimic BK channels and convert spiking cells to bursting cells [8]. The dynamic mechanism underlying this type of activity is called pseudo-plateau bursting because the plateau is built out of transient spikes. As a consequence, it requires calcium to be slower than membrane potential but not too slow, or the trajectory would be captured by the non-oscillatory plateau voltage. For detailed analysis of this interesting and subtle activity pattern, see [9, 10].

As the input stimulus is increased (moving left to right in figures 10.1(b) and (c)), the frequency of spikes or bursts increases, as does the mean whole-cell calcium. Regardless of activity pattern, calcium oscillates in phase with voltage. However, spiking activity elicits calcium oscillations with very small amplitude about the mean level, while bursting produces whole-cell calcium oscillations with large amplitude. The amplitudes of voltage, domain calcium, whole-cell calcium, and oscillation frequency as a function of input strength are summarized in figure 10.2 for both spiking and bursting cases. The spiking cell effectively encodes input strength into the mean level of whole-cell calcium (an example of amplitude coding), which may be sufficient to drive exocytosis. Bear in mind that exocytosis is governed mainly by calcium in proximity to secretory granules rather than whole-cell calcium. For more detailed models that account for calcium diffusion in the three-dimensional cytosolic space and calculate the relative capabilities of spiking and bursting to drive secretion see [11, 12]. A bursting cell, however, encodes the signal not only into mean calcium concentration, but also into oscillation pattern. This could be harnessed to drive downstream processes sensitive to oscillatory calcium, such as calcium-dependent gene transcription [13]. A possible advantage of calcium oscillations for regulating gene transcription is that it makes it possible to synchronize expression of target genes that have different affinities to calcium. In contrast, the amplitude coding exemplified by the spiking cell could lead to excessive expression of high-affinity genes and insufficient expression of low-affinity genes [14]. As such, the calcium influx due to bursting activity offers more diverse signaling possibilities than simple spiking. This BK channel-dependent mechanism for bursting, characteristic of somatotrophs, lactotrophs, and corticotrophs, therefore provides these cells with a mechanism for generating whole-cell calcium oscillations, while also encoding input strength into mean calcium levels and oscillation frequency.

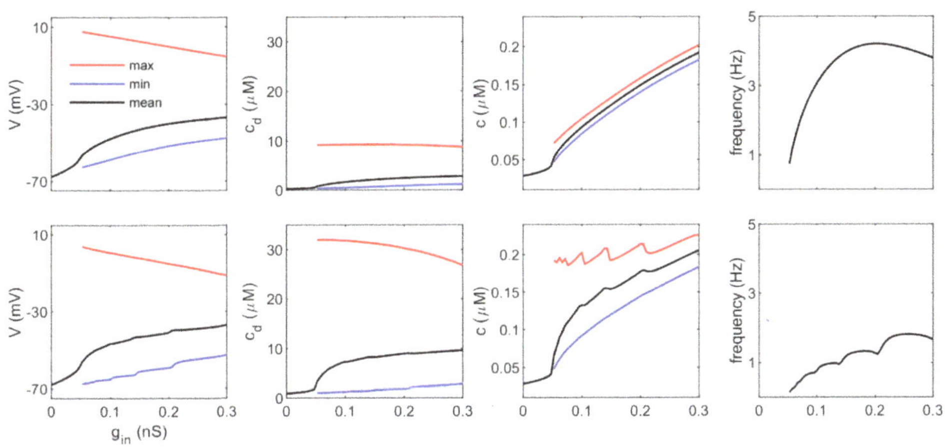

Figure 10.2. Summary of oscillation properties of the pituitary endocrine cell model. In the spiking cell with large r_d (top row, corresponding to the top row in figure 10.1(c)) domain calcium (c_d) does not reach sufficient levels to activate BK channels (second panel; the red curve is below $K_{CaBK} = 16.6$ μM). Despite oscillation frequency rising with increasing input (fourth panel), this cell encodes increasing input primarily as an increase in mean whole-cell calcium, with only very small amplitude oscillations about the mean (third panel; the red and blue curves remain close to the black curve). The bursting cell (bottom row) has a smaller r_d and thus c_d reaches sufficient levels for BK activation. This configuration generates large-amplitude c oscillations that increase in frequency with increasing input.

10.3 IP3-dependent calcium oscillations

Unique among anterior pituitary endocrine cells, gonadotrophs employ a different strategy to generate whole-cell calcium oscillations. These cells, lacking BK channels, typically display spontaneous spiking activity, and yet they also transition to bursting activity upon stimulation by gonadotropin-releasing hormone (GnRH). GnRH acts via the Gq-coupled GnRH receptor, which upon binding GnRH triggers production of inositol trisphosphate (IP3) and diacyl glycerol (DAG) by phospholipase C.

IP3 opens IP3 receptor (IP3R) calcium channels in the ER membrane, triggering release of calcium into the cytosol. The IP3 receptor itself is also sensitive to calcium, which provides the possibility for oscillatory calcium release from the ER. A classical model for these channels is based on a rapid activation and a slower inhibition due to calcium [15]. IP3-induced calcium flux through the channel into the surrounding cytosol promotes further opening. This calcium-induced calcium release (CICR), much like the upstroke of an action potential, may result in a large release of calcium. The delayed inhibitory feedback can eventually close the channel, resulting in a spike of calcium release. Increased cytosolic calcium in turn may activate protein kinase C (PKC), which along with DAG can trigger a GnRH-activated inward current to enhance excitability of the plasma membrane. PKC also increases the availability of secretory vesicles, which synergizes with the increased calcium to promote secretion.

We now modify our model pituitary endocrine cell to demonstrate this type of calcium signal encoding and its possible dependence on localization of calcium signaling. We consider a new calcium domain in the cytosol, which in this case need not be small. This calcium domain represents a region of the cytosol into which IP3Rs release calcium and is connected to the remaining cytosol via calcium diffusion, represented in the model with a diffusive coupling strength parameter, k_{diff} (figure 10.3(a)). Importantly, the remaining set of components included in the model are unchanged, except that the BK current is removed because it is poorly expressed in the gonadotroph.

Stimulation with GnRH increases IP3 concentration, which in turn opens IP3Rs and generates flux of calcium into the IP3R domain. If the domain is well connected to the cytosol (i.e., the diffusive coupling strength is large), domain calcium cannot build up and trigger CICR, resulting in a single transient calcium spike followed by a stable plateau of elevated intracellular calcium (figure 10.3(b), top panel). Such a response to Gq-coupled inputs is observed in other pituitary cells, such as arginine vasopressin responses in corticotrophs, or thyroid-stimulating hormone responses in lactotrophs. In contrast, if the domain represents a region of cytosol sufficiently isolated from the whole-cell calcium pool (low diffusive coupling), CICR can be activated and repetitive IP3R-mediated calcium spikes can occur (figure 10.3(b), bottom panel). The periodic calcium release from intracellular stores activates KCa channels, interrupting spiking to produce a bursting pattern of activity. This mechanism underlies the characteristic response of gonadotrophs to GnRH, in which voltage and calcium oscillate out of phase with each other. Mean calcium rises appreciably only when IP3-dependent calcium oscillations are triggered, which occurs for sufficiently low diffusive coupling of calcium between the cytosol and IP3R domain (figure 10.3(c)).

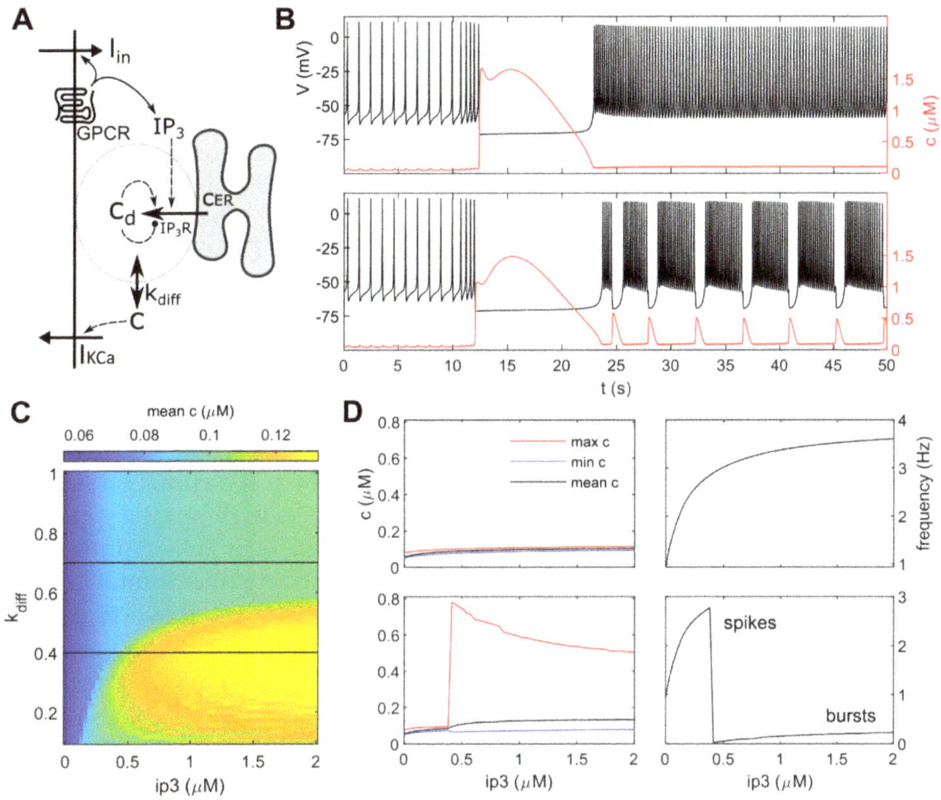

Figure 10.3. Calcium signal encoding via IP3-dependent release from intracellular stores. (a) Diagram of the components added to the endocrine cell model to support the generation of IP3-dependent calcium oscillations. The BK current was removed, resulting in a spiking cell much like the one shown in the previous section. (b) Response of the spiking cell to a smooth increase in IP3, beginning t = 10 s (not shown). A domain with high k_{diff} yields a transient calcium response followed by spiking and a slightly elevated calcium plateau (top trace, $k_{\text{diff}} = 0.7$). When k_{diff} is smaller, repetitive calcium pulses hyperpolarize the plasma membrane (bottom trace, $k_{\text{diff}} = 0.4$). (c) Increases in IP3 trigger increased mean calcium for sufficiently small k_{diff}. The black lines represent the parameters used for the two trajectories in panel (b). Simulations were done for a grid of 64 by 64 combinations of k_{diff} and IP3. (d) Summary of steady-state calcium oscillation amplitude and frequency in response to IP3 for the two selected values of k_{diff}.

Furthermore, the calcium oscillations have large amplitude, and their frequency, though lower than in the absence of IP3, increases with IP3 (figure 10.3(d)). The concomitant activation of a small inward current in response to IP3 gives rise to the increase in spike frequency (figure 10.3(d)); it has been suggested that the bursting membrane potential serves to provide calcium influx to replenish the ER calcium stores [16].

IP3-dependent encoding of Gq-coupled input signals into calcium oscillations is also observed in non-excitable cell types, such as hepatocytes [17], salivary gland acinar cells [18], pancreatic acinar cells [19], and T-lymphocytes [20]. An early model for frequency dependence of CICR-based calcium oscillations that triggered a wave of further modeling was [21]. The details of the mechanisms vary. An IP3R calcium domain is not strictly required, depending on the context of calcium handling in the cell, but it is likely that IP3R do in fact sense calcium in domains [22]; for an application to ryanodine receptors see [23]. Non-excitable cells also need calcium influx to replenish calcium stores. That role is often played by a store-operated current (SOC), an inward voltage-insensitive calcium current activated by ER depletion and not dependent on membrane potential oscillations [24].

The model described here generates oscillations despite IP3 being constant, while in other cells, IP3 oscillations are integral to the mechanism. IP3-dependent oscillations of both types have been the focus of detailed theoretical study [25, 26]. For a comprehensive, didactic overview of calcium oscillation models see [27].

10.4 Calcium oscillations in pancreatic beta cells

We now turn to another role the ER calcium can play in encoding calcium signals, and another way IP3 signaling can enhance calcium signals generated by electrical oscillations. Pancreatic beta cells are endocrine cells responsible for secreting insulin in response to increases in circulating glucose. As with pituitary and other endocrine cells, the primary signal driving secretion in beta cells is intracellular calcium, and beta cells generate this signal via bursting membrane potential activity. In terms of electrical activity and calcium signaling, they are equipped with a very similar complement of ion channels and calcium handling machinery to their pituitary cousins (compare figures 10.4(a), 10.1(a), and 10.3(a)). Their primary mechanism for spike generation and calcium influx similarly involves voltage-gated calcium and potassium channels. As with pituitary cells, calcium compartmentalization is important for their calcium signaling, but the role of ER calcium differs. Furthermore, the pattern of bursting is distinct; beta cells can produce much longer bursts and burst periods on the order of minutes. We will use the model of Bertram and Sherman [28] to illustrate a mechanism for this type of bursting as seen in mice; human beta cells have more varied and more complex patterns, but the principles are broadly the same [29].

Unlike pituitary and other endocrine cells, the main input is a nutrient, glucose, the concentration of which is transduced to the calcium signal by the rate of metabolism. Receptor-mediated signals take a back seat, acting as modulators of the glucose signal rather than primary triggers for secretion. Beta cells have several specializations that enable them to encode glucose input into calcium signals. Glucose sensing begins with glycolysis; the expression of the low affinity glucose transporter, GLUT2, and the low affinity glucokinase enzyme, responsible for the first step of glycolysis, means the cell's metabolism is driven by extracellular glucose concentration instead of intracellular demand. Glycolysis and subsequent oxidative phosphorylation in the mitochondria increase the ATP/ADP ratio, which in turn closes the ATP-sensitive potassium (KATP) channel and depolarizes the plasma membrane.

At low glucose, corresponding to or below the *in vivo* fasting glucose level, the KATP conductance dominates and maintains the beta cells at a low membrane potential (ca. −60 mV). This prevents insulin secretion, which would be not only inappropriate but could lead to life-threaten hypoglycemia.

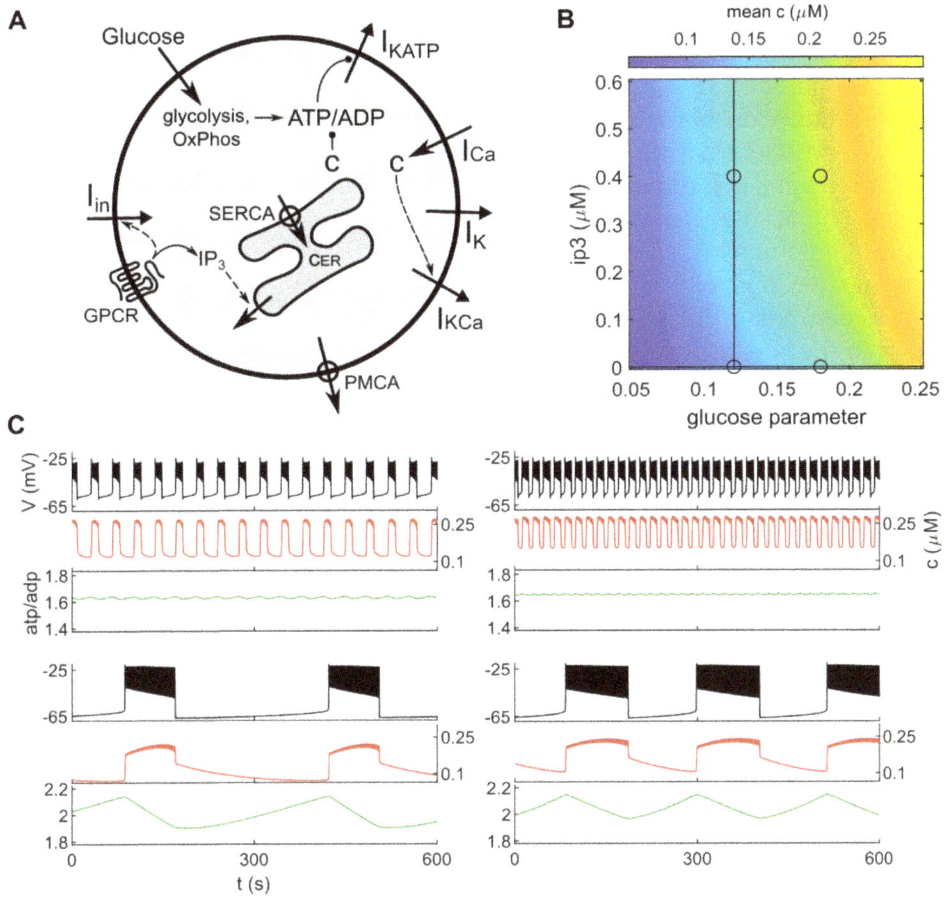

Figure 10.4. Calcium signal encoding in a model of pancreatic beta cells. (a). Diagram of the components included in the beta-cell model [BMB04]. ATP/ADP ratio is increased due to glucose metabolism and decreased in response to cytosolic calcium. I_{KATP}, ATP-sensitive potassium current; OxPhos, oxidative phosphorylation. (b). Summary of changes in mean calcium in response to the glucose parameter (r) and IP3. Simulations were done for a grid of 64 by 64 combinations of IP3 and r. (c) Membrane potential and calcium oscillations corresponding to the four points indicated by circles in panel B (combinations of ip3 = 0, 0.4 and r = 0.12, 0.18).

As glucose is raised to post-prandial levels, most of the KATP channels close, depolarizing the cells and allowing the other ion channels to become activated. Depolarization initiates a burst of action potentials and drives influx of calcium through VGCCs. Although KCa currents of the SK family are also present in beta cells, burst termination depends less on these currents and more on the much reduced but still substantial remaining KATP currents. In the model presented here, much of the calcium is sequestered in the ER by the SERCA pump and removed from the cell by the plasma membrane calcium ATPase. This precludes sufficient activation of KCa channels to terminate spiking, especially if KCa channel density is low, and prolongs the burst (note the plateau shape of the calcium oscillations in figure 10.4(c), bottom left panel). However, ATP is consumed by these processes, and eventually the ATP/ADP ratio drops enough to reopen a sufficient fraction of KATP channels and repolarize the membrane in combination with KCa channels. Calcium influx then abruptly stops, along with ATP consumption, allowing the ATP/ADP ratio to recover.

For this reason, it is primarily the rates of ATP production and consumption that control the burst characteristics in this model. At higher glucose concentration, ATP levels rise more quickly, thereby shortening the silent phase between bursts and/or prolonging the high-calcium active phase. Burst period may either increase or decrease, depending on which of these effects is larger, but the main effect is an increase in the burst duty cycle or plateau fraction. This is the main way that mean calcium is increased as a function of glucose (figure 10.5).

Beta cells are also stimulated by acetylcholine (ACh) via Gq-coupled muscarinic ACh receptors. We use the same model of IP3Rs and IP3-activated inward current described above, but without a calcium domain, to demonstrate how these input signals synergize. In this case, the main action of IP3R opening is to make the ER leakier with respect to calcium, making SERCA pumps less effective at sequestering cytosolic calcium into the ER. The result is higher cytosolic calcium, and therefore more KCa current activation occurs. In turn, less KATP current is needed to terminate the burst. This means ATP levels need to rise and fall by a smaller amount to start and stop each burst, and the burst frequency increases. The activation of an

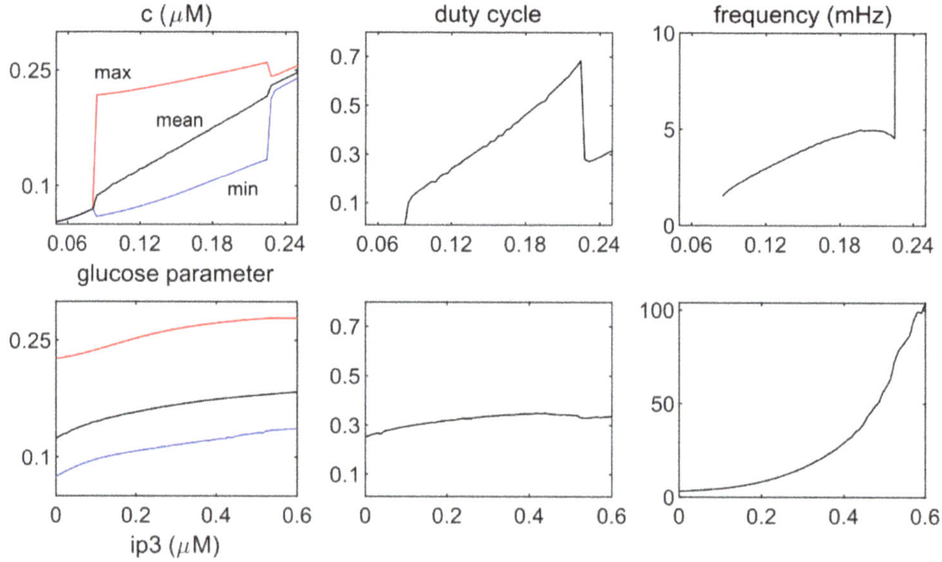

Figure 10.5. Summary of oscillation properties in the beta-cell model. Increasing glucose (top row, corresponding to the horizontal line in figure 10.4(b)) while keeping ip3 = 0 results in a steady rise in mean calcium and burst duty cycle, with only modest changes in burst frequency. Increasing IP3 with glucose parameter $r = 0.12$ (bottom row, vertical line in figure 10.4(b)) results in a modest increase in calcium and duty cycle, but a significant increase in burst frequency via a transition to fast, pituitary endocrine cell-like bursting.

IP3-stimulated inward current, although not necessary for the changes in burst period, promotes a greater increase in calcium by increasing firing frequency during bursts, increasing calcium influx, and further increasing the burst frequency. Both active and silent phases shrink together, resulting in a relatively unchanged duty cycle (figure 10.4(b), figure 10.5, bottom row).

If IP3 is increased enough, the ER becomes sufficiently poor at sequestering calcium that KCa currents become the primary mechanism for burst termination, much like the first case of pituitary cell bursting we examined. This is evidenced by the very small amplitude of ATP/ADP oscillations (figure 10.4(c), top panels); the fast bursting pattern would persist even if ATP/ADP were clamped to its mean value. Thus, in beta cells, the ER mediates a dialogue between KATP and KCa currents as burst termination mechanisms; as the ER becomes more labile, the KATP mechanism loses importance and the KCa mechanism is unmasked. We also note that in beta-cell models, increasing the KCa conductance, and therefore the relative strength of that mechanism, also reliably increases burst frequency [28].

A subtle point is that the increase in mean calcium due to increased IP3 is limited by the fact that increased calcium leads to increased ATP consumption by SERCA and PMCA pumps, which reduces ATP and thereby reopens some KATP channels. This countervailing effect is greater when KATP-channel density is larger and KCa density is smaller. The model predicts that beta cells that have faster oscillations will tend to have relatively more KCa channels compared to KATP channels and will respond to ACh with a greater increase in mean calcium [28].

In summary, the primary stimulus to beta cells, glucose, acts to increase calcium levels by increasing the burst duty cycle with modest changes in frequency, while IP3 significantly increasing burst frequency with little change in duty cycle (figure 10.5). When both IP3 and glucose are increased, the two effects synergize, leading to a robust increase in mean calcium (figures 10.4(b) and (c)). We can view the beta-cell calcium handling system as being a classical endocrine system wrapped in a KATP-channel blanket, which suppresses insulin secretion at basal glucose and mediates a graded increase in insulin secretion as glucose rises after meals.

Another physiologically important modulator of beta-cell calcium oscillations that feeds into this integrated package is glucagon-like peptide 1 (GLP-1), a hormone released by the L cells of the small intestine when exposed to glucose in the intestinal lumen. GLP-1 has similar effects to ACh on both calcium and vesicle availability but is mediated by Gs-coupled receptors, which increase cAMP production. Like ACh, GLP-1 has little effect on calcium, and hence insulin secretion, at low glucose but amplifies secretion at high glucose. This means GLP-1 based drugs pose much less risk of hypoglycemia than an earlier generation of oral diabetes medications that directly close KATP channels (e.g., the sulfonylureas), circumventing glucose metabolism. This feature, along with separate effects of GLP-1 receptors to inhibit secretion of the glucose-elevating hormone glucagon, secreted by pancreatic alpha cells, and suppress appetite in the central nervous system, has made drugs based on GLP-1 a multi-billion dollar industry. Their development represents a triumph of decades of basic research on beta-cell calcium and signaling mechanisms.

10.5 Conclusion

We have illustrated, using simple mathematical models, how a common set of calcium signaling components can encode different inputs into a variety of calcium signals with temporal and spatial specificity. We have focused mainly on calcium oscillations in endocrine pituitary and pancreatic cells, which may be generated by either plasma membrane or ER calcium channels.

Both classes of mechanisms follow the ubiquitous pattern of fast positive feedback to initiate each calcium rise, combined with slow negative feedback to restore basal calcium. In the plasma membrane mechanisms, negative feedback is provided by potassium channels activated directly by calcium (e.g., SK channels) in pituitary lactotrophs, somatotrophs and corticotrophs and pancreatic beta cells, or indirectly by calcium, i.e., the ATP-sensitive potassium (KATP) channel in beta cells, which is activated when rising calcium leads to ATP consumption by calcium pumps. In the ER driven mechanism, positive and negative feedback inhere in different calcium binding sites on the IP3R.

We showed that the size of the microdomain around L-type calcium channels combined with co-localization of BK channels determines whether cells exhibit large-amplitude voltage spikes, which can produce only small amplitude calcium oscillations, or bursting oscillations, which build a sustained calcium plateau out of small amplitude depolarized voltage spikes. Bursting depends on both adequate expression of BK channels and tight proximity to the calcium channels (i.e., small microdomains), which allows them to compete with the slower voltage-gated potassium channels and stretch the would-be spike into a burst. Both spiking and bursting transduce increased input signal strength to increased mean calcium, but the bursting mode also generates large-amplitude oscillations, which may be important for downstream targets.

We showed that calcium oscillations driven by release through ER calcium channels (IP3R), rather than plasma membrane calcium channels, similarly may depend on calcium domains around the IP3R. If diffusion away from the calcium release channels is sufficiently restricted, calcium can feed back on to the channel to drive oscillations; otherwise only a transient release of calcium is seen.

We contrasted the above pituitary patterns with those exhibited by their cousins, the pancreatic beta cells. Beta cells are actually capable of manifesting similar SK-driven bursting to that seen in a subclass of pituitary cells, but this is ordinarily suppressed by the much larger conductance of the KATP channels. In fact, the first model for bursting in beta cells [30] envisioned a starring role for calcium-activated potassium channels, which had already been proposed as a mechanism for other bursting cell types, such as the R-15 neuron [31]. That model had to be discarded when the new technique of calcium imaging showed that calcium did not rise progressively during the active phase of the burst (contrast figures 10.1(c) and 10.4(c)). When the KCa model ran aground, it was realized that KATP conductance could oscillate and drive the very slow bursting of beta cells. The fast SK mechanism in this context plays a secondary but important new role of potentiating the KATP mechanism, modulating burst frequency and mean calcium levels. The ER also was coopted into this KATP-dominated regime to serve as a mediator to regulate the balance between KATP and KCa mechanisms, enhancing the contribution of KCa when ER-depleting agonists, such as acetylcholine and glucagon-like peptide 1 are present. Combined, the potentiating mechanisms, which also tie into vesicle trafficking, are responsible for half of the total insulin secretion.

We have focused on production of calcium signals for brevity, but the sensitivity of downstream targets to mean calcium and calcium oscillation frequency also play important roles [13, 32, 33]. Thus, calcium can be targeted specifically to particular outcomes based on the patterns we have described.

References

[1] Goodman H M 2009 *Basic Medical Endocrinology* (Amsterdam/Boston, MA: Elsevier/Academic)

[2] Rorsman P and Ashcroft F M 2018 Pancreatic beta-cell electrical activity and insulin secretion: of mice and men *Physiol. Rev.* **98** 117–214

[3] Stanley C A 2016 Perspective on the genetics and diagnosis of congenital hyperinsulinism disorders *J. Clin. Endocrinol. Metab.* **101** 815–26

[4] Stojilkovic S S, Tabak J and Bertram R 2010 Ion channels and signaling in the pituitary gland *Endocr. Rev.* **31** 845–915

[5] Fletcher P A, Sherman A and Stojilkovic S S 2018 Common and diverse elements of ion channels and receptors underlying electrical activity in endocrine pituitary cells *Mol. Cell. Endocrinol.* **463** 23–36

[6] Montefusco F, Tagliavini A, Ferrante M and Pedersen M G 2017 Concise whole-cell modeling of BK_{Ca}–CaV activity controlled by local coupling and stoichiometry *Biophys. J.* **112** 2387–96

[7] Van Goor F, Li Y X and Stojilkovic S S 2001 Paradoxical role of large-conductance calcium-activated K^+ (BK) channels in controlling action potential-driven Ca^{2+} entry in anterior pituitary cells *J. Neurosci.* **21** 5902–15

[8] Tabak J, Tomaiuolo M, Gonzalez-Iglesias A E, Milescu L S and Bertram R 2011 Fast-activating voltage- and calcium-dependent potassium (BK) conductance promotes bursting in pituitary cells: a dynamic clamp study *J. Neurosci.* **31** 16855–63

[9] Stern J V, Osinga H M, LeBeau A and Sherman A 2008 Resetting behavior in a model of bursting in secretory pituitary cells: distinguishing plateaus from pseudo-plateaus *Bull. Math. Biol.* **70** 68–88

[10] Vo T, Bertram R, Tabak J and Wechselberger M 2010 Mixed mode oscillations as a mechanism for pseudo-plateau bursting *J. Comput. Neurosci.* **28** 443–58

[11] Montefusco F and Pedersen M G 2019 From local to global modeling for characterizing calcium dynamics and their effects on electrical activity and exocytosis in excitable cells *Int. J. Mol. Sci.* **20** 6057

[12] Tagliavini A, Tabak J, Bertram R and Pedersen M G 2016 Is bursting more effective than spiking in evoking pituitary hormone secretion? A spatiotemporal simulation study of calcium and granule dynamics *Am. J. Physiol. Endocrinol. Metab.* **310** E515–25

[13] Dolmetsch R E, Xu K and Lewis R S 1998 Calcium oscillations increase the efficiency and specificity of gene expression *Nature* **392** 933–6

[14] Cai L, Dalal C K and Elowitz M B 2008 Frequency-modulated nuclear localization bursts coordinate gene regulation *Nature* **455** 485–90

[15] Li Y X and Rinzel J 1994 Equations for InsP3 receptor-mediated $[Ca^{2+}]_i$ oscillations derived from a detailed kinetic model: a Hodgkin–Huxley like formalism *J. Theor. Biol.* **166** 461–73
[16] Li Y X, Stojilkovic S S, Keizer J and Rinzel J 1997 Sensing and refilling calcium stores in an excitable cell *Biophys. J.* **72** 1080–91
[17] Woods N M, Cuthbertson K S and Cobbold P H 1986 Repetitive transient rises in cytoplasmic free calcium in hormone-stimulated hepatocytes *Nature* **319** 600–2
[18] Ambudkar I S 2014 Ca^{2+} signaling and regulation of fluid secretion in salivary gland acinar cells *Cell Calcium* **55** 297–305
[19] Petersen O H 2014 Calcium signalling and secretory epithelia *Cell Calcium* **55** 282–9
[20] Lewis R S 2003 Calcium oscillations in T-cells: mechanisms and consequences for gene expression *Biochem. Soc. Trans.* **31** 925–9
[21] Dupont G, Berridge M J and Goldbeter A 1991 Signal-induced Ca^{2+} oscillations: properties of a model based on Ca^{2+}-induced Ca^{2+} release *Cell Calcium* **12** 73–85
[22] Smith G 2002 An extended DeYoung–Keizer-like IP3 receptor model that accounts for domain Ca^{2+}-mediated inactivation *Rec. Res. Dev. Biophys. Chem.* **2** 37–55
[23] Huertas M A, Smith G D and Gyorke S 2010 Ca^{2+} alternans in a cardiac myocyte model that uses moment equations to represent heterogeneous junctional SR Ca^{2+} *Biophys. J.* **99** 377–87
[24] Putney J W, Steinckwich-Besancon N, Numaga-Tomita T, Davis F M, Desai P N, D'Agostin D M, Wu S and Bird G S 2017 The functions of store-operated calcium channels *Biochim. Biophys. Acta, Mol. Cell. Res.* **1864** 900–6
[25] Dupont G, Combettes L, Bird G S and Putney J W 2011 Calcium oscillations *Cold Spring Harb. Perspect. Biol.* **3** a004226
[26] Politi A, Gaspers L D, Thomas A P and Hofer T 2006 Models of IP3 and Ca^{2+} oscillations: frequency encoding and identification of underlying feedbacks *Biophys. J.* **90** 3120–33
[27] Dupont G, Falcke M, Kirk V and Sneyd J 2016 *Models of Calcium Signaling* (Cham: Springer International Publishing)
[28] Bertram R and Sherman A 2004 A calcium-based phantom bursting model for pancreatic islets *Bull. Math. Biol.* **66** 1313–44
[29] Pedersen M G 2010 A biophysical model of electrical activity in human beta-cells *Biophys. J.* **99** 3200–7
[30] Chay T R and Keizer J 1983 Minimal model for membrane oscillations in the pancreatic beta-cell *Biophys. J.* **42** 181–90
[31] Plant R E and Kim M 1976 Mathematical description of a bursting pacemaker neuron by a modification of the Hodgkin–Huxley equations *Biophys. J.* **16** 227–44
[32] Fletcher P A, Clement F, Vidal A, Tabak J and Bertram R 2014 Interpreting frequency responses to dose-conserved pulsatile input signals in simple cell signaling motifs *PLoS One* **9** e95613
[33] Yildirim V and Bertram R 2017 Calcium oscillation frequency-sensitive gene regulation and homeostatic compensation in pancreatic beta-cells *Bull. Math. Biol.* **79** 1295–324

Part IV

Calcium channels in organelles

Chapter 11

Characterization of endo-lysosomal cation channels using calcium imaging

Christian Wahl-Schott, Marc Freichel, Volodymyr Tsvilovsky and Hristo Varbanov

11.1 Introduction

Membranes of intracellular organelles harbor a large number of transmembrane proteins, including many ion channels. In fact, just as the total area of membranes of intracellular organelles is much larger than the area of the plasma membrane, the number of organellar ion channels and transporters is also larger than those of the plasma membrane. Recently, ion channels localized in endo-lysosomal membranes have gained particular attention. They control a variety of fundamental physiological functions, including endo-lysosomal ion and pH homeostasis, regulation of endo-lysosomal resting membrane potential, catabolic export of amino acids, vesicle trafficking and vesicle fusion (Cang *et al* 2013, 2015, Dong *et al* 2008, 2010, Grimm *et al* 2012, 2014, Jentsch and Pusch 2018). Loss of function or dysfunction of endo-lysosomal ion channels underlie human metabolic diseases, such as nonalcoholic fatty liver disease and hyperlipoproteinemia, affect the development of certain infectious diseases and are implicated in the development of congenital lysosomal storage disorders and neurodegenerative diseases, such as Alzheimer's disease and Parkinson's disease (Cang *et al* 2013, Grimm *et al* 2012, Hockey *et al* 2015). The strong link between dysfunction or loss of function of these channels and the pathophysiology of human disease indicates that these ion channels are clinically relevant. In addition, the recent identification of small molecules activating or inhibiting these channels revealed that these channels not only represent useful tools for basic research but also undermine the very high potential of these channels to become clinically relevant drug targets. In addition, organelle proteomic studies suggest that there is a plethora of other putative ion channels and transporters (>70) in endo-lysosomes that remain to be characterized (Chapel *et al* 2013, Schwake *et al* 2013).

doi:10.1088/978-0-7503-2009-2ch11

A summary of endo-lysosomal ion channels and transporters which are known so far is given in figure 11.1. It can be clearly seen that many ion channels, such as five members of the transient receptor potential (TRP) family, the so called mucolipins TRPML 1–3 channels, two-pore-loop cation channels TPC1 and 2 channels as well as P2X receptors are permeable for Ca^{2+}. Importantly, alterations in endo-lysosomal Ca^{2+} signaling have been associated with lysosomal storage diseases, such as Niemann–Pick type C disease and the mucolipidosis type IV disease (Morgan *et al* 2011). The function of these channels can be investigated by Ca^{2+} imaging using the technical approaches and the methodology outlined below. Here, we will focus on the specific challenges of these Ca^{2+} imaging techniques, which are related to the endo-lysosomal localization of the ion channels. For general aspects of Ca^{2+} imaging involving Ca^{2+} indicators or genetically encoded Ca^{2+} sensors we refer to other chapters of the book.

The luminal Ca^{2+} concentrations of endo-lysosomal vesicles range from ~200 to 500 μM and are thus higher than Ca^{2+} concentrations of the cytosol (~100 nM). There is a large body of recent evidence suggesting that endo-lysosomes, even though small in size, play a prominent role as intracellular Ca^{2+} stores able to release

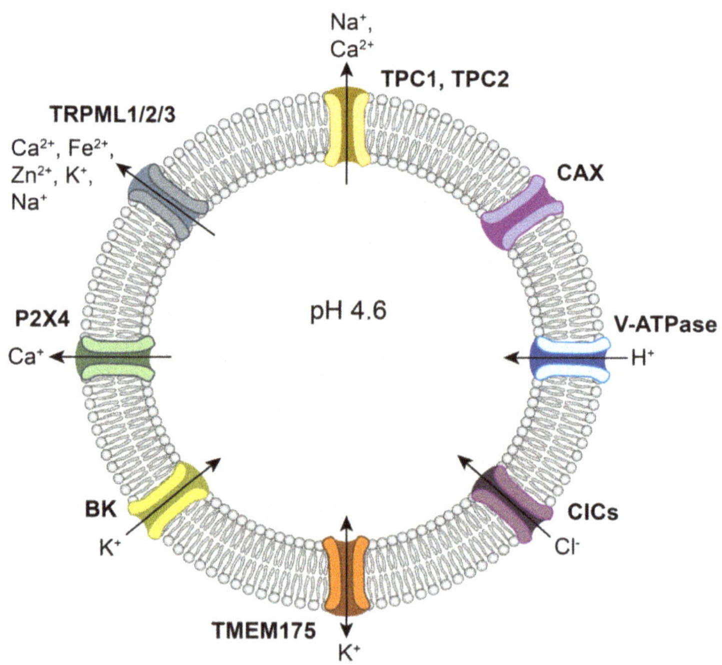

Figure 11.1. A schematic drawing illustrating the types of ion channels and transporters expressed in the lysosomal membrane. Please note that there are several ion channels that play a role in the Ca^{2+} release from endo-lysosomes, including TPCs, P2X4 and TRPMLs. Abbreviations: BK, large conductance Ca^{2+}- and voltage-activated potassium channel; CAX, vacuolar Ca^{2+}/H^+ exchanger; ClC, Cl^- channel; TMEM, TMEM175, transmembrane protein 175; TPC, two-pore channel; TRPML, mucolipin TRP channel; P2X4, P2X purinoreceptor subunit 4; V-ATPase, vacuolar-type H^+-ATPase (modified from Chen *et al* (2017b), copyright 2017, Nature Publishing Group, a division of Macmillan Publishers Limited. All Rights Reserved.).

and buffer Ca^{2+}, similar to the endoplasmic reticulum (ER) (Christensen *et al* 2002, Lloyd-Evans *et al* 2008, Morgan *et al* 2011). Notably, the Ca^{2+} uptake of lysosomes is largely driven by the H^+ gradient across the lysosomal membrane (Morgan *et al* 2011). While the cytosol shows neutral pH values in the range of 7.0–7.5, the luminal pH in lysosomes has been reported to be ~3.8–5 (Grabe and Oster 2001, Johnson *et al* 2016). This acidic luminal pH is maintained by the activity of several ion transporters, which include mainly the vacuolar-type H^+-ATPase (V-ATPase), the Na^+/H^+ exchanger, passive H^+ leaks, ClC (Cl^-) channels, K^+ channels, and the Na^+/K^+ ATPase (Grabe and Oster 2001, Morgan *et al* 2011).

In mammalian cells, the exact mechanisms of Ca^{2+} uptake in lysosomes are not yet understood. Several modes of Ca^{2+} exchange have been suggested, including a simple Ca^{2+}/H^+ exchange and a form of coupled exchange, in which a Na^+/H^+ exchanger functions together with the Na^+/Ca^{2+} exchanger or alternatively, with a Na^+/Ca^{2+}–K^+ exchanger (Morgan *et al* 2011). In the case of coupled exchangers, the Na^+/H^+ exchanger requires the H^+ gradient to translocate Na^+ into the lumen, thus leading to a Na^+ gradient that in turn promotes the uptake of Ca^{2+} via the Na^+/Ca^{2+} exchanger.

Moreover, two different types of endo-lysosomal Ca^{2+} pumps have been implicated in Ca^{2+} uptake in acidic stores. The mammalian plasma membrane-like Ca^{2+}-ATPase and the SERCA3-like Ca^{2+}-ATPase translocate Ca^{2+} into the lysosome and transport H^+ into the cytosol at the expense of ATP hydrolysis. One major difference between these two pumps is their sensitivity to selective inhibitors of the sarco-endoplasmatic reticulum Ca^{2+}-ATPase (SERCA). The PMCA-like Ca^{2+}-ATPase shows no sensitivity to the SERCA inhibitor thapsigargin and low sensitivity to the Ca^{2+}-ATPase inhibitor vanadate, which is a membrane-impermeant drug (Lytton *et al* 1992). In contrast, the SERCA3-like Ca^{2+}-ATPase is highly sensitive to vanadate and to tert-butylhydroquinone (tBHQ) (Lopez *et al* 2005). It is noteworthy that tBHQ specifically inhibits this type of Ca^{2+}-ATPase on acidic stores sensitive to NAADP in human platelets (López *et al* 2006), and that NAADP-evoked Ca^{2+} release is abolished in platelets from $Serca3^{-/-}$ mice (Feng *et al* 2020).

Several pharmacological studies have revealed that the intracellular messenger Nicotinic Acid Adenine Dinucleotide Phosphate (NAADP) elicits a functionally relevant release of Ca^{2+} from endo-lysosomal organelles (Galione 2006, Morgan *et al* 2011). Other NAADP targets include type 1 ryanodine receptors (RYR1) (Guse 2012, Wolf *et al* 2015) as well as members of the TRP family, such as TRPML1 (Zhang *et al* 2011), the (structurally) related TPC1 and TPC2 channels. The amount of Ca^{2+} released by endo-lysosomes is small in the first place. However, it has been shown that local Ca^{2+} release from endo-lysosomes in turn activates inositol 1,4,5-trisphosphate receptor channels (IP_3Rs) and ryanodine receptor channels (RyRs) on membranes of the endoplasmic reticulum (Morgan *et al* 2011, Roderick *et al* 2003) and thus induces a strong amplification of cytosolic Ca^{2+}, also known as Ca^{2+}-induced Ca^{2+}-release (CICR) via the coupling of endo-lysosomal Ca^{2+} to Ca^{2+}-release from the sarco-endoplasmic reticulum (SR), which is called the 'trigger hypothesis' (Cancela *et al* 1999).

Further evidence for the release of Ca^{2+} from lysosomes could be provided by recent studies using genetically encoded Ca^{2+} indicators that are directly coupled to the lysosomal channel or lysosomal membrane proteins of interest (McCue *et al* 2013, Shen *et al* 2012). Interestingly, electron microscopy studies have revealed that Ca^{2+} coupling between lysosomes and the ER relies on specialized ER-lysosomal membrane contact sites (~20 nm separation) enabling the transmission of Ca^{2+} oscillations between organelles (Kilpatrick *et al* 2013). In 2009, three seminal studies independently showed that the NAADP-triggered Ca^{2+}-release is mediated by two-pore channels (TPCs), in consistence with the trigger hypothesis (Brailoiu *et al* 2009, Calcraft *et al* 2009, Zong *et al* 2009). Furthermore, Ca^{2+} release and refilling of endo-lysosomes and the ER involve Ca^{2+} permeable and K^+ permeable cation channels, such as TRPML and K^+ channels (Chen *et al* 2017b, Xu and Ren 2015). In order to comprehensively and efficiently characterize Ca^{2+} signals arising from endo-lysosomes and the Ca^{2+}-dependent crosstalk of these organelles with the ER and other organelles, novel Ca^{2+} imaging concepts are necessary.

Here, we will present useful state-of-the-art approaches well suited to characterizing the function of endo-lysosomal cation channels using *four* different Ca^{2+} imaging approaches: (1) global cytosolic Ca^{2+} measurements (figure 11.2(A)); (2) peri-endo-lysosomal Ca^{2+} imaging using genetically encoded Ca^{2+} sensors, which are directed to the cytosolic endo-lysosomal membrane surface (figure 11.2(B)); (3) Ca^{2+} imaging of endo-lysosomal cation channels, which are engineered in order to redirect them to the plasma membrane in combination with approaches 1 and 2 (figures 11.2(C) and (D)); and (4) Ca^{2+} imaging by directing Ca^{2+} indicators to the endo-lysosomal lumen (figure 11.2(E)). Finally, we will describe a set of two complementary assays to investigate reloading of endo-lysosomal Ca^{2+} stores. In the following, we will present these methods, provide methodological details and discuss these approaches in comparison with the current literature. Furthermore, we will provide information on useful small molecules which can be used as valuable tools for endo-lysosmal Ca^{2+} imaging, most importantly novel compounds which activate or inhibit endo-lysosomal cation channels, small molecules which differentially enlarge individual endo-lysosomal subpopulations as well as substances that deplete endo-lysosomal Ca^{2+} stores. Rather than providing a complete protocol, we will discuss specific issues which are related to endo-lysosomal Ca^{2+} imaging. For details of more general Ca^{2+} imaging application we refer to individual chapters.

11.1.1 Global cytosolic Ca^{2+} measurements

This technique is designed to quantify Ca^{2+} signals generated by endo-lysosomal ion channels and to dissect these signals from cytosolic Ca^{2+} signals. There are multiple possible Ca^{2+} sources and a variety of ion channels that potentially could simultaneously contribute to these signals (figure 11.1). In order to dissect the input to these signals, an activator of an endo-lysosomal Ca^{2+} channel (NAADP, $PI(3,5)P_2$; table 11.1) is applied in order to release Ca^{2+} from the endo-lysosome to the cytosol

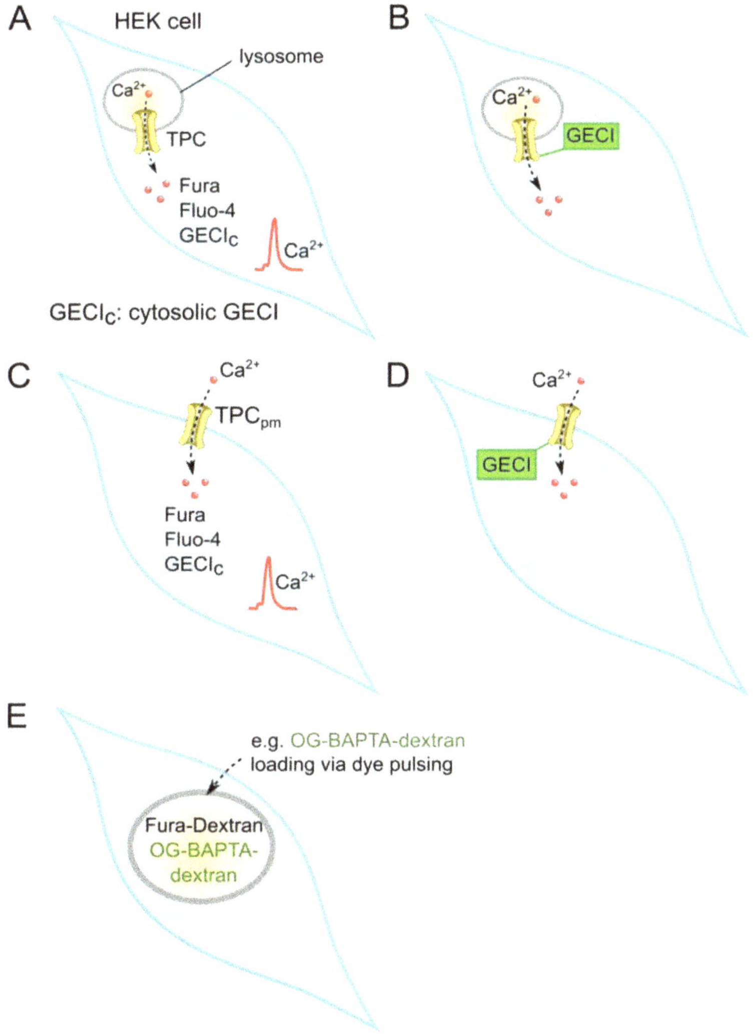

Figure 11.2. Configurations for recording Ca^{2+} signals from endo-lysosomal Ca^{2+} channels. (A) Native endo-lysosomal Ca^{2+} channels. (B) and (D) Genetically encoded Ca^{2+} sensors (GECIs) fused to endo-lysosomal ion channels in endo-lysosomes (B) or directed to the plasma membrane (D). (C) Endo-lysosomal ion channels targeted to the plasma membrane by disrupting the lysosomal targeting motives. (E) Monitoring of intra-lysosomal Ca^{2+} using chase-pulsing of dextran-coupled dyes, e.g. Fura-Dextran or Oregon 488 BAPTA-1 dextran. Abbreviations: c, cytosolic; GECI, genetically encoded Ca^{2+} indicators; HEK, human embryonic kidney 293 cell; TPC, two-pore channel; pm, plasma membrane.

and the resulting cytosolic signal is monitored in the absence and presence of pharmacological blockade of a given Ca^{2+} source (e.g., Ca^{2+} store depletion, channel blockers, inhibitor of V-ATPase, NAADP antagonists) or by genetic approaches aiming to knock down/delete essential release channels. Given that

Table 11.1. Overview of Ca^{2+}-releasing drugs, agonists and antagonists of lysosomal cation channels, and genetically encoded Ca^{2+} sensors, which have been used in assays for imaging of lysosomal Ca^{2+}. Abbreviations: EE, early endosomes; ES, endosomes; LE, late endosomes; LY, lysosomes; RE, recycling endosomes; LAMP1, lysosomal-associated membrane protein 1.

Ca^{2+}-releasing drugs	**Organelle**	**Target/Mechanism**	**Citation**
Thapsigargin	ER	SERCA	Thastrup *et al* (1990), Yagodin *et al* (1999)
GPN	LY		Berg *et al* (1994), Churchill *et al* (2002), Haller *et al* (1996)
Bafilomycin A1	ES/LY	V-ATPase	Bowman *et al* (1988), Dröse and Altendorf (1997)
Nigericine, Monensin	ES/LY	Ionophores	Churchill *et al* (2002), Morgan *et al* (2015), Srinivas *et al* (2002)
Small molecules	**Target**	**Mode of action**	**Citation**
NAAPD	TPCs	Agonist	Brailoiu *et al* (2009), Calcraft *et al* (2009), Gerndt *et al* (2020), Grimm *et al* (2014), Schieder *et al* (2010), Zong *et al* (2009)
$PI(3,5)P_2$	TPCs	Agonist	Chao *et al* (2017), Gerndt *et al* (2020), She *et al* (2018), She *et al* (2019)
TPC2-A1-N	TPCs	Agonist	Gerndt *et al* (2020)
TPC2-A1-P	TPCs	Agonist	Gerndt *et al* (2020)
ML-SA1	TRPMLs	Agonist	Garrity *et al* (2016), Shen *et al* (2012)
ML-SI1	TRPMLs	Antagonist	Garrity *et al* (2016)
Small molecules	**Organelle selectivity**	**Target/Mechanism**	**Citation**
Vacuolin-1	Enlarges EE, RE, and LE/LY	Blocks Ca^{2+}-dependent exocytosis of LYs	Cerny *et al* (2004), Chao *et al* (2017), Chen *et al* (2017b), Dong *et al* (2008), Gerndt *et al* (2020)
YM201636	Enlarges LE, LY	PIKfyve inhibitor	Chen *et al* (2017a), Garrity *et al* (2016), Jefferies *et al* (2008)
Apilimod	induces exosome secretion	PIKfyve Inhibitor; decreases PI(3,5)P2 levels	Cai *et al* (2013), Garrity *et al* (2016)
Latrunculin	EE	Disrupts actin polymerization	Chen *et al* (2017a), Spector *et al* (1983)
Wortmannin	EE	PI3K inhibitor	Chen *et al* (2017a), Wymann *et al* (1996)
Chloroquine	exosomes	increases exosome release	Ortega *et al* (2019)

Genetically encoded Ca^{2+} sensors	Location	Target	Citation
GCaMP	peri-vesicular	TPCs, TRPML1	Cheng *et al* (2014), Garrity *et al* (2016), Medina *et al* (2015), Samie *et al* (2013), Shen *et al* (2012), Zhang *et al* (2016)
GCaMP3 fused to the cytosolic N-terminus of TRPML1	peri-vesicular	TRPML1, MCOLN1	Shen *et al* (2011) later used by Medina *et al* (2015), Tsunemi *et al* (2019a), Garrity *et al* (2016)
LAMP1 fused to the YCaM3.6 cameleon	peri-vesicular	LAMP1	McCue *et al* (2013)
GCaMP6m fused with C-term. of TPC2	peri-vesicular	TPC2	Ambrosio *et al* (2015)
human TPC2 C-terminally tagged with GCaMP6s	peri-vesicular	TPC2	Gerndt *et al* (2020)
D1 connected to tissue plasminogen activator	inside of secretory granules	tissue plasminogen activator	Dickson *et al* (2012)
GEM-GECO1 fused with TIVAMP	inside of endosomes	TIVAMP	Albrecht *et al* (2015)

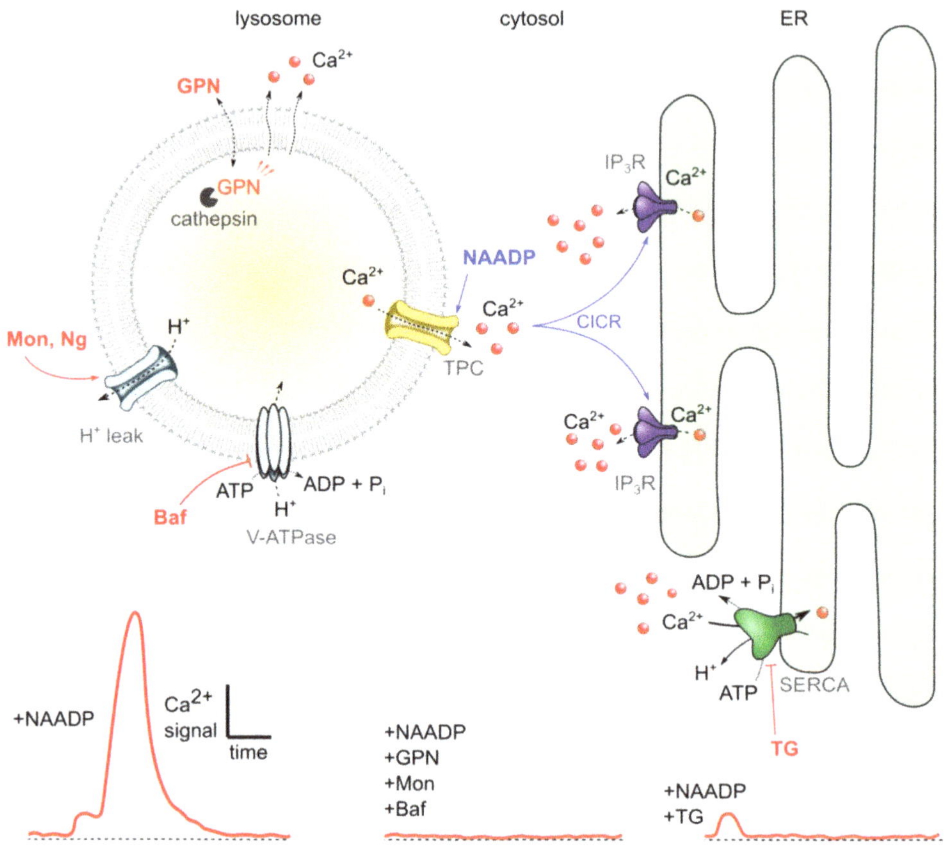

Figure 11.3. Dissection of endo-lysosomal Ca^{2+} signals from global Ca^{2+} signals generated by Ca^{2+} permeable ion channels in the plasma membrane and the endo-sarcoplasmic reticulum. Different mechanisms of drug effects are indicated. For details see text. Abbreviations: Baf, bafilomycin A1; CICR, Ca^{2+}-induced Ca^{2+}-release; ER, endoplasmic reticulum; GPN, glycyl-L-phenylalanine-naphthylamide; IP_3Rs, inositol 1,4,5-trisphosphate receptor; NAADP, nicotinic acid adenine dinucleotide phosphate; Mon, monensin; Ng, nigericin; TG, thapsigargin; SERCA, sarco-endoplasmic reticulum Ca^{2+}-ATPase; TPC, two-pore channel; V-ATPase, vacuolar-type H^+-ATPase;

both agonists, NAADP and $PI(3,5)P_2$, are membrane impermeable, we usually apply these compounds via the patch pipette. By comparing the signals in the absence and presence of inhibitors of a distinct Ca^{2+} source, the contribution of a given store to the signal can be calculated. For example, the observed cytosolic Ca^{2+} signal in figure 11.3 is a compound signal containing a Ca^{2+} signal generated by an endo-lysosomal Ca^{2+} store and a second component that is generated by the subsequent Ca^{2+} release from the endoplasmic reticulum, which in turn is triggered by the endo-lysosomal Ca^{2+} signal. Application of thapsigargin leads to a Ca^{2+} depletion of the ER. Application of NAADP after depletion of the ER-mediated Ca^{2+} store by thapsigargin therefore reveals the endo-lysosomal contribution to the cytosolic Ca^{2+} signal, but it needs to be considered that NAADP-evoked Ca^{2+}

release is characterized by a bell-shaped dose response curve leading to small response at high NAADP concentrations (Cancela *et al* 1999, Masgrau *et al* 2003, Zong *et al* 2009). On the other hand, depletion of the endo-lysosomal Ca^{2+} store eliminates the endo-lysosomal and the ER-dependent Ca^{2+} signal.

The approach described can also be used to indirectly assess the Ca^{2+} content of endo-lysosomes. Here, rapid and complete release of the Ca^{2+} content from an endo-lysosomal store is important. This can be achieved by activation of endo-lysosomal ion channels. Alternatively, a set of pharmacological tools, most of which act through direct or indirect inhibition of Ca^{2+} uptake into lysosomes or simply disrupt endo-lysosomal membranes could be used (Morgan *et al* 2011). Subsequently, the time course and the amplitude of the cytosolic Ca^{2+} signal is monitored using conventional Ca^{2+} dyes. The more Ca^{2+} is contained inside an organelle, the higher is the amplitude of the cytosolic Ca^{2+} spike and the more rapidly the rising phase of the signal occurs.

There are several technical issues that need to be considered for endo-lysosomal Ca^{2+} imaging.

(1) Firstly, Ca^{2+} signals can be recorded using fluorescent Ca^{2+} indicators, such as Fura-2, Fluo-4 (figures 11.2(A) and (C)) or by expressing a cytosolic genetically encoded Ca^{2+} sensor such as $GCaMP_c$ (= cytosolic GCamP; figures 11.2(B) and (D)).

(2) Secondly, it is important to empirically determine for any cell type that the Ca^{2+} indicator is exclusively loaded into the cytosol as the compartmentation of dye into other compartments will result in artifacts. Consequently, routine loading of cells by AM-ester at room temperature, which reduces compartmentation, is recommended. The exact concentration of the dye and the loading times must be determined empirically for each cell type and could differ considerably.

(3) The readout of the method is qualitative and is suitable in order to investigate whether endo-lysosomal ion channels contribute to Ca^{2+} storage and release $Ca^{2+}$from endo-lysosomal organelles.

(4) Ca^{2+} indicator-based experiments can be formatted to different scales. They can be applied to either cell populations (plate-readers, FACS or cuvettes) or to single-cell imaging using fluorescence microscopes (epifluorescence or confocal laser scanning). The only prerequisite is to have at hand the appropriate excitation and emission light wavelengths and appropriate filter cubes for the fluorophore(s).

(5) The method is ideal to compare the endo-lysosomal Ca^{2+} signals and content under different conditions, e.g. in the presence of a drug, altered protein expression or disease state.

(6) A potential drawback of the method is that depleting Ca^{2+} stores represent a major perturbation and may have unexpected additional effects. For a valid interpretation of the results this drawback should be considered. Along the same line, different stores frequently communicate with one

another and targeting one organelle can unavoidably affect a neighboring organelle.

(7) One potential drawback of the method is that its results cannot be calibrated and converted into a luminal [Ca^{2+}].

There are two main native activators described for TPCs, NAADP and PI(3,5)P_2. Recently, in a drug screen involving 80 000 small molecule compounds two TPC agonists have been identified. For TRPML1, several agonists and antagonists have been described. They are summarized in table 11.1.

Beside drugs which selectively activate or inhibit endo-lysosomal Ca^{2+} channels, a set of compounds for Ca^{2+} release from endo-lysosomal stores is available, comprising Glycyl-L-Phenylalanine 2-Naphthylamide (GPN), bafilomycin A1, nigericin and monensin. GPN has been extensively used to deplete lysosomal Ca^{2+} stores by disrupting lysosomes (Berg *et al* 1994, Churchill *et al* 2002, Coen *et al* 2012, Davis *et al* 2012, Dionisio *et al* 2011, Fois *et al* 2015, Garrity *et al* 2016, Haller *et al* 1996, Jadot *et al* 1984, Kilpatrick *et al* 2013, Li *et al* 2012, Melchionda *et al* 2016, Morgan and Galione 2007, Morgan *et al* 2011, Penny *et al* 2014, 2015, Ruas *et al* 2015). GPN is a synthetic membrane-permeable di-peptide and represents a pseudo-substrate that is degraded by the lysosome-specific enzyme Cathepsin C, which is also known as dipeptidyl peptidase 1 (Rao *et al* 1997). It has been shown that GPN is able to enter the lysosomal membrane by diffusion and is subsequently hydrolyzed by intra-luminal cathepsin C to produce free amino acids. Because of their polarity, these amino acids accumulate in the lysosomal lumen, leading to reversible permeabilization of the lysosomal membrane by osmotic swelling (Berg *et al* 1994, Haller *et al* 1996). Notably, GPN treatment results in membrane pores in lysosomes allowing leaks of small molecules with a molecular mass <10 kD (Penny *et al* 2014), which explains the GPN-evoked Ca^{2+} release from lysosomes. Nevertheless, the specificity of the Ca^{2+} source, from which GPN-evoked Ca^{2+} rise originates, has been questioned as experiments of a recent study lead to the conclusion that GPN does not selectively target lysosomes but leads to increases in cytosolic pH and stimulates Ca^{2+} release from the ER in a manner that is independent of IP_3 and ryanodine receptors (Atakpa *et al* 2019).

The macrolide antibiotic bafilomycin A1 acts as a specific inhibitor of the lysosomal vacuolar-type H^+-ATPase (V-ATPase), which is required for the acidification of endo-lysosomal lumen (Bowman *et al* 1988, Dröse and Altendorf 1997, Huss *et al* 2011). Bafilomycin A1 inhibits V-ATPases with a high affinity, at nanomolar concentrations (Bowman *et al* 1988). It has been suggested that the pharmacological inhibition of H^+ uptake into the lumen via the V-ATPase leads to passive loss of H^+ through leaks, which in turn dissipates the H^+ gradient across the lysosomal membrane, leading to luminal alkalization. Since this H^+ gradient is of key importance for the proper functioning of Ca^{2+} uptake via Ca^{2+}/H^+ exchangers, bafilomycin and related inhibitors of the V-ATPase indirectly block the Ca^{2+} filling of endo-lysomes, thus promoting the release of Ca^{2+} in the cytosol (Brailoiu *et al* 2009, Christensen *et al* 2002, Churchill *et al* 2002,

Gerasimenko *et al* 2006, Kelu *et al* 2017, Kinnear *et al* 2004, Lloyd-Evans *et al* 2008, Morgan *et al* 2015, Rah *et al* 2010, Vasudevan *et al* 2010, Yamasaki *et al* 2004).

Ca^{2+} release from acidic Ca^{2+} stores can also be induced by application of electroneutral cation-exchanging ionophores, such as nigericin and monensin (Churchill *et al* 2002, Ramos *et al* 2010, Srinivas *et al* 2002, Yagodin *et al* 1999). Nigericin represents a K^+/H^+ exchanger that allows the transport of H^+ out of the lysosome at the expense of K^+ influx, thus collapsing the H^+ gradient across the lysosomal membrane. In consequence of this luminal alkalinization, the Ca^{2+} filling via the lysosomal Ca^{2+}/H^+ exchanger is reduced, and the Ca^{2+} leak is left unmasked. Monensin, a polyether antibiotic, works in a similar fashion, but it translocates Na^+ instead of K^+ from the cytosol into the lysosome (Inabayashi *et al* 1995). Moreover, monensin has been shown to reverse the mode of the Na^+/Ca^{2+} exchanger (Asano *et al* 1995). It has to be considered that nigericin can act also in mitochondria, where application of nigericin hyperpolarizes the mitochondrial inner membrane (Robb-Gaspers *et al* 1998).

11.1.2 Peri-vesicular Ca^{2+} imaging

In the past years a variety of genetically encoded Ca^{2+} indicators (GECIs) have emerged (table 11.1). One significant advantage of these indicators is that they can be selectively expressed and retained in organelles by fusing organelle-specific targeting sequences to the indicator molecule (for review see (Suzuki *et al* 2016)). This approach has been used for example to design Ca^{2+} sensors targeted to the endoplasmatic reticulum (ER) and to endo-lysosomes. The ER-targeted sensor 'cameleon' was generated by adding the calreticulin signal sequence 5′ to CFP, and an ER-retention sequence was added to the 3′ end of citrine (Palmer *et al* 2004). Likewise, the low affinity GCaMP3 variant (GCaMPer10.19) coding sequence was fused downstream of calreticulin and an ER-retention sequence was fused to the carboxy-terminus for imaging of the ER Ca^{2+} stores (Henderson *et al* 2015). GCaMP-type GECIs were successfully subcellularly targeted in neurons (Mao *et al* 2008). Beside the ER, endo-lysosomal Ca^{2+} stores are an important part of Ca^{2+}-signaling in numerous cell types. However, the low pH environment in the lysosome lumen and high pH-sensitivity of most GECIs (reflected by their relatively high pKa values, e.g. in the range between 6 and 7 for GCamP3 to GCaMP6 variants (Chen *et al* 2013)) makes it very challenging to measure $[Ca^{2+}]$ accurately in acidic organelles. Thus, by placing GECIs to the cytosolic surface of endo-lysosomal vesicles elegant assays can be developed to detect Ca^{2+} release from endo-lysosomes more locally around the peri-endo-lysosomal surface of the vesicle itself without interfering with lysosomal pH. To record peri-vesicular Ca^{2+} release Shen *et al* (2012) fused GCaMP3 to the cytosolic N-terminus of TRPML1 in order to tether the sensor to the cytosolic surface of vesicles. The rationale of the assay is to monitor cytosolic Ca^{2+} and in parallel to activate endo-lysosomal channels. Alternatively, intracellular Ca^{2+} stores can be rapidly discharged with agents that selectively target the endo-lysosomal stores described above (GPN, bafilomycin A1, monensin and nigericin). Using this approach, these authors demonstrated that

GCaMP3 fluorescence responded preferentially and reliably to reagents, which selectively mobilize Ca^{2+} from endo-lysosomal vesicles, including GPN and bafilomycin A1 (Shen *et al* 2012). In the majority of the cells thapsigargin failed to significantly increase GCaMP3 fluorescence, suggesting that slow and small ER Ca^{2+} release was not detected by GCaMP3-ML1. This construct was also used by other researchers to study lysosomal Ca^{2+} signalling mechanisms (Gerndt *et al* 2020, Medina *et al* 2015, Tsunemi *et al* 2019b). Based on the same construct, Garrity *et al* (2016) developed a robust lysosomal Ca^{2+} refilling assay (see section (11.1.5)) to study release and uptake mechanisms of acidic Ca^{2+} stores (for a review see Yang *et al* 2019). Gerndt *et al* fused TPC2 to the genetically encoded Ca^{2+} indicator, GCaMP6(s) both with (TPC2GCaMP) and without (TPC2GCaMP/L265P) an intact pore in order to measure global Ca^{2+} signals. TPC2-A1-N and TPC2-A1-P evoked Ca^{2+} signals in cells expressing intracellular TPC2GCaMP but not TPC2GCaMP/L265P. Another researcher group generated a lysosomally targeted Ca^{2+}-sensitive FRET probe consisting of the lysosomal-resident membrane protein LAMP1 fused to the YCaM3.6 cameleon (LAMP1–YCaM) (McCue *et al* 2013). They demonstrated that this genetically encoded ratiometric sensor efficiently targets and detects Ca^{2+} in close proximity to the cytoplasmic surface of lysosomes. Furthermore, they have shown in HeLa cells that responses to the physiological agonist histamine persist in cells with depleted ER-Ca^{2+} content. Some progress was achieved in the development of Ca^{2+}-sensors working at low pH values observed in secretory granules and endosomes. Dickson *et al* adapted the D1-endoplasmic reticulum probe (Palmer *et al* 2004) to develop a unique probe (D1-SG) to measure Ca^{2+} in secretory granules by attaching of D1 to tissue plasminogen activator (Dickson *et al* 2012). For Ca^{2+} imaging in the endosomes, the pH-resistant ratiometric Ca^{2+}-biosensor GEM-GECO1 was fused with tetanus-insensitive vesicle-associated membrane protein (Albrecht *et al* 2015). TPC2 channel proteins fused to GCaMP6m were also used for the visualization of Ca^{2+} spikes around dense granules of platelets, which represent endo-lysosome-related acidic Ca^{2+} stores in platelets (Ambrosio *et al* 2015). Using this approach these researchers were able to visualize Ca^{2+} release from platelet dense granules and formation of perigranular Ca^{2+} nanodomains in real time and reveal organelle 'kiss-and-run' events. They observed also the presence of membranous tubules transiently connecting PDGs and demonstrated that this process was dramatically enhanced by TPC2 in a mechanism that requires ion flux through TPC2.

We used this Ca^{2+}-sensor attached to the C-terminus of TPC2 channels and exposed to the cytoplasmic face of acidic organelles to create transgenic mice ubiquitously expressing this construct. Our measurements were carried out in isolated acinar clusters expressing TPC2-GCaMP6m. In highly polarized pancreatic acinar cells, TPC2-GCaMP6m was localized exclusively at the apical side as shown by visualization of GCamp6 by immunostaining using anti-GFP antibody (figure 11.4(A)-left, right). Cells were stimulated by application of an analogue of Cholecystokinin CCK-8, which is known to generate NAADP. An application of CCK-8 (2 pM), in the absence of extracellular Ca^{2+} ions led to the generation of Ca^{2+}-oscillations (figure 11.4(B)). Targeting the Ca^{2+} sensor GCamP6m to the

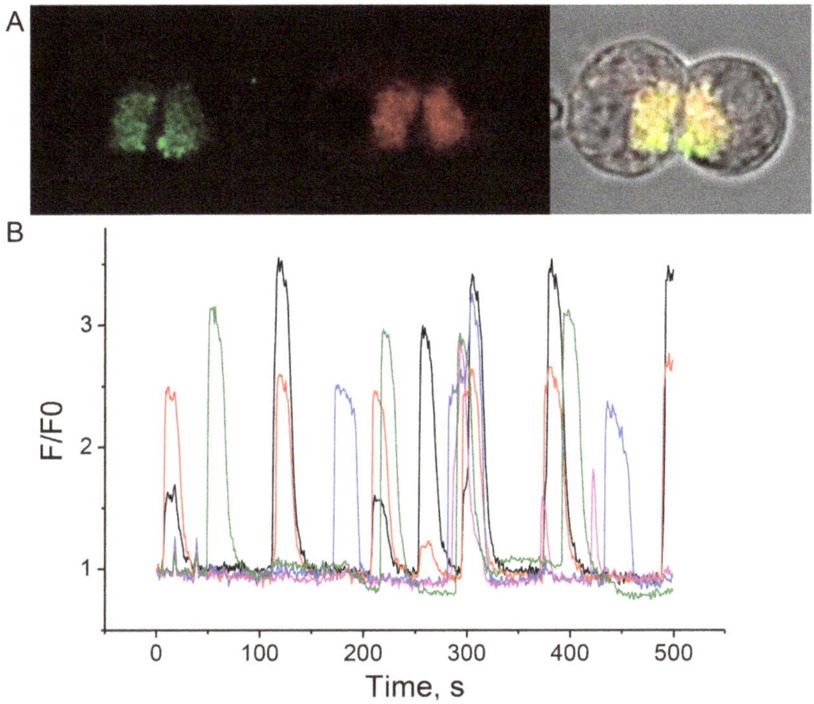

Figure 11.4. (A) Confocal images of fixed and permeabilized pancreatic acinar cells cluster co-stained with an anti-GFP antibody (green, left panel); the right panel demonstrates an overlay picture of the transmission light and the GFP channel. (B) Time course of the TPC2-GCaMP6m fluorescence (normalized to fluorescent values observed at zero time point) measured in five representative pancreatic acinar cells stimulated with 2 pM of Cholecystokinin in the absence of extracellular calcium ions. Black, red, green, blue and magenta traces represent the Ca^{2+} signal of the five independent measured cells.

C-terminal end of TPC2 channel proteins (TPC2-GCaMP6m) suggests that these Ca^{2+} oscillations are arising at the cytosolic face of organelles containing TPC2 channels, which are localized in endosomes and lysosomes in many cells types investigated so far (Galione *et al* 2009, Grimm *et al* 2017, Patel and Kilpatrick 2018). Our experiments suggest that targeting GECIs such as GCamP6m to endo-lysosomal TPC2 channels is a suitable approach to study Ca^{2+} signals triggered by NAADP-dependent Ca^{2+} release from primary acinar cell clusters. Whether Ca^{2+} signals measured by the TPC2-GCaMP6m sensors are specific to stimulation by the NAADP-generating CCK or can also be evoked by stimuli triggering Ca^{2+} release from independent Ca^{2+} release channels in other Ca^{2+} stores, e.g. via engagement of IP3 receptors following stimulation of muscarinic receptors needs to be studied in detail in future studies. In this case, sensitivity of the targeted Ca^{2+} sensor in endo-lysosomes could be varied by the use of GCaMP variants with lower Ca^{2+} affinity (Henderson *et al* 2015) or by GECO proteins (Wu *et al* 2014).

Overall, targeting GECIs to endo-lysosomal Ca^{2+} channels bear the advantage that potentially small or local Ca^{2+} signals can be detected, which are otherwise buried in global cytosolic signals. Alternatively, this approach could be used, when signal-to-noise ratio is too small using Ca^{2+} indicators. This issue is quite important, since the endo-lysosomal Ca^{2+} pool is small compared to the huge Ca^{2+} reservoirs of the ER or the extracellular milieu. Local peri-endo-lysosomal Ca^{2+} signals are small in size, but they have a high functional relevance for vesicle trafficking, vesicle fusion and exocytosis. Other situations where genetically encoded Ca^{2+} sensors can be used are whenever loading of chemical Ca^{2+} dyes is poor or too compartmentalized in a given cell type, or in case a drug of interest interferes with the fluorescence of the chemical indicator. Finally, tethering also overcomes distortions introduced by the rapid diffusion of mobile Ca^{2+} indicators.

Recordings should be performed in brief runs and in Ca^{2+}-free medium to eliminate complications of transmembrane Ca^{2+} influx.

Controls and other important issues:

(1) It is important to confirm that these fusion proteins co-localize well, in healthy cells, with lysosomal-associated membrane protein-1 (Lamp1), but not with markers for the ER, mitochondria, or early endosomes (Garrity *et al* 2016).
(2) Another important point is the Ca^{2+} affinity of the GECIs: to selectively measure local [Ca^{2+}] around endo-lysosomal channels (10–100 μM) rather than global [Ca^{2+}] (nM–μM), the probe must exhibit a low micromolar affinity, otherwise the tethered probe simply acts as cytosolic Ca^{2+} indicator. To date, the GECIs that have been tethered to vesicles manifest fairly high affinities to Ca^{2+} (in the sub- to low micromolar range). This does not make them ideal for selectively monitoring peri-vesicular Ca^{2+}, but they could be useful for recordings of small changes when cytosolic Ca^{2+} signals are different from these local signals.
(3) Concerning the assessment of vesicular Ca^{2+} content, it should be stated that these probes provide an indirect measure of endo-lysosomal Ca^{2+} content, which however, is better than those obtained by global cytosolic recordings.
(4) One also should keep in mind that pH on the peri-endo-lysosmal surface could be subjected to dynamic fluctuations. Fluctuations of the pH are known for endo-lysosomal vesicles and could, if present, confound Ca^{2+} measurements.
(5) In control experiments, it should be confirmed that fluorescent signals are endo-lysosome- and Ca^{2+}-specific. For example, pre-treatment with GPN or BAPTA-AM should completely abolish the response to an induction of Ca^{2+} release.

11.1.3 Ca^{2+} imaging of endo-lysosomal Ca^{2+} channels redirected to the plasma membrane

Endo-lysosomal cation channels frequently contain conserved di-leucine motifs, which target these channels to endo-lysosomal compartments. Mutation of these leucine residues to alanine or deletion of the lysosomal targeting sequence efficiently target these channels to the plasma membrane. For example, mutation of the N-terminal endo-lysosomal targeting motif in human TPC2 (TPC2L11A/L12A) redirects it to the plasma membrane. This approach was validated by our own group and by Brailoiu *et al* (2009). In particular, Brailoiu *et al* (2009) reported that TPC2, redirected to the plasma membrane mediates robust Ca^{2+} entry upon activation with NAADP. Our own group used this approach to screen a cell line stably expressing (TPC2L11A/L12A) with a library of 80 000 natural and synthetic small molecules using a FLIPR-based Ca^{2+} assay. Using this approach new TPC2 activators could successfully be identified (Gerndt *et al* 2020). In order to increase the robustness of the assay, as control a 'pore-dead' channel variant, in which the function of the pore was disrupted (TPC2L11A/L12A/L265P) was used in control experiments. While it is more efficient to obtain results using this approach, results need to be confirmed using Ca^{2+} imaging approaches to characterize Ca^{2+} release from the endo-lysosomes (approaches 1 and 2). Furthermore, results can be validated using electrophysiological characterization of ion channels redirected to the plasma membrane and by endo-lysosomal patch-clamp. Similar approaches have been used for TRPML channels. In line with this, deletion of N- and C-terminus including lysosomal targeting motifs of human TRPML1 redirects this channel to the plasma membrane (TRPML1ΔNC).

11.1.4 Fura-Dextran or Oregon-Green 488 BAPTA-1 dextran imaging

Intra-endo-lysosomal Ca^{2+} can be directly monitored using intra-luminal Ca^{2+} indicators that are dextran-coupled and can be pulse-chased into endo-lysosomal compartments. Among possible indicators are Fura-Dextran and Oregon 488 BAPTA-1 dextran (OG-BAPTA-dextran) (Morgan *et al* 2015). For pulse/chasing of these indicators, they are applied for a given time (usually $\leqslant$15 min) at 37 °C and then washed out. During this time interval, dextran-coupled indicators are taken up into cells via endocytosis, and subsequently the dyes enter the endo-lysosome lumen (Christensen *et al* 2002). These indicators can be pulsed/chased into HEK293 cells stably or transiently expressing TPC2-mCherry or TRPML1-mCherry-tagged fusion proteins. It should be considered that both dyes are pH sensitive. For reproducible and robust detection of intra-lysosomal Ca^{2+} ($[Ca^{2+}]_{LY}$) changes, intra-lysosomal pH should remain constant below pH 5.0. This could be confirmed using lysotracker or specific pH sensors. A convincing and clear demonstration of this approach also in connection with the refilling assay described below (sections (11.1.5) and (11.1.6)) has been performed by Garrity *et al* (2016). The authors demonstrated in the Fura-Dextran-loaded ML1-mCherry-transfected HEK-293T cells, activation of TRPML1 by application of ML-SA1-induced Ca^{2+} release from the lysosome lumen.

11.1.5 Endo-lysosomal Ca^{2+} refilling assay to investigate reloading of endo-lysosomal Ca^{2+} stores

Recently, two complementary assays that allow for studying dynamics of refilling and depletion of endo-lysosomal Ca^{2+} stores have been developed (Garrity *et al* 2016,

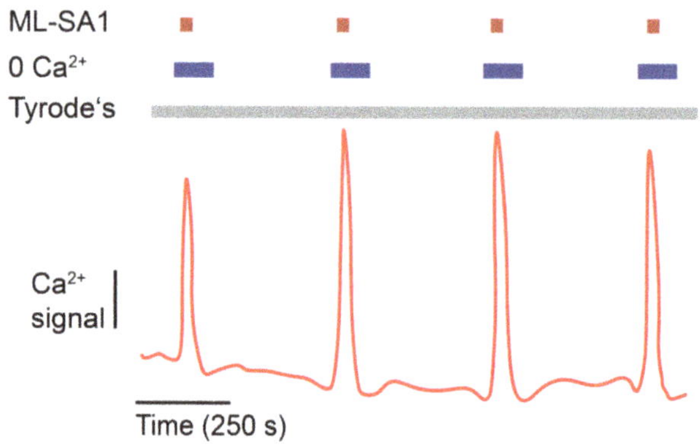

Figure 11.5. Schematic drawing of Ca^{2+} refilling assay for assessing the refilling of endo-lysosomal Ca^{2+} stores. When performed in cells expressing the genetically encoded Ca^{2+} indicator GCAMP3 coupled to the lysosomal Mucolipin TRP channel 1 (ML1) (in Ca^{2+}-free external solution), application of the ML1 channel agonist ML-SA1 evokes an increase in GCaMP3 Ca^{2+} fluorescence, which indicates the Ca^{2+} release from lysosomes trough the ML1 channel. The washout of the ML1 agonist for 5 min allows for Ca^{2+} refilling of the lysosomes; then, ML-SA1 is applied and washed out for three more times at time interval of 5 min. Please note that the increase in Ca^{2+} signal upon the second ML-SA1 application is more pronounced than the first one. Grey, blue and red bars denote the bath application of Tyrode's external solution, Ca^{2+}-free external solution and ML-SA1, respectively. Modified from Garrity *et al* (2016), CC-BY 4.0.

Wang *et al* 2017). These assays are particularly valuable to dissect the dependence of individual ER stores and other stores in refilling of endo-lysosomes. For these assays, approaches 2 (peri-endo-lysosomal Ca^{2+} imaging; figure 11.2(B)) and 4 (luminal endo-lysosomal Ca^{2+} imaging; figure 11.2(E)) described above are combined. The principle of this assay is to repetitively apply a membrane-permeable specific activator of the endo-lysosomal channel, which directly activates the endo-lysosomal Ca^{2+} channel and repeatedly induces Ca^{2+} release. For both, TPC2 and TRPML1 channels specific activators are available. In the following we describe the assay specifically for TRPML1, as performed by Garrity *et al* (2016) and Shen *et al* (2012). To activate TRPML1, a specific, membrane-permeable synthetic agonist of endo-lysosomal TRPML1 channels (ML-SA1) was used (Shen *et al* 2012). Figure 11.5 shows the rationale and the design of the endo-lysosomal Ca^{2+} refilling assay. In HEK293 cell lines stably-expressing GCaMP3-ML1 (HEK-GCaMP3-ML1 cells), bath application of ML-SA1 (30 s) in a 'zero' (nominally Ca^{2+} free; low; free $[Ca^{2+}]$ <10 nM) Ca^{2+} external solution produced robust lysosomal Ca^{2+} release measured by GCaMP3 fluorescence (4F/F0 >0.5; figure 11.5). After the initial Ca^{2+} release from endo-lysosomal store upon first application of ML-SA1, lysosomal Ca^{2+} stores are depleted. Refilling requires 3–5 min. Increasing the time interval between consecutive applications completely restores the lysosomal ML-SA1-induced responses. By contrast, immediate re-application of ML-SA1 after the first application evokes much smaller or no responses.

In a series of control experiments, it needs to be confirmed that lysosomal Ca^{2+} stores are refilled between first and second application of ML-SA1 (Garrity *et al* 2016). Furthermore, it is important to confirm that these release events are not dependent on the pH, not on mitochondria, but selectively depend on IP3-sensitive ER stores, but not on RyR-dependent ER stores.

To validate the assay, the following control experiments need to be performed:

(1) Ca^{2+} specificity of the assay can be demonstrated using the BAPTA-AM, which is the membrane-permeable form of the fast Ca^{2+} chelator BAPTA. If the process is Ca^{2+}-dependent, BAPTA-AM should completely block the ML-SA1-induced response.

(2) Correct localization of GCaMP3-ML1 can be confirmed by co-localization experiments. For correct localization, GCaMP3-ML1 and lysosomal marker proteins, such as LAMP1 and Rab7 could co-localize. Furthermore, GCaMP3-ML1-tagged lysosomes should co-localize with LysoTracker-positive lysosomes, and the pH of these lysosomes should be similar to LysoTracker-positive lysosomes without GCaMP3-ML1.

(3) It should be ruled out that channel desensitization underlies a reduction in the amplitude of the second response. This can be done by using Ca^{2+}-containing (2 mM) external solution. In this case, surface-expressed TRPML1 mutant (TRPML1-4A (Shen *et al* 2012)) should show repeated Ca^{2+} entry (Garrity *et al* 2016).

(4) To ensure that observed ML-SA1-induced Ca^{2+} responses are exclusively intracellular and lysosomal, ML-SA1 responses need to be measured either in the nominally Ca^{2+}-free external solution or in the presence of La^{3+}, a membrane-impermeable TRPML channel blocker (Dong *et al* 2008). Application of La^{3+} should completely inhibit surface-expressed TRPML1 channels.

(5) To confirm that Ca^{2+} release can be completely blocked, antagonists specifically inhibiting TRPML channels, such as ML-SI1 or ML-SI3 could be applied.

(6) To confirm that observed responses are specific for lysosomes, cells should be pretreated with the lysosome-disrupting reagent GPN. In a positive case, this pre-treatment should completely abolish the refilling either in nominally Ca^{2+} free external solution or in the presence of La^{3+}. This response should be reversible, and washout of GPN should lead to gradual recovery of ML-SA1 responses.

(7) Responses should be robust in independent cell systems, such as GCaMP3-ML1-transfected human fibroblasts, COS-7 cells, primary mouse macrophages, mouse myoblasts, and DT40 chicken B cells.

GPN assay

Lysosomal Ca^{2+} content and lysosome-specificity of the assay can be tested, independently of activation of lysosomal channels, using GPN as pointed out above. GPN can also be used for Ca^{2+} refilling experiments. In this case, GPN is repetitively applied, and refilling of lysosomal Ca^{2+} stores is evident as a Ca^{2+}

signal of similar magnitude to the first GPN-induced response. One potential problem of GPN, which needs to be accounted for by further control experiments is that permeabilization of lysosomal membranes by GPN induces leakage of Ca^{2+} and H^+ into the cytosol. This could significantly change pH in both the lysosome lumen and the peri-lysosomal (juxta-lysosomal) cytosol (Appelqvist *et al* 2012, Berg *et al* 1994, Kilpatrick *et al* 2013). It is well known that Ca^{2+} dyes and GFP-based Ca^{2+} indicators are sensitive to pH (Appelqvist *et al* 2012, Rudolf *et al* 2003). Therefore, Ca^{2+}-specificity of GPN-induced increases of the Fura-2 and GCaMP3 signals should be tested. In such control experiments, ER-mediated Ca^{2+} refilling of endo-lysosomes is abolished by pre-treatment with BAPTA-AM. GPN-induced responses are monitored using a Ca^{2+} indicator or peri-lysosomal Ca^{2+} sensor such as GCaMP3-ML1. Under these conditions, BAPTA-insensitive GCaMP3 and residual Fura-2 signals could be caused by GPN-induced H^+ release into the peri-lysosomal space and by changes in lysosomal and peri-lysosomal pH. Therefore, BAPTA-insensitive GCaMP3 and residual Fura-2 signals are caused by contaminating non-Ca^{2+} H^+ signals. In order to confirm this hypothesis, cells could be pretreated with bafilomycin-A (Baf-A), a specific inhibitor of the V-ATPase (Morgan *et al* 2011). This manipulation should abolish GPN-induced increases in GCaMP3 and Fura-2 responses. Furthermore, the extent of pH sensitivity of GCaMP3 fluorescence in comparison to the Ca^{2+} sensitivity can be confirmed and quantified using isolated single lysosomes (Garrity *et al* 2016). Finally, relative Ca^{2+}/pH-sensitivity of Ca^{2+} indicators and genetically encoded Ca^{2+} sensors could be compared with each other. Ratiometric dyes such as Fura-2 are less susceptible to pH changes (Morgan *et al* 2015). Therefore, in Fura-2-based assays, GPN may induce a large Ca^{2+} signal, but also a pH-mediated contamination by a non-Ca^{2+} signal. By contrast, both signals might be considerable in case of using non-ratiometric indicators or genetically encoded Ca^{2+} sensors.

Dissection of the mechanisms underlying Ca^{2+} refilling

Using specific compounds, the mechanisms underlying Ca^{2+} refilling of lysosomes can be further dissected. Experiments 1–3 could discriminate whether the secretory and endocytic pathways are involved in Ca^{2+} refilling, experiment 4 tests for dependence of effects on the generation of endo-lysosomal membrane lipid PI(3,5) P_2, and experiment 5 tests for pH dependence.

(1) The small molecule dynamin inhibitor dynasore (von Kleist *et al* 2011), inhibits endocytosis. If Ca^{2+} refilling depends on endocytosis, dynasore should block refilling.

(2) The microtubule inhibitor nocodazole (Webb *et al* 2004) and the potent histone deacetylase inhibitor trichostatin A (Vigushin *et al* 2001) inhibit cytoskeleton and organelle mobility. If Ca^{2+} refilling depends on organelle mobility, the two compounds should block refilling.

(3) Brefeldin-A disrupts Golgi function. If Ca^{2+} refilling depends on Golgi function, it should affect refilling.

(4) $PI(3,5)P_2$ is a lysosome-specific phosphoinositide that regulates multiple lysosomal channels and transporters including TRPML1 (Xu and Ren 2015). $PI(3,5)P_2$ levels can be pharmacologically decreased using two small molecule PIKfyve inhibitors: YM201636 (Jefferies *et al* 2008) and Apilimod (Cai *et al* 2013). If Ca^{2+} refilling depends on TRPMLs or TPCs, these PI $(3,5)P_2$ levels and thus the two inhibitors should prevent lysosomal Ca^{2+} refilling.

(5) Baf-A and concanamycin-A (Con-A) inhibit the V-ATPase and increase the pH of the lysosome (Morgan *et al* 2011). This can be demonstrated by abolishing LysoTracker staining within minutes after application of Baf-A. If Ca^{2+} refilling depends on pH, application of these compounds should affect Ca^{2+} refilling.

Specific experiments can be used to further confirm that Ca^{2+} refilling of endo-lysosomes are driven by ER-dependent mechanisms. Evidence suggesting that the ER is involved in lysosomal Ca^{2+} refilling are:

(1) Reduction of lysosomal refilling upon removal of extracellular Ca^{2+} during the refilling time. The ER stores are passively and slowly depleted in Ca^{2+}-free medium (Wu *et al* 2006). This is in line with the role of extracellular Ca^{2+} in ER store refilling.

(2) Insensitivity of refilling on blocking Ca^{2+} entry using the generic cation channel blocker La^{3+}.

(3) Thapsigargin (TG), a specific inhibitor of the ER SERCA pump (Thastrup *et al* 1990), should rapidly and completely abolish Ca^{2+} refilling of lysosomes.

(4) TG should not affect the first ML-SA1 response or lysosomal pH.

(5) The GPN and Fura-2 assays provide a reasonable (but not perfect; see above) measurement of lysosomal Ca^{2+} release independent of ML1. TG application also largely reduced the second GPN response, which could be further reduced or abolished by Baf-A pre-treatment.

(6) A rapid and complete block of Ca^{2+} refilling should be observed for another SERCA pump inhibitor CPA.

(7) N,N,N′,N′-Tetrakis (2-pyridylmethyl) ethylenediamine (TPEN), a membrane-permeable metal ion chelator with a low affinity for Ca^{2+} (Hofer *et al* 1998) can be used to chelate ER $[Ca^{2+}]$, but not cytosolic Ca^{2+}. Given that Ca^{2+}-binding affinity in the acidic pH (pH LY = 4.6) is critically reduced (>100-fold less), chelation of lysosomal Ca^{2+} can be considered minimal, even though TPEN enters the lysosomal lumen. TPEN blocks Ca^{2+} release from ER, and Ca^{2+} refilling, but not acute Ca^{2+} release from lysosomes.

(8) To confirm that lysosomal Ca^{2+} signals depend on the ER, a genetically encoded IP3-sponge can be used. To this end, a fusion protein containing the IP3R-ligand-binding domain fused to an ER targeting sequence (IP3R-LBD-ER) (Várnai *et al* 2005) can be expressed. IP3R-LBD-ER expression acts as IP3-sponge, which binds IP3 and chronically reduces ER Ca^{2+}store without raising intracellular Ca^{2+} levels. It also should decrease acute ATP-induced IP3R-mediated Ca^{2+} release. In case endo-lysosomal Ca^{2+} levels depend on ER Ca^{2+} levels, this could be tested by a GPN assay. GPN-induced lysosomal Ca^{2+} release should be reduced (WT cells). Furthermore, in GCaMP3-ML1 expressing cells transfected with IP3R-LBD-ER, peri-lysosomal Ca^{2+} release should be significantly reduced.

(9) Ca^{2+} release induced by SERCA inhibition of the ER can be detected using endo-lysosomal GCaMP3-ML1 probes. Close membrane contacts between the ER and lysosomes (Eden 2016) are a prerequisite for a positive experimental result. Similar detection of ER Ca^{2+} release by a genetically-encoded, lysosomally targeted chameleon Ca^{2+} sensor utilizing the lysosome membrane protein Lamp1 has also been reported (McCue *et al* 2013).

(10) Structural association can be confirmed using time-lapse confocal imaging. Is this case, the majority of lysosomes, marked by Lamp1-mCherry, should move and traffic together with ER tubules, labeled with CFP-ER. Thus, the ER could be the direct source of Ca^{2+} to lysosomes by forming nano-junctions with them (Eden 2016).

Experiments to dissect whether IP3 receptors or ryanodine receptors on the ER are required for Ca^{2+} refilling/signaling of lysosomes.

In cases where IP3Rs are involved:

(1) Xestospongin-C, a relatively specific IP3R blocker (Peppiatt *et al* 2003) should block Ca^{2+} refilling of lysosome Ca^{2+} store.

(2) Instead of using ML-SA1, GPN could be used to induce endo-lysosomal Ca^{2+} release in Fura-2 or GCaMP assays. Blocking IP3 receptor by Xestospongin-C in these assays should cause attenuated lysosomal Ca^{2+} refilling.

(3) Acute application of Xestospongin-C after allowing lysosomal Ca^{2+} stores to refill for 5 min (hence stores are completely refilled and functionally equivalent to 'naıve' ones) also slowly (up to 10 min) reduced lysosomal Ca^{2+} release, suggesting that constitutive lysosomal Ca^{2+} release under resting conditions may gradually deplete lysosome Ca^{2+} stores if refilling is prevented.

(4) Likewise, long-term (20 min) treatment with the aforementioned ER Ca^{2+} manipulators including TG and TPEN should abolish lysosomal Ca^{2+} release.

(5) 2-APB, a non-specific IP3R antagonist (Peppiatt *et al* 2003), should block Ca^{2+} refilling.

(6) Basal production of IP3 may be essential for Ca^{2+} refilling of lysosomes. To test for this possibility, U73122, a PLC inhibitor that blocks the constitutive production of IP3 (Cárdenas *et al* 2010) and prevents ATP-induced IP3R-mediated Ca^{2+} release, could be applied. In case basal IP3 is essential for refilling, application of U73122 should completely prevent Ca^{2+} refilling of lysosomes.

(7) In case that RYR are involved in Ca^{2+} refilling of endo-lysosomes, blocking the RyRs with high (>10 mM) concentrations of ryanodine, or with the RyR antagonist 1,1'-diheptyl-4,4'-bipyridinium (DHBP) (Berridge 2012), should affect Ca^{2+} refilling.

11.1.6 Endo-lysosomal Ca^{2+} refilling assay using Fura-Dextran or Oregon-Green 488 or BAPTA-1 dextran imaging

Intra-endo-lysosomal Ca^{2+} can be directly monitored using Ca^{2+} indicators, which are coupled to dextran and can be pulse-chased into endo-lysosomal compartments, as described in section 11.1.4. Among possible indicators are Fura-Dextran and Oregon 488 BAPTA-1 dextran (OG-BAPTA-dextran) (Morgan *et al* 2015). For pulse/chasing of these indicators (see section (11.1.4)), they are applied for a given time, usually 15 min and then washed out. During this time interval, dextran-coupled indicators are taken up into cells via endocytosis, and subsequently the dyes enter the endo-lysosome lumen (Christensen *et al* 2002). It should be considered that both dyes are pH sensitive. For reproducible and robust detection of intra-lysosomal Ca^{2+} ($[Ca^{2+}]_{LY}$) changes, intra-lysosomal pH should remain constant below pH 5.0 (Morgan *et al* 2015). It has been shown that in the Fura-Dextran-loaded ML1-mCherry-transfected HEK-293T cells, ML-SA1 application induced Ca^{2+} release from the lysosome lumen (Garrity *et al* 2016).

In order to confirm ER-dependence of endo-lysosomal refilling, IP3R-dependent Ca^{2+} release from the ER can be inhibited by Xestospongin-C. Alternatively, ER Ca^{2+} loading can be blocked by TG. Both manipulations should abolish the ML-SA1-induced decrease in $[Ca^{2+}]_{LY}$. In addition to Fura-Dextran, Oregon-Green BAPTA can be loaded more efficiently into endo-lysosomes. To control for pH changes induced in control experiments, pH can be evaluated using LysoTracker staining. This staining should reveal no changes in pH due to Ca^{2+} release. The results by this complementary refilling assay should be compared to the refilling assay outlined above and should provide consistent results (aforementioned ML-SA1 and GCaMP3-ML1 assay) and the GPN and Fura-2 assay.

11.2 Conclusions

Together the set of methods outlined in the current chapter should represent a comprehensive toolbox from which appropriate methods can be used to characterized Ca^{2+} signals generated by endo-lysosomal Ca^{2+} channels. From this toolbox, individual methods can be picked and used in an appropriate format ranging from single-cell microscopy to FACS and plate-reader-based assays to confocal microscopy.

References

Albrecht T, Zhao Y, Nguyen T H, Campbell R E and Johnson J D 2015 Fluorescent biosensors illuminate calcium levels within defined beta-cell endosome subpopulations *Cell Calcium* **57** 263–74

Ambrosio A L, Boyle J A and Di Pietro S M 2015 TPC2 mediates new mechanisms of platelet dense granule membrane dynamics through regulation of Ca^{2+} release *Mol. Biol. Cell* **26** 3263–74

Appelqvist H, Johansson A C, Linderoth E, Johansson U, Antonsson B, Steinfeld R, Kågedal K and Ollinger K 2012 Lysosome-mediated apoptosis is associated with cathepsin D-specific processing of bid at Phe24, Trp48, and Phe183 *Ann.. Clin Lab. Sci.* **42** 231–42

Asano S, Matsuda T, Takuma K, Kim H S, Sato T, Nishikawa T and Baba A 1995 Nitroprusside and cyclic GMP stimulate Na^{+}–Ca^{2+} exchange activity in neuronal preparations and cultured rat astrocytes *J. Neurochem.* **64** 2437–41

Atakpa P, van Marrewijk L M, Apta-Smith M, Chakraborty S and Taylor C W 2019 GPN does not release lysosomal Ca^{2+} but evokes Ca^{2+} release from the ER by increasing the cytosolic pH independently of cathepsin C *J. Cell Sci.* **132** 1–12

Berg T O, Strømhaug E, Løvdal T, Seglen O and Berg T 1994 Use of glycyl-L-phenylalanine 2-naphthylamide, a lysosome-disrupting cathepsin C substrate, to distinguish between lysosomes and prelysosomal endocytic vacuoles *Biochem. J.* **300** 229–36

Berridge M J 2012 Calcium signalling remodelling and disease *Biochem. Soc. Trans.* **40** 297–309

Bowman E J, Siebers A and Altendorf K 1988 Bafilomycins: a class of inhibitors of membrane ATPases from microorganisms, animal cells, and plant cells *Proc. Natl Acad. Sci. USA* **85** 7972–76

Brailoiu E *et al* 2009 Essential requirement for two-pore channel 1 in NAADP-mediated calcium signaling *J. Cell Biol.* **186** 201–9

Cai X *et al* 2013 PIKfyve, a class III PI kinase, is the target of the small molecular IL-12/IL-23 inhibitor apilimod and a player in Toll-like receptor signaling *Chem. Biol.* **20** 912–21

Calcraft P J *et al* 2009 NAADP mobilizes calcium from acidic organelles through two-pore channels *Nature* **459** 596–600

Cancela J M, Churchill G C and Galione A 1999 Coordination of agonist-induced Ca^{2+}-signalling patterns by NAADP in pancreatic acinar cells *Nature* **398** 74–6

Cang C, Aranda K, Seo Y J, Gasnier B and Ren D 2015 TMEM175 Is an organelle K^{+} channel regulating lysosomal function *Cell* **162** 1101–12

Cang C, Zhou Y, Navarro B, Seo Y J, Aranda K, Shi L, Battaglia-Hsu S, Nissim I, Clapham D E and Ren D 2013 MTOR regulates lysosomal ATP-sensitive two-pore Na^{+} channels to adapt to metabolic state *Cell* **152** 778–90

Cárdenas C *et al* 2010 Essential regulation of cell bioenergetics by constitutive InsP3 receptor Ca^{2+} transfer to mitochondria *Cell* **142** 270–83

Cerny J, Feng Y, Yu A, Miyake K, Borgonovo B, Klumperman J, Meldolesi J, McNeil P L and Kirchhausen T 2004 The small chemical vacuolin-1 inhibits Ca^{2+}-dependent lysosomal exocytosis but not cell resealing *EMBO Rep* **5** 883–88

Chao Y K *et al* 2017 TPC2 polymorphisms associated with a hair pigmentation phenotype in humans result in gain of channel function by independent mechanisms *Proc. Natl Acad. Sci. USA* **114** E8595–602

Chapel A *et al* 2013 An extended proteome map of the lysosomal membrane reveals novel potential transporters *Mol. Cell Proteom.* **12** 1572–88

Chen C C, Butz E S, Chao Y K, Grishchuk Y, Becker L, Heller S, Slaugenhaupt S A, Biel M, Wahl-Schott C and Grimm C 2017a Small molecules for early endosome-specific patch clamping *Cell Chem. Biol.* **24** 907–16

Chen C C, Cang C, Fenske S, Butz E, Chao Y K, Biel M, Ren D, Wahl-Schott C and Grimm C 2017b Patch-clamp technique to characterize ion channels in enlarged individual endolysosomes *Nat. Protoc.* **12** 1639–58

Chen T W *et al* 2013 Ultrasensitive fluorescent proteins for imaging neuronal activity *Nature* **499** 295–300

Cheng X *et al* 2014 The intracellular Ca^{2+} channel MCOLN1 is required for sarcolemma repair to prevent muscular dystrophy *Nat. Med.* **20** 1187–92

Christensen K A, Myers J T and Swanson J A 2002 PH-dependent regulation of lysosomal calcium in macrophages *J. Cell Sci.* **115** 599–607

Churchill G C, Okada Y, Thomas J M, Genazzani A A, Patel S and Galione A 2002 NAADP mobilizes Ca^{2+} from reserve granules, lysosome-related organelles, in sea urchin eggs *Cell* **111** 703–8

Coen K *et al* 2012 Lysosomal calcium homeostasis defects, not proton pump defects, cause endo-lysosomal dysfunction in PSEN-deficient cells *J. Cell Biol.* **198** 23–35

Davis L C *et al* 2012 NAADP activates two-pore channels on T cell cytolytic granules to stimulate exocytosis and killing *Curr. Biol.* **22** 2331–7

Dickson E J, Duman J G, Moody M W, Chen L and Hille B 2012 Orai-STIM-mediated Ca^{2+} release from secretory granules revealed by a targeted Ca^{2+} and pH probe *Proc. Natl Acad. Sci. USA* **109** E3539–48

Dionisio N, Albarrán L, López J J, Berna-Erro A, Salido G M, Bobe R and Rosado J A 2011 Acidic NAADP-releasable Ca^{2+} compartments in the megakaryoblastic cell line MEG01 *Biochim. Biophys. Acta* **1813** 1483–94

Dong X P, Cheng X, Mills E, Delling M, Wang F, Kurz T and Xu H 2008 The type IV mucolipidosis-associated protein TRPML1 is an endolysosomal iron release channel *Nature* **455** 992–6

Dong X P *et al* 2010 PI(3,5)P(2) controls membrane trafficking by direct activation of mucolipin Ca^{2+} release channels in the endolysosome *Nat. Commun.* **1** 38

Dröse S and Altendorf K 1997 Bafilomycins and concanamycins as inhibitors of V-ATPases and P-ATPases *J. Exp. Biol.* **200** 1–8

Eden E R 2016 The formation and function of ER-endosome membrane contact sites *Biochim. Biophys. Acta* **1861** 874–79

Feng M, Elaïb Z, Borgel D, Denis C V, Adam F, Bryckaert M, Rosa J P and Bobe R 2020 NAADP/SERCA3-dependent Ca^{2+} stores pathway specifically controls early autocrine ADP secretion potentiating platelet activation *Circ. Res.* **127** e166–83

Fois G, Hobi N, Felder E, Ziegler A, Miklavc P, Walther P, Radermacher P, Haller T and Dietl P 2015 A new role for an old drug: ambroxol triggers lysosomal exocytosis via pH-dependent Ca^{2+} release from acidic Ca^{2+} stores *Cell Calcium* **58** 628–37

Galione A 2006 NAADP, a new intracellular messenger that mobilizes Ca^{2+} from acidic stores *Biochem. Soc. Trans.* **34** 922–6

Galione A, Evans A M, Ma J, Parrington J, Arredouani A, Cheng X and Zhu M X 2009 The acid test: the discovery of two-pore channels (TPCs) as NAADP-gated endolysosomal Ca^{2+} release channels *Pflugers Arch.* **458** 869–76

Garrity A G, Wang W, Collier C M, Levey S A, Gao Q and Xu H 2016 The endoplasmic reticulum, not the pH gradient, drives calcium refilling of lysosomes *Elife* **5** e15887

Gerasimenko J V, Sherwood M, Tepikin A V, Petersen O H and Gerasimenko O V 2006 NAADP, cADPR and IP3 all release Ca^{2+} from the endoplasmic reticulum and an acidic store in the secretory granule area *J. Cell Sci.* **119** 226–38

Gerndt S *et al* 2020 Agonist-mediated switching of ion selectivity in TPC2 differentially promotes lysosomal function *Elife* **9** e54712

Grabe M and Oster G 2001 Regulation of organelle acidity *J. Gen. Physiol.* **117** 329–44

Grimm C, Chen C C, Wahl-Schott C and Biel M 2017 Two-pore channels: catalyzers of endolysosomal transport and function *Front. Pharmacol.* **8** 45

Grimm C, Hassan S, Wahl-Schott C and Biel M 2012 Role of TRPML and two-pore channels in endolysosomal cation homeostasis *J. Pharmacol. Exp. Ther.* **342** 236–44

Grimm C *et al* 2014 High susceptibility to fatty liver disease in two-pore channel 2-deficient mice *Nat. Commun.* **5** 4699

Guse A H 2012 Linking NAADP to ion channel activity: a unifying hypothesis *Sci. Signal* **5** pe18

Haller T, Dietl P, Deetjen P and Völkl H 1996 The lysosomal compartment as intracellular calcium store in MDCK cells: a possible involvement in InsP3-mediated Ca^{2+} release *Cell Calcium* **19** 157–65

Henderson M J *et al* 2015 A low affinity GCaMP3 variant (GCaMPer) for imaging the endoplasmic reticulum calcium store *PLoS One* **10** e0139273

Hockey L N, Kilpatrick B S, Eden E R, Lin-Moshier Y, Brailoiu G C, Brailoiu E, Futter C E, Schapira A H, Marchant J S and Patel S 2015 Dysregulation of lysosomal morphology by pathogenic LRRK2 is corrected by TPC2 inhibition *J. Cell Sci.* **128** 232–38

Hofer A M, Fasolato C and Pozzan T 1998 Capacitative Ca^{2+} entry is closely linked to the filling state of internal Ca^{2+} stores: a study using simultaneous measurements of ICRAC and intraluminal $[Ca^{2+}]$ *J. Cell Biol.* **140** 325–34

Huss M, Vitavska O, Albertmelcher A, Bockelmann S, Nardmann C, Tabke K, Tiburcy F and Wieczorek H 2011 Vacuolar H^+-ATPases: intra- and intermolecular interactions *Eur. J. Cell Biol.* **90** 688–95

Inabayashi M, Miyauchi S, Kamo N and Jin T 1995 Conductance change in phospholipid bilayer membrane by an electroneutral ionophore, monensin *Biochemistry* **34** 3455–60

Jadot M, Colmant C, Wattiaux-De Coninck S and Wattiaux R 1984 Intralysosomal hydrolysis of glycyl-L-phenylalanine 2-naphthylamide *Biochem. J.* **219** 965–70

Jefferies H B *et al* 2008 A selective PIKfyve inhibitor blocks PtdIns(3,5)P(2) production and disrupts endomembrane transport and retroviral budding *EMBO Rep.* **9** 164–70

Jentsch T J and Pusch M 2018 CLC chloride channels and transporters: structure, function, physiology, and disease *Physiol. Rev.* **98** 1493–590

Johnson D E, Ostrowski P, Jaumouillé V and Grinstein S 2016 The position of lysosomes within the cell determines their luminal pH *J. Cell Biol.* **212** 677–92

Kelu J J, Webb S E, Parrington J, Galione A and Miller A L 2017 Ca^{2+} release via two-pore channel type 2 (TPC2) is required for slow muscle cell myofibrillogenesis and myotomal patterning in intact zebrafish embryos *Dev. Biol.* **425** 109–29

Kilpatrick B S, Eden E R, Schapira A H, Futter C E and Patel S 2013 Direct mobilisation of lysosomal Ca^{2+} triggers complex Ca^{2+} signals *J. Cell Sci.* **126** 60–6

Kinnear N P, Boittin F X, Thomas J M, Galione A and Evans A M 2004 Lysosome-sarcoplasmic reticulum junctions. a trigger zone for calcium signaling by nicotinic acid adenine dinucleotide phosphate and endothelin-1 *J. Biol. Chem.* **279** 54319–26

Li S, Hao B, Lu Y, Yu P, Lee H C and Yue J 2012 Intracellular alkalinization induces cytosolic Ca^{2+} increases by inhibiting sarco/endoplasmic reticulum Ca^{2+}-ATPase (SERCA) *PLoS One* **7** e31905

Lloyd-Evans E, Morgan A J, He X, Smith D A, Elliot-Smith E, Sillence D J, Churchill G C, Schuchman E H, Galione A and Platt F M 2008 Niemann-pick disease type C1 is a sphingosine storage disease that causes deregulation of lysosomal calcium *Nat. Med.* **14** 1247–55

Lopez J J, Camello-Almaraz C, Pariente J A, Salido G M and Rosado J A 2005 Ca^{2+} accumulation into acidic organelles mediated by Ca^{2+}- and vacuolar H^{+}-ATPases in human platelets *Biochem. J.* **390** 243–52

López J J, Redondo P C, Salido G M, Pariente J A and Rosado J A 2006 Two distinct Ca^{2+} compartments show differential sensitivity to thrombin, ADP and vasopressin in human platelets *Cell Signal.* **18** 373–81

Lytton J, Westlin M, Burk S E, Shull G E and MacLennan D H 1992 Functional comparisons between isoforms of the sarcoplasmic or endoplasmic reticulum family of calcium pumps *J. Biol. Chem.* **267** 14483–89

Mao T, O'Connor D H, Scheuss V, Nakai J and Svoboda K 2008 Characterization and subcellular targeting of GCaMP-type genetically-encoded calcium indicators *PLoS One* **3** e1796

Masgrau R, Churchill G C, Morgan A J, Ashcroft S J and Galione A 2003 NAADP: a new second messenger for glucose-induced Ca^{2+} responses in clonal pancreatic beta cells *Curr. Biol.* **13** 247–51

McCue H V, Wardyn J D, Burgoyne R D and Haynes L P 2013 Generation and characterization of a lysosomally targeted, genetically encoded Ca^{2+}-sensor *Biochem. J.* **449** 449–57

Medina D L *et al* 2015 Lysosomal calcium signalling regulates autophagy through calcineurin and TFEB *Nat. Cell Biol.* **17** 288–99

Melchionda M, Pittman J K, Mayor R and Patel S 2016 Ca^{2+}/H^{+} exchange by acidic organelles regulates cell migration *in vivo J. Cell Biol.* **212** 803–13

Morgan A J, Davis L C and Galione A 2015 Imaging approaches to measuring lysosomal calcium *Methods Cell. Biol.* **126** 159–95

Morgan A J and Galione A 2007 NAADP induces pH changes in the lumen of acidic Ca^{2+} stores *Biochem. J.* **402** 301–10

Morgan A J, Platt F M, Lloyd-Evans E and Galione A 2011 Molecular mechanisms of endolysosomal Ca^{2+} signalling in health and disease *Biochem. J.* **439** 349–74

Ortega F G, Roefs M T, de Miguel Perez D, Kooijmans S A, de Jong O G, Sluijter J P, Schiffelers R M and Vader P 2019 Interfering with endolysosomal trafficking enhances release of bioactive exosomes *Nanomedicine* **20** 102014

Palmer A E, Jin C, Reed J C and Tsien R Y 2004 Bcl-2-mediated alterations in endoplasmic reticulum Ca^{2+} analyzed with an improved genetically encoded fluorescent sensor *Proc. Natl Acad. Sci. USA* **101** 17404–9

Patel S and Kilpatrick B S 2018 Two-pore channels and disease *Biochim. Biophys. Acta, Mol. Cell. Res.* **1865** 1678–86

Penny C J, Kilpatrick B S, Eden E R and Patel S 2015 Coupling acidic organelles with the ER through Ca^{2+} microdomains at membrane contact sites *Cell Calcium* **58** 387–96

Penny C J, Kilpatrick B S, Han J M, Sneyd J and Patel S 2014 A computational model of lysosome-ER Ca^{2+} microdomains *J. Cell Sci.* **127** 2934–43

Peppiatt C M, Collins T J, Mackenzie L, Conway S J, Holmes A B, Bootman M D, Berridge M J, Seo J T and Roderick H L 2003 2-Aminoethoxydiphenyl borate (2-APB) antagonises inositol 1,4,5-trisphosphate-induced calcium release, inhibits calcium pumps and has a use-dependent and slowly reversible action on store-operated calcium entry channels *Cell Calcium* **34** 97–108

Rah S Y, Mushtaq M, Nam T S, Kim S H and Kim U H 2010 Generation of cyclic ADP-ribose and nicotinic acid adenine dinucleotide phosphate by CD38 for Ca^{2+} signaling in interleukin-8-treated lymphokine-activated killer cells *J. Biol. Chem.* **285** 21877–87

Ramos I B, Miranda K, Pace D A, Verbist K C, Lin F Y, Zhang Y, Oldfield E, Machado E A, De Souza W and Docampo R 2010 Calcium- and polyphosphate-containing acidic granules of sea urchin eggs are similar to acidocalcisomes, but are not the targets for NAADP *Biochem. J.* **429** 485–95

Rao N V, Rao G V and Hoidal J R 1997 Human dipeptidyl-peptidase I. Gene characterization, localization, and expression *J. Biol. Chem.* **272** 10260–65

Robb-Gaspers L D, Rutter G A, Burnett P, Hajnóczky G, Denton R M and Thomas A P 1998 Coupling between cytosolic and mitochondrial calcium oscillations: role in the regulation of hepatic metabolism *Biochim. Biophys. Acta* **1366** 17–32

Roderick H L, Berridge M J and Bootman M D 2003 Calcium-induced calcium release *Curr. Biol.* **13** R425

Ruas M *et al* 2015 Expression of Ca^{2+}-permeable two-pore channels rescues NAADP signalling in TPC-deficient cells *EMBO J.* **34** 1743–58

Rudolf R, Mongillo M, Rizzuto R and Pozzan T 2003 Looking forward to seeing calcium *Nat. Rev. Mol. Cell Biol.* **4** 579–86

Samie M *et al* 2013 TRP channel in the lysosome regulates large particle phagocytosis via focal exocytosis *Dev. Cell* **26** 511–24

Schieder M, Rötzer K, Brüggemann A, Biel M and Wahl-Schott C A 2010 Characterization of two-pore channel 2 (TPCN2)-mediated Ca^{2+} currents in isolated lysosomes *J. Biol. Chem.* **285** 21219–22

Schwake M, Schröder B and Saftig P 2013 Lysosomal membrane proteins and their central role in physiology *Traffic* **14** 739–48

She J, Guo J, Chen Q, Zeng W, Jiang Y and Bai X C 2018 Structural insights into the voltage and phospholipid activation of the mammalian TPC1 channel *Nature* **556** 130–34

She J, Zeng W, Guo J, Chen Q, Bai X C and Jiang Y 2019 Structural mechanisms of phospholipid activation of the human TPC2 channel *Elife* **8** e45222

Shen D *et al* 2011 Lipid storage disorders block lysosomal trafficking by inhibiting a TRP channel and lysosomal calcium release *Nat. Commun.* **3** 731

Shen D *et al* 2012 Lipid storage disorders block lysosomal trafficking by inhibiting a TRP channel and lysosomal calcium release *Nat. Commun.* **3** 731

Spector I, Shochet N R, Kashman Y and Groweiss A 1983 Latrunculins: novel marine toxins that disrupt microfilament organization in cultured cells *Science* **219** 493–95

Srinivas S P, Ong A, Goon L, Goon L and Bonanno J A 2002 Lysosomal Ca^{2+} stores in bovine corneal endothelium *Invest. Ophthalmol. Vis. Sci.* **43** 2341–50

Suzuki J, Kanemaru K and Iino M 2016 Genetically encoded fluorescent indicators for organellar calcium imaging *Biophys. J.* **111** 1119–31

Thastrup O, Cullen P J, Drøbak B K, Hanley M R and Dawson A P 1990 Thapsigargin, a tumor promoter, discharges intracellular Ca^{2+} stores by specific inhibition of the endoplasmic reticulum Ca^{2+}-ATPase *Proc. Natl Acad. Sci. USA* **87** 2466–70

Tsunemi T *et al* 2019a Increased lysosomal exocytosis induced by lysosomal Ca^{2+} channel agonists protects human dopaminergic neurons from alpha-synuclein toxicity *J. Neurosci.* **39** 5760–72

Tsunemi T *et al* 2019b Increased lysosomal exocytosis induced by lysosomal Ca^{2+} channel agonists protects human dopaminergic neurons from α-synuclein toxicity *J. Neurosci.* **39** 5760–72

Várnai P, Balla A, Hunyady L and Balla T 2005 Targeted expression of the inositol 1,4,5-triphosphate receptor (IP3R) ligand-binding domain releases Ca^{2+} via endogenous IP3R channels *Proc. Natl Acad. Sci. USA* **102** 7859–64

Vasudevan S R, Lewis A M, Chan J W, Machin C L, Sinha D, Galione A and Churchill G C 2010 The calcium-mobilizing messenger nicotinic acid adenine dinucleotide phosphate participates in sperm activation by mediating the acrosome reaction *J. Biol. Chem.* **285** 18262–69

Vigushin D M, Ali S, Pace P E, Mirsaidi N, Ito K, Adcock I and Coombes R C 2001 Trichostatin A is a histone deacetylase inhibitor with potent antitumor activity against breast cancer *in vivo Clin. Cancer Res.* **7** 971–6

von Kleist L *et al* 2011 Role of the clathrin terminal domain in regulating coated pit dynamics revealed by small molecule inhibition *Cell* **146** 471–84

Wang W *et al* 2017 A voltage-dependent K^+ channel in the lysosome is required for refilling lysosomal Ca^{2+} stores *J. Cell Biol.* **216** 1715–30

Webb J L, Ravikumar B and Rubinsztein D C 2004 Microtubule disruption inhibits autophagosome-lysosome fusion: implications for studying the roles of aggresomes in polyglutamine diseases *Int. J. Biochem. Cell Biol.* **36** 2541–50

Wolf I M *et al* 2015 Frontrunners of T cell activation: initial, localized Ca^{2+} signals mediated by NAADP and the type 1 ryanodine receptor *Sci Signal* **8** ra102

Wu J *et al* 2014 Red fluorescent genetically encoded Ca^{2+} indicators for use in mitochondria and endoplasmic reticulum *Biochem. J.* **464** 13–22

Wu M M, Buchanan J, Luik R M and Lewis R S 2006 Ca^{2+} store depletion causes STIM1 to accumulate in ER regions closely associated with the plasma membrane *J. Cell Biol.* **174** 803–13

Wymann M P, Bulgarelli-Leva G, Zvelebil M J, Pirola L, Vanhaesebroeck B, Waterfield M D and Panayotou G 1996 Wortmannin inactivates phosphoinositide 3-kinase by covalent modification of Lys-802, a residue involved in the phosphate transfer reaction *Mol. Cell. Biol.* **16** 1722–33

Xu H and Ren D 2015 Lysosomal physiology *Annu. Rev. Physiol.* **77** 57–80

Yagodin S, Pivovarova N B, Andrews S B and Sattelle D B 1999 Functional characterization of thapsigargin and agonist-insensitive acidic Ca^{2+} stores in drosophila melanogaster S2 cell lines *Cell Calcium* **25** 429–38

Yamasaki M, Masgrau R, Morgan A J, Churchill G C, Patel S, Ashcroft S J and Galione A 2004 Organelle selection determines agonist-specific Ca^{2+} signals in pancreatic acinar and beta cells *J. Biol. Chem.* **279** 7234–40

Yang J, Zhao Z, Gu M, Feng X and Xu H 2019 Release and uptake mechanisms of vesicular Ca^{2+} stores *Protein Cell* **10** 8–19

Zhang F, Xu M, Han W Q and Li P L 2011 Reconstitution of lysosomal NAADP-TRP-ML1 signaling pathway and its function in TRP-ML1(−/−) cells *Am. J. Physiol. Cell Physiol.* **301** C421–30

Zhang X *et al* 2016 MCOLN1 is a ROS sensor in lysosomes that regulates autophagy *Nat. Commun.* **7** 12109

Zong X, Schieder M, Cuny H, Fenske S, Gruner C, Rotzer K, Griesbeck O, Harz H, Biel M and Wahl-Schott C 2009 The two-pore channel TPCN2 mediates NAADP-dependent Ca^{2+}-release from lysosomal stores *Pflugers Arch.* **458** 891–9

IOP Publishing

Calcium Signals
From single molecules to physiology
Leslie S Satin, Manu Ben-Johny and Ivy E Dick

Chapter 12

The structural era of the mitochondrial calcium uniporter

Enrique Balderas, Salah Sommakia, David Eberhardt, Sandra Lee and Dipayan Chaudhuri

Mitochondria accumulate calcium (Ca^{2+}) via the mitochondrial Ca^{2+} uniporter, an ion channel embedded in the mitochondrial inner membrane. This channel faces a daunting task within the cell, selecting and transporting Ca^{2+} ions present at micromolar levels in a sea of other cations that are up to six orders of magnitude more concentrated. Yet, despite knowing of its existence for over half a century, the genes encoding the transport machinery have only been identified in the past decade. Since then, a substantial number of ongoing investigations are defining uniporter physiology, culminating in the recent publication of atomic structures of the pore-forming and accessory gatekeepers. Here, we provide a focused overview of the composition of the various essential and regulatory subunits of the channel, detailing the structure-function relationships that produce channel activity. We also review the latest evidence on how the channel is regulated in a species- and tissue-specific manner, producing unique Ca^{2+} uptake profiles.

12.1 Introduction

Ca^{2+} signaling allows the coupling of electrical phenomena to biochemical regulation. A ubiquitous form of such signaling occurs within each cell, where Ca^{2+} influx into mitochondria potently regulates metabolism and cell survival. Here we review the molecular composition and function of the mitochondrial calcium uniporter, the main pathway for Ca^{2+} entry into mitochondria. The contribution of uniporter activity to cellular physiology in health and disease has been reviewed elsewhere [1–4].

12.1.1 The mitochondrial Ca^{2+} uniporter is the main portal for Ca^{2+} influx into the matrix

Rapid mitochondrial accumulation of Ca^{2+} was demonstrated nearly six decades ago [5, 6]. Such uptake required the driving force provided by the electrochemical gradient [7, 8]. Further studies established that entry of Ca^{2+} and other divalent cations (Sr^{2+}, Mn^{2+}, Ba^{2+}) could be separated from other ion transport mechanisms pharmacologically using inhibitors, such as ruthenium red, leading to the supposition that Ca^{2+} influx required a uniporter [9–11]. Besides the uniporter, other pathways for Ca^{2+} influx have been suggested [12–14]. In all tissues, however, Ca^{2+} uptake through the uniporter far exceeds other pathways.

Several factors suggest that Ca^{2+} influx through the uniporter could be regulated. Within mitochondria, Ca^{2+} has varied effects. For example, it can control the choice of substrate entering the citric acid cycle, stimulate ATP synthesis by speeding up the citric acid cycle, alter mitochondrial architecture, or trigger cell death [15–19]. Moreover, after activation, Ca^{2+} influx ceases long before reaching equilibrium distributions. Therefore, the uniporter can be gated on or off, with little, if any, transport at baseline cytoplasmic [Ca^{2+}]. This sensitivity of uniporter activation to cytoplasmic [Ca^{2+}] revealed that the uniporter-mediated Ca^{2+}-activated Ca^{2+} influx [20, 21].

Early efforts to determine the nature of the uniporter used a variety of techniques to measure Ca^{2+} uptake, including assaying ^{45}Ca transport, volume changes from osmotic shifts due to ion movement, or fluorescence changes from small molecule and genetically-encoded Ca^{2+} sensors. These revealed a very fast rate of Ca^{2+} translocation, but did not definitively prove whether a carrier or ion channel was responsible [22].

12.1.2 The uniporter is a highly-selective Ca^{2+} channel

The seminal discovery that the uniporter was an ion channel came from the application of voltage-clamp electrophysiology to isolated mitoplasts (mitochondria stripped of their outer membranes) [23]. This study opened a new era of investigation into the uniporter, with several fundamental features of channel activity becoming evident. Most importantly, the channel was exquisitely selective for Ca^{2+}—more so than any other Ca^{2+} channel discovered so far. Even ~10 nM Ca^{2+} was sufficient to block conduction of monovalent cations, orders of magnitude more selective than other Ca^{2+} channels (~1–10 μM) [24–26]. This property is critical to avoiding a futile depolarization of the inner membrane, as K^+ is ~six orders of magnitude more abundant in the cytoplasm. In addition, the uniporter had substantial Ca^{2+} carrying capacity, unitary channel conductance (2.6–5.2 pS, in 105 mM [Ca^{2+}]) similar to other Ca^{2+} channels, but minimal inactivation compared to other Ca^{2+} channels.

12.1.3 The uniporter channel is composed of multiple subunits

Efforts to identify the genes encoding the ion channel were stymied by the lack of highly-specific ligands, and lack of homology to plasma membrane Ca^{2+} channels. One advance was the development of mitochondria-targeted genetically-encoded Ca^{2+} sensors, allowing higher throughput screening [27]. The key advance came from integrative genomics, with the establishment of a high-confidence compendium of genes encoding mitochondrial proteins (MitoCarta) [28]. Using MitoCarta, along with the finding that yeast possesses no uniporter activity for Ca^{2+} uptake, led the

Mootha laboratory to identity the first component of the uniporter complex, *mitochondrial calcium uptake 1* (*MICU1*) [29]. Shortly thereafter, investigations from the Mootha and Rizzuto labs converged on the pore-forming subunit, *mitochondrial calcium uniporter* (*MCU*) [30, 31]. Subsequent studies defined the remaining key components of the channel, *MCUB, MICU2–3* and *essential MCU regulator* (*EMRE*) [32–34]. From here onwards, we will refer to the multi-subunit complex as the uniporter, reserving MCU to designate the pore-forming subunit.

12.2 MCU

12.2.1 The MCU gene encodes the channel pore

MCU was identified by a combination of *in silico* and RNA interference screens [30, 31]. The basic architecture of the pore-forming MCU protein comprises an N-terminal domain (NTD) of ~100 residues that faces the mitochondrial matrix and regulates Ca^{2+} uptake, two coiled-coil domains (CC1, CC2) that are structurally necessary for Ca^{2+} uptake, and two transmembrane spanning domains TM1 and TM2 [35, 36] (figure 12.1(A)). The linker between TM1 and TM2 leads into a highly-conserved WDXXEP (Trp-Asp-X-X-Glu-Pro, where X denotes hydrophobic residues) motif, which is essential for Ca^{2+} selectivity. This basic MCU topology is conserved in homologs expressed across the eukaryotic lineage, though the gene has been lost in some protozoans and fungal species such as *Saccharomyces cerevisiae* [37, 38].

MCU was shown to be the pore-forming subunit by several functional studies [30, 31, 39]. Mutating residues in the putative selectivity filter altered channel conduction. In particular, replacement of a conserved serine residue with alanine (S259A) abolished the sensitivity to the inhibitors ruthenium red and Ru360. Moreover, the reconstitution of the protein into lipid bilayers appeared to produce channel-like behavior, though the conductance (6–7 pS in symmetrical 100 mM Ca^{2+}) and open probability were somewhat discordant from the native channel. Additionally, in a variety of cell types, reduction of *MCU* correlated with inhibited Ca^{2+} uptake, whereas overexpression increased it. Finally, expression of *Dictyostelium discoideum* MCU, which operates independently of accessory subunits, conferred electrophoretic Ca^{2+} uptake in yeast, which do not have an endogenous uniporter [40]. Ultimate confirmation came from channel structures recently described.

12.2.2 Most evidence suggests the channel is a tetramer of MCU subunits

Although it is well established that the MCU can form oligomers [30], the subunit stoichiometry between MCU and the remaining components is an ongoing area of research. For MCU itself, the focus has been in defining the architecture of the pore. The first structure was determined by nuclear magnetic resonance and electron microscopy (EM) from a truncated *Caenorhabditis elegans* MCU (cMCU-ΔNTD). In this structure, MCU formed pentamers, with the WDXXEP motif located within the conduction pathway of the channel. Subsequently, several laboratories used cryo-EM to solve structures of full-length fungal MCUs, which do not require

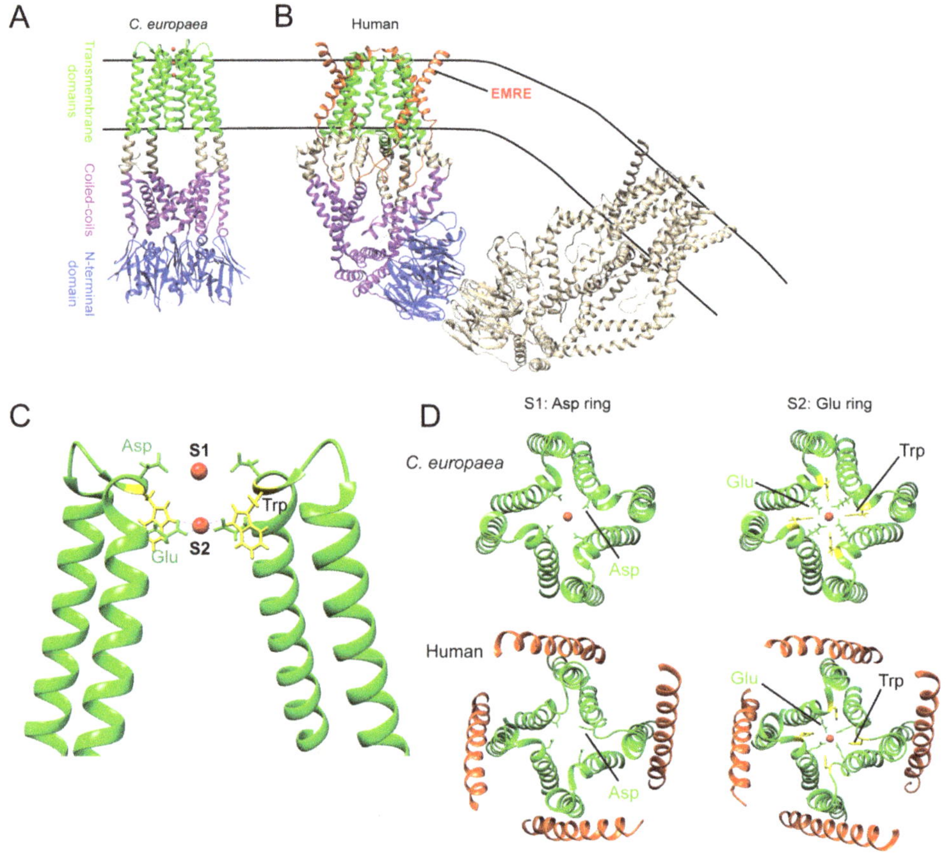

Figure 12.1. (A) Uniporter channel formed by MCU tetramer from *C. europaea.* MCU domains are color-coded and labeled. (B) Human uniporter dimer, showing additionally the EMRE subunit. Dimer formation would require membrane curvature as shown. The human NTDs break symmetry and arrange in a side-by-side crescent-like conformation. (C) Side view of the selectivity filter with two of four subunits removed. Ca at the upper (S1) and lower (S2) sites is shown, bound by a ring of Asp (S1) and Glu (S2), shown in green. The key Trp residue is shown in yellow. (D) Top view of the S1 (left) and S2 (right) sites with side chains from Asp and Glu/Trp residues shown. PDB identifiers for *C. europaea* structure: 6DNF [44], human structure: 6O58 [45]. Molecular graphics and analyses performed with UCSF Chimera, developed by the Resource for Biocomputing, Visualization, and Informatics at the University of California, San Francisco, with support from NIH P41-GM103311 (http://www.rbvi.ucsf.edu/chimera/) [118].

EMRE for expression and Ca^{2+} uptake, unlike metazoans. The MCU structures from *Cyphellophora europaea*, *Fusarium graminearum*, *Metarhizium acridum*, *Neosartorya fischeri*, and *Neurospora crassa* all displayed homo-tetrameric stoichiometry [41–44] (figure 12.1(A)). All four structures were highly similar, showing four-fold symmetry in the transmembrane domains. In these, TM1 is the outer transmembrane domain, and interacts closely with the pore-lining, inner TM2 of an adjacent subunit, but only weakly with the TM2 from the same subunit. In the soluble portion, four NTDs form a well-packed tetrameric ring in a dimer-of-dimers configuration, connected to TM1 through an elongated CC1. Compared to the cMCU-ΔNTD NMR structure, the fungal structures show a tetrameric rather than pentameric structure, a selectivity filter that spans one alpha-helical turn at the pore mouth rather than an unstructured loop, and a more extended relationship between the transmembrane and soluble portions.

The tetrameric channel likely represents the physiological oligomer, as to create the cMCU-ΔNTD protein, a truncated construct lacking the NTD was used, which failed to produce Ca^{2+} uptake in reconstitution experiments. Moreover, the truncated protein required extraction from inclusion bodies using harsher detergents. Finally, a recent cryo-EM structure of the human MCU in complex with the EMRE protein showed a similar tetrameric structure, mimicking the fungal structures extensively in the transmembrane domains [45] (figure 12.1(B)).

12.2.3 Ca^{2+} selectivity relies on concentric rings of negatively-charged residues and an extremely rigid bottleneck

The highly-conserved WDXXEP motif forms the selectivity filter. A slightly less widely-conserved sequence (Trp-Asp-X-X-Glu-Pro-X-Thr-Tyr or WDXXEPXTY) is shared by the fungal and human selectivity filters, with the negatively-charged Asp and Glu residues essential for Ca^{2+} transport [30, 31] (figure 12.1(C)). It was presumed that these residues conferred selectivity by binding Ca^{2+} at the pore mouth, in a manner similar to other Ca^{2+} channels [25]. Structural studies have since confirmed this hypothesis, with the fungal and human structures showing the WDXXEPVTY motif forms the first complete turn of the alpha-helical TM2 at the mouth of the pore [41–45]. The critical residue is the Glu. Its carboxylate group is rigidly constrained to optimally coordinate dehydrated Ca^{2+} at the narrowest portion of the pore.

One feature of the WDXXEPXTY motif that allows preferential binding of divalent cations over monovalents is the high charge density conferred by the acidic Asp and Glu residues. Because of their orientation in the alpha helix, the carboxyl side chains from both these residues point into the solvent, creating stacked concentric rings of negative charge (figures 12.1(C) and (D)). This leads to Ca^{2+} binding sites separated by ~6 Å along the pore [44]. The solvent-exposed Asp ring forms the upper binding site (S1), possessing a diameter of approximately 4–8 Å in the fungal and human structures, and likely coordinates a hydrated Ca^{2+} ion. In the structure with the lowest diameter at this site (~4 Å, [44]), a density corresponding to Ca^{2+} was visible, whereas it was absent and the Asp side chain was least resolved in the structure with largest diameter [43]. This suggests that there may be some flexibility in this ring, with coordination of a hydrated Ca^{2+} ion locking it into a narrow configuration. Consistent with this interpretation, the Asp is more tolerant of substitutions, and can even increase Ca^{2+} transport when mutated to Glu [30, 42, 46]. The second binding site (S2) is composed of a ring of carboxyl side chains from the Glu residue pointing towards the center of the conduction pathway, producing substantial concentration of charge in a very small volume. In all the fungal and human structures, the pore diameter within this ring is quite small, with a diameter of ~2 Å. A Ca^{2+} ion is seen directly coordinated by opposing Glu carboxylate side chains, with ion-ligand distances ~2.5 Å, optimal for Ca^{2+} coordination. Such coordination is also evident in plasma membrane Ca^{2+} channel structures [47, 48]. With these negative Glu carboxylate groups constrained to be so close to each other, this site may have Ca^{2+} constitutively bound, which may explain the exquisite sensitivity of the channel to Ca^{2+}. At cytoplasmic nanomolar Ca^{2+} concentrations, a bound but non-transported Ca^{2+} ion may block the permeation of cations with far greater cytoplasmic concentrations. A third possible binding site (S3) was resolved in the CeMCU structure, 6 Å below the second site [44]. Here, the density, which may be Ca^{2+} or another molecule, is centered ~6 Å from surrounding Thr and Tyr hydroxyl groups that terminate the WDXXEPXTY motif.

Beyond charge density, the highly-conserved Trp residue in the WDXXEPVTY motif adds substantial rigidity to the S2 binding site via an extensive network of interactions (figures 12.1(C) and (D)). Although the Trp precedes the Asp in the TM2 alpha helix, it is oriented such that its indole side chain dips into the plane of the S2 ring. Here, the Trp indoles are oriented parallel to the pore axis, sandwiching the Glu residues, and stabilizing them. Additionally, the Trp phenyl ring interacts with the highly-conserved Pro residue of an adjacent subunit, stacking against it to further stiffen the orientation of the critical Glu residue. By constraining the Glu residues, this grid of interactions maintains both an optimal architecture for direct Ca^{2+} coordination as well as creating a rigid, narrow ring that likely excludes larger cations such as K^+. Finally, the Trp may also contribute to the stability of the S3 binding site, as it interacts with the Tyr moiety contributing to that ring. In support of their crucial stabilizing function, replacement of the Trp or Pro residues produces non-conducting channels [41, 43, 44].

12.2.4 A wide vestibule beyond the selectivity filter allows rapid Ca^{2+} conduction

To support the rapid conduction rates with high-selectivity, the two- or three-site single-file pathway mediated by the WDXXEPVTY motif in the constricted portion of the pore gives way to a wide, likely water-filled vestibule immediately below. Thus, the pore is divided into two sections, a short, narrow portion comprising the selectivity filter residing in the outer leaflet of the membrane, and a much wider portion leading to the matrix in the inner leaflet. Widening the vestibule immediately below the narrow selectivity filter for diffusion of hydrated ions is a general strategy for rapid conduction, seen also in K^+ channels [49]. To create this widening, whereas the upper portion of the pore is lined only by TM2, the vestibule is lined by TM1, TM2, and a gap between these presumably sealed off by lipids.

12.2.5 The linker between the second transmembrane and coiled-coil domains may serve as a channel gate in human MCU

Although uniporter gating strongly depends on the interaction between MCU and auxiliary MICU1 and MICU2 subunits, how this alters the channel pore remains unknown. A current hypothesis is that flexibility at the end of the TM2 domain, where it is connected to CC2 by a short juxtamembrane loop, may produce a conduction gate at the junction between the pore and mitochondrial matrix [45, 50]. In the fungal structures, the linker between the C-terminus of the TM2 and CC2 is rather disordered, but a focused subclassification revealed channels where TM2 was bent towards the central pore axis, suggesting a closed conformation [43]. Such a constriction is also seen at low resolution in the human structure, leading the authors to propose that the metazoan-specific EMRE subunit opens the channel by hooking this loop and CC2 away from the pore, pulling TM2 along with it [45]. Whether the constriction produced in this 'closed' conformation by TM2 is sufficient to prevent the transport of hydrated Ca^{2+} remains undetermined.

12.2.6 Regulation of MCU occurs through its N-terminal domain

The N-terminal domain resides in the mitochondrial matrix and adopts a structure resembling a β-grasp fold [35, 51]. This evolutionarily-ancient structural motif has been adapted for many uses, including protein–protein interactions and small molecule binding, and the uniporter NTD similarly appears to be a nexus of channel regulation [52]. Despite little homology in sequence between the fungal and human NTDs, their three-dimensional structure is very similar. The NTD structure comprises six antiparallel β-sheets surrounding a central α-helix, with a variable number of additional α-helices and β-sheets depending on the species.

In the fungal uniporter tetramer, four NTDs form a ring in a dimer or dimers configuration [41–44] (figure 12.1(A)). The human NTD, despite being structural similar to the fungal NTDs, has a substantially different architecture in the channel tetramer [45] (figure 12.1(B)). Alpha-helical linker domains between the NTD and CC1 break the four-fold symmetry and swing the four NTDs into a side-by-side crescent-like configuration, seen also in low-resolutions structures of the zebrafish MCU [44]. Most surprisingly, two sets of NTD tetramers mediate a dimerization of the channel itself. Dimerized channels are not in the same plane, requiring a membrane curvature of ~50° for embedding both sets of transmembrane domains. It is currently unclear if such an architecture occurs under physiological conditions in cells. In favor of channel dimerization is evidence of potential dimers after blue-native PAGE isolation, and oligomerization of channels after oxidative stress [33, 53]. In addition, the residues producing both NTD oligomerization within a single channel as well as channel dimerization are among the most conserved in metazoans. However, the membrane curvature required for dimerization, which might suggest that MCU is localized to cristae junctions or other highly-curved spots, is at odds with the homogenous distribution of MCU within the inner membrane [54, 55]. Further investigation will be required to establish the relevance of this intriguing architecture.

The NTD appears to be regulatory rather than essential for Ca^{2+} flux through the uniporter [119]. Elimination of the NTD entirely does not prevent the formation of uniporter channels nor their Ca^{2+} uptake [35, 46]. Nevertheless, the NTD has been implicated in channel regulation in several reports, and NTD-mediated oligomerization appears to be a common theme. Purified NTD binds Ca^{2+}

and Mg^{2+} with low affinity (K_D ~1–2 mM for both) [51]. Adding Mg^{2+} inhibits MCU subunit oligomerization, and blunts the rate of Ca^{2+} uptake in a manner similar to mutating the Mg^{2+}-coordinating residues in the NTD [51, 56]. Similarly, the NTD appears to mediate channel aggregation during oxidative stress, leading to persistent Ca^{2+} accumulation [53]. This mechanism is transduced by a highly-conserved cysteine residue (Cys-97) located in the NTD interaction interface, and requires glutathionylation of its solvent-exposed thiol group. The NTD has also been implicated in channel phosphorylation. Several kinases may alter uniporter activity, including Pyk2, CaMKII, and LKB1 [57–59]. An initial report suggested that CaMKII-mediated phosphorylation of residues on the NTD increased channel activity [57]. However, later studies put this in doubt, as neither a direct effect of CaMKII on uniporter currents, nor an effect of CaMKII knockout on mitochondrial Ca^{2+} uptake was seen [60, 61]. The residues affected by the other kinases have not yet been isolated. Thus, it remains to be established if NTD phosphorylation may be a target for channel regulation. Finally, the MCU regulator MCUR1 modulates MCU function through an interaction with the NTD (see section 12.6.1).

Two substantial questions are still unanswered regarding NTD-mediated channel regulation. First, the remarkable structural similarity in the NTD across species with minimal sequence-level homology suggests that the overall architecture of this domain is more important than any particular residue, for channel regulation. What necessary behaviors are gained by this overall structure? Second, given that NTD modifications alter Ca^{2+} uptake, investigators need to establish whether such effects depend on emergent cooperativity due to channel oligomerization or are directly transduced within the same channel by a structural rearrangement.

12.3 MCUB

MCUB remains the least understood of the uniporter subunits. It possesses close sequence homology to MCU (~50%), with similar domain organization and an identical WDIMEPVTY motif and surrounding sequence. It can form hetero-oligomers with MCU, and such channels appear to have diminished Ca^{2+} uptake, whereas reducing its expression leads to enhanced Ca^{2+} uptake. Such evidence has led to the hypothesis that this subunit may act in a dominant-negative manner [34]. It interacts with the accessory subunit EMRE (see section 12.4) [62]. Its expression in mammals may be regulated to reduce Ca^{2+} uptake in the setting of injury, though in Trypanosomes, MCUB appears necessary for Ca^{2+} uptake and survival [63]. MCUB expression has been shown to be protective in cardiac ischemic injury but may limit metabolic flexibility in the diabetic heart [62, 64, 65].

12.4 EMRE

EMRE was discovered through a proteomic screen for MCU interactors [33]. This small ~10 kDa protein is a metazoan-specific component of the uniporter, and is required for Ca^{2+} uptake. Deletion of the subunit entirely abolishes uniporter Ca^{2+}

currents, though uniporter protein remains. This led to the concept that the EMRE subunit is essential for MCU activation [66, 67].

The EMRE protein was resolved in the recent human MCU–EMRE cryo-EM structure [45] (figure 12.1(B)). As expected from sequence analysis, the protein has a single transmembrane domain (TMD). This sits at an angle compared to the MCU transmembrane domains, allowing it to make contacts with adjacent MCU subunits. The soluble N-terminal portion of EMRE consists of a β-hairpin with a linker connecting it to the TMD. These sit on the matrix side of the inner membrane, confirmed by several functional studies [50, 68–70]. The N-terminal portion of the EMRE TMD contacts TM2 and the juxtamembrane loop from one MCU subunit, whereas the central portion of the EMRE TMD makes hydrophobic contacts with the outer TM1 of the neighboring MCU subunit.

Mutation of residues to disrupt the MCU–EMRE interface prevents uniporter activity, both in the EMRE TMD as well as in the soluble domains [50, 69, 70]. As mentioned above, the β-hairpin and linker may mediate gating of the uniporter via the juxtamembrane loop and CC2 of MCU (see section 12.2.5) [45]. For this purpose, the linker appears more necessary, as a rigid, highly-conserved Pro-Lys-Pro sequence is essential for EMRE function, whereas deletion of the β-hairpin prior to this motif fails to impair Ca^{2+} uptake. Taking advantage of EMRE-induced MCU opening, electrogenic Ca^{2+} uptake can be reconstituted in yeast, which have no endogenous uniporter, by co-expressing human MCU and EMRE [40, 69, 71]. Thus, the EMRE N-terminus and transmembrane domain are essential for maintaining the uniporter in an open conformation.

The solvent-exposed EMRE C-terminus, on the other hand, is essential for Ca^{2+}-induced activation of the uniporter. The C-terminus is disordered in the cryo-EM structure, but contains a highly-conserved polyaspartate tail. This stretch of acidic residues interacts with the polybasic domain of MICU1, the uniporter gatekeeper (see section 12.5) [72]. Thus, EMRE bridges MCU and MICU1. In fact, deletion of EMRE abolishes the interaction between MCU and MICU1, unless these are overexpressed [33]. Moreover, deletion of the aspartate tail leads to a channel that is open at even low Ca^{2+} levels, consistent with a loss of MICU1 gatekeeping [70]. Acting as an adaptor for the MCU-MICU1 interaction, precise control of EMRE levels is essential. Excess EMRE creates MCU–EMRE complexes without MICU1 bound, which are constitutively open even at basal cytoplasmic Ca^{2+} levels, leading to mitochondrial uncoupling and Ca^{2+} overload. To prevent this, robust activity of the inner membrane-resident *m*-AAA proteases degrades excess EMRE, keeping MCU and MICU1 levels well coupled [68]. Consistent with this mechanism, depletion of MCU leads to a similar decrease in EMRE levels [33].

12.5 The MICU proteins (MICU1, MICU2, MICU3)

MICU1 was initially identified as a calcium-binding autoantigen involved in atopic dermatitis [73]. It was the first protein identified to be part of the uniporter complex [29]. As with the MCU gene, an RNAi screen was performed on candidates selected

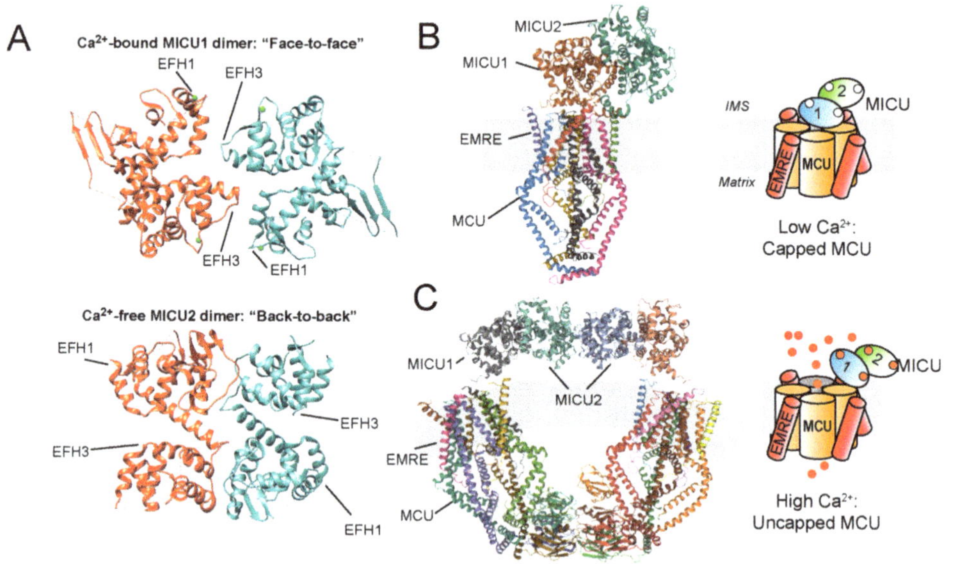

Figure 12.2. (A) Top, Ca^{2+}-bound MICU1 dimer in 'face-to-face' conformation, with EFH1 from one subunit interacting with EFH3 from another. Bottom, MICU2 dimer in 'back-to-back' apo form, with EF1 and EFH3 on opposite sides of the dimer no longer interacting. (B) In Ca^{2+} free conformation, MCU is capped by a MICU1–MICU2 heterodimer that occludes the pore via interactions with the aspartate ring. (C) Ca^{2+}-bound MICU1–MICU2 heterodimers tilt away, uncapping the pore for Ca^{2+} permeation. The rearranged MICU1–MICU2 dimers can now form a bridge across two channel complexes via back-to-back interactions. PDB identifiers: for human MICU1 dimer: 4NSD [74], human MICU2 dimer: 6AGH [85], Low Ca^{2+} MCU–EMRE–MICU1–MICU2 complex: 6WDN [88], High Ca^{2+} MCU–EMRE–MICU1–MICU2 complex: 6WDO [88]. IMS, intermembrane space. Molecular graphics and analyses performed with UCSF Chimera, developed by the Resource for Biocomputing, Visualization, and Informatics at the University of California, San Francisco, with support from NIH P41-GM103311 (http://www.rbvi.ucsf.edu/chimera/).

from the MitoCarta database that were present in mammals but not yeast. MICU1 was further identified as an inner membrane associated protein resident in the intermembrane space. It possesses four EF-hands, two of which possess Ca^{2+} binding residues and two of which have lost these and appear to be structural [74] (figure 12.2(A)). Within humans, MICU1 possessed two paralogs with ~25% homology, subsequently named MICU2 and MICU3, and these were found to interact with MICU1 [32]. Homologs of the MICU proteins are found in most eukaryotes, but have been lost in certain protozoan and fungal lineages [37].

12.5.1 The gatekeeper MICU proteins set the threshold for Ca^{2+} activation and alter the cooperativity of Ca^{2+} uptake

Studies prior to establishing the identity of uniporter genes had shown that Ca^{2+} uptake was sensitive to Ca^{2+} binding at an extra-mitochondrial site [20, 21]. Functionally, MICU1 was found to act as a gatekeeper of MCU-mediated mitochondrial calcium uptake, preventing Ca^{2+} uptake at low concentrations (<~1 μM) [75, 76]. In its absence, limited Ca^{2+} uptake can occur even at basal cytoplasmic levels and can lead to mitochondrial Ca^{2+} overload. The Ca^{2+} binding sites were shown to be extra-mitochondrial and therefore sensitive to cytoplasmic Ca^{2+} [77]. The MICU2 and MICU3 paralogs were found to reside in the same uniporter complex and help cross-stabilize MICU1 and MCU in a cell-type dependent manner, with MICU2 being the version present in most tissues, and MICU3 enriched in brain and skeletal muscle [32]. The MICU protein complex thus makes the uniporter a Ca^{2+}-activated Ca^{2+}-transporting channel.

MICU1 and MICU2 form a regulatory heterodimer that enables fine-tuning of the uniporter Ca^{2+} threshold [78–80]. The EF hands of MICU1 bind Ca^{2+} with an affinity of 200–400 nM, the EF hands of MICU2 bind Ca^{2+} with an affinity of 700–900 nM, whereas the MICU1–MICU2 heterodimer has an intermediate K_D of ~600 nM. Thus, at basal cytoplasmic concentrations, Ca^{2+} does not bind the EF hands, and MICU1 and MICU2 cooperate to inhibit the channel, preventing Ca^{2+} uptake and overload. In support of this, depletion of either protein reduces the threshold for Ca^{2+} uptake, leading to mitochondrial Ca^{2+} accumulation even under basal conditions [66, 75, 78–82]. Mutation of the Ca^{2+}-binding residues on the EF hands keeps the gates closed, inhibiting uniporter permeation during Ca^{2+} signals. This Ca^{2+} binding gatekeeping effect not only prevents Ca^{2+} transport, but also prevents the uptake of manganese or other potentially toxic divalents that can permeate the channel [83].

During active Ca^{2+} signaling, such as during IP_3R-mediated ER Ca^{2+} release, the MICU1–MICU2 heterodimer also allows cooperative activation of the uniporter [75, 79, 82]. Under these larger Ca^{2+} loads, the high gain of uniporter activation leads to enhanced Ca^{2+} uptake. This enhancement depends on MICU1, as inhibition of this subunit prevents such cooperative activation. In HeLa, HEK-293T, and hepatocyte cells, the initial Ca^{2+} uptake rate is much faster with higher Ca^{2+} loads (a slope of ~3 on double logarithmic plots), but becomes much less sensitive to the Ca^{2+} load (slope of 1.4–2) upon MICU1 knockdown. Conversely, inhibition of MICU2 has a minimal effect on the slope, though it shifts the curve such that cooperative activation is evident a much lower Ca^{2+} loads [82]. Thus, MICU2 increases the Ca^{2+} required for cooperative behavior, while MICU1 increases its gain. The mechanism by which cooperativity occurs, however, is unknown.

The MICU1–MICU2 heterodimerization is mediated by the essential oxidoreductase Mia40, which introduces intermolecular disulfide bonds that link MICU1 and MICU2 [84]. Such crosslinking is essential for MICU2 regulation of the uniporter, as it does not directly interact with MCU. Thus, MICU2 activity and association with the pore requires MICU1 while the converse is not true, indicating nonredundant roles in setting the calcium threshold of the uniporter [78]. MICU2 protein stability also depends on MICU1. MICU2 protein levels decrease after depleting MICU1, whereas depleting MICU2 has no effect on MICU1 levels [32, 78].

12.5.2 Structures of the MICU proteins show Ca^{2+}-dependent rearrangements and oligomerization

Structurally, MICU1-3 resemble each other, possessing an N-terminal domain, two EF-hand containing lobes (N- and C-lobes), and a C-terminus [74, 85–87].

The N-terminal domains all contain a mixture of α-helices and antiparallel β-sheets. The MICU1 N-terminal domain contains a polybasic region composed of several lysine and arginine residues that mediates the interaction with EMRE, likely via electrostatic interactions with the poly-asparate EMRE tail [70, 72]. The N-terminal domains in MICU2 and MICU3 are similar to each other, suggesting similar function. They are more compact and packed closer to the N-lobe than in MICU1. In all the MICU structures, the N- and C-lobes each contain a Ca^{2+} binding and a structural EF hand. The Ca^{2+}-binding EF hands are EF1 and EF4. The C-termini also differ between the structures, with the MICU1 structure having a helical segment (C-helix), whereas the MICU2 C-helix is bent into a plane orthogonal to the rest of the crystal, where it can bind other MICU2 molecules, but not MICU1 [86]. The C-helices contain the critical Cys for MICU1–MICU2 oligomerization. Functionally, C-terminal deletion impairs both the binding to MCU and the Ca^{2+} gatekeeping activity [74, 78].

In isolation, calcium-free MICU1 forms a hexamer, packed as a trimer of dimers [74]. MICU2 and MICU3 use this EF-hand interface to dimerize as well [85–87]. Notably, however, MICU2 and MICU3 can also dimerize in an entirely different manner in their Ca^{2+}-free form [85]. In this alternate dimer, the EF-hands are far from the interaction surface, though the N-lobe of one subunit still interacts with the C-lobe of the opposing one (figure 12.2(A)). MICU1 is apparently prevented from this arrangement due to steric hindrances caused by its N-terminal domain. This alternate arrangement has been called the 'back-to-back' dimer, to contrast with the EF1-EF3 interaction comprising the 'face-to-face' dimer. In the Ca^{2+}-bound forms, however, MICU2 and MICU3 adopt the 'face-to-face' alignment [85–87]. How such a large domain swap occurs with Ca^{2+} binding during physiological gating is unclear, particularly as the 'back-to-back' conformation was seen in dimers with truncated C-helices. As with the MICU1 C-helix, though the MICU2 C-helix is dispensable for MICU2 dimerization, it is essential for uniporter gatekeeping [86].

Given their structural similarity, MICU1–MICU2 heterodimerization likely follows the same general principles for each homodimer, involving 'face-to-face' contacts. The unique feature of the heterodimer is the requirement for the C-helix in the absence of Ca^{2+} [86]. In this situation, the MICU1–MICU2 heterodimer is stabilized by the C-helix disulfide bond, as noted previously in section 12.5.1 [84].

How the MICU1–MICU2 heterodimer interacts with MCU–EMRE is not entirely resolved. In recent holocomplex structures containing MCU–EMRE–MICU1–MICU2, the MICU1 N-terminus polybasic region is close to the poly-aspartate EMRE C-tail and parallel to the membrane, though the spatial characteristics of a potential salt-bridge interaction remain unresolved [88–90]. In addition, a second point of contact appears to be present, though this varies between the three sets of structures, possibly involving the α11 MICU1 helix contacting the poly-aspartate tail of an adjacent EMRE in one structure and hydrophobic interactions with conserved EMRE and MCU transmembrane residues in the others [88–90]. Finally, both MICU1 and MICU2 were found to specifically bind cardiolipin, suggesting an independent mechanism for inner membrane association [80].

12.5.3 Holocomplex structures reveal MICU1–MICU2 uncap the MCU pore upon Ca^{2+} binding

Very recent structures by three groups show the novel mechanism of MCU Ca^{2+} activation by the MICU1–MICU2 heterodimer. These structures report a 4:1 MCU: MICU1–MICU2 stoichiometry. This symmetry allows an MCU tetramer to bind a single MICU1. The bottom of MICU1's N-lobe docks directly onto the IMS surface of MCU such that a lysine/arginine ring (K126, R129, R259, R261 and R263) of MCU1 forms a 'cap' that seals the D261 ring at the MCU pore entrance, with all D261 residues engaged in electrostatic interactions (figure 12.2(B)). Such a mechanism was predicted in prior functional studies, where mutagenesis of several basic Arg and Lys residues altered MICU1-dependent Ca^{2+} gating, though the residues predicted to interact with the MCU asparate ring differed between these functional assays and the recent structures [91, 92].

The relative instability and mobility of the MICU1–MICU2 contacts with MCU may be key to its gating function. Ca^{2+} binding to the EF-hands sites on MICU1–MICU2 activate MCU by dislodging MICU1 from the top of the pore, and tilting its position relative to MICU2, 'uncapping' the pore and leaving it open for Ca^{2+} transport [88–90] (figure 12.2(C)). In both Ca^{2+}-bound and Ca^{2+}-free forms, the MCU–EMRE structure is largely unchanged, suggesting that rearrangements within the channel itself are unnecessary for Ca^{2+} gating into a permeable state. Surprisingly, the structures also reveal how Ca^{2+} binding may lead to the cooperativity in Ca^{2+} uptake [75]. Ca^{2+} binding to a single MICU subunit causes steric hindrance at the heterodimer interface, while Ca^{2+} binding to both MICU1–MICU2 subunits allows reciprocal conformational changes that favor dissociation of MICU1–MICU2 from the pore [88]. In addition, it is intriguing that a large fraction of structures in high Ca^{2+} revealed MCU dimerization involving the MCU N-terminal domains, occurring even in structures without MICU1–MICU2 heterodimers. It remains an open question whether the formation of such MCU–EMRE dimers in high Ca^{2+} has a physiological advantage over the monomeric counterpart.

Structures in high Ca^{2+} still show MCU–EMRE–MICU1–MICU2 heteroligomers. In these, full dissociation of the MICU1–MICU2 heterodimer from MCU is prevented by the creation of the MICU1–MICU2 bridged holocomplex, forming an O-like structure (figure 12.2(C)). Maintaining the Ca^{2+}-bound MICU1–MICU2 heterodimer close to the channel would presumably allow rapid 'recapping' upon Ca^{2+} dissociation from the MICU1–MICU2 EF hands when cytoplasmic levels drop.

12.5.4 Variation in MICU protein expression and post-translational modification can further regulate mitochondrial Ca^{2+} uptake

Mitochondrial Ca^{2+} uptake profiles vary substantially between tissues, consistent with heterogenous energy requirements [93]. An additional level of tissue-specific variations occurs in the threshold and cooperativity for Ca^{2+} uptake. As noted above, cardiac muscle appears to have no threshold for Ca^{2+} uptake, and also has minimal cooperative behavior, whereas skeletal muscle and liver appear to have both [94, 95]. An important mechanism contributing to this appears to be the ratio of MICU1 to MCU protein levels. As MICU1 helps set the threshold and cooperativity gain for the uniporter, cells with larger amounts of MICU1 relative to MCU have mitochondria with evident thresholds and high gains, such as hepatocytes, whereas those with low ratios, such as cardiomyocytes, appear to have less evident thresholds and minimal gain [94].

Regulatory modifications of MICU1 also modulate Ca^{2+} uptake. In skeletal muscle, an alternatively spliced isoform of the MICU1 N-terminal domain appears to activate the uniporter at lower Ca^{2+} concentrations [96]. Similarly, Akt-mediated phosphorylation of the N-terminus impairs MICU1 stability, leading to increased mitochondrial Ca^{2+} levels and culminating in elevated ROS production and tumor progression [97]. The C-helix may also be subject to regulation, as methylation on an Arg decreases Ca^{2+} sensitivity [98]. Finally, synthesis and degradation of the MICU1 protein may also be regulated [99–101].

12.5.5 MICU1 regulates cristae morphology in a non-canonical, MCU-independent manner

Although no pathophysiological variants in MCU have been described in humans, multiple MICU1 and MICU2 mutations have been associated with severe pathology, including encephalopathy, myopathy, fatigue, and hepatopathy [102–105]. In animal models, elimination of MCU produces relatively mild phenotypes, whereas deletion of MICU1 proteins appears to be far more severe [66, 67, 81]. Although initially this pathology was thought to be due solely to mitochondrial Ca^{2+} overload, more recently effects independent of Ca^{2+} uptake have been suggested. In *Drosophila*, for example, MICU1 deletion results in developmental lethality that is not rescued by deletion of MCU or EMRE, suggesting a Ca^{2+}-independent role of MICU1 in development [67].

In one potential mechanism for such non-canonical effects, MICU1 appears to be involved in the control of cristae structure, independent of its association with MCU [55, 106]. Unlike MCU, which is distributed throughout the inner membrane, MICU1 appears to be localized to the inner boundary membrane and excluded from cristae [54, 55]. This effect appears to be mediated by MICU1 interactions with the Mitochondrial Contact Site and Cristae Organizing System (MICOS) proteins. MICU1 depletion remodels cristae junctions, removing a bottleneck for cytochrome c release, and such pro-apoptotic signaling may explain the Ca^{2+}-independent effects of MICU1 loss. Interestingly, upon Ca^{2+} stimulation, MCU and EMRE appear to move from a homogenous distribution to one that is restricted to the inner boundary membrane, mimicking MICU1 localization [55]. Whether the inner boundary membrane residency of MICU1 and associated MCU redistribution is truly occurring physiologically remains somewhat unclear. If not associated with MICU1, it is unclear how MCU–EMRE channels within cristae would be prevented from conducting Ca^{2+} under basal conditions and leading to Ca^{2+} overload. More importantly, multiple studies examining MICU1 function, including those that show differential localization, are performed by tagging MICU1 at the C-terminus. If the oligomeric structure of MICU1 is correct, the C-terminal helices face into the narrow center of the structure. Adding tags such as fluorescent proteins at the C-terminus is therefore very likely to disrupt higher-order oligomer formation, given the limited space and likely steric hindrance produced. Thus, measurement of sub-organellar distributions using C-terminal fluorescent or other tags may be assaying MICU1 monomers or dimers, rather than the physiological oligomers controlling MCU gating. Nevertheless, these analyses still reveal that MICU1 controls cristae morphology in a Ca^{2+}-independent manner.

12.6 Additional regulatory proteins

A range of other molecules have been shown to interact with the uniporter in a variety of contexts, including UCP2/3, SLC25A23, and TRPC3 [98, 107, 108]. We discuss below the regulatory protein most studied in this context, MCUR1.

12.6.1 MCUR1 is a regulator of uniporter-mediated Ca^{2+} uptake

An RNAi screen for proteins affecting mitochondrial Ca^{2+} uptake identified an inner membrane protein with coiled-coil domains, MCUR1 [109]. MCUR1 was found to bind to MCU and serve as a potential uniporter scaffold [110, 111]. A partial MCUR1 structure reveals a head-neck-stalk-anchor architecture, in which a membrane-anchored coiled-coil stalk projects an N-terminal head domain via a β-layer neck. The head interacts directly with the NTD of MCU [35, 110–112].

MCUR1 appears to enhance uniporter activity. MCUR1 loss disrupts the formation of heterooligomeric MCU complexes, impairs mitochondrial Ca^{2+} uptake and downstream bioenergetics [109, 111]. Direct assessment of the mitochondrial Ca^{2+} current via mitoplast patch clamp, revealed that MCUR1 knockdown reduced uniporter Ca^{2+} current [113].

MCUR1 also affects Ca^{2+}-dependent bioenergetics. A comparison of human and *Drosophila* cells revealed that MCUR1 regulates the Ca^{2+} threshold necessary to induce toxic Ca^{2+}-overload (the mitochondrial permeability transition) [110]. In this system, the protection against Ca^{2+} overload induced by MCUR1 inhibition improved cell survival. In several studies of hepatocellular carcinoma cells, overexpression of MCUR1 increases mitochondrial Ca^{2+} uptake that is, however, accompanied by improved cell survival and tumor invasion [114, 115]. One study suggested that MCUR1 is a cytochrome oxidase assembly factor, and that the impaired Ca^{2+} uptake produced by suppression of MCUR1 is due to the bioenergetic impairment [116]. Given this apparent link between the uniporter and the electron transport chain, it is also notable that previous studies also established a link between MCU (when it was known as CCDC109A) and the Complex I subunit NDUFA9 [117].

12.7 Summary

Our understanding of the mitochondrial Ca^{2+} uniporter has increased considerably in the past decade, culminating recently with the fungal and human structures.

Located on the inner membrane of the mitochondria, the uniporter is made up of the MCU pore-forming subunit, and MICU1-3, MCUb, and EMRE accessory subunits. It is likely that MCU is a well-packed tetrameric ring with a selectivity filter, formed by a highly-conserved WDXXEP motif, that is exquisitely precise in coordinating Ca^{2+}. Beyond the pore, though its amino acid sequence diverges, the MCU NTD shows great structural homology between species. What crucial regulatory function the NTD β-grasp fold structure provides needs to be established. The MICU proteins have been established as both the gatekeepers making the uniporter a Ca^{2+}-activated Ca^{2+}-channel, as well as the mediators of highly cooperative Ca^{2+} uptake. They gate the channel by a capping mechanism that occludes Ca^{2+} access to the selectivity filter and pore. We are well into the structural era of the mitochondrial Ca^{2+} uniporter, and look forward to when this recent knowledge is exploited for clinical use in treating mitochondrial dysfunction.

Acknowledgment

Molecular graphics using PDB data files performed with UCSF Chimera, developed by the Resource for Biocomputing, Visualization, and Informatics at the University of California, San Francisco, with support from NIH P41-GM103311. This work was supported in part by the Nora Eccles Treadwell Foundation; the NIH Heart, Lung, and Blood Institute under awards R01HL141353 (D C), and T32HL007576 (S S).

References

[1] Tarasova N V, Vishnyakova P A, Logashina Y A and Elchaninov A V 2019 Mitochondrial calcium uniporter structure and function in different types of muscle tissues in health and disease *Int. J. Mol. Sci.* **20** 4823

[2] Vultur A, Gibhardt C S, Stanisz H and Bogeski I 2018 The role of the mitochondrial calcium uniporter (MCU) complex in cancer *Pflugers Archiv. : Eur. J. Physiol.* **470** 1149–63

[3] Mammucari C, Raffaello A, Vecellio Reane D, Gherardi G, De Mario A and Rizzuto R 2018 Mitochondrial calcium uptake in organ physiology: from molecular mechanism to animal models *Pflugers Archiv: Eur. J. Physiol.* **470** 1165–79

[4] Nemani N, Shanmughapriya S and Madesh M 2018 Molecular regulation of MCU: implications in physiology and disease *Cell Calcium* **74** 86–93

[5] Vasington F D and Murphy J V 1962 Ca ion uptake by rat kidney mitochondria and its dependence on respiration and phosphorylation *J. Biol. Chem.* **237** 2670–7

[6] Deluca H F and Engstrom G W 1961 Calcium uptake by rat kidney mitochondria *PNAS* **47** 1744–50

[7] Selwyn M J, Dawson A P and Dunnett S J 1970 Calcium transport in mitochondria *FEBS Lett.* **10** 1–5

[8] Scarpa A and Azzone G F 1970 The mechanism of ion translocation in mitochondria. 4. coupling of K^+ efflux with Ca^{2+} uptake *Eur. J. Biochem./FEBS* **12** 328–35

[9] Lehninger A L 1974 Role of phosphate and other proton-donating anions in respiration-coupled transport of Ca^{2+} by mitochondria *PNAS* **71** 1520–24

[10] Vasington F D, Gazzotti P, Tiozzo R and Carafoli E 1972 The effect of ruthenium red on Ca^{2+} transport and respiration in rat liver mitochondria *Biochim. Biophys. Acta* **256** 43–54

[11] Puskin J S, Gunter T E, Gunter K K and Russell P R 1976 Evidence for more than one Ca^{2+} transport mechanism in mitochondria *Biochemistry* **15** 3834–42

[12] Ryu S Y, Beutner G, Dirksen R T, Kinnally K W and Sheu S S 2010 Mitochondrial ryanodine receptors and other mitochondrial Ca^{2+} permeable channels *FEBS Lett.* **584** 1948–55

[13] Beutner G, Sharma V K, Giovannucci D R, Yule D I and Sheu S S 2001 Identification of a ryanodine receptor in rat heart mitochondria *J. Biol. Chem.* **276** 21482–88

[14] Michels G, Khan I F, Endres-Becker J, Rottlaender D, Herzig S, Ruhparwar A, Wahlers T and Hoppe U C 2009 Regulation of the human cardiac mitochondrial Ca^{2+} uptake by 2 different voltage-gated Ca^{2+} channels *Circulation* **119** 2435–43

[15] Glancy B and Balaban R S 2012 Role of mitochondrial Ca^{2+} in the regulation of cellular energetics *Biochemistry* **51** 2959–73

[16] Zoratti M and Szabo I 1995 The mitochondrial permeability transition *Biochim. Biophys. Acta* **1241** 139–76

[17] Sommakia S, Houlihan P R, Deane S S, Simcox J A, Torres N S, Jeong M Y, Winge D R, Villanueva C J and Chaudhuri D 2017 Mitochondrial cardiomyopathies feature increased uptake and diminished efflux of mitochondrial calcium *J. Mol. Cell. Cardiol.* **113** 22–32

[18] Gherardi G, Nogara L, Ciciliot S, Fadini G P, Blaauw B, Braghetta P, Bonaldo P, De Stefani D, Rizzuto R and Mammucari C 2019 Loss of mitochondrial calcium uniporter rewires skeletal muscle metabolism and substrate preference *Cell Death Diff.* **26** 362–81

[19] Nemani N *et al* 2018 MIRO-1 determines mitochondrial shape transition upon GPCR activation and Ca^{2+} stress *Cell Rep.* **23** 1005–19

[20] Igbavboa U and Pfeiffer D R 1988 EGTA inhibits reverse uniport-dependent Ca^{2+} release from uncoupled mitochondria. possible regulation of the Ca^{2+} uniporter by a Ca^{2+} binding site on the cytoplasmic side of the inner membrane *J. Biol. Chem.* **263** 1405–12

[21] Kroner H 1986 Ca^{2+} ions, an allosteric activator of calcium uptake in rat liver mitochondria *Arch. Biochem. Biophys.* **251** 525–35

[22] Gunter T E and Pfeiffer D R 1990 Mechanisms by which mitochondria transport calcium *Am. J. Physiol.* **258** C755–86

[23] Kirichok Y, Krapivinsky G and Clapham D E 2004 The mitochondrial calcium uniporter is a highly selective ion channel *Nature* **427** 360–4

[24] Bakowski D and Parekh A B 2002 Monovalent cation permeability and Ca^{2+} block of the store-operated Ca^{2+} current I(CRAC)in rat basophilic leukemia cells *Pflugers Archiv : Eur. J. Physiol.* **443** 892–902

[25] Yang J, Ellinor P T, Sather W A, Zhang J F and Tsien R W 1993 Molecular determinants of Ca^{2+} selectivity and ion permeation in L-type Ca^{2+} channels *Nature* **366** 158–61

[26] Kuo C C and Hess P 1993 Ion permeation through the L-type Ca^{2+} channel in rat phaeochromocytoma cells: two sets of ion binding sites in the pore *J. Physiol.* **466** 629–55

[27] Rizzuto R, Simpson A W, Brini M and Pozzan T 1992 Rapid changes of mitochondrial Ca^{2+} revealed by specifically targeted recombinant aequorin *Nature* **358** 325–7

[28] Pagliarini D J *et al* 2008 A mitochondrial protein compendium elucidates complex I disease biology *Cell* **134** 112–23

[29] Perocchi F, Gohil V M, Girgis H S, Bao X R, McCombs J E, Palmer A E and Mootha V K 2010 MICU1 encodes a mitochondrial EF hand protein required for Ca^{2+} uptake *Nature* **467** 291–6

[30] Baughman J M *et al* 2011 Integrative genomics identifies MCU as an essential component of the mitochondrial calcium uniporter *Nature* **476** 341–5
[31] De Stefani D, Raffaello A, Teardo E, Szabo I and Rizzuto R 2011 A forty-kilodalton protein of the inner membrane is the mitochondrial calcium uniporter *Nature* **476** 336–40
[32] Plovanich M *et al* 2013 MICU2, a paralog of MICU1, resides within the mitochondrial uniporter complex to regulate calcium handling *PLoS One* **8** e55785
[33] Sancak Y *et al* 2013 EMRE is an essential component of the mitochondrial calcium uniporter complex *Science* **342** 1379–82
[34] Raffaello A, De Stefani D, Sabbadin D, Teardo E, Merli G, Picard A, Checchetto V, Moro S, Szabo I and Rizzuto R 2013 The mitochondrial calcium uniporter is a multimer that can include a dominant-negative pore-forming subunit *EMBO J.* **32** 2362–76
[35] Lee Y *et al* 2015 Structure and function of the N-terminal domain of the human mitochondrial calcium uniporter *EMBO Rep.* **16** 1318–33
[36] Yamamoto T *et al* 2019 Functional analysis of coiled-coil domains of MCU in mitochondrial calcium uptake *Biochim. Biophys. Acta, Bioenerg.* **1860** 148061
[37] Bick A G, Calvo S E and Mootha V K 2012 Evolutionary diversity of the mitochondrial calcium uniporter *Science* **336** 886
[38] Li Y, Calvo S E, Gutman R, Liu J S and Mootha V K 2014 Expansion of biological pathways based on evolutionary inference *Cell* **158** 213–25
[39] Chaudhuri D, Sancak Y, Mootha V K and Clapham D E 2013 MCU encodes the pore conducting mitochondrial calcium currents *eLife* **2** e00704
[40] Kovacs-Bogdan E, Sancak Y, Kamer K J, Plovanich M, Jambhekar A, Huber R J, Myre M A, Blower M D and Mootha V K 2014 Reconstitution of the mitochondrial calcium uniporter in yeast *PNAS* **111** 8985–90
[41] Yoo J, Wu M, Yin Y, Herzik M A Jr., Lander G C and Lee S Y 2018 Cryo-EM structure of a mitochondrial calcium uniporter *Science* **361** 506–11
[42] Nguyen N X, Armache J P, Lee C, Yang Y, Zeng W, Mootha V K, Cheng Y, Bai X C and Jiang Y 2018 Cryo-EM structure of a fungal mitochondrial calcium uniporter *Nature* **559** 570–4
[43] Fan C *et al* 2018 X-ray and cryo-EM structures of the mitochondrial calcium uniporter *Nature* **559** 575–9
[44] Baradaran R, Wang C, Siliciano A F and Long S B 2018 Cryo-EM structures of fungal and metazoan mitochondrial calcium uniporters *Nature* **559** 580–4
[45] Wang Y, Nguyen N X, She J, Zeng W, Yang Y, Bai X C and Jiang Y 2019 Structural Mechanism of EMRE-dependent gating of the human mitochondrial calcium uniporter *Cell* **177** 1252–61
[46] Oxenoid K *et al* 2016 Architecture of the mitochondrial calcium uniporter *Nature* **533** 269–73
[47] Hou X, Pedi L, Diver M M and Long S B 2012 Crystal structure of the calcium release-activated calcium channel Orai *Science* **338** 1308–13
[48] Wu J, Yan Z, Li Z, Qian X, Lu S, Dong M, Zhou Q and Yan N 2016 Structure of the voltage-gated calcium channel $Ca_v1.1$ at 3.6 a resolution *Nature* **537** 191–96
[49] Tao X, Hite R K and MacKinnon R 2017 Cryo-EM structure of the open high-conductance Ca^{2+}-activated K^+ channel *Nature* **541** 46–51
[50] MacEwen M J S, Markhard A L, Bozbeyoglu M, Bradford F, Goldberger O, Mootha V K and Sancak Y 2019 Molecular basis of EMRE-dependence of the human mitochondrial calcium uniporter *bioRxiv* 637918

[51] Lee S K, Shanmughapriya S, Mok M C Y, Dong Z, Tomar D, Carvalho E, Rajan S, Junop M S, Madesh M and Stathopulos P B 2016 Structural insights into mitochondrial calcium uniporter regulation by divalent cations *Cell Chem. Biol.* **23** 1157–69

[52] Burroughs A M, Balaji S, Iyer L M and Aravind L 2007 Small but versatile: the extraordinary functional and structural diversity of the beta-grasp fold *Biol. Direct* **2** 18

[53] Dong Z *et al* 2017 Mitochondrial Ca^{2+} uniporter is a mitochondrial luminal redox sensor that augments MCU channel activity *Mol. Cell* **65** 1014–28

[54] De La Fuente S *et al* 2016 Strategic positioning and biased activity of the mitochondrial calcium uniporter in cardiac muscle *J. Biol. Chem.* **291** 23343–62

[55] Gottschalk B *et al* 2019 MICU1 controls cristae junction and spatially anchors mitochondrial Ca^{2+} uniporter complex *Nat. Commun.* **10** 3732

[56] Blomeyer C A, Bazil J N, Stowe D F, Dash R K and Camara A K 2016 Mg^{2+} differentially regulates two modes of mitochondrial Ca^{2+} uptake in isolated cardiac mitochondria: implications for mitochondrial Ca^{2+} sequestration *J. Bioenerg. Biomembr.* **48** 175–88

[57] Joiner M L *et al* 2012 CaMKII determines mitochondrial stress responses in heart *Nature* **491** 269–73

[58] O-Uchi J *et al* 2014 Adrenergic signaling regulates mitochondrial Ca^{2+} uptake through Pyk2-dependent tyrosine phosphorylation of the mitochondrial Ca^{2+} uniporter *Antiox. Redox Signal.* **21** 863–79

[59] Kwon S K, Sando R 3rd, Lewis T L, Hirabayashi Y, Maximov A and Polleux F 2016 LKB1 regulates mitochondria-dependent presynaptic calcium clearance and neurotransmitter release properties at excitatory synapses along cortical axons *PLoS Biol.* **14** e1002516

[60] Fieni F, Johnson D E, Hudmon A and Kirichok Y 2014 Mitochondrial Ca^{2+} uniporter and CaMKII in heart *Nature* **513** E1–2

[61] Nickel A G *et al* 2019 CaMKII does not control mitochondrial Ca^{2+} uptake in cardiac myocytes *J. Physiol.* **598** 1361–76

[62] Lambert J P, Luongo T S, Tomar D, Jadiya P, Gao E, Zhang X, Lucchese A M, Kolmetzky D W, Shah N S and Elrod J W 2019 MCUB regulates the molecular composition of the mitochondrial calcium uniporter channel to limit mitochondrial calcium overload during stress *Circulation* **140** 1720–33

[63] Chiurillo M A, Lander N, Bertolini M S, Storey M, Vercesi A E and Docampo R 2017 Different roles of mitochondrial calcium uniporter complex subunits in growth and infectivity of trypanosoma cruzi *MBio* **8** e00574-17

[64] Cividini F *et al* 2021 Ncor2/PPARalpha-dependent upregulation of MCUb in the type 2 diabetic heart impacts cardiac metabolic flexibility and function *Diabetes* **70** 665–79

[65] Huo J, Lu S, Kwong J Q, Bround M J, Grimes K M, Sargent M A, Brown M E, Davis M E, Bers D M and Molkentin J D 2020 MCUb induction protects the heart from postischemic remodeling *Circ. Res.* **127** 379–90

[66] Liu J C *et al* 2016 MICU1 serves as a molecular gatekeeper to prevent *in vivo* mitochondrial calcium overload *Cell Rep.* **16** 1561–73

[67] Tufi R, Gleeson T P, von Stockum S, Hewitt V L, Lee J J, Terriente-Felix A, Sanchez-Martinez A, Ziviani E and Whitworth A J 2019 Comprehensive genetic characterization of mitochondrial Ca^{2+} uniporter components reveals their different physiological requirements *in vivo Cell Rep.* **27** 1541–50

[68] Konig T *et al* 2016 The m-AAA protease associated with neurodegeneration limits MCU activity in mitochondria *Mol. Cell* **64** 148–62

[69] Yamamoto T *et al* 2016 Analysis of the structure and function of EMRE in a yeast expression system *Biochim. Biophys. Acta* **1857** 831–39

[70] Tsai M F, Phillips C B, Ranaghan M, Tsai C W, Wu Y, Willliams C and Miller C 2016 Dual functions of a small regulatory subunit in the mitochondrial calcium uniporter complex *eLife* **5** e15545

[71] Arduino D M *et al* 2017 Systematic identification of MCU modulators by orthogonal interspecies chemical screening *Mol. Cell* **67** 711–23

[72] Hoffman N E *et al* 2013 MICU1 motifs define mitochondrial calcium uniporter binding and activity *Cell Rep.* **5** 1576–88

[73] Aichberger K J *et al* 2005 Hom s 4, an IgE-reactive autoantigen belonging to a new subfamily of calcium-binding proteins, can induce Th cell type 1-mediated autoreactivity *J. Immunol.* **175** 1286–94

[74] Wang L, Yang X, Li S, Wang Z, Liu Y, Feng J, Zhu Y and Shen Y 2014 Structural and mechanistic insights into MICU1 regulation of mitochondrial calcium uptake *EMBO J.* **33** 594–604

[75] Csordas G *et al* 2013 MICU1 controls both the threshold and cooperative activation of the mitochondrial Ca^{2+} uniporter *Cell Metab.* **17** 976–87

[76] Mallilankaraman K *et al* 2012 MICU1 Is an essential gatekeeper for MCU-mediated mitochondrial Ca^{2+} uptake that regulates cell survival *Cell* **151** 630–44

[77] Waldeck-Weiermair M, Malli R, Parichatikanond W, Gottschalk B, Madreiter-Sokolowski C T, Klec C, Rost R and Graier W F 2015 Rearrangement of MICU1 multimers for activation of MCU is solely controlled by cytosolic Ca^{2+} *Sci. Rep.* **5** 15602

[78] Kamer K J and Mootha V K 2014 MICU1 and MICU2 play nonredundant roles in the regulation of the mitochondrial calcium uniporter *EMBO Rep.* **15** 299–307

[79] Patron M, Checchetto V, Raffaello A, Teardo E, Vecellio Reane D, Mantoan M, Granatiero V, Szabo I, De Stefani D and Rizzuto R 2014 MICU1 and MICU2 finely tune the mitochondrial Ca^{2+} uniporter by exerting opposite effects on MCU activity *Mol. Cell* **53** 726–37

[80] Kamer K J, Grabarek Z and Mootha V K 2017 High-affinity cooperative Ca^{2+} binding by MICU1–MICU2 serves as an on-off switch for the uniporter *EMBO Rep.* **18** 1397–411

[81] Antony A N *et al* 2016 MICU1 regulation of mitochondrial Ca^{2+} uptake dictates survival and tissue regeneration *Nat. Commun.* **7** 10955

[82] Payne R, Hoff H, Roskowski A and Foskett J K 2017 MICU2 restricts spatial crosstalk between InsP3R and MCU channels by regulating threshold and gain of MICU1-mediated inhibition and activation of MCU *Cell Rep.* **21** 3141–54

[83] Kamer K J, Sancak Y, Fomina Y, Meisel J D, Chaudhuri D, Grabarek Z and Mootha V K 2018 MICU1 imparts the mitochondrial uniporter with the ability to discriminate between Ca^{2+} and Mn^{2+} *PNAS* **115** E7960–9

[84] Petrungaro C, Zimmermann K M, Kuttner V, Fischer M, Dengjel J, Bogeski I and Riemer J 2015 The Ca^{2+}-dependent release of the Mia40-induced MICU1–MICU2 dimer from MCU regulates mitochondrial Ca^{2+} uptake *Cell Metab.* **22** 721–33

[85] Xing Y, Wang M, Wang J, Nie Z, Wu G, Yang X and Shen Y 2019 Dimerization of MICU proteins controls Ca^{2+} influx through the mitochondrial Ca^{2+} uniporter *Cell Rep.* **26** 1203–2

[86] Kamer K J, Jiang W, Kaushik V K, Mootha V K and Grabarek Z 2019 Crystal structure of MICU2 and comparison with MICU1 reveal insights into the uniporter gating mechanism *PNAS* **116** 3546–55

[87] Wu W, Shen Q, Lei Z, Qiu Z, Li D, Pei H, Zheng J and Jia Z 2019 The crystal structure of MICU2 provides insight into Ca^{2+} binding and MICU1–MICU2 heterodimer formation *EMBO Rep.* **20** e47488

[88] Fan M, Zhang J, Tsai C W, Orlando B J, Rodriguez M, Xu Y, Liao M, Tsai M F and Feng L 2020 Structure and mechanism of the mitochondrial Ca^{2+} uniporter holocomplex *Nature* **582** 129–33

[89] Wang C, Jacewicz A, Delgado B D, Baradaran R and Long S B 2020 Structures reveal gatekeeping of the mitochondrial Ca^{2+} uniporter by MICU1–MICU2 *eLife* **9** e59991

[90] Wang Y, Han Y, She J, Nguyen N X, Mootha V K, Bai X C and Jiang Y 2020 Structural insights into the Ca^{2+}-dependent gating of the human mitochondrial calcium uniporter *eLife* **9** e60513

[91] Phillips C B, Tsai C W and Tsai M F 2019 The conserved aspartate ring of MCU mediates MICU1 binding and regulation in the mitochondrial calcium uniporter complex *eLife* **8** e41112

[92] Paillard M, Csordas G, Huang K T, Varnai P, Joseph S K and Hajnoczky G 2018 MICU1 interacts with the D-ring of the MCU pore to control Its Ca^{2+} flux and sensitivity to Ru360 *Mol. Cell* **72** 778–85

[93] Fieni F, Bae Lee S, Jan Y N and Kirichok Y 2012 Activity of the mitochondrial calcium uniporter varies greatly between tissues *Nat. Commun.* **3** 1317

[94] Paillard M *et al* 2017 Tissue-specific mitochondrial decoding of cytoplasmic Ca^{2+} signals is controlled by the stoichiometry of MICU1/2 and MCU *Cell Rep.* **18** 2291–300

[95] Wescott A P, Kao J P Y, Lederer W J and Boyman L 2019 Voltage-energized calcium-sensitive ATP production by mitochondria *Nat. Metab.* **1** 975–84

[96] Vecellio Reane D, Vallese F, Checchetto V, Acquasaliente L, Butera G, De Filippis V, Szabo I, Zanotti G, Rizzuto R and Raffaello A 2016 A MICU1 splice variant confers high sensitivity to the mitochondrial Ca^{2+} uptake machinery of skeletal muscle *Mol. Cell* **64** 760–73

[97] Marchi S *et al* 2019 Akt-mediated phosphorylation of MICU1 regulates mitochondrial Ca^{2+} levels and tumor growth *EMBO J.* **38** e99435

[98] Madreiter-Sokolowski C T *et al* 2016 PRMT1-mediated methylation of MICU1 determines the UCP2/3 dependency of mitochondrial Ca^{2+} uptake in immortalized cells *Nat. Commun.* **7** 12897

[99] Stoll S, Xi J, Ma B, Leimena C, Behringer E J, Qin G and Qiu H 2019 The valosin-containing protein protects the heart against pathological Ca^{2+} overload by modulating Ca^{2+} uptake proteins *Toxicol. Sci.* **171** 473–84

[100] Matteucci A, Patron M, Vecellio Reane D, Gastaldello S, Amoroso S, Rizzuto R, Brini M, Raffaello A and Cali T 2018 Parkin-dependent regulation of the MCU complex component MICU1 *Sci. Rep.* **8** 14199

[101] Shanmughapriya S *et al* 2018 FOXD1-dependent MICU1 expression regulates mitochondrial activity and cell differentiation *Nat. Commun.* **9** 3449

[102] Logan C V *et al* 2014 Loss-of-function mutations in MICU1 cause a brain and muscle disorder linked to primary alterations in mitochondrial calcium signaling *Nat. Genet.* **46** 188–93

[103] Musa S *et al* 2019 A middle eastern founder mutation expands the genotypic and phenotypic spectrum of mitochondrial MICU1 deficiency: a report of 13 patients *JIMD Rep.* **43** 79–83

[104] Shamseldin H E *et al* 2017 A null mutation in MICU2 causes abnormal mitochondrial calcium homeostasis and a severe neurodevelopmental disorder *Brain: J. Neurol.* **140** 2806–13
[105] Lewis-Smith D *et al* 2016 Homozygous deletion in MICU1 presenting with fatigue and lethargy in childhood *Neurol. Genet.* **2** e59
[106] Tomar D, Thomas M, Garbincius J F, Kolmetzky D W, Salik O, Jadiya P, Carpenter A C and Elrod J W 2019 MICU1 regulates mitochondrial cristae structure and function independent of the mitochondrial calcium uniporter channel *bioRxiv* 803213
[107] Hoffman N E *et al* 2014 SLC25A23 augments mitochondrial Ca^{2+} uptake, interacts with MCU, and induces oxidative stress-mediated cell death *Mol. Biol. Cell* **25** 729–964
[108] Feng S, Li H, Tai Y, Huang J, Su Y, Abramowitz J, Zhu M X, Birnbaumer L and Wang Y 2013 Canonical transient receptor potential 3 channels regulate mitochondrial calcium uptake *PNAS* **110** 11011–6
[109] Mallilankaraman K *et al* 2012 MCUR1 is an essential component of mitochondrial Ca^{2+} uptake that regulates cellular metabolism *Nat. Cell Biol.* **14** 1336–43
[110] Chaudhuri D, Artiga D J, Abiria S A and Clapham D E 2016 Mitochondrial calcium uniporter regulator 1 (MCUR1) regulates the calcium threshold for the mitochondrial permeability transition *PNAS* **113** E1872–80
[111] Tomar D *et al* 2016 MCUR1 is a scaffold factor for the MCU complex function and promotes mitochondrial bioenergetics *Cell Rep.* **15** 1673–85
[112] Adlakha J, Karamichali I, Sangwallek J, Deiss S, Bar K, Coles M, Hartmann M D, Lupas A N and Hernandez Alvarez B 2019 Characterization of MCU-binding proteins MCUR1 and CCDC90B – representatives of a protein family conserved in prokaryotes and eukaryotic organelles *Structure* **27** 464–75
[113] Vais H, Tanis J E, Muller M, Payne R, Mallilankaraman K and Foskett J K 2015 MCUR1, CCDC90A, is a regulator of the mitochondrial calcium uniporter *Cell Metab.* **22** 533–5
[114] Ren T *et al* 2018 MCUR1-mediated mitochondrial calcium signaling facilitates cell survival of hepatocellular carcinoma via reactive oxygen species-dependent P53 degradation *Antiox. Redox Signal.* **28** 1120–36
[115] Jin M, Wang J, Ji X, Cao H, Zhu J, Chen Y, Yang J, Zhao Z, Ren T and Xing J 2019 MCUR1 facilitates epithelial-mesenchymal transition and metastasis via the mitochondrial calcium dependent ROS/Nrf2/Notch pathway in hepatocellular carcinoma *J. Exp. Clin. Cancer Res.* **38** 136
[116] Paupe V, Prudent J, Dassa E P, Rendon O Z and Shoubridge E A 2015 CCDC90A (MCUR1) is a cytochrome c oxidase assembly factor and not a regulator of the mitochondrial calcium uniporter *Cell Metab.* **21** 109–16
[117] Guarani V, Paulo J, Zhai B, Huttlin E L, Gygi S P and Harper J W 2014 TIMMDC1/C3orf1 functions as a membrane-embedded mitochondrial complex I assembly factor through association with the MCIA complex *Mol. Cell. Biol.* **34** 847–61
[118] Pettersen E F, Goddard T D, Huang C C, Couch G S, Greenblatt D M, Meng E C and Ferrin T E 2004 UCSF Chimera – a visualization system for exploratory research and analysis *J. Comput. Chem.* **25** 1605–12
[119] Balderas E *et al* 2022 Mitochondrial calcium uniporter stabilization preserves energetic homeostasis during Complex I impairment *Nat. Commun.* **13** 2769

Part V

Calcium channels and disease

Chapter 13

Voltage-gated calcium channelopathies

John W Hussey, Kevin G Herold and Ivy E Dick

Voltage-gated Ca^{2+} channels (VGCCs) control Ca^{2+} entry in electrically excitable cells in numerous tissues, including the heart, brain, smooth muscle, skeletal muscle, and endocrine system. As such, they are responsible for numerous critical physiological functions, and genetic mutations which affect the actions of these channels can cause a myriad of pathologies, known as Ca^{2+} channelopathies. These mutations contribute to disease states through their diverse impacts on channel properties including channel gating, expression, trafficking, and downstream cellular processes. Often the effects of these mutations are characterized as either gain-of-function (GOF) or loss-of-function (LOF); however, the biophysical effects of mutations on channel function can be complex, making such categorization challenging. Here, we review the currently known mutations in the pore-forming subunits of VGCCs and their impact on channel function and pathophysiology. As the number of identified mutations continues to grow, so too will the pathogenic mechanisms and clinical phenotypes associated with these channelopathies.

13.1 Introduction

Voltage-gated calcium channels (VGCCs) are critical conduits for Ca^{2+} influx in all excitable cells, as well as some non-excitable cells [1, 2]. They play a major role in a myriad of physiological processes including gene transcription, hormone secretion, neurotransmitter release, synaptic excitability, synaptic plasticity, and cardiac, smooth, and skeletal muscle contraction [3–5]. Accordingly, Ca^{2+} entry through these channels is precisely regulated, and mutations which disrupt the function of these channels can result in severe disease phenotypes known as channelopathies [6, 7]. Most VGCC channelopathies result from changes in channel gating, implicating a complex array of pathogenic mechanisms.

The primary subunit of VGCCs is the α_1 pore-forming subunit [8, 9], which consists of four homologous domains each containing six transmembrane α helices,

doi:10.1088/978-0-7503-2009-2ch13

designated S1–S6. The S5 and S6 helices line the channel pore, with the S5–6 linker forming the pore loop containing the selectivity filter [10, 11]. The S1-S4 segments contain the voltage sensing domain (VSD), with S4 harboring the positive gating charges which enable the channel to respond to changes in membrane voltage [12]. The activation gate of the channel is formed by the intracellular side of the S6 helices [13], and the C-terminus of each channel is known to serve as a locus for numerous cytosolic protein interactions [14–16]. There are ten different α_1 subunits within the VGCC family. The Ca_V1 subfamily contains four members: $Ca_V1.1$, $Ca_V1.2$, $Ca_V1.3$, and $Ca_V1.4$, corresponding to the α_{1s}, α_{1c}, α_{1D}, and α_{1F} pore-forming subunits encoded by the genes *CACNA1S, CACNA1C, CACNA1D* and *CACNA1F*, respectively [17]. These channels are also known as L-type Ca^{2+} channels (LTCCs). $Ca_V1.2$ and $Ca_V1.3$ channels display a wide expression pattern, existing in nearly all excitable cells. These channels play important roles in excitation–contraction (EC) coupling, excitation–transcription coupling, synaptic regulation, hormone release, and cardiac rhythmogenesis. On the other hand, $Ca_V1.1$ and $Ca_V1.4$ have more selective expression in skeletal muscle and retinal cells, respectively. Thus, Ca_V1 channels are critical for normal cardiac, visual, neuronal, endocrine, and muscular function, and mutations in these channels can result in a wide array of clinical phenotypes. The Ca_V2 subfamily contains three members: $Ca_V2.1$, $Ca_V2.2$, and $Ca_V2.3$, corresponding to the α_{1A}, α_{1B}, and α_{1E} pore-forming subunits encoded by the genes *CACNA1A, CACNA1B* and *CACNA1E*, respectively [17]. $Ca_V2.1$ channels are also known as P/Q-type channels, $Ca_V2.2$ channels are known as N-type channels, and $Ca_V2.3$ are designated as R-type. Ca_V2 channels are primarily expressed in the nerve terminals and dendrites of neurons and in neuroendocrine cells [3, 18]. These channels are essential for neurotransmitter release.

Together, Ca_V1 and Ca_V2 channels comprise the high voltage activated (HVA) channels. In addition to the α_1 subunit, HVA channels are known to interact with multiple auxiliary subunits [19]. A β subunit is required for proper expression and gating of the channel [20], and an auxiliary $\alpha_2\delta$ subunits is also associated with mature VGCCs [21, 22]. In many HVA channels, a γ subunit also associates with the channel [23]. These three auxiliary subunits are known to modulate various aspects of channel function, including regulating trafficking, modifying biophysical gating properties, and serving as the target for second messengers or drugs. This variety of channel composition and regulation can lead to challenges when evaluating the effect of mutations, as multiple mutations have demonstrated differential behavior in the context of various channel splice variants and auxiliary subunits [24–26]. Moreover, mutations in any of the four VGCC subunits can cause significant changes to channel properties and lead to disease phenotypes; however, here we focus selectively on those mutations occurring within the main α_1 subunit of each channel.

In addition to modification by auxiliary subunits, HVA channels are modulated by numerous second messengers and protein partners. Calmodulin (CaM) is of particular relevance and is constitutively attached to the C-terminus of many HVA channels where it serves as the sensor for calcium [27–29, 577, 578]. Upon Ca^{2+} binding, CaM can impart two distinct forms of feedback regulation on VGCCs, either Ca^{2+} dependent inactivation (CDI), or Ca^{2+} dependent facilitation (CDF). While the majority of HVA channels exhibit robust CDI, Ca^{2+}-CaM mediated CDF is relevant primarily in $Ca_V2.1$ channels. Importantly, multiple Ca^{2+} channel splice variants contain components which modulate CDI or CDF, enabling specialization of Ca^{2+}-dependent feedback regulation within different cell types. Interestingly, numerous channelopathic mutations in VGCCs disrupt the normal Ca^{2+}-dependent regulation of the channel.

Ca_V3 channels (or T-type channels) represent the low-voltage activated (LVA) Ca^{2+} channel subfamily. There are three Ca_V3 channel isoforms, each with unique electrophysiological and pharmacological characteristics [3]. α_{1G} [30], α_{1H} [31], and α_{1I} [32] are the pore-forming α subunits encoded by the genes *CACNA1G*, *CACNA1H*, *CACNAI*, respectively, and correspond to $Ca_V3.1$, $Ca_V3.2$, and $Ca_V3.3$. These channels are characterized by fast inactivation, small unitary conductance, and fast activation at more hyperpolarized membrane potentials as compared to HVA channels [3, 33–35]. T-type channels show a wide expression pattern and can be found in the cardiovascular, neuroendocrine, and nervous systems, as well as in some non-excitable cell types, including osteocytes, sperm, and immune cells [36]. In neurons, the low-voltage activation of these channels depolarizes the membrane, which promotes the opening of voltage-gated sodium channels, thus contributing to action potential (AP) firing [37]. Ca_V3 channels are also subject to alternative splicing, further diversifying the functional properties of the channels. This alternative splicing can determine channel expression at the membrane and alter the biophysical effects of known mutations [32, 38–44].

13.2 $Ca_V1.1$

CACNA1S encodes for the α_{1S} pore-forming subunit of the $Ca_V1.1$ VGCC, which is often also referred to as the dihydropyridine (DHP) receptor [45]. $Ca_V1.1$ channels are almost exclusively expressed in skeletal muscle, where they sit within the T-tubule and play a central role in EC coupling [46–50]. Unlike other VGCCs, the primary function of $Ca_V1.1$ in muscle does not appear to be controlling the flux of Ca^{2+} through the plasma membrane. Rather, $Ca_V1.1$ serves as the voltage sensor for Ca^{2+} release from the sarcoplasmic reticulum (SR). This is accomplished through a physical coupling between $Ca_V1.1$ and the ryanodine receptor (RyR), such that movement of the $Ca_V1.1$ voltage sensor initiates Ca^{2+} release from the SR through RyR1, thus initiating muscle contraction. Moreover, RyR1 may also serve as a feedback regulator of the $Ca_V1.1$ channel [51–53].

Complete loss of $Ca_V1.1$ causes paralysis and death at birth due to respiratory failure, as demonstrated in dysgenic $Ca_V1.1$ null mice [50, 54]. In addition, numerous functionally relevant pathogenic mutations have been described in $Ca_V1.1$ (figure 13.1 [55–89]), resulting in muscle dysfunction [47]. Interestingly, because $Ca_V1.1$ does not require Ca^{2+} to initiate EC coupling, loss-of-function (LOF) mutations that disrupt the entry of Ca^{2+} through the channel are unlikely to cause an overt phenotype, as evidenced by the fact that ablation of Ca^{2+} flux through $Ca_V1.1$ appears to be of minimal consequence to the skeletal muscle

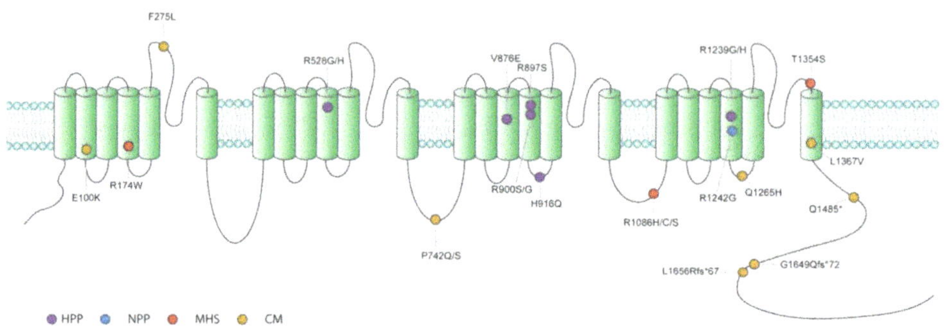

Figure 13.1. The location of mutations in *CACNA1S* ($Ca_V1.1$), with the associated phenotype indicated by the corresponding color. HPP, hypokalemic periodic paralysis; NPP, normokalemic periodic paralysis; MHS, malignant hyperthermia susceptibility; CM, congenital myopathy.

performance in mice [90]. Even so, disruption of the $Ca_V1.1$ selectivity filter can precipitate changes in muscle metabolism fiber composition [91, 92]. However, while the gating properties of the adult form of $Ca_V1.1$ do not support appreciable channel opening in response to an action potential (AP), an embryonic variant of $Ca_V1.1$ harbors a hyperpolarized activation curve permissive of Ca^{2+} entry [93]. Persistence of this embryonic splice variant into adulthood results in muscle dysfunction, demonstrating a potential pathogenicity of gain-of-function (GOF) $Ca_V1.1$ mutations [94].

13.2.1 Hypokalemic periodic paralysis (HPP)

HPP is an autosomal-dominant disorder in which muscle excitability is reduced [95]. Symptoms usually manifest in the second decade of life and are characterized by episodes of hypotonia, muscle weakness or paralysis, and are associated with a dramatic decrease in blood potassium levels. Episodes often occur in response to cold or heat, rest after exercise, insulin intake or after a carbohydrate-rich meal. In rare cases there is involvement of respiratory muscles which may be life threatening, and HPP may progress to myopathy or muscle degeneration. In ~60% of cases, the cause of HPP is a mutation in $Ca_V1.1$, while mutations within $Na_V1.4$ account for ~20% of cases [47, 96].

The majority of HPP mutations within $Ca_V1.1$ occur within the S4 voltage sensing domain (VSD). R528G/H [55–60, 97], R897S [61, 62] and R900S/G [63, 64] and R1239H/G [65–73] all result in neutralization of an S4 gating charge, while the mutation V876E [74–76] resides in the S3 helix, replacing a hydrophobic residue with a negative charge (figure 13.1). R528H/G and R1239H/G have been shown to cause a broadening of the AP and reduced AP amplitude. However, functional characterization of these mutations demonstrated a general LOF effect including slower activation, reduced open probability, decreased current amplitude and a left shift in inactivation, all of which cannot explain the effects on the AP. Instead, it appears likely that these mutations result in a leak current through the VSD known as an ω-current [98–101]. This current appears to be formed due to disruption of the

S4 charged residue, enabling the passage of protons or small ions through the region. For the V876E mutation, a similar mechanism appears likely, such that the extra negative charge in the S3 region produced by this mutation may disrupt the stabilization of the positive residues within the VSD. The existence of this ω-current has been demonstrated for multiple mutations. Muscle fibers from patients and mice carrying the R528H mutation exhibit inward cation leak currents characteristic of an ω-current [102, 103], while ω-currents have been recorded from mouse muscle in response to R1239H and V876E mutations [69, 77]. In addition, co-expression of $Ca_V1.1$ with STAC3 protein enabled the recording of ω-currents resulting from the R528H mutation in oocytes [104]. The ω-currents produced by either the R528H or V876E mutations appear to be carried by sodium ions, while those due to R1239H are carried by protons. These ω-currents are thought to shift the resting membrane potential of the cell to a more depolarized state. Skeletal muscle cells exhibit a bistable resting potential and are generally held at a potential of approximately −80 mV by a balance between a hyperpolarizing potassium current and a depolarizing leak current. The added depolarizing ω-current alters this balance, resulting in a shift to a second stable membrane voltage around −60 to −50 mV, causing a temporary loss of muscle cell excitability. However, a recent functional study has demonstrated that not all HPP mutations utilize this same pathogenic mechanism. A charge conserving lysine mutation at the R897 locus did not cause detectable ω-current when expressed in oocytes [105]. Finally, a single mutation, H916Q [78], has been identified within the IIIS4–5 linker that connects the VSD with the pore domain. However, this mutation causes incomplete penetrance and only affects males, distinguishing it from other HPP mutations.

13.2.2 Normokalemic periodic paralysis (NPP)

The mutation R1242G [79, 80] in the domain IV VSD also results in a neutralization of one of the S4 gating charges, and has been shown to cause complicated NPP, where patients suffer from myopathy, progressive permanent muscle weakness and exercise induced contracture in addition to NPP. Unlike the HPP mutations, which occur primarily within the outer two arginines of the VSD, R1242 represents the third arginine of the VSD, resulting in an ω-current which conducts within the activated state of the channel. When evaluated in dysgenic myotubes this mutation reduces current amplitude and left shifts inactivation, thus producing a LOF effect which is unlikely to be the pathogenic mechanism of this mutation [79]. Instead, the outward ω-current observed at positive potentials explains a slowing of the AP upstroke and hypo-excitability of the muscle, ultimately leading to permanent weakness and myopathy. In addition, a negative ω-current was found in response to long depolarizations, predicted to depolarize the fiber enough to trigger Ca^{2+} release and inappropriate muscle contracture.

13.2.3 Myotonic dystrophy type 1 (MD1)

MD1 causes muscle weakness and slow muscle relaxation known as myotonia. It can be caused by splicing abnormalities in *CACNA1S* which result in the exclusion of exon 29 and expression of the embryonic form of the $Ca_V1.1$ channel. The lack of exon 29 in the embryonic $Ca_V1.1$ splice variant results in a hyperpolarizing shift in the channel activation curve, resulting in a GOF effect and atypical flux of Ca^{2+} through $Ca_V1.1$ when inappropriately expressed in adults [106]. This effect, however, may not be sufficient on its own to produce the full MD1 phenotype, as evidenced by relatively normal motor function in mice constitutively expressing the embryonic form of $Ca_V1.1$ [94]. However, these mice did display reduced muscle strength and altered fiber type composition, demonstrating a pathogenic role for the excess Ca^{2+} flux through the aberrantly spliced $Ca_V1.1$ channel.

13.2.4 Malignant hyperthermia (MH) susceptibility

MH is a potentially fatal autosomal-dominant disorder presented in otherwise healthy individuals. The underlying mechanism of MH generally includes uncontrolled Ca^{2+} release from the SR, which results in severe muscle contractures and ATP depletion in response to volatile anesthetics and depolarizing muscle relaxants [47]. This causes muscle rigidity, acidosis, hypercapnia, hypoxemia, lactic acidosis, and hyperthermia which leads to muscle breakdown and can be fatal if untreated with dantrolene [107]. In many cases, MH is caused by a mutation in RyR1, which destabilizes the closed state of the channel making it hypersensitive and leaky [108, 109]. However, in some cases the cause of MH is a mutation within $Ca_V1.1$, which acts as the voltage sensor for RyR1. The highly conserved arginine R1086 within the III–IV cytoplasmic loop of $Ca_V1.1$ may be mutated to either histidine, cysteine, or serine in patients with MH [81–85]. When characterized in myotubes, R1086H caused a dramatically increased sensitivity of SR Ca^{2+} release to depolarization and caffeine [82], as well as decreased channel conductance. In addition, a T1354S mutation located in the outer pore of $Ca_V1.1$ has been linked to MH, and causes increased activation kinetics of the channel and increased sensitivity of SR Ca^{2+} release to caffeine [86]. Thus, these mutations seem to mimic the functional effects of RyR1 MH mutations. Finally, the MH mutation R174W in domain I VSD abolishes the $Ca_V1.1$ current without altering EC coupling in cultured dysgenic myotubes [87, 110]. Nonetheless, this mutation still results in an increased sensitivity to caffeine [87]. Thus, it appears that R174W impedes a slow conformational transition required for pore opening while, possibly, altering conformational coupling with RyR1 resulting in a leaky SR channel. However, the specific mechanisms by which these $Ca_V1.1$ mutations sensitize RyR1 Ca^{2+} release has yet to be established.

13.2.5 Congenital myopathy (CM)

CMs are a heterogeneous set of rare muscle diseases characterized by muscle weakness and loss of muscle tone. They primarily result from disrupted EC coupling and altered Ca^{2+} handling. Often, CMs are the result of mutations within RyR1, however multiple $Ca_V1.1$ mutations have also been linked to the disorder [88]. Most missense $Ca_V1.1$ mutations associated with myopathy have yet to be characterized functionally. However, several mutations occur within a locus with predicted function. The F275L mutation alters a conserved amino acid within the pore-forming loop of $Ca_V1.1$ near the selectivity filter, while the variant Q1265H [89] is predicted to disrupt channel splicing. An E100K mutation [88] results in the replacement of a conserved negative residue within the IS2 region, which is predicted to eliminate the counter charge for the R174 VSD residue which was previously described as a locus for MH [87]. Two myopathy associated mutations P742Q and P742S were identified within the II–III loop [88] of $Ca_V1.1$ at a locus previously shown to be critical for EC coupling [111]. Mutations at residues Q1485, L1656 [88] and G1649 [89] cause a truncation of the C-terminus of the channel, which has known regulatory function and phosphorylation targets. Finally, myotubes from patients harboring either the L1656Rfs*67 truncation mutation or an L1367V missense mutation exhibited reduced Ca^{2+} transients in response to depolarization [88], demonstrating a functional disruption due to these mutations.

13.2.6 Non-skeletal muscle diseases

Despite the selective localization of $Ca_V1.1$ in skeletal muscle, there have been reports of low expression of the channel in non-skeletal muscle tissue, including postsynaptic expression in ON-bipolar cell dendrites in the photoreceptors [112]. Moreover, a handful of reports describe $Ca_V1.1$ mutations associated with non-skeletal muscle diseases including autism spectrum disorder [113, 114], schizophrenia [115, 116] and aberrant tooth morphogenesis [117]. However, little functional characterization has been performed in these studies, making a causative link to $Ca_V1.1$ tenuous.

13.3 $Ca_V1.2$

α_{1C} comprises the pore-forming subunit of $Ca_V1.2$ and is encoded by the *CACNA1C* gene [118]. $Ca_V1.2$ is the most widespread of the L-type calcium channels and is robustly expressed in heart, brain, smooth muscle and the immune system [5, 119, 120]. In the brain, $Ca_V1.2$ is known to be critical for excitation–transcription coupling, long-term potentiation and memory, and *CACNA1C* is commonly identified in genome-wide association studies of schizophrenia and bipolar disorder [121, 122]. In the heart, it is critical for EC coupling, where Ca^{2+} entry through $Ca_V1.2$ initiates Ca^{2+} release from the SR through Ca^{2+} induced Ca^{2+} release (CICR) [5, 123]. As a result, mutations within $Ca_V1.2$ can impact a myriad of physiological functions. For some time, the critical role of $Ca_V1.2$ in the heart, coupled with the lack of identified mutations in patients, seemed to indicate that significant mutations which altered the function of $Ca_V1.2$ would likely be embryonic lethal [124]. However, in 2004, Splawski and Keating identified the underlying cause of Timothy syndrome (TS) as a single point mutation within the $Ca_V1.2$ channel [125]. Since this first discovery, numerous mutations have been identified within $Ca_V1.2$ (figure 13.2 [114, 116, 125–160]), some with multisystem effects and some with a more selective impact on a specific tissue. While the overall number and variety of mutations in $Ca_V1.2$ continues to grow, many of the mutations are identified within very few patients, often only a single proband.

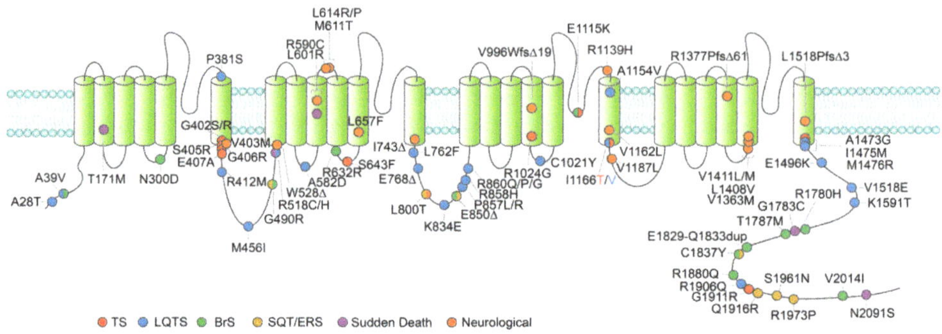

Figure 13.2. The location of mutations in *CACNA1C*, ($Ca_V1.2$), with the associated phenotype indicated by the corresponding color. TS, Timothy syndrome; LQTS, long-QT syndrome; BrS, Brugada syndrome; SQT/ERS, short-QT/early repolarization syndrome.

As such, $Ca_V1.2$ channelopathies remain exceedingly rare, likely due to the severity of the phenotype. Overall, the myriad of symptoms identified in $Ca_V1.2$ channelopathy patients has expanded our understanding of the role of $Ca_V1.2$ in both excitatory and non-excitatory cells [7, 161].

13.3.1 Timothy syndrome (TS)

TS was first described in 1993 as a multisystem syndrome presenting with severe cardiac symptoms, neurodevelopmental deficits, craniofacial abnormalities and syndactyly [125, 162, 163]. The cardiac phenotype of these patients includes severe long-QT syndrome (LQTS), AV block, torsades de pointes, early after depolarizations, ventricular tachycardia and ventricular fibrillation [139, 164–166]. The QT interval of TS patients ranged from 615 to 690 ms, making it one of the most severe forms of LQTS. As a result, TS patients had a life expectancy of only two and a half years. In addition to the cardiac features, most TS patients also suffer from dysmorphic facial features, syndactyly, developmental delays, autism spectrum disorder (ASD) and immune deficits [125, 139]. In 2004, the cause of TS was identified as a single, heterozygous point mutation, G406R, within the IS6 region of $Ca_V1.2$ [125]. The mutation occurred within the mutually exclusive exon 8a, which is known to be expressed within only 20% of cardiac channels but is also widely expressed in other tissues including brain and muscle, accounting for the multisystem nature of TS. Patients harboring this mutation are now classified as TS type 1 (TS1). Previous reports have used conflicting nomenclature for this channel exon as described here, in brief. While the original description of TS used nomenclature which places exon 8 upstream of exon 8a, previous electrophysiological studies referred to the upstream exon as 8a and the downstream exon as 8 [167–169]. However, as the majority of recent papers has adopted the Splawski *et al* nomenclature (exon 8 upstream) it will be used for the remainder of this chapter. The functional effect of the G406R mutation on $Ca_V1.2$ includes a loss of voltage-dependent inactivation (VDI), decreased Ca^{2+}-dependent inactivation (CDI) and a hyperpolarizing shift in the voltage dependence of channel activation [125, 139, 170–172]. These GOF effects fit well with the prolongation of the cardiac AP, which results from an increase in Ca^{2+} entry through $Ca_V1.2$. Moreover, TS1 represents one of the most penetrant monogenic forms of ASD. As such, understanding the biophysical disruptions caused by TS may shed light on the pathogenesis of ASD.

Within a year of the first description of the G406R mutation in exon 8a of $Ca_V1.2$, two additional mutations were described as underlying a similar set of phenotypes known as TS type 2 (TS2) [139]. TS2 patients lacked the syndactyly which was shared by all TS1 patients, and seemed to have more severe cardiac deficits, with QT intervals extending up to 730 ms [139]. This is explained largely by the locus of TS2 mutations within the alternate exon 8, which is found in the majority of $Ca_V1.2$ channels in the heart, and has a distinct expression pattern as compared to exon 8a [139, 169]. In addition to the original G406R mutation, a nearby G402S mutation was also found in TS2 patients. Like G406R, G402S also causes a significant loss of VDI, and a disruption of CDI, however, the voltage dependence of activation exhibited a depolarizing shift, demonstrating a challenge in defining channel deficits as simple GOF *versus* LOF effects. Interestingly, the mutation G402R has recently been identified at this same site, resulting in cardiac phenotypes including LQT and torsades de pointes as well as developmental delay, however, the majority of the other systemic TS symptoms were absent.

Mouse models of TS have primarily focused on the G406R mutation. A TS2–Neo mouse was made in which the G406R mutation was inserted within exon 8a, along with the neomycin cassette which enabled survival of the mice to adulthood [173]. These mice demonstrated ASD phenotypes including repetitive and perseverative behavior, social impairment, reduced sensorimotor learning and altered vocalizations. In another study, oligodendrocyte progenitor cells isolated from the cortex of these mice demonstrated enhanced maturation and myelination [174]. In addition, these mice exhibited altered activity of the ascending serotonin system, a feature which may correlate with the ASD phenotype [175]. Finally, dissociated cortical neurons from these mice demonstrated activity-dependent dendritic retraction, a result that was replicated in induced pluripotent stem cells (iPSCs) derived from a TS patient [176–178]. In addition, a mouse overexpressing rabbit $Ca_V1.2$ channels harboring the equivalent of the G406R mutation demonstrated abnormal coupling of the channel to AKAP150 [179], increased SR Ca^{2+} leak, higher diastolic Ca^{2+} and increased SR Ca^{2+} load and Ca^{2+} spark activity in ventricular myocytes [180]. Moreover, overexpression of the G406R $Ca_V1.2$ channel in mice has been demonstrated to alter mandibular development [181], consistent with the dysmorphic features displayed by TS patients and demonstrative of a role for $Ca_V1.2$ in non-excitable cells. Interestingly, expression of the G406R mutation has also been linked to altered splicing of the channel, a feature which may alter the development of the cortex [182].

Since the initial description of TS, a myriad of mutations have been discovered within $Ca_V1.2$, often occurring in constitutively expressed exons. However, only a subset of these mutations result in the full multisystem TS phenotype. In 2012 a single patient was identified with an A1473G mutation [153]. Similar to G406R and G402S, the mutation is located within a distal S6 region, this time occurring within domain IV of the channel. Like the original TS mutations, the patient exhibited a severe multisystem disorder, including LQTS, dysmorphic facial features, joint contractures, syndactyly, hypoglycemia and severe developmental delay.

The mutation I1166T has been identified in at least two patients [140, 142, 183], resulting in a severe cardiac and neuronal phenotype. The mutation again occurs within the distal S6 region of the channel, this time in domain III. However, a I1166V mutation at the same locus was identified within the same study but produced a cardiac specific phenotype, indicating that S6 mutations do not inevitably produce the full TS phenotype. Nonetheless, I1166T resulted in a severe manifestation of TS, including LQTS, AV block, facial dysmorphisms, joint hypermobility, hypotonia and syndactyly. In one case, the patient suffered cardiac death at the age of 1 year and the mutation was identified postmortem, while the second patient died at age 3 years, 8 months despite the implantation of an internal cardiac defibrillator. The severity of this mutation correlates with a large GOF effect on channel gating in which channel activation is shifted to more hyperpolarized potentials, inactivation is reduced and current amplitude is increased [140, 184].

It is often ambiguous as to whether to define a patient as having TS as opposed to a cardiac only phenotype. For example, a patient identified with a E407A mutation presented with primarily cardiac symptoms and convulsions which occurred during syncopal episodes [151]. The patient suffered from LQTS, AV block, ventricular tachycardia, but lacked the full multisystem disorder. Unlike most TS patients, this patient was diagnosed as an adult. However, despite the lack of a full multisystem phenotype, the patient was diagnosed with TS [151]. Interestingly, despite being located just next to the canonical G406 locus, this mutation falls within the constitutively expressed exon 9. The patient was implanted with a pacemaker at the age of six and an ICD at the age of 7.

Many multisystem TS patients display a subset of the initially described TS1 phenotype. Ozawa identified an S643F mutation in a patient exhibiting LQTS, developmental delay, ASD, dysmorphic features, but no syndactyly [158]. The mutation is located within the S4–5 linker of domain II and was identified in a 14 year old boy. Channels harboring the S643F mutation had a deficit in inactivation, activation was shifted in a hyperpolarized direction, and current amplitude appeared to be decreased. Thus, this mutation exhibited both GOF and LOF effects on channel gating. Moreover, R1024G has been shown to cause developmental delay, joint contractures and gastroenteritis, but not LQTS [155, 185]. The patient did, however, display multiple cardiac phenotypes including fever-induced tachycardia, enlarged right ventricle, impaired left ventricular contraction and pulmonary artery hypertension [155]. Likewise, the mutation S405R [149] produced syndactyly and LQTS as well developmental delay in one patient, but not in another. Further, patients with this mutation lack facial dysmorphisms, immune deficits and hypoglycemia, demonstrating the variability of symptoms among TS patients. Finally, the mutation G1911R was identified within the C-terminus of the channel, resulting in a multisystem phenotype including LQTS, seizures, dental abnormalities and facial dysmorphisms [154]. Like other TS mutations, the reported effects on the channel include a hyperpolarizing shift in channel activation and a decrease in VDI. Finally, the mutation, E1115K, was identified within the selectivity filter of $Ca_V1.2$ channels. Unlike the biophysical changes described for other mutations, this mutation caused a change in the selectivity of the channel, making it nonselective or monovalent cations. Interestingly, this mutation has been linked to both a multisystem phenotype including LQTS, bradycardia, and ASD in one patient [186], and Brugada syndrome in other patients [129, 131, 187].

While the number and variety of mutations continues to grow, to date the majority of *CACNA1C* mutations which have been associated with the multisystem TS phenotype have been identified within an S6 region of the $Ca_V1.2$ channel (figure 13.2), and result in a GOF effect via a hyperpolarizing shift in channel activation and a decrease in channel inactivation [184]. This often causes an increase in the window current of the channel, which may be of particular relevance to a neuronal AP, where it will enable Ca^{2+} influx at subthreshold potentials [579].

13.3.2 Long-QT syndrome (LQTS)

LQTS resulting from a mutation within *CACNA1C* has been designated as LQT type 8. $Ca_V1.2$ mutations may result in LQT8 either as part of the full multisystem TS phenotype, or as a cardiac selective phenotype. This variability in symptoms has led to a somewhat inconsistent naming system, with the terms cardiac only TS (COTs), non-syndromic long-QT and atypical TS used to describe mutations which produce non-syndromic forms of TS. As the variety of phenotypes has continued to grow, sometimes including a smaller subset of the original TS symptoms, the term atypical TS has been increasingly applied.

In addition to the TS mutations described above, a myriad of missense mutations in *CACNA1C* have been associated with LQTS (figure 13.2). While many of these mutations remain to be functionally characterized, those that have point towards common mechanisms leading to LQTS. In particular, studies in heterologous expression systems have implicated a loss of channel inactivation (either CDI, VDI or both) in producing LQT8. The LQT8 mutations A28T [140], A582D [115, 152], L762F [144], R860G [140, 145], I1475M [140], E1496K [140] and R518C/H [143, 152, 188] mutations all produce a decrease or slowing of channel inactivation. Computational modeling bears out a mechanism whereby a loss of either CDI or VDI can prolong the cardiac AP due to excessive Ca^{2+} entry during phase 2 of the AP, thus delaying repolarization [125, 172]. In a similar manner, the LQT8 mutations A28T, I1166V, P857R and R858H have been shown to increase the current amplitude when overexpressed in heterologous cells. Such a GOF effect is predicted to act in a similar manner to the loss of inactivation, increasing Ca^{2+} entry during the plateau phase of AP and thus extending the AP duration. On the other hand, while numerous mutations have been shown to alter channel activation, the LQT phenotype has been observed in response to mutations which cause either a hyperpolarizing shift (G406R [125, 172, 189], I1166T [140, 142, 184], G1911R [154, 190], S643F [152, 158], I1475M [140], E1496K [140]), a depolarizing shift (G402S [139, 172]) or no shift (A28T [140], P381S [133, 152], M456I [133, 152], A582D [133, 152],

L762F [144], R860G [140, 145], and R518H [143, 152]) in channel activation, making this feature unlikely to be fully responsible for the cardiac features of $Ca_V1.2$ channelopathies.

In addition to this GOF effect during the cardiac AP, mutations in *CACNA1C* have the potential to disrupt the interaction of the channel with other proteins. Of particular note, the LQT8 mutations P857R/L, R858H and R860Q/P/G fall within a region known to be important for the binding of STAC (SH3 and cysteine rich domain), a protein which has been shown to alter the inactivation properties of $Ca_V1.2$ [145, 191–194]. However, as STAC proteins are not known to be highly expressed in the heart [195], the mechanisms underlying the pathogenicity of LQTS due to these mutations remains uncertain. However, given the abundance of proteins in the cardiac myocyte which may interact with $Ca_V1.2$, it would not be surprising if several of the mutations identified within cytosolic regions of the channel alter cardiac function through the disruption of a protein binding site.

13.3.3 Short-QT (SQT) syndrome

SQT syndrome is a rare cardiac disorder in which the QT interval is abnormally short. Patients with SQT can suffer from life-threatening ventricular arrhythmias, believed to be due to heterogeneity in the AP duration across the cardiac tissue [196]. SQT syndrome may be the result of a mutation in multiple ion channels, including $Ca_V1.2$. The mechanism is essentially the opposite as that of LQTS. Mutations which cause a LOF effect in $Ca_V1.2$ result in a shortening of the plateau phase of the AP and early repolarization. These mutations tend to decrease the inward Ca^{2+} current during cell repolarization, while increasing the transmural and epicardial dispersion of repolarization, resulting in a combined phenotype of both SQT and Brugada syndromes (BrS) [127, 196]. Thus, in the case of $Ca_V1.2$, SQT and BrS generally form an overlapping phenotype. Nevertheless, several mutations have been identified as associated with SQT or early repolarization syndrome (ERS), without mention of BrS, with at least one of these mutations identified within victims of sudden unexplained death syndrome (SUDS) [134]. The majority of SQT mutations which have been biophysically characterized cause a reduction in current amplitude. This is true for the mutations Q1916R [156], R1973P [197], S1961N [138], all located within the C-terminus of the $Ca_V1.2$ channel. In addition, R1973P and S1961N appear to increase VDI, enhancing the LOF effect. Finally, mutation L800T serves to expand the SQT phenotype in that the patient harboring this mutation exhibited not only SQT, but also ASD, dental defects, and affective disorder [160]. Like other SQT $Ca_V1.2$ mutations, L800T caused a marked reduction in current amplitude, likely due to a trafficking defect. Thus, as more mutations are identified the range of clinical phenotypes will likely continue to expand.

13.3.4 Brugada syndrome (BrS)

BrS is characterized by a unique ECG pattern featuring ST-segment elevation in the right precordial leads. Patients harbor a high susceptibility for ventricular arrhythmias and sudden cardiac death. The altered electrical pattern in BrS patients can result from an imbalance between inward and outward currents during the plateau phase of the AP [196]. The reentrant arrhythmias displayed by BrS patients may be the result of mutations causing distinct effects in select populations of myocytes, resulting in heterogeneity in AP duration across the cardiac tissue, however, the exact mechanism underlying the arrhythmogenicity of BrS remains a subject of continued investigation [196]. In the case of $Ca_V1.2$, BrS seems to result primarily from LOF mutations, and generally includes SQT features. The majority of BrS mutations which have been characterized thus far (E850Del [129, 159], A39V [127], G490R [127], V2014I [129], E1829-Q1833dup [129], T1787M [128] and N300D [198]) cause a decrease in $Ca_V1.2$ current amplitude. In addition to this effect on current amplitude, mutation T1787M also appears to increase VDI [128]. Finally, the LOF effect of some mutations may be caused by a deficit in channel trafficking, however, these results are not consistent. The A39V mutation resulted in a loss of channel trafficking in CHO cells [127], but no loss in surface expression of the channel was detected in rat neurons [199]. Likewise, G490R displayed normal trafficking. Thus, the BrS phenotype appears to track primarily with a loss of current flux through the channel.

13.3.5 Sudden cardiac death

The severe cardiac phenotypes suffered by many of the patients harboring *CACNA1C* mutations are well known to be substrates for sudden cardiac death. Multiple studies have implicated mutations in *CACNA1C* in causing sudden unexplained death or cardiac arrest in both young and adult patients [127, 134, 143, 157, 159, 200, 201]. In some cases, these mutations are known to be associated with LQTS or BrS, while other studies identified the mutations postmortem, making it difficult to assign a specific mechanism to each mutation. Nonetheless, it is likely that *CACNA1C* channelopathies represent a significant subset of unexplained death due to the critical role the channel plays in the heart.

13.3.6 Autism spectrum disorder (ASD), developmental delay and epilepsy

While the majority of $Ca_V1.2$ channelopathy patients exhibiting ASD or developmental delay are currently diagnosed as having TS, recent studies highlight the likelihood that select mutations in $Ca_V1.2$ can specifically affect neurological function. The majority of *CACNA1C* mutations have been identified by screening LQTS patients, making it challenging to identify such neuro-specific patients. However, as genetic screening has become more widespread, patients with primarily neurological features have emerged. Several mutations have been linked to ASD [114, 126, 135, 136, 202], intellectual disability [203] and schizophrenia [116], however, the majority of these studies include no functional data, making it difficult to assign causation. Nonetheless, the mutation V1162L was identified in three separate ASD studies [114, 135, 136] pointing to a high likelihood that this mutation participates in the ASD pathogenesis. In addition, epilepsy has emerged as a prominent feature in some patients [580]. Moreover, a recent study by Rodan *et al* demonstrated predominately neurological phenotypes among 25 patients harboring *CACNA1C* mutations [581]. Patients phenotypes included developmental delay, intellectual disability, ASD, hypotonia, epilepsy and ataxia; all without significant cardiac phenotypes. Thus, the known impact of *CACNA1C* mutations on neurological function continues to grow.

13.3.7 Hypoxic injury

It is worth noting that the identification of neurological symptoms among patients harboring $Ca_V1.2$ mutations can be complicated by the existence of severe cardiac phenotypes which may result in hypoxia. As an example, the G402S mutation was initially identified as causative of neurologic deficits in a TS2 patient [139], however, this presentation was subsequent to sudden cardiac arrest, leaving hypoxic injury as a possible cause of the neurological phenotype. This possibility appears more likely when a more recent case is considered [204]. In this report, two siblings inherited the G402S mutation from a mosaic father. The elder of the two children was successfully resuscitated following cardiac arrest at the age of 2 months and required an implantable defibrillator at the age of 4 years. The child developed generalized epilepsy and did not learn to walk until age 7, and was unable to speak at the age of 8. However, the younger sibling experienced only a single reported episode of cardiac arrest, which was immediately remedied though the use of the older sibling's external defibrillator. Immediately following this event, an internal cardiac defibrillator was implanted. Unlike his older sister, the younger child met all developmental milestones and exhibited no neurological symptoms. It therefore appears likely that the developmental delay suffered by the elder sibling was a result of hypoxic injury rather than a direct effect of the mutation. In light of these findings, it appears that the G402S mutation is not linked to ASD or developmental delay as originally described. This potential for hypoxic injury due to cardiac arrest remains a challenge for evaluating the neurological impact of $Ca_V1.2$ mutations which cause severe cardiac deficits.

13.4 $Ca_V1.3$

$Ca_V1.3$ channels are comprised of the α_{1D} pore-forming subunit, encoded by the *CACNA1D* gene. The tissue distribution of $Ca_V1.3$ is widespread and often overlaps with that of $Ca_V1.2$. $Ca_V1.3$ channels play an important role in cardiac, auditory, neuroendocrine, and neurological function [120, 205]. In the brain, $Ca_V1.3$ channels are primarily expressed postsynaptically, where they shape neuronal activity and contribute to Ca^{2+}-dependent gene expression, and the like *CACNA1C, CACNA1D* has been associated with bipolar disorder and ASD [122], and appears to be involved with the neurodegenerative mechanisms associated with Parkinson's disease [120]. In the heart, $Ca_V1.3$ is primarily expressed in the sinoatrial (SA) and atrioventricular (AV) nodes, where it contributes to cardiac pacemaking [206–208]. $Ca_V1.3$ also plays an important role in the endocrine system, where it participates in Ca^{2+}-induced aldosterone production in the adrenal gland [209]. Finally, $Ca_V1.3$ channels play an important role in cochlear hair cells, where they provide the Ca^{2+} required to trigger neurotransmitter release at synaptic ribbons, and thus are required for normal auditory

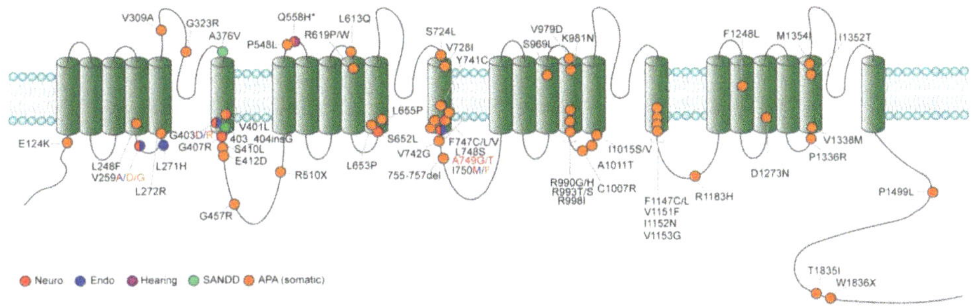

Figure 13.3. The location of mutations in *CACNA1D* ($Ca_V1.3$), with the associated phenotype indicated by the corresponding color. Neuro., neurodevelopmental disorder; Endo., endocrine disorder; Hearing, hearing impairment; SANDD, sinoatrial node dysfunction and deafness; APA, aldosterone producing adenoma.

function [206, 210]. Knockout of $Ca_V1.3$ in mice has demonstrated the importance of the channel in auditory and cardiac function. Knockout mice display SA node dysfunction, resulting in bradycardia, as well as congenital deafness [206, 208]. In addition, these mice display subtle defects in brain development, including decreased number of dopamine-producing neurons in the substantia nigra, a decreased number of neurons in the auditory brainstem and dentate gyrus, and behavioral changes consistent with mood disorders [211]. Thus, mutations in $Ca_V1.3$ have the potential to impact auditory, neuronal, endocrine, and cardiac function (figure 13.3 [205, 209, 211–222]).

13.4.1 Primary aldosteronism (PA)

Aldosterone is important for the proper regulation of blood pressure, and excess aldosterone production results in hypertension. $Ca_V1.3$ channels play a key role in driving aldosterone production in the zona glomerulosa of the adrenal cortex. Numerous somatic mutations in $Ca_V1.3$ have been identified in aldosterone producing adenomas (APA), resulting in PA and treatment-resistant hypertension (figure 13.3) [209, 211]. These mutations primarily cause a GOF effect, either through a dramatic loss of channel inactivation (V259D [209], F747L [223], I750M, V1153G [220]), a hyperpolarizing shift in the voltage dependence of activation (G403D [221], G403R [221], V401L [212], S652L [216]), slower or incomplete inactivation over prolonged depolarization (P1336R [209]), or the introduction of an ω-current (R990H [223, 224]); these effects have been characterized as types 1–4, respectively [216, 223]. Thus, the biophysical effects of these mutations support a role for increased Ca^{2+} entry through $Ca_V1.3$, driving excessive aldosterone production. It should also be noted that the majority of these mutations (V401L [212], G403D, G403R [221], G407R [222, 223], A749G [223], I750M [209, 221]) reside within the distal S6 region of $Ca_V1.3$, a locus also identified as significant in $Ca_V1.2$ channelopathies. In addition, A749G, I750M, G403D, V401L have been found as both somatic and germline mutations. When occurring in the germline, these mutations produce a much wider spectrum of symptoms including intellectual disability, ASD, seizures, and hypotonia. Interestingly, G403D and I750M, but not V401L, include PA in the germline phenotype.

13.4.2 Neurodevelopmental and multisystem disorders

When mutations in $Ca_V1.3$ occur in the germline they produce a wide array of symptoms, including intellectual disability and ASD. These patients often suffer from a multisystem disorder which may also include hypoglycemic hyperinsulinism, hypotonia, seizures, limb spasticity, facial dysmorphism, cardiac effects, and/or PA, with ASD being the most common recurring phenotype [211]. The first reported systemic $Ca_V1.3$ mutation was A749G, located within the distal IIS6 region. This mutation caused ASD and intellectual impairment with no report of effects on other systems [218, 222]. When expressed in heterologous systems, $Ca_V1.3$ channels harboring the A749G mutation exhibited a significant hyperpolarizing shift in the voltage dependence of activation, reduced CDI, and enhanced VDI, in a β subunit dependent manner [222, 223, 225]. A second mutation (A749T), was identified at the same locus, resulting in ASD, hypotonia, and intellectual disability in at least one patient. In addition, mutation L271H has been identified as causing hyperinsulinemic hypoglycemia, PA, developmental delay, muscle hypotonia, and facial dysmorphisms [213], and mutation V259D has been reported to cause PA, seizures, and developmental delay [219]. Like the somatic mutations, germline $Ca_V1.3$ mutations also tend to cause GOF effects which can be categorized into the same four main types described above. The ASD-associated G407R mutation, which occurs within the mutually exclusive exon 8A, causes a significant loss of channel inactivation [222]. A similar loss of channel inactivation is apparent in $Ca_V1.3$ channels harboring a G403D mutation within the mutually exclusive exon 8B [223], however, patients harboring this mutation exhibit a broader phenotype, including hyperinsulinaemic hypoglycemia, heart defects, intellectual disability, seizures, and hypotonia [214]. The mutation V401L within the mutually exclusive exon 8A causes ASD, intellectual disability, and seizures; and displays a hyperpolarizing shift in the voltage dependence of both activation and inactivation [212]. Likewise, the mutation I750M also hyperpolarizes channel activation [221] and is clinically associated with intellectual disability, PA, limb spasticity, and seizures [211]. Mutation S652L also causes a left shift in channel activation [216] and was associated with intellectual disability in ASD in patients. Thus, neurodevelopmental disease appears to be well correlated with GOF effects on $Ca_V1.3$. Yet, the biophysical characteristics are not always so clearly defined. Mutation Q558H causes intellectual disability, seizures, and hearing impairment, and has been shown to decrease current amplitude of $Ca_V1.3$ channels through a loss of membrane expression [215]. However, this mutation also disrupts VDI, consistent with other neurodevelopmental $Ca_V1.3$ mutations. Thus, like other channels, the mechanisms underlying $Ca_V1.3$ channelopathies are likely more complex than simple GOF descriptions. Additional sequencing studies have identified several *CACNA1D* mutations associated with ASD [226–228], however, without functional characterization, direct causation is difficult to ascertain. Nonetheless, it seems probable that a growing number of patients harboring *CACNA1D* mutations are likely to be identified as genetic studies continue. Moreover, it would not be unexpected for the phenotype of $Ca_V1.3$ channelopathies to grow. In fact, it has already been shown that a rare variant of *CACNA1D* (A1791P) segregates with bipolar disorder [229]; however, a causative link remains unidentified.

13.4.3 Sinoatrial node dysfunction and deafness (SANDD)

SANDD is the result of a LOF of $Ca_V1.3$ channels, which causes congenital deafness and SA node dysfunction, resulting in bradycardia. The disorder was first identified in a patient harboring the insertion mutation 403_404insG [205]. This mutation occurs within the IS6 region of the channel, selectively within mutually exclusive exon 8B. This $Ca_V1.3$ splice variant is expressed preferentially in inner hair cells and in the SA node, consistent with the patient phenotype. Introduction of the 403_404insG mutation into $Ca_V1.3$ channels in tsA-201 cells resulted in a near-complete loss of current. Thus, the phenotype of this mutation is entirely consistent with that of the $Ca_V1.3$ knockout mice which displayed deafness and bradycardia. In addition, mutation A376V in $Ca_V1.3$ also resulted in SANDD [217]; however, no functional characterization of this mutant has yet been carried out. Interestingly, the GOF mutation G403D also produced sinus bradycardia, and the presentation of the GOF mutation I750M included ventricular hypertrophy; however, these cardiac features may not be a direct result of these mutations.

13.4.4 Alternative splicing

$Ca_V1.3$ is known to undergo extensive alternative splicing, resulting in channels with distinct biophysical characteristics. Mutations in $Ca_V1.3$ have been shown to display divergent effects on the channel within different splice variant backgrounds. As these different splice variants have variable distribution across tissues, splice variant-specific effects may have a significant impact on the pathophysiology of each mutation. In particular, $Ca_V1.3$ exists as both long ($Ca_V1.3_L$) and short ($Ca_V1.3_{43S}$) splice variants [230], such that the short splice variant contains a truncated C-terminus causing increased CDI, enhanced channel opening, and a hyperpolarizing shift in channel activation as compared to the long variant. This short $Ca_V1.3_{43S}$ is the most abundant variant expressed in brain. Mutations A749G, G407R, and R990H have all been shown to have splice variant-specific effects on $Ca_V1.3$ function [223, 225]. A749G causes the loss of CDI and a hyperpolarizing shift in channel activation in both short and long $Ca_V1.3$ splice variants, although to differing extents [225]. In addition, a significantly larger fraction of A749G $Ca_V1.3$ channels were unavailable to open at −50 mV in the short as compared to the long splice variant [223]. Further, the left shift in channel activation for G407R was only significant in the short splice variant as compared to the long variant [223]. Finally, the R990H mutation caused a significant hyperpolarizing shift in steady-state inactivation in the short $Ca_V1.3$ splice variant, but not in the long channel variant [223, 224]. However, not all mutations exhibit, splice specific effects, and mutation V115G displayed similar biophysical effects in both long and short splice variations [220]. Thus, splice variation may play an important role in the pathogenesis of $Ca_V1.3$ channelopathies.

13.4.5 $Ca_V1.4$

$Ca_V1.4$ channels are primarily expressed in rod and cone photoreceptors within the retina, and localize to the synaptic active zone within the outer plexiform layer, where they play a role in synaptic transmission between photoreceptors and adjacent bipolar cells. Photoreceptors are rather unique cells. In the dark, these cells sit at a somewhat depolarized resting potential of around −40 mV and exhibit sustained Ca^{2+} influx and tonic glutamate release. Upon absorption of light, the photoreceptor membrane hyperpolarizes in a graded fashion, thus reducing Ca^{2+} influx and decreasing neurotransmitter release. $Ca_V1.4$ channels possess several unique properties which make them ideally suited to function within this scheme. The VDI of $Ca_V1.4$ is slow and CDI is essentially eliminated by the presence of a CDI inhibiting element (inhibitor of CDI: ICDI [231]; also called C-terminal modulatory domain; CTM[14]) within the C-terminus of the channel. Moreover, these channels activate at relatively depolarized membrane potentials. The combination of these features results in a large window current permissive of Ca^{2+} entry at the rest potential of a photoreceptor in the dark. This chronic Ca^{2+} influx contributes to tonic glutamate release at the synapse between the photoreceptor and bipolar cell, which is alleviated upon hyperpolarization of the cell and closing of the Ca^{2+} channels.

The importance of $Ca_V1.4$ in visual function has been demonstrated by the generation of multiple $Ca_V1.4$-deficient transgenic mice. Complete knockout of $Ca_V1.4$ results in immature photoreceptors and abolishment of visual function [112, 232, 233], making the phenotype of these animals more severe than similar human $Ca_V1.4$ mutations. Nonetheless, the use of transgenic mice has demonstrated that the $Ca_V1.4$ protein is required for the formation of rod photoreceptor synapses, even in the absence of Ca^{2+} flux [233]. Ca^{2+} influx through $Ca_V1.4$, however, is necessary for the proper recruitment of postsynaptic partners [233]. Thus, $Ca_V1.4$ channels appear to play a dual role in the ribbon synapse, serving as a source for Ca^{2+} influx and playing a scaffolding role in synaptic organization. In humans, over 140 different mutations have been identified within $Ca_V1.4$, consisting of nonsense, frameshift, and missense mutations, as well as deletions and insertions (figure 13.4 [234–246]), all of which impair vision.

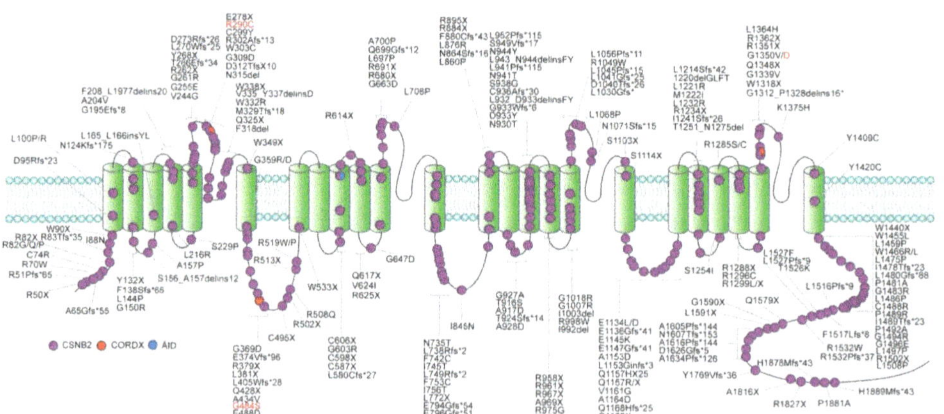

Figure 13.4. The location of mutations in *CACNA1F* ($Ca_V1.4$), with the associated phenotype indicated by the corresponding color. CSNB2, congenital stationary night blindness type 2; CORDX, cone-rod dystrophy; ÅID, Åland Island Eye Disease.

13.4.6 Incomplete congenital stationary night blindness type 2 (CSNB2)

CSNB2 is an X-linked inherited retinal disorder in which patients suffer from varying degrees of low visual acuity, myopia, nystagmus, light sensitivity, and night blindness. However, all symptoms may not be present in every patient, and in many cases patients present with little or no night vision deficits [234, 247, 248]. Electroretinography (ERG) recordings indicate that both rod and cone function is reduced in CSNB2 patients. CSNB2 is primarily caused by mutations in the *CACNA1F* gene, which encodes for the pore-forming subunit of $Ca_V1.4$, although mutations in the Ca^{2+} channel auxiliary subunit $\alpha_2\delta4$ and calcium binding protein (CaBP) also account for a small fraction of cases. However, as both these proteins are known to interact with $Ca_V1.4$, the underlying pathogenic mechanisms are likely the same. Loss of $Ca_V1.4$ channels would be predicted to decrease neurotransmitter release from photoreceptors, and the disease mechanism of CSNB2 appears to include a decrease in neurotransmission efficiency between photoreceptors and second-order neurons [96].

In many CSNB2 cases, $Ca_V1.4$ is truncated, and unlikely to form a functional channel (e.g. R50X, R82X, E312X, Q428X, R614X, R625X, R680X, R830X, R1234X—see figure 13.4) [234]. In some cases the premature stop codon may lead to nonsense-mediated mRNA decay [96, 238], and in as many as 4% of cases, disease causing variants may be intronic or synonymous, leading to splicing defects [237]. Because *CACNA1F* resides on the X chromosome, these mutations lead to complete loss of $Ca_V1.4$ channels only in male patients. This LOF feature of $Ca_V1.4$ CSNB2 mutations has been validated for multiple mutations using heterologous expression systems. Mutations S229P, W1440X and L1068P displayed almost no current amplitude when expressed in *Xenopus* oocytes, with L1068P producing currents only in the presence of a Ca^{2+} channel agonist Bay K8644 [235]. In the case of W1440X, this loss of current appears to be due to the absence of channel expression, while L1068P appears to reduce channel opening, increase the rate of channel inactivation, and reduce the rate of recovery from inactivation. Mutation L860P was also shown to decrease $Ca_V1.4$ current density by reducing the number of channels at the plasma membrane [249], and mutations R508Q and L1364H were shown to reduce channel expression in a temperature-dependent manner when expressed in tsA-201 cells [250]. Importantly, the temperature dependence is such that a decrease in protein expression would be predicted at physiological temperatures. However, it should be noted that R508Q has been suggested to represent a SNP, rather than a true pathogenic missense mutation [251]. In addition, the L1364H mutation also caused a small increase in the rate of channel inactivation and an increase in the rate of recovery from inactivation; however, the authors conclude that this biophysical effect may not be relevant to the pathogenesis of this mutation [250]. Likewise, introduction of the mutations G1007R and R1049W into $Ca_V1.4$ resulted in full-length proteins which targeted to the membrane when expressed in tsA-201 cells, but did not produce any current [252]. Overall, LOF of $Ca_V1.4$ appears to be a common mechanism leading to CSNB2.

This pattern, however, does not hold true for all CSNB2 mutations. Some truncation mutations lead to functional $Ca_V1.4$ channels with altered biophysical properties which can be categorized as both LOF and GOF. For example, the truncation mutation K1591X eliminates the ICDI/CTM module on the C-terminus of the channel, resulting in the introduction of a substantial amount of CDI, and also produces a left shift in channel activation and inactivation [14]. Likewise, the truncation mutation R1827X has been shown to unmask significant CDI [249], while also producing a hyperpolarizing shift in channel activation and increased current density. Thus, many $Ca_V1.4$ C-terminal truncation mutations may enable $Ca_V1.4$ CDI through the loss of ICDI/CTM, and produce a mixed GOF/LOF effect on the channel.

Consideration of additional missense mutations increases this mechanistic complexity. Mutation G369D occurs within the IS6 region of $Ca_V1.4$, resulting in an overall GOF effect due to a hyperpolarizing shift in channel activation and a decrease in channel inactivation [250]. However, an alternate study demonstrated a slight depolarizing shift in channel activation and an increase in VDI [253], pointing to a mixed LOF/GOF effect. In addition, mutation F742C, located within the IIS6 region of $Ca_V1.4$, caused a large hyperpolarizing shift in channel activation and inactivation, decreased inactivation and reduced current size [252]. Likewise, the nearby I745T [254] mutation, also located within the IIS6 region, produces a significant hyperpolarizing shift in the voltage dependence of activation and a reduction in inactivation [255]. Thus, I745T represents a clear GOF CSNB2 mutation. Interestingly, the magnitude of the biophysical effects of I745T depend on the splice variant in which the mutation is expressed [256]. Moreover, the mutation was shown to alter the ion selectivity of the channel, resulting in a hyperpolarizing shift in the reversal potential of $Ca_V1.4$ in a manner dependent upon the auxiliary subunits associated with the channel. These rare GOF effects, which lead to an overlapping clinical phenotype with the LOF mutations, complicate elucidation of the pathogenic mechanisms underlying CSNB2. It has been suggested that the increased Ca^{2+} entry through the $Ca_V1.4$ channel results in a decreased dynamic range for membrane potential-induced changes in Ca^{2+} and consequently neurotransmitter release [96, 257]. To gain additional mechanistic understanding of the effects of this mutation, an I745T knockin mouse was generated [257]. This mouse displayed a reduction in the size of the retinal outer nuclear layer, disorganization of the outer plexiform layer and immature photoreceptors. Thus, the causative nature of the I745T mutation on retinal dysfunction has been confirmed.

Not all $Ca_V1.4$ mutations resulted in significant functional changes in $Ca_V1.4$. Mutations G647D, A928D, and W1459X produced no detectable biophysical changes in channel properties when expressed in heterologous cells [253]. In addition to direct biophysical effects, it has been suggested that some of the mutations within the C-terminus of the channel the alter the ability of the channel to bind CaBP4 [238]. Changes in trafficking or protein interactions at the ribbon synapse may also play a role in the pathogenesis of some of the mutations which display limited biophysical changes. Thus, the full pathogenic mechanisms underlying CSNB2 remain under investigation.

13.4.7 Cone-rod dystrophy (CORDX3)

CORDX3 is a progressive X-linked visual disorder which causes reduced visual acuity, impaired color vision, fundus abnormalities, blind spots, photophobia and myopia, and has been associated with mutations in $Ca_V1.4$. A CORDX3 mutation described as IVS28-1 occurs within the splice acceptor site of intron 28 of *CACNA1F* and is predicted to cause either a premature stop codon or deletion [242]. In addition, a sizable in-frame deletion of *CACNA1F* exons 18 through 26 was identified within a large family displaying CORDX3 phenotypes across three generations [240]. Moreover, the *CACNA1F* mutations G484S [241], R290C [244], G1350D [243] were also associated with CORDX3, however, their effects on the channel have yet to be evaluated. Finally, a 2019 study suggested that mutations within $Ca_V1.4$ may be responsible for nearly 20% of X-linked cone-rod dystrophy cases [239].

13.4.8 Åland island eye disease

Åland Island disease, also known as Forsius–Eriksson syndrome, causes myopia, astigmatism, nystagmus, dyschromatopsia, foveal hypoplasia, and fundus hyperpigmentation. Åland Island disease has been linked to a mutation in *CACNA1F* which causes exclusion of exon 30 from the channel protein [245]. In addition, the mutation G603R in *CACNA1F* has been reported to cause both Åland Island disease and CSNB2 within the same family [246]. As Åland Island disease and CSNB2 share overlapping symptoms, it may be that these two disorders are not entirely distinct.

13.5 $Ca_V2.1$

$Ca_V2.1$ is a voltage-gated Ca^{2+} channel, and the pore-forming α_{1A} subunit is encoded by the *CACNA1A* gene [258–260]. This channel subtype gives rise to P/Q-type Ca^{2+} currents [22, 258, 261]. These channels are widely expressed in both the central and peripheral nervous system, and are particularly enriched in the cerebellum [22, 262, 263], as well as the hippocampus [264] and neuromuscular junctions [265]. *CACNA1A*, located on chromosome 19p13, contains 47 exons [266]. Many of these exons are the subject of alternative splicing, giving rise to numerous splice variants [266, 267]. Further, some of these variants exhibit differential functional properties, gating behavior, trafficking, and expression patterns [266–268]. This flexibility is believed to help optimize Ca^{2+} signaling to specific cellular needs [266, 269, 270]. $Ca_V2.1$ channels are critical components of the cellular mechanisms within the presynaptic terminal that underlie neurotransmitter release [17, 271–273]. Additionally, $Ca_V2.1$ channels have been shown to be important determinants of proper synaptic architecture in developing cerebellar Purkinje cells [274]. The complete knockout of these channels in mice results in animals with rapid-onset progressive neurological impairments, with specific deficits that resemble ataxia and dystonia, followed by shortly by death, while heterozygotes are relatively normal [275].

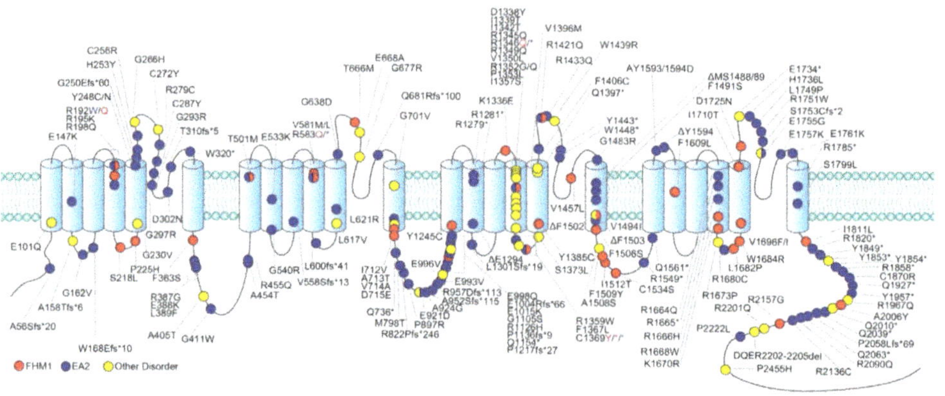

Figure 13.5. The location of mutations in *CACNA1A* ($Ca_V2.1$), with the associated phenotype indicated by the corresponding color. FHM1, familial hemiplegic migraine type 1; EA2, Episodic Ataxia Type 2.

Numerous mutations within *CACNA1A* have been associated with multiple identified disorders of the nervous system, including familial hemiplegic migraine type 1 (FHM1) [276], episodic ataxia type 2 (EA2) [266], and spinocerebellar ataxia type 6 (SCA6) [276, 277]. Additionally, emerging research continues to implicate rare *CACNA1A* variants in a variety of neurological dysfunctions, often with features of epilepsy and developmental delay [278–284] (figure 13.5 [276, 278, 281, 282, 285–368]). Most of the documented human variants of *CACNA1A* are the result of missense mutations, but a number of deletions, insertions, and repeat expansions have also been identified [288, 328, 369]. Much work has been done to characterize the functional consequences of these mutations, with complimentary results from heterologous expression systems, primary cells, and *in vivo* models. Broadly speaking, the effects of $Ca_V2.1$ mutations have been classified as either GOF or LOF. A large number of missense $Ca_V2.1$ mutations, however, exhibit more complex effects that go beyond the traditional GOF/LOF dichotomy [279]. These complex changes most commonly manifest as alterations to channel activation and inactivation, and can be due to intrinsic channel properties, or altered interactions with accessory channel subunits and channel binding proteins.

13.5.1 Familial hemiplegic migraine type 1 (FHM1)

FHM1 is a rare, autosomal-dominant disorder that is characterized by migraines with an early onset, and is produced by mutations within $Ca_V2.1$ channels encoded by the *CACNA1A* gene [276, 370]. FHM1 symptoms include unilateral headaches, hemiparesis, auras, and in some cases progressive cerebellar atrophy and ataxia. It is thought that migraine pain is initiated following activation of nociceptive sensory afferents [336]. A phenomenon called cortical spreading depression (CSD) is an element in migraine symptoms and may play a role in migraine initiation [371, 372]. CSD is characterized as a self-propagating wave of neuronal and glial depolarizations that travel throughout the cortex and is followed by a period of neuronal silencing [373–375], and susceptibility to CSD can be conferred by a number of *CACNA1A* mutations [371, 373, 376–378].

Using HEK293 cells and *Xenopus* oocytes, Hans *et al* (1999) and Kraus *et al* (1998) performed some of the first quantifications of the impact of FHM1-associated mutations on channel function [379, 380]. As the number of identified $Ca_V2.1$ mutations grew, researchers began examining FHM1 mutations in neuronal model systems [378, 381–386]. In the course of this work, a consensus emerged that FHM1 mutants within $Ca_V2.1$ were broadly GOF, primarily through a hyperpolarizing shift in channel activation [370, 380, 383, 387–390]. These kinds of GOF effects have been linked to the mutant channels' ability to contribute to CSD susceptibility [385, 386], tying the biophysical changes observed in channel function into a proposed mechanism of migraine pathology.

Most FHM1 missense mutations lie within the S4 voltage sensors (e.g. R192Q [276], R195K [285], V581L [290], R583Q [292], R1668W [285]), the S4-S5 linkers (e.g. S218L [286], C1396Y [296], W1684R [285]), or the S6 activation gates (e.g. V714A [276], D715E [391], F1506S [392]) [279, 393] (figure 13.5). This localization is broadly consistent with other GOF channelopathic mutations [393]. Single-channel patch clamp experiments in neurons and HEK293 cells expressing $Ca_V2.1$ channels harboring different FHM1 mutations (T666M, V1457L, R192Q, S218L, V714A, and I1811L) revealed an increase in their open probabilities and single-channel Ca^{2+} flux over a broad range of voltages, primarily by exhibiting hyperpolarizing shifts in activation and deficits in inactivation [380, 383, 387]. Results from knockin mice expressing human $Ca_V2.1$ harboring FHM1 mutations also provide evidence for broad GOF effects. Investigation of the S218L mutation using a knockin mouse model revealed strong GOF channel properties, consistent with data from heterologous expression systems [385, 394]. In this neuronal model, S218L causes a hyperpolarizing shift in $Ca_V2.1$ voltage-dependent activation [336, 394]. Further, S218L $Ca_V2.1$ channels exhibit a similar shift in a complimentary *drosophila* model [395]. Thus, the most consistent finding reported for FHM1 mutations is a hyperpolarizing shift in activation [336, 395].

There is also evidence that FHM1 mutations can confer a GOF effect on channel activity through the disruption of channel regulation normally imparted by other proteins. The R192Q and Y1245C mutations have been shown to increase Ca^{2+} flux by disrupting G-protein-mediated inhibition of the channel [381, 390]. Additionally, the E1015K mutation imparts a GOF effect on $Ca_V2.1$ primarily through its impact on channel inactivation. This mutation sits in the cytoplasmic II–III loop within the synaptic protein interaction (synprint) site of $Ca_V2.1$ [294, 396, 397]. This site is known to be involved in protein-protein interactions between $Ca_V2.1$ and t-SNARES [294, 398]. It is thought that diminished t-SNARE channel regulation plays a significant role in the observed disruptions to channel inactivation seen in E1015K channels [294]. Normally, interaction between $Ca_V2.1$ channels and SNARE proteins shifts the voltage dependence of inactivation to hyperpolarized potentials [399], but in the presence of E1015K, this form of channel regulation is impaired, leading to a net GOF effect [294].

In addition, FHM1 mutations can alter the Ca^{2+}-dependent regulation of $Ca_V2.1$. R192Q as well as S218L substantially reduce $Ca_V2.1$ CDI [400], a negative feedback mechanism thought to be important for neuronal functions including synaptic plasticity [401, 402]. Further, R192Q and S218L both impair $Ca_V2.1$ Ca^{2+}-dependent facilitation (CDF) [52, 400], a positive feedback mechanism thought to play a role in synaptic facilitation [403]. The precise balance of $Ca_V2.1$ CDI and CDF in Purkinje cells has been demonstrated to be important for stable currents during repetitive firing [404]. R192Q, and to a larger extent S218L, shift $Ca_V2.1$ into a chronically facilitated state [394, 400]. Additional experiments revealed S218L channels have impaired inactivation over sustained depolarizations, and faster recovery from inactivation [387]. It was theorized that these alterations to inactivation were in part due to a mechanism that is sensitive to resting-state Ca^{2+}, potentially implicating calmodulin, the Ca^{2+} sensor that confers Ca^{2+}-dependent channel regulation to $Ca_V2.1$ [385, 394]. Interestingly, synaptic strength was increased in calyx of Held neurons and synaptic transmission was increased at the neuromuscular junctions of the S218L neurons, effects which were proposed to be due to the loss of Ca^{2+}-sensitivity of the channels [394].

While most studies of FHM1 mutations report similar GOF features, this is not universal [336, 382, 383, 405, 406]. Inconsistent alterations in unitary conductance have been observed in select mutations, including decreased unitary conductance in channels harboring T666M [380], V714A [380, 387], and increased unitary conductance in D1725N channels [387]. Additional variability has been observed in the impact of mutations on inactivation parameters [26, 407]. Experiments in heterologous expression systems using a variety of FHM1 mutations showed either increased (K1336E, W1684R, V1696I) [26] or decreased (I1811L, V714A) [379] VDI in response to 1 Hz stimulation trains. In addition, multiple studies reported no change in the voltage dependence of channel activation for T666M [380, 408], while other work [383] demonstrates hyperpolarizing shifts in activation for this same mutation. Reports on the effects of FHM1 mutations on current density are also varied. In most cases, FHM1 mutations appear to cause reductions in whole-cell current densities [380, 383, 405, 408], however, this effect is not universal. For example, a number of studies report that the mutation R192Q decreases whole-cell current densities [382, 383], while other studies that indicate that R192Q can increase current density [378, 380]. In a study that examined the impact of FHM1 mutations on synaptic transmission, the authors noted that R192Q, T666M, V714A, and 1I811L all decreased the $Ca_V2.1$ current density when expressed in cultured mouse neurons [382]. The contribution of the FHM1 mutants to excitatory postsynaptic currents (EPSCs) was diminished in a pattern that matched the magnitude of current density reductions, and the authors concluded that these mutations operated *in vivo* through a LOF mechanism by impairing inhibitory neurotransmission [382]. In another study utilizing knockin mice harboring human $Ca_V2.1$ mutations, S218L resulted in a decrease in current density at the calyx of Held [387]. A study using a similar knockin animal model harboring $Ca_V2.1$ R192Q found that the mutation is capable of altering Ca^{2+} currents, but only in the context of certain AP waveforms [409]. This suggests that biophysical changes to channel function may present different physiological consequences depending on the cellular environment. Moreover, it has been proposed that some of the discrepancies in the biophysical parameters reported may arise from differences in the particular $Ca_V2.1$ splice variant used, the model system employed, the species of origin of the $Ca_V2.1$ channels, or the specific auxiliary subunits present [383, 407, 410].

Taken together, these results indicate that typical FHM1 mutations impart a GOF effect on $Ca_V2.1$ by increasing the open probability of the channel, shifting the activation voltage to more negative potentials, altering current density, and/or diminishing channel inactivation at negative potentials. The activation shifts observed for most FHM1 mutations are thought to be sufficient to confer a GOF effect [336]. The physiological impact of these alterations to channel function is thought to be dependent on the particular mutation, as well as interactions with the distinct cellular environment of different neuronal subtypes [393, 407].

13.5.2 Episodic ataxia type 2 (EA2)

EA2 is a rare neurological disorder, characterized primarily by sudden episodes of ataxia, as well as nystagmus and vertigo [411]. Numerous mutations of $Ca_V2.1$ have been identified in patients suffering from EA2. These mutations can be found throughout the channel, with the majority residing in the pore loop or neighboring S5/S6 helices [276, 289, 313, 411] (figure 13.5). The majority of EA2 mutations display LOF effects, often by disrupting the open reading frame and causing a premature stop codon and loss of Ca^{2+} current [266, 313, 332, 336, 411].

It has been suggested that haploinsufficiency of $Ca_V2.1$ may explain some of the observed effects of a number of EA2 mutations. The frameshift mutation L1301Sfs*19, for example, results in a premature stop codon, and pathogenicity is believed to arise due to selective nonsense-mediated mRNA decay of the mutant transcript, leaving neurons with an inadequate supply of functional $Ca_V2.1$ channels [333]. In addition, the frameshift mutation A56Sfs*20 has also been predicted to lead to pathogenic haploinsufficiency through nonsense-mediated decay [289, 412]. Similar results have been observed for other EA2 mutations, including E147K [307]. However, haploinsufficiency cannot explain all documented effects of EA2 mutations [274, 330]. Some mutations produce a dominant negative effect such that co-expression of wildtype and mutant channels results in suppression of even the wildtype current. This is due, in part, to the suppression of wildtype channel translation in a process known as unfolded protein response (UPR) [330], a mechanism which can lead to cell death. Thus, even in patients possessing a healthy copy of *CACNA1A*, gross deficits in $Ca_V2.1$ currents and cell death may be caused by a shared mechanism downstream of a single mutated *CACNA1A* [413]. In particular, it has been shown that three human $Ca_V2.1$ EA2 mutations, F1406C, E1761K, and the single nucleotide deletion P1217fs*27, exert dominant negative effects on $Ca_V2.1$ currents when expressed alongside wildtype channels in *Xenopus* oocytes [276, 330, 413, 414]. Of note, E1761 is an essential residue of the selectivity filter, thus E1761 may also impair channel function by permitting nonselective monovalent ion flow [279, 411]. In addition to the UPR hypothesis, it has also been speculated that competition for synaptic 'slots' reserved for $Ca_V2.1$ channels in the membrane contribute to the dominant negative effect of some EA2 mutants [410, 415].

Though dominant negative effects and haploinsufficiency of $Ca_V2.1$ currents may explain the majority of pathogenic impacts of EA2 mutations, other potential LOF effects have been identified. H1736L reduces, but does not eliminate, $Ca_V2.1$ currents when compared to wildtype channels expressed in HEK293 cells [339]. This reduction in current was due, in part, to a slight but significant depolarizing shift in activation voltage. This depolarizing shift, coupled with an increase in the rate of VDI, imparted a moderate LOF effect on the mutated channel. In addition, H1736L exhibits some GOF effects, including a significant reduction in current decay in response to a train of stimuli as compared to wildtype channels [339], resulting in increased reliance on the mutant channel in response to repetitive firing. Similarly, G293R and AY1593/1594D $Ca_V2.1$ mutations produce functional channels that undergo membrane insertion in heterologous cells [341], yet these mutations both also result in LOF via a significant depolarizing shift in activation and accelerated VDI [341]. Interestingly, current inactivation of both G293R and AY1593/1594D was enhanced during rapid pulse trains, in contrast to the reduced inactivation observed in channels harboring H1736L [339, 341].

In addition, alternative splicing of *CACNA1A* can affect and be affected by a number of EA2 mutations [266, 289, 346]. A nucleotide change of A>G was identified in a patient with EA2 at position −2 within the acceptor splice site of the mutually exclusive exon 37a [266, 416]. This change is thought to shift splicing to favor the mutually exclusive exon 37b [266, 416]. Importantly, exons 37a and 37b control the inclusion of the functionally relevant motifs EFa or EFb, respectively [258, 266, 270, 417]. $Ca_V2.1$[EFa] has a number of different biophysical properties as compared to $Ca_V2.1$[EFb]. EFa, but not EFb, demonstrates robust CDF in channels where exon 47 is also included [417, 418], and thus EA2 mutations that shift splicing in favor of 37b may impair this form of channel regulation. It has been proposed that $Ca_V2.1$[EFa] reliably participates in the mechanisms underlying neurotransmitter release, while $Ca_V2.1$[EFb] only engages in neurotransmitter secretion following repetitive stimulation [266, 267, 419]. Notably, $Ca_V2.1$[EFa] is selectively expressed at high levels in cerebellar Purkinje cells [258, 346], and altered activity of these cells are thought to be involved in the pathology of ataxia [420–422]. Multiple predicted LOF mutations from EA2 patients have been identified within exon 37a, including mutations that lead to the introduction of premature stop codons, namely Y1849* [313], Y1854* [346], R1858* [346], as well as the missense mutation C1870R [289]. Notably, the three mutations resulting in premature stop codons are located within the EF hand motif, which is critical for proper CDF of the channel [346]. In contrast, no pathogenic mutations to date have been observed in exon 37b, suggesting that the $Ca_V2.1$[EFa] isoform may play a role in EA2 pathophysiology. These kinds of projected impacts of EA2 mutations on $Ca_V2.1$ splicing have become targets for potential therapeutic interventions such as antisense oligonucleotides [266].

In some cases, $Ca_V2.1$ mutations may be associated with more than one possible disease phenotype. Some FHM1 mutations can also lead to ataxic symptoms, such as T666M [423], S218L [424], and ΔF1502 [365]. T666M has also been identified in a patient diagnosed with episodic ataxia, without comorbid migraine [425]. This overlapping of symptoms between FHM1 and EA2 can complicate the mechanistic understanding of each disease. Whereas EA2 is considered to result generally from LOF mutations in *CACNA1A*, FHM1 mutations that cause comorbid ataxia have generally been found to exhibit GOF effects [365, 423, 424]. In addition, select mutations have been identified in patients diagnosed with either FHM1 or EA2. R1421Q [298] and C1369* [363] can be found in patients exhibiting either hemipelagic migraine or EA2, and T501M results in changes to activation and inactivation properties consistent with a GOF effect, despite being identified in patients with either EA2 or hemipelagic migraine [313, 426]. In light of results like these demonstrating the ability of ataxia- and migraine-associated $Ca_V2.1$ mutations to overlap in their functional impacts, it has been proposed that there exists a 'critical bandwidth' for proper Ca^{2+} flux through $Ca_V2.1$, outside of which pathogenicity can arise [411].

13.5.3 Spinocerebellar ataxia type 6 (SCA6)

Patients with SCA6 exhibit slowly progressive cerebellar ataxia, loss of coordination, speech impairments and nystagmus [393, 427, 428]. The disorder is caused by a trinucleotide repeat expansion of a native polyglutamine (poly-Q) encoding CAG sequence within *CACNA1A* [25, 277, 428, 429]. Typically, healthy individuals possess between 4 and 18 glutamines in this tract, whereas SCA6 patients harbor between 20 and 33 [428, 429]. This expansion causes degeneration of Purkinje cells within the cerebellum [428], and neurons within the inferior olivary nucleus [430, 431]. Remaining Purkinje cells often display aberrant cellular morphology, with diminished arborization and cytosolic organelles [428].

It is thought that alternative splicing of $Ca_V2.1$ plays a role in the genetic events underlying SCA6 pathology. Alternative splicing at exon 47 of *CACNA1A* can generate two channel isoforms, often called MPI and MPc, both of which are observed in human neurons [22, 269, 432]. The MPI splice variant results in a full-length channel harboring the poly-Q tract within which the pathogenic expansion occurs. In contrast, the MPc splice variant results in a channel isoform which is truncated after exon 46, thus excluding the region containing the poly-Q expansion [22, 269, 277]. SCA6 is relatively unique among SCA disorders in that the pathogenic $Ca_V2.1$ protein is functional, producing channels capable of membrane insertion [433]. However, it remains unclear how direct biophysical changes to channel properties impact the pathogenesis in SCA6, in light of the fact that mutated channel fragments also contribute to the disease state [25, 433–435].

Interrogating the mechanisms underlying SCA6 pathogenicity has proven difficult, with apparently contradictory results reported [22, 269, 428, 436]. Early studies utilizing $Ca_V2.1$ heterologously expressed in BHK and HEK293 cells suggested that

SCA6 mutations may confer a LOF effect through moderate hyperpolarizing shifts in VDI, which was theorized to result in a net decrease in Ca^{2+} influx [437, 438]. These studies found that this effect was only present when certain β subunits were co-expressed, and the severity of the impact on channel inactivation was dependent on splice variation in exon 31 [25, 438]. In contrast, studies utilizing heterologous expression of $Ca_V2.1$ in *Xenopus* oocytes observed that channels with extreme expansions of 30 glutamines (Q30) co-expressed with β_4, but not β_2 or β_3, demonstrated hyperpolarizing shifts in the voltage dependence of activation and a slowing of inactivation [24, 428], indicating a GOF effect. Moreover, G-protein-mediated inhibition of $Ca_V2.1$ was shown to be reduced in oocytes expressing channels harboring the SCA6 poly-Q expansions, again selectively in the context of β_4 co-expression [24]. Additional work carried out in HEK293 cells expressing β_1 and $Ca_V2.1$ harboring Q23 and Q27 expansions seemed to indicate an increase in the functional surface expression of the channel, doubling the Ca^{2+} current density as compared to wildtype channels [439]. This GOF effect was suggested to be due to increased stability of the channel at the membrane. This is consistent with other poly-Q expansion disorders, such as in Huntington's disease, where alterations to protein stability are thought to underly some of the associated pathogenic effects [440–443].

Additional experimental evidence suggests that frequency-dependent $Ca_V2.1$ channel properties may be altered by SCA6 mutations in a generally GOF pattern [444]. In these experiments, Ca^{2+}-dependent and Ca^{2+}-independent activation, inactivation, and facilitation kinetics were unchanged between the wildtype CAG_{11} and pathogenic CAG_{23} channels when utilizing square-pulse depolarization protocols [444]. However, the authors noted a significant decrease in frequency-dependent inactivation and an increase in frequency-dependent facilitation when utilizing high-frequency stimulation reminiscent of neuronal burst firing [444]. The authors note these frequency-dependent effects could be physiologically relevant in the context of Purkinje cells, which exhibit relatively high firing rates [444–446].

Taken together, these experiments present a complex array of somewhat contradictory findings. Some results indicate that $Ca_V2.1$ channels harboring SCA6 mutations have LOF characteristics, while others display GOF effects. Some of these differences may be explained by the use of different $Ca_V2.1$ isoforms, auxiliary subunits, or model systems employed in each study [22, 428]. Still, other studies appear to indicate that $Ca_V2.1$ channels harboring an expanded poly-Q tract do not display significant alterations in channel properties *in vivo* [430, 447]. Electrophysiological interrogation of the artificial hyperexpanded poly-Q tract $SCA6^{84Q}$ in knockin mice demonstrated no alteration to the intrinsic membrane properties of Purkinje cells [430, 448]. Findings from $SCA6^{84Q}$ and MPI-118Q mice, which recapitulate many of the symptomatic features of SCA6, suggest that current densities, voltage-dependence of activation and inactivation, and Ca^{2+}-dependent firing properties independent of cellular degeneration were all unchanged *in vivo* [430, 447]. Thus, the functional impact of biophysical changes observed *in vitro* remain unclear.

Beyond its ability to encode the $Ca_V2.1$ ion channel, *CACNA1A* is capable of producing peptide products which act independently of the channel pore. It has been shown that the poly-Q-containing C-terminus of $Ca_V2.1$ can be subject to proteolytic cleavage, resulting in a stable peptide fragment [448–450]. Kordasiewicz *et al* demonstrated that such a peptide can localize to the nucleus in both mouse and human Purkinje cells [434]. Critically, C-terminal peptides containing expanded poly-Q tracts were found to be toxic to cultured cells, with nuclear translocation of these peptides being necessary for this cytotoxicity [434, 448]. Additionally, a gene product of *CACNA1A* derived from an internal ribosome entry site, termed α1ACT, includes many C-terminal elements of $Ca_V2.1$ but functions independently as a transcription factor within the cell nucleus [435, 451]. Experiments utilizing *in vivo* mouse models and virally delivered wildtype or SCA6 α1ACT revealed that animals injected with SCA6 α1ACT showed behavioral characteristics of ataxia, as well as degeneration of cerebellar tissue [451]. Thus, regardless of the potential impact on $Ca_V2.1$ channel gating and kinetics, it is likely that a significant portion of SCA6 pathogenicity is conferred by C-terminal channel fragments and/or α1ACT.

13.5.4 Developmental epileptic encephalopathies (DEEs)

Encephalopathies are considered DEEs when the epileptic activity contributes directly to developmental impairments, with the most severe DEEs falling under the umbrella of Lennox–Gastaut syndrome (LGS) [278, 283]. The causes of DEE are diverse, and can include brain malformations, perinatal brain lesions, neurometabolomic disorders, and chromosomal rearrangements [278]. Despite this heterogeneity, a number of likely pathogenic *CACNA1A* mutations have been identified in patients suffering from non-lesional DEEs [278]. Symptoms in these patients include neurocognitive deficits, ASD, refractory generalized seizures, severe intellectual disability, as well as mild-moderate ataxia [278, 326].

Experiments using a heterologous expression system were carried out to examine the functional impact of these mutations, and immunocytochemical labelling in both HEK293 cells and in mouse neurons revealed normal channel localization. *De novo* mutations from patients with DEEs had a diverse range of impacts on channel functionality. Both G230V and I1357S substitutions resulted in a general LOF effect, with greatly diminished current densities in HEK293 cells that was not fully rescued with isomolar co-expression of the wildtype channel alongside the mutant channels [278]. The G230V mutation did not impact the time constants of activation or inactivation, whereas the I1357S channels demonstrated an increased time constant of inactivation, while activation was unchanged compared to wildtype channels [278]. While both mutations did not impact the voltage dependence of activation, they did decrease the slope of the activation curve. In addition, A713T and V1396M increased whole-cell currents and resulted in a hyperpolarized shift in activation in HEK293 cells [278]. A713T also resulted in slower current decay [278]. Thus, it appears that $Ca_V2.1$ variants that result in either GOF or LOF in *in vitro* environments may contribute to DEE pathogenesis.

13.5.5 Progressive myoclonic epilepsy (PME)

The spectrum of disorders associated with $Ca_V2.1$ mutations has grown to include a form of PME. PME is characterized by refractory myoclonic seizures, neurological deterioration, and cognitive decline [452]. A novel homozygous poly-Q mutation within $Ca_V2.1$, similar in type to those associated with SCA6, was documented in PME patients from a single family with no previously identified PME genetic risk factors [452]. The abnormal CAG repeats were localized to residues 2319, which is normally followed by CAG repeats until residue 2331. In contrast, CAG repeats were extended in the afflicted individuals to 2344 [452]. Further investigation in SH-SY5Y cells revealed that overexpression of $Ca_V2.1$ channels containing the CAG repeats led to increased nuclear translocation of the protein [453]. Additionally, there is evidence that this poly-Q mutation may result in truncated channel fragments that activate apoptotic pathways and induce cell death, similar to mechanisms proposed to underly SCA6 pathology [451, 453]. The effects of this poly-Q expansion on channel properties such as gating, conductance, and inactivation have yet to be fully investigated.

13.5.6 Trigeminal neuralgia

Whole-exome sequencing has revealed the existence of a *CACNA1A* mutation within patients suffering from trigeminal neuralgia, a common condition characterized by unilateral, paroxysmal pain [454, 455]. This missense mutation, P2455H, revealed a slowing of activation and inactivation kinetics, as well as a depolarizing shift in the voltage dependence of activation and inactivation when evaluated in tsA-201 cells. No changes in cell surface expression or current densities were observed [368]. These effects resulted in a mixed functional impact, with both GOF and LOF features. P2455H was also shown to impair normal $Ca_V2.1$ CDI, which represents an additional GOF effect [368]. This mutation shows a heterogeneous impact on the biophysical properties of $Ca_V2.1$, and it has been suggested that the GOF effects will likely predominate in neurons with faster average firing rates, whereas the reverse is expected for neurons that demonstrate slower rates of activity. Thus, the authors suggest that altered $Ca_V2.1$ activity may contribute to the disease state in trigeminal neuralgia [368].

13.5.7 Dravet syndrome susceptibility

$Ca_V2.1$ mutations have also been described as having an interaction with another channelopathy, namely, Dravet syndrome. Dravet syndrome is a severe form of myoclonic epilepsy in infancy [456]. Approximately 70%–80% of patients suffering from Dravet syndrome exhibit mutations within *SCN1A*, which encodes the $Na_V1.1$ channel, and these mutations are thought to be the primary source of pathogenicity. However, mutations within $Ca_V2.1$ have also been identified in patients suffering from Dravet syndrome. Notably, these mutations were typically present alongside mutations within *SCN1A*, making interpretation of the mutational effects more complex. Pairs of $Ca_V2.1$ mutations have been shown to occur simultaneously within individual patients, including a mutation pair E921D and E996V, as well as a pair of mutations R1126H and R2201Q [327]. E921D and E996V were both individually associated with EA2 in previous studies [322]. While *SCN1A* mutations in Dravet syndrome patients are largely *de novo*, the *CACNA1A* mutations found in the same patients are more often inherited [327]. The $Ca_V2.1$ mutations identified in Dravet syndrome patients demonstrated a general trend towards increased current density when expressed in HEK293 cells. The R1126H and R2201Q paired mutations, as well as a $Ca_V2.1$ deletion DQER2202–2205, displayed large increases in current density. Interestingly, individually mutating R1126H or R2201Q did not significantly alter channel functionality [327]. Clinically, these $Ca_V2.1$ mutations were significantly associated with increased susceptibility to epilepsy, and were predictive of the age-of-onset of Dravet syndrome symptoms [327]. These results indicate how $Ca_V2.1$ variants that may not generate clinically significant outcomes in and of themselves can interact with other mutations inside $Ca_V2.1$ as well as other voltage-gated ion channel classes to potentially contribute to disease states.

13.5.8 The widening spectrum of $Ca_V2.1$ channelopathies

Diseases like DEE, PME, Trigeminal Neuralgia, and Dravet Syndrome susceptibility are representative of a widening spectrum of $Ca_V2.1$ channelopathies. As this spectrum continues to grow, additional phenotypes continue to be identified. Patients harboring channel variants often exhibit multifaceted conditions which overlap in terms of their symptomology, and combine elements of the previously described $Ca_V2.1$ channelopathies [282, 393, 411]. In the past decade and a half, advances in whole-exome sequencing have permitted the identification of *CACNA1A* variants in patients exhibiting disorders with symptoms that do not precisely match those of FHM1, EA2, or SCA6 [280–282, 326, 342, 358, 359, 361, 393, 411]. Most of these variants are exceedingly rare, and as is often the case with spectrum disorders, these rarer variants are frequently associated with more extreme pathology [411]. Further, there is a broad range of documented and predicted functional impacts. This expanding spectrum of $Ca_V2.1$ channelopathies has yet to be fully classified, and functional characterization of many of these variants is ongoing [279, 280, 358, 361, 411, 452].

These more recently identified mutations are found throughout the channel, indicating that there are multiple potential mechanisms underlying these disorders. For instance, the A405T mutation, identified in a single adolescent patient [281], rests within the I–II linker region. The patient with the A405T substitution demonstrated similar, but more extreme symptoms as compared to EA2 patients harboring the nearby mutation A454T, including ataxia in addition to dysmetria, hypotonia, and developmental delay. Multiple hypotheses exist to explain this mutation's impact on channel function. It may have similar consequences to A454T [319], which is known to functionally uncouple the effect of the β subunit on the voltage-dependence of $Ca_V2.1$ steady-state inactivation, and impair the interplay between $Ca_V2.1$ channels and plasma membrane SNARE proteins [457, 458].

Alternatively, the A405T mutation may disrupt the binding of the $Ca_V2.1$ α_{1A} subunit with the auxiliary β subunit. This would be similar to what was observed when an artificial substitution within the I–II loop, Y392S, was introduced in $Ca_V2.1$ channels expressed in *Xenopus* oocytes [411, 459]. This mechanism would likely impact proper channel trafficking and potentially result in a LOF effect via haploinsufficiency [411].

A number of identified mutations within $Ca_V2.1$ S4 segments result in diverse biophysical impacts [393, 411]. The S4 mutation R1673P, identified in a patient with ataxia, hypotonia, cerebellar atrophy, and developmental delay, precipitates a strong LOF effect when expressed in tsA-201 cells, primarily through a ~25 mV depolarizing shift in the voltage dependence of activation [279, 280]. Despite the important role that highly-conserved S4 residues play in $Ca_V2.1$'s voltage-sensitivity [460], a number of these mutations have effects that impact channel properties beyond this functionality. Another S4 mutation that produces a LOF effect on channel properties, P1353L, was identified as a *de novo* mutation in a single pediatric patient suffering from pronounced language, developmental, and motor delay, as well as hypotonia [361]. Investigation of the biophysical properties of P1353L channels in tsA-201 cells indicated that the mutation was capable of nearly eliminating ion flux through $Ca_V2.1$ channels, with a reduction in current density of over 95% at approximately physiological Ca^{2+} concentrations. From this, it was concluded that it is likely the P1353L mutation operates through a dominant negative LOF mechanism, rather than haploinsufficiency. In contrast to the P1353L mutation is the nearby R1349Q mutation, first identified in a pediatric patient with cerebellar ataxia and developmental delay [282]. In the *tottering* mouse model, a mutation equivalent to R1349Q causes a significant hyperpolarizing shift in activation [461], which, coupled with the fact that another patient with the same mutation had their symptoms alleviated by Ca^{2+} channel blockers [359], indicates that R1349Q likely represents a GOF effect [411]. The heterogeneous effects of mutations within the S4 segments are thought to be the result of the specific charges, polarity, and positions of the affected amino acids [411]. The wide range of documented impacts of rare *CACNA1A* variants in this spectrum will likely continue to grow, and characterization of $Ca_V2.1$ functionality in these disorders will present specific challenges on a case-by-case basis.

13.6 $Ca_V2.2$

$Ca_V2.2$ channels contain the pore-forming α_{1B} subunit, which is encoded by *CACNA1B*. $Ca_V2.2$, or N-type, channels are expressed throughout the central nervous system in regions including the cerebellum, cerebral white matter, cortex, hippocampus, and the basal ganglia [462–465]. $Ca_V2.2$ channels contribute to fast neurotransmitter release alongside $Ca_V2.1$ and are similarly capable of interacting with synaptic vesicular machinery including t-SNARE proteins [397, 466–468]. While $Ca_V2.2$ channels contribute to synaptic maturation, gene expression, and neuronal migration and survival [462, 468, 469], it is believed that their role in synaptic transmission is eventually eclipsed during maturation by $Ca_V2.1$ in neurons

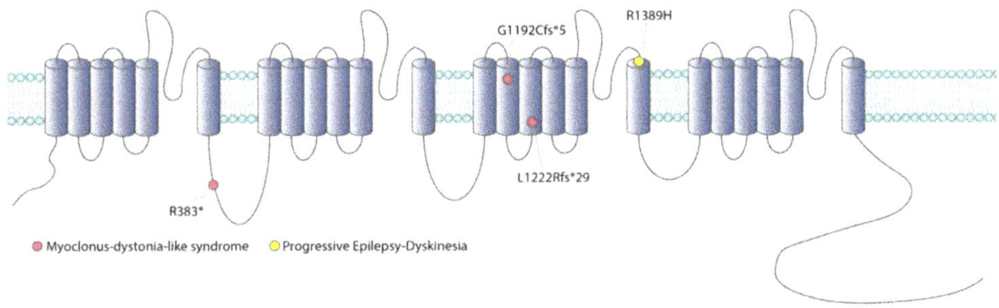

Figure 13.6. The location of mutations in *CACNA1B* ($Ca_V2.2$), with the associated phenotype indicated by the corresponding color.

of the thalamus, cerebellum, and auditory brainstem [462, 470]. Whereas animals with artificially induced mutations in $Ca_V2.1$ demonstrate severe ataxic phenotypes, equivalent artificial mutations in $Ca_V2.2$ typically result in much milder phenotypes [393, 471]. $Ca_V2.2$ knockout mice are hyposensitive, with significantly reduced responses to inflammatory and neuropathic pain [468, 472–474], consistent with the role these channels play in synaptic transmission in sensory fibers. These animals also demonstrate decreased voluntary ethanol intake and reduced responses to alcohol [475], as well as deficits in heart rate control and blood pressure [476]. Additionally, knockout animals display significant hyperactivity [477]. Given the relatively mild phenotype of animals harboring $Ca_V2.2$ channel mutations as well as the developmental expression profile, it is perhaps unsurprising that $Ca_V2.2$ variants were not identified in clinical patients until recently (figure 13.6 [462, 478]).

A number of rare $Ca_V2.2$ variants have been identified in individuals without any apparent clinical disorder, despite the fact that that these mutations occur within a channel locus predicted to be involved in gating functionality [393]. Moreover, analogous mutations in other Ca_V channels produce a clear clinical phenotype. For example, V351E within $Ca_V2.2$ is found in the gnomAD database corresponding to healthy individuals, yet in $Ca_V1.3$, a mutation V401L at an analogous locus is associated with a severe neurodevelopmental disorder due to a GOF effect on the $Ca_V1.3$ channel [211]. Even more strikingly, S215L within $Ca_V2.2$ does not appear to be associated with any described disorder, yet it is analogous to the severe FHM1 $Ca_V2.1$ mutation S218L which displays known GOF effects [393, 479]. This lack of severe impact from $Ca_V2.2$ variants may help explain the relatively small number of identified pathogenic mutations to date.

13.6.1 Myoclonus-dystonia-like syndrome

Myoclonus-dystonia is characterized by sudden jerk-like movements, and is typically produced by mutations within the *SGCE* gene [478, 480]. Psychiatric problems are often comorbid, and symptoms have been noted to respond positively to alcohol ingestion [481]. A $Ca_V2.2$ missense variant, R1389H, was identified in a family of patients suffering from a form of myoclonus-dystonia-like syndrome [478, 480]. Functionally, R1389H demonstrated a minimal impact on whole-cell current properties when compared to wildtype $Ca_V2.2$ channels in tsA-201 cells, though a slight increase in current density was reported [478]. Analysis of single-channel currents revealed an impact of the R1389H mutation on channel function, resulting in an apparent LOF effect [478]. However, the pathogenicity of this mutation remains unclear, as the variant is relatively common and is found in healthy individuals at similar rates to patients [393, 482]. Other work has identified two variants within $Ca_V2.2$ in patients suffering a superficially similar disorder, adult-onset isolated focal dystonia, but complete characterization of these mutant channels has not yet been carried out [393, 483].

13.6.2 Progressive epilepsy-dyskinesia

In contrast to the more common mono-allelic mutations within Ca_V proteins, a set of autosomal recessive, bi-allelic truncating mutations within $Ca_V2.2$ (R383*, G1192Cfs*5, and L1222Rfs*29) was identified in a sample of patients suffering from developmental epileptic encephalopathies associated with dyskinesia [462]. This severe and progressive neurological disorder is complex, consisting of treatment-resistant epileptic encephalopathy, hypotonia, microcephaly, and hyperkinetic movements [393, 462]. The bi-allelic nature of these $Ca_V2.2$ mutations may explain the severity of symptoms. These mutations are predicted to result in LOF effects [462], however, functional characterization has not yet been carried out.

13.7 $Ca_V2.3$

CACNA1E encodes the pore-forming α_{1E} subunit of the voltage-gated $Ca_V2.3$ Ca^{2+} channel, which are highly expressed in the central and peripheral nervous systems [484]. Additionally, histological and pharmacological evidence demonstrates $Ca_V2.3$ channels are present and functional in the endocrine [485, 486] and cardiovascular systems [487], and these channels are involved in pulmonary development [488]. In neurons [489] and glia [444], $Ca_V2.3$ channels influence the development of cellular morphology. $Ca_V2.3$ channels mediate R-type Ca^{2+} currents [490], and are found in presynaptic terminals in neurons, notably within hippocampal mossy fibers [491], the globus pallidus [492], and neuromuscular junctions [493]. Within the presynaptic terminal, $Ca_V2.3$ channels are less closely associated with the active zone exocytosis machinery as compared to $Ca_V2.1$, and are more predominant in the peripheral region of the terminal [494, 495]. Dendritic expression of $Ca_V2.3$ channels is complex, being evident in select neuronal subtypes such as those found in the CA1 region of the hippocampus [496]. Additionally, $Ca_V2.3$ channels are highly expressed in substantia nigra dopaminergic neurons [497]. Substantia nigra neurons from $Ca_V2.3$ knockout mice differ only slightly in their spontaneous firing, however, Ca^{2+} oscillations and Ca^{2+}-dependent after-hyperpolarizations are significantly affected [497]. Animal models lacking $Ca_V2.3$ channels have a reduced susceptibility to chemically induced seizures, altered pain responses, and are at increased risk for hyperglycemia [498–500]. *CACNA1E* has been identified as a potential locus of disease-associated variants in developmental disorders [501] based on a meta-analysis of whole-exome sequencing studies.

13.7.1 Developmental and epileptic encephalopathies (DEEs)

DEEs are typically characterized by intractable seizures with an early onset, developmental impairment, hyperkinetic movement disorders, and hypotonia [393, 502]. Severe verbal and gross motor deficits are also common in these patients. *CACNA1E* missense variants have been associated with DEEs, and a number of these variants have been shown to lead to GOF effects within $Ca_V2.3$ channels [502]. Notably, clustering of these variants was observed within S6 segments of the channel (figure 13.7 [502, 503]). In the patient population identified, a plurality of cases were associated with the recurrent variants G352R and A702T [393, 502]. G352R is found within the GX_9GX_3G motif of the IS6 domain, and is thought to stabilize the open state of $Ca_V2.3$ by diminishing inactivation and introducing a pronounced hyperpolarizing shift in the voltage dependence of activation [171]. The glycine rich GX_9GX_3G motif of domain IS6 is found in multiple high-voltage activated Ca_V channels, and mutational disruption of this motif in both $Ca_V1.2$ and $Ca_V2.3$ slows inactivation kinetics, although the specific residues mutated to produce these inactivation deficits differ between the two [171]. The $Ca_V2.3$ G352R mutation selectively impairs VDI. In addition the $Ca_V2.3$ mutations A702T, I701V, and F698S, all found within IIS6, demonstrate a hyperpolarizing shift in the voltage dependence of activation when expressed in tsA-201 cells, as well as a slowing of the fast component of inactivation [502]. Thus, these S6 mutations result in strong GOF effects on $Ca_V2.3$ channel activity. Finally, the $Ca_V2.3$ IIS4-S5 mutation I603L resulted in a large increase in current density when expressed in tsA-201 cells, to such an extent that it made precise biophysical characterization of the functional impact of the mutation difficult [502]. These results indicate that GOF missense mutations of $Ca_V2.3$, especially those within the S6 regions, can significantly alter channel function and contribute to disease states.

13.8 $Ca_V3.1$

The $Ca_V3.1$ channel is composed of the pore-forming α_{1G} subunit and is encoded by *CACNA1G*. Genetic knockout studies have revealed important physiological roles of the channel. Abolishing $Ca_V3.1$ in knockout mice resulted in bradycardia, delays in atrioventricular conduction, slowed *in vivo* heart rate, prolonged sinoatrial node

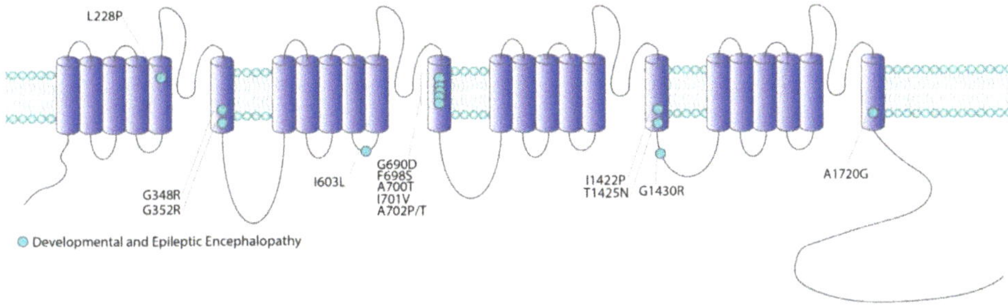

Figure 13.7. The location of mutations in *CACNA1E* ($Ca_V2.3$), with the associated phenotype indicated by the corresponding color.

recovery time, and slowed pacemaker activity [504]. Moreover, $Ca_V3.1$ knockout mice have revealed a role of the channel in the development of trigeminal neuropathic pain (TNP). In response to a TNP protocol, the mice showed diminished mechanical hypersensitivity and altered activity in the primary somatosensory cortex as compared to wildtype [505]. In addition, mice that lacked $Ca_V3.1$ showed attenuated spontaneous pain responses and thermal hyperalgesia, as well as an increased paw withdrawal threshold in response to mechanical stimulation, showcasing the role of $Ca_V3.1$ in pain perception [506]. In addition, knockout mice display increased nitric oxide levels in senescence, demonstrating a potential role for $Ca_V3.1$ in the pathology of aging [507]. Finally, there is evidence for a role of $Ca_V3.1$ in absence seizures. Mice deficient in $Ca_V3.1$ lack burst firing in thalamocortical relay neurons and were resistant to absence seizures [508]. Thus, $Ca_V3.1$ is critical for the normal function of the heart and brain.

13.8.1 Cerebellar ataxia

Cerebellar ataxias are a heterogeneous set of disorders that affect the cerebellum and typically result in impaired coordination. Though they can be the consequence of an acquired condition or neurodegenerative disorder, they can also have a genetic component including mutations in *CACNA1G* (figure 13.8 [509–519]). R1715H has been identified as one such $Ca_V3.1$ mutation which is associated with autosomal-dominant cerebellar ataxia [513–516]. This particular mutation resides in the IVS4 voltage-sensing region of the channel (figure 13.8) and results in a depolarizing shift in the voltage dependence of channel activation, as well as a higher slope factor in the inactivation curve, thus giving rise to an overall LOF effect [513]. Computer modeling in the same study also predicts that the LOF effects of $Ca_V3.1$ channels harboring R1715H would result in decreased neuronal excitability in deep cerebellar nuclei neurons. In addition, iPSC-derived Purkinje neurons generated from a patient with R1715H showed no differences in cellular morphology [514]. It has also been suggested that mutations such as R1715H that reside within voltage sensing domains may result in new ion conducting pathways, similar to the ω-currents described in $Ca_V1.1$ channelopathies [47]. Thus, some $Ca_V3.1$ mutations may alter the function of the channel without altering Ca^{2+} current through the α pore [393].

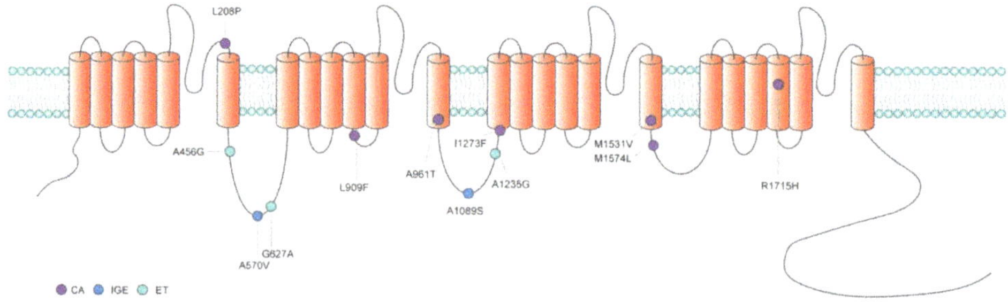

Figure 13.8. The location of mutations in *CACNA1G* ($Ca_V3.1$), with the associated phenotype indicated by the corresponding color. CA, cerebellar ataxa; IGE, idiopathic generalized epilepsy; ET, essential tremor.

The cerebellar ataxia mutation L208P causes axial hypotonia, developmental delay, and severe cognitive impairment [510]. Functionally, $Ca_V3.1$ channels harboring this mutation exhibit a hyperpolarizing shift in the voltage dependence of activation and inactivation, as well as slower inactivation and deactivation kinetics, comprising a mix of both GOF and LOF effects. The $Ca_V3.1$ L909F mutation causes a similar but somewhat less severe phenotype. When heterologously expressed in HEK293 cells, the mutation result in a hyperpolarizing shift in the voltage dependence of activation and slower deactivation. Thus, this mutation has a more purely GOF effect [510]. Mutations A961T and M1531V, identified in a genetic screen of patients with childhood-onset cerebellar atrophy [511], also cause a mixed GOF/LOF effect. Channels harboring these mutations display significantly impaired channel inactivation with slower kinetics and a hyperpolarizing shift in VDI when expressed in HEK293 cells. In a computational model of a cerebellar neuron, these two mutations increased neuronal firing and resulted in a persistent background Ca^{2+} current compared to wildtype neurons. This is consistent with the increased window current seen in biophysical studies. Finally, whole-exome sequencing of spinocerebellar ataxia type 42 patients revealed that three individuals within a single family harbor an M1574L mutation within their $Ca_V3.1$ channels [517]. However, no functional data on the effects of this mutation have yet been described. Taken together, these mutations associated with cerebellar ataxia mostly result in either mixed or GOF effects on channel gating with a single exception (R1715H).

13.8.2 Idiopathic generalized epilepsy (IGE)

IGE is inclusive of a wide range of syndromes which include childhood absence, juvenile absence, juvenile myoclonic, myoclonic astatic, and febrile seizures, as well as temporal lobe epilepsy [37]. Several lines of evidence indicate that IGE may be linked to mutations in $Ca_V3.1$. As previously mentioned, thalamic $Ca_V3.1$ currents are increased in absence epilepsy rodent models [508]. Moreover, when $Ca_V3.1$ channels are overexpressed in rodents, the animals develop epilepsy [520]. Finally, if $Ca_V3.1$ channels are inhibited pharmacologically, it leads to the suppression of both thalamic burst firing and seizures [521–523]. Taken together, this data indicates that $Ca_V3.1$ may contribute to the development of IGE. Consistent with this, a number of mutations in $Ca_V3.1$ have been identified in patients with IGE (figure 13.8) [512, 518]. One study identified five mutations in $Ca_V3.1$ among a cohort of 123 IGE patients [518]. Of those, A570V and A1089S, were analyzed for biophysical effects, however, no significant alterations in biophysical properties were identified [512]. In addition, a I1273F mutation was identified among individuals with early onset epileptic encephalopathy. Like A570V and A1089S, I1273F also displayed no change in electrophysiological properties compared to the wildtype channel [512]. This particular *de novo* mutation was identified in a patient suffering from both epilepsy and developmental delay. As such, further work is required to determine the pathogenicity of these mutations. Recent studies noted that elevated $Ca_V3.1$ expression correlated with seizure frequency in a *SCN2A* epilepsy mouse model and a *SCN1A* Dravet syndrome mouse model [524, 525], indicating that *CACNA1G* may play a role in the pathogenesis of other channelopathies

13.8.3 Essential tremor (ET)

ET is a highly prevalent movement disorder characterized by rhythmic shaking, particularly of the arms, hands, and eventually the cranial region [526]. $Ca_V3.1$ has been implicated as having a role in neuronal autorhythmicity and ET, and recently mutations in *CACNA1G* have been associated with the disorder (figure 13.8). Two of these mutations, A456G and G627A are located within the I–II loop, however biophysical studies did not identify any changes in channel properties [519]. As the I–II loop is associated with channel trafficking [527], further examination of the effect of these mutations on this property may be warranted. A third mutation, A1235G, also showed no significant changes in the biophysical properties of the channel, though there was a trend towards a more hyperpolarized voltage dependence of activation and a depolarizing shift in deactivation [519]. Thus, additional studies are necessary to correlate altered function of $Ca_V3.1$ with ET.

13.9 $Ca_V3.2$

The $Ca_V3.2$ channel is composed of the pore-forming subunit α_{1H} and is encoded by *CACNA1H*. The channel is expressed in smooth and cardiac muscle, the kidneys, liver, brain, pancreas, placenta, lung, skeletal muscle, and the adrenal cortex [36]. $Ca_V3.2$ knockout mice display constitutively constricted coronary arterioles, focal myocardial fibrosis, and reduced relaxation responses to acetylcholine and nitroprusside. These results indicate that $Ca_V3.2$ has an important role in the relaxation of coronary arteries [528, 529]. The deletion of $Ca_V3.2$ can also affect vasodilation [530], conduction of vascular responses in cremaster arterioles [531], relaxation of cerebral vascular smooth muscles [532], and contractility during oxidative stress [533]. Moreover, $Ca_V3.2$ impacts renal hemodynamics as the $Ca_V3.2$ knockout mice display an increase in the contractile response of the efferent arteriole, which in turn increases the glomerular filtration rate [529, 534]. In addition to this effect, it has been elucidated that $Ca_V3.2$ is important for cardiac hypertrophic growth, as $Ca_V3.2$ knockout mice show suppressed hypertrophy in response to pressure overload and angiotensin II [535]. $Ca_V3.2$ channels have been found to be associated with the control of neuronal firing patterns through the regulation of HCN channels and K_V7 [36]. Furthermore, mice that lack $Ca_V3.2$ show a reduction in sensitivity to painful stimuli, increased anxiety, and impaired memory [536, 537]. Thus $Ca_V3.2$ channels are a critical component of normal vascular, kidney, cardiac, and neuronal function.

13.9.1 Idiopathic generalized epilepsy (IGE)

In addition to the $Ca_V3.1$ mutations underlying IGE, mutations within $Ca_V3.2$ may also confer a similar phenotype. The extent of the contribution of $Ca_V3.2$ mutations to the pathogenesis of IGE remains a topic of ongoing investigation. It is predicted that many of these mutations are likely only a small contributing factor acting in combination with other genetic and environmental components, and may not

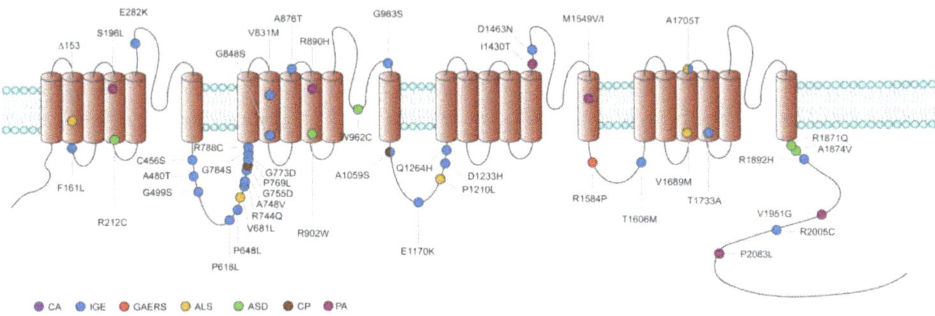

Figure 13.9. The location of mutations in ($Ca_V3.2$), with the associated phenotype indicated by the corresponding color. CA, cerebellar ataxa; IGE, idiopathic generalized epilepsy; GAERS, genetic absence epilepsy rat from Strasburg; ALS, amyotrophic lateral sclerosis; ASD, autism spectrum disorder; CP, chronic pain; PA, primary aldosteronism.

represent a monogenic cause of epilepsy [37, 538]. Despite this, it is known that $Ca_V3.2$ currents are more pronounced in some rodent models of epilepsy [539–541], and multiple mutations within $Ca_V3.2$ have been associated with IGE (figure 13.9 [189, 542–557]). In particular, the genetic absence epilepsy rat from Strasburg (GAERS) showcases that the mutation R1584P in $Ca_V3.2$ segregates with the development of seizures. Expression of $Ca_V3.2$ channels harboring R1584P display faster recovery from inactivation and increased channel expression at the membrane [542], demonstrating an overall GOF effect. Likewise, F161L, E282K, and C456S mutations result in a hyperpolarizing shift in the voltage dependence of activation, also contributing to a GOF effect [543]. Further study of the C456S mutation revealed that expression of the mutant channel significantly increases the excitability of cultured neurons as seen by an elevated firing rate and a reduction in the threshold for rebound firing [544]. Moreover, the mutations G499S, P648L, R788C, and V831M were all found to inactivate more slowly than wildtype. In addition, some mutations (G773D, V831M, G848S and R788C) showed slowed deactivation kinetics [543]. In computer models of thalamic neurons and circuits, the mutations described above (with the exception of V831M and E282K) resulted in overall increases in neuronal firing properties including spike initiation threshold and number of spikes per burst. Finally, another mutation, D1463N, showed GOF effects on channel activity in the form of faster activation kinetics compared to wildtype [543]. However, in another study, no effect on biophysical properties of the channel were seen in response to either D1463N or C456S [545], adding uncertainty to the impact of these mutations.

Not all $Ca_V3.2$ IGE mutations display a GOF effect. P618L is an example of a mutation showing mixed effects as it causes an accelerated time course of activation at more negative potentials and faster inactivation at more positive potentials [546]. Likewise, the mutation G773D displays a shift in the voltage dependence of activation to more depolarized potentials compared to wildtype, but also inactivates more slowly [543]. In some rare instances, some mutations had LOF effects on channel activity. The G983S mutation for example, which is located in the pore loop of domain II (figure 13.9) results in a slowing of the recovery from inactivation and a hyperpolarization of the voltage dependence of inactivation. Another mutation with a LOF effect is A1059S, which results in a hyperpolarized voltage dependence of activation, and smaller current densities [547]. Further, G755D displays faster inactivation at depolarizations of around 20mV, similar to the P618L mutation [546].

In general, when human $Ca_V3.2$ mutations associated with IGE are expressed in a heterologous expression system and assessed for changes in biophysical properties, a majority of these mutations either result in mild GOF effects on channel gating or no effect at all. In fact, nearly a third of the known $Ca_V3.2$ mutations associated with IGE (7 mutations out of 27 in total at the time of writing) show no effect on channel gating. It is possible that these mutations may affect channel splicing, which has been documented in one study [558]. An alternative explanation is that the surface expression of the channel may be altered, as many of these mutations exist in the I–II loop, which is known to impact trafficking [37]. In a recent study, Eckle *et al* found that alterations in the I–II loop can increase the surface expression of $Ca_V3.2$ channels in neurons, though the distribution of the channels in the dendrites was unchanged [544].

13.9.2 Primary aldosteronism (PA)

Aldosterone production is triggered by Ca^{2+} release and is increased in PA, the leading cause of secondary hypertension. PA is primarily the result of bilateral adrenal hyperplasia and unilateral adrenal adenoma. $Ca_V3.2$ channels have previously been implicated in aldosterone secretion from the adrenal zona glomerulosa [551]. Through whole-exome sequencing of patients with PA, several $Ca_V3.2$ mutations that segregate with the disease have been identified (figure 13.9). A number of these mutations cause altered biophysical properties of $Ca_V3.2$ channels when expressed in heterologous systems, and are implicated in PA pathogenesis [549–552]. Multiple lines of evidence suggest that the majority of PA mutations confer a GOF effect. One study identified four different germline mutations using whole-exome sequencing and examined the biophysical properties of the mutant channels in tsA-201 cells [549]. The M1549I mutation caused a significant hyperpolarizing shift in the voltage-dependence of activation compared to the wildtype channel. The other three identified mutations in the study did not show any change in the voltage dependence of activation. In addition, M1549I and S196L in the same study displayed slower activation and inactivation kinetics. It was observed that M1549I also caused a larger window current at lower voltages. The other three mutants (S196L, V1951G, and P2083L) showed significantly faster recovery from inactivation. It has been suggested that this alteration to inactivation could favor larger Ca^{2+} entry in response to repeated electrical activity. These three mutants also showed increased voltage-dependent facilitation compared to wildtype. This indicates that the activity of these channels could be enhanced during repetitive action potentials [549]. Moreover, M1549V was identified in another study using whole-exome sequencing and was shown to have a GOF effect, such that channel inactivation was significantly attenuated at hyperpolarized potentials [550]. These GOF effects on channel gating could produce an increase in intracellular Ca^{2+}, a signal for aldosterone production, which may then contribute to PA pathogenesis.

In contrast to the GOF effects described above, the mutation R890H, which was identified through whole-exome sequencing [552], results in a pronounced decrease in channel activity. Though the activation and inactivation mechanics were identical compared to wildtype, the channel displayed a marked decrease in whole-cell current in heterologous cells. In the patient, this mutation correlated with increased aldosterone production and cellular proliferation, similar to the previously described mutations associated with primary aldosteronism [552]. Given that this mutation was identified within a single patient and is the only described LOF mutation for PA, direct causation remains to be confirmed.

13.9.3 Chronic pain

$Ca_V3.2$ channels are essential for the processing of peripheral nociception and can contribute to the development and maintenance of chronic pain [37]. Altered expression of $Ca_V3.2$ channels has been identified in a number of chronic pain conditions including diabetic neuropathy [559], nerve injury [560], and irritable bowel syndrome [561]. The involvement of $Ca_V3.2$ in these conditions is mainly due to aberrant post-translational modifications of the channel. Recent work has raised the possibility that mutations within the channel itself may play a role in chronic pain conditions (figure 13.9). Two missense mutations within the $Ca_V3.2$ channel sequence, P769L and A1059S, have been identified in a patient with chronic pain [555]. In tsA-201 cells the mutation resulted in decreased current densities and depolarizing shifts in channel activation. However, when the mutations were expressed in neuronal CAD cells, the current densities were larger compared to wildtype channels and there was no difference in the voltage dependence of activation [555]. It was suggested that this discrepancy could be due to differences in experimental conditions.

13.9.4 Amyotrophic lateral sclerosis (ALS)

ALS is a neuromuscular disease that is characterized by the gradual loss of cortical, brainstem, and spinal motor neurons, which results in muscle weakness, paralysis, and death. Whole-exome sequencing has uncovered two heterozygous recessive $Ca_V3.2$ mutations (figures 13.9, V1689M and A1705T) identified in patients with ALS [562].

Functional analysis of these mutations showed that they produce a mild LOF effect on channels [556]. The mutations caused a significant reduction in $Ca_V3.2$ currents, shifted the voltage dependence of activation to more positive potentials, and accelerated activation kinetics when compared to wildtype channels. In addition, the voltage dependence of inactivation was shifted to more negative potentials, though there was no difference in the inactivation kinetics compared to wildtype. Another mutation recently identified in patients with ALS is P1210L (figure 13.9), which results in reduced current density and diminished expression of $Ca_V3.2$ at the cell surface [557]. However, this mutation has a high occurrence in the general population, leaving its contribution to ALS pathogenicity uncertain [557]. The study also identified a ΔI153 in-frame deletion (figure 13.9) which results in a complete loss of channel function and has a dominant negative effect on the wildtype channel when co-expressed. Thus, most of the known mutations in $Ca_V3.2$ associated with ALS appear to result in LOF effects on channel function.

13.9.5 Autism spectrum disorder (ASD)

ASD is a neurodevelopmental disorder characterized by repetitive stereotypic behavior, impaired communication, and deficits in social interaction [563] and $Ca_V3.2$ mutations have been discovered in patients with the disorder (figure 13.9). Two of these mutations are in the VSD (R212C and R902W) and one is located in the pore-loop (W962C). The two voltage sensor mutations result in decreased current density and a depolarizing shift in the voltage dependence of activation compared to wildtype, as detected through whole-cell voltage clamp in HEK293T cells. The mutation in the pore-loop (W962C) resulted in a significant reduction in current. Two other mutations identified in the same study (R1871Q and A1874V) are located in the C-terminus of the channel (figure 13.9) and resulted in slower activation kinetics, and a depolarizing shift in the voltage dependence of activation. Thus, the known ASD-associated $Ca_V3.2$ mutations tend to result in LOF effects on channel activity.

13.10 $Ca_V3.3$

The *CACNA1I* gene encodes the pore-forming α_{1I} subunit of $Ca_V3.3$, and is present in a number of brain regions including the thalamus, olfactory bulb, striatum, cerebral cortex, hippocampus, reticular nucleus, lateral habenula, and the cerebellum [564]. *CACNA1I* has also been detected at low levels in the adrenal glands [36, 565]. Through transgenic knockout mice, $Ca_V3.3$ channels have been found to be essential for proper sleep rhythmogenesis [357, 566, 567]. These mice demonstrate alterations in neuronal excitability, resulting in abnormal oscillatory wave activity generated from thalamocortical and nucleus reticularis thalami (nRT) cells. Furthermore, $Ca_V3.3$ deficient mice are more susceptible to altered cortical activity characteristic of absence seizures [568]. In addition $Ca_V3.3$ contributes to smooth muscle cell contraction in human cerebral arteries [569]. Finally, mutations within *CACNA1I* have been predicted to be a risk factor for ASD, and several other neurological disorders [570, 571]. Thus, $Ca_V3.3$ has an important role in a number of systems including the nervous and vascular systems.

13.10.1 Schizophrenia

Schizophrenia is a neuropsychiatric disorder characterized by altered mental states, including hallucinations, delusions, and disorganized thinking. Genetics are a considerable risk factor, and both common and rare mutations are thought

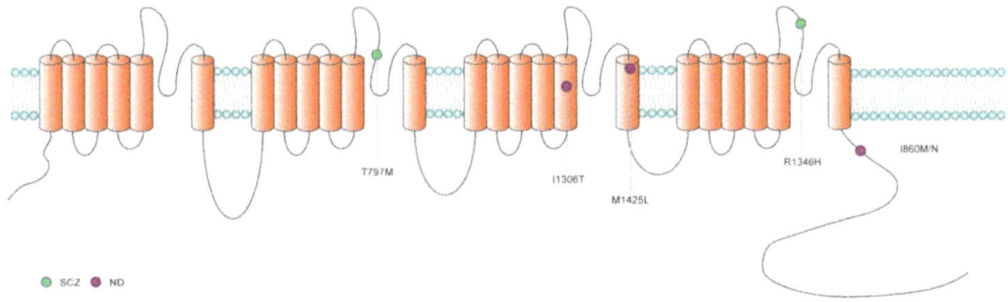

Figure 13.10. The location of mutations in *CACNA1I* ($Ca_V3.3$), with the associated phenotype indicated by the corresponding color. SCZ, schizophrenia; ND, neurodevelopmental disorders.

to contribute to the disease pathology [572]. There are only a handful of known *CACNA1I* mutations associated with schizophrenia, and only two of these have been functionally characterized (figure 13.10 [573–575]). The $Ca_V3.3$ mutations T797M and R1346H, identified in schizophrenia patients, are both located in the pore-loops of domains II and III (figure 13.10) and neither alter channel kinetics or change the voltage dependence of activation or inactivation [573, 574]. However, R1346H results in a decrease in current density of $Ca_V3.3$ when compared to wildtype channels, and it has been suggested that this is due to changes in glycosylation. Recently a R1346H knockin mouse model was generated and had altered excitability in the thalamic reticular nucleus, where *CACNA1I* is highly expressed [357]. The mutation also resulted in sleep deficits. Finally, a computer model of thalamic reticular nucleus neurons demonstrated that the effects of R1346H on the channel are sufficient to alter neuronal excitability [576].

13.10.2 Neurodevelopmental disorders

Mutations in $Ca_V3.3$ have been associated with various neurodevelopmental disorders including developmental delay and cognitive impairment. A $Ca_V3.3$ mutation, I860M, has recently been identified in a family of patients who exhibit varying levels of cognitive dysfunction [575] and is located near the cytoplasmic end of the IIS6 region (figure 13.10). A second mutation (I860N) at the same site results in severe developmental delay and seizures. Both mutations cause a significant slowing of activation and inactivation kinetics, though this effect is less severe in I860M. In addition to these effects, the voltage dependence of activation was shifted towards hyperpolarizing potentials for both I860M/N. Finally, the same study identified two additional mutations in the IIIS5 and IIIS6 regions, I1306T and M1425L, respectively (figure 13.10). Each mutation occurred within a single patient who displayed developmental delay, hypotonia, and epilepsy. These mutations each cause a slowing of activation and inactivation kinetics, as well as a hyperpolarizing shift in the voltage dependence of activation. Thus, the authors concluded that the four mutations have a mostly GOF effect on channel activity.

13.11 Summary

VGCCs are critical for numerous physiological processes, and are required for normal function of heart, brain, smooth muscle, skeletal muscle and endocrine systems. As a result, mutations within any of these channels can result in a myriad of clinical pathologies. Channelopathic mutations can impact the channel in a variety of ways including altered gating properties, changes in trafficking and disrupted downstream signaling. As the number and variety of known VGCC mutations continues to grow, so will our knowledge of the pathogenic mechanisms underlying these channelopathies.

References

[1] Catterall W A 2011 Voltage-gated calcium channels. *Cold Spring Harb. Perspect. Biol.* **3** a003947

[2] Hille B 2001 Ion channels of excitable membranes. 3rd edn (Sunderland, MA: Sinauer)

[3] Zamponi G W, Striessnig J, Koschak A and Dolphin A C 2015 The physiology, pathology, and pharmacology of voltage-gated calcium channels and their future therapeutic potential *Pharmacol. Rev.* **67** 821–70

[4] Ma H, Cohen S, Li B and Tsien R W 2012 Exploring the dominant role of Ca_V1 channels in signalling to the nucleus. *Biosci. Rep.* **33** 97–101

[5] Harvey R D and Hell J W 2013 $Ca_V1.2$ signaling complexes in the heart. *J. Mol. Cell. Cardiol.* **58** 143–52

[6] Felix R 2006 Calcium channelopathies. *Neuromol. Med.* **8** 307–18

[7] Liao P and Soong T W 2010 $Ca_V1.2$ channelopathies: from arrhythmias to autism, bipolar disorder, and immunodeficiency. *Pflugers Arch.* **460** 353–9

[8] Catterall W A 1995 Structure and function of voltage-gated ion channels *Annu. Rev. Biochem.* **64** 493–531

[9] Pinto A, Gillard S and Moss F *et al* 1998 Human autoantibodies specific for the α1A calcium channel subunit reduce both P-type and Q-type calcium currents in cerebellar neurons *Proc. Natl. Acad. Sci. USA* **95** 8328–33

[10] Stephens R F, Guan W, Zhorov B S and Spafford J D 2015 Selectivity filters and cysteine-rich extracellular loops in voltage-gated sodium, calcium, and NALCN channels *Front. Physiol.* **6** 153

[11] Sather W A, Yang J and Tsien R W 1994 Structural basis of ion channel permeation and selectivity *Curr. Opin. Neurobiol.* **4** 313–23

[12] Flucher B E 2016 Specific contributions of the four voltage-sensing domains in L-type calcium channels to gating and modulation *J. Gen. Physiol.* **148** 91–5

[13] Xie C, Zhen X G and Yang J 2005 Localization of the activation gate of a voltage-gated Ca^{2+} channel *J. Gen. Physiol.* **126** 205–12

[14] Singh A, Hamedinger D and Hoda J C *et al* 2006 C-terminal modulator controls Ca^{2+}-dependent gating of $Ca_V1.4$ L-type Ca^{2+} channels. *Nat. Neurosci.* **9** 1108–16

[15] Lee A, Zhou H, Scheuer T and Catterall W A 2003 Molecular determinants of Ca^{2+}/calmodulin-dependent regulation of $Ca_V2.1$ channels *Proc. Natl. Acad. Sci. USA* **100** 16059–64

[16] Van Petegem F, Chatelain F C and Minor D L Jr. 2005 Insights into voltage-gated calcium channel regulation from the structure of the $Ca_V1.2$ IQ domain-Ca^{2+}/calmodulin complex *Nat. Struct. Mol. Biol.* **12** 1108–15
[17] Catterall W A, Perez-Reyes E, Snutch T P and Striessnig J 2005 International union of pharmacology. XLVIII. nomenclature and structure-function relationships of voltage-gated calcium channels *Pharmacol. Rev.* **57** 411–25
[18] Catterall W A 1998 Structure and function of neuronal Ca^{2+} channels and their role in neurotransmitter release *Cell Calcium* **24** 307–23
[19] Dolphin A C 2016 Voltage-gated calcium channels and their auxiliary subunits: physiology and pathophysiology and pharmacology *J. Physiol.* **594** 5369–90
[20] Buraei Z and Yang J 2010 The β subunit of voltage-gated Ca^{2+} channels *Physiol. Rev.* **90** 1461–506
[21] Dolphin A C 2018 Voltage-gated calcium channel α2δ subunits: an assessment of proposed novel roles *F1000Res* **7** 1–14
[22] Menard C, Charnet P, Rousset M, Vignes M and Cens T 2020 $Ca_V2.1$ C-terminal fragments produced in *Xenopus laevis* oocytes do not modify the channel expression and functional properties *Eur. J Neurosci.* **51** 1900–13
[23] Zhang J F, Ellinor P T, Aldrich R W and Tsien R W 1996 Multiple structural elements in voltage-dependent Ca^{2+} channels support their inhibition by G proteins *Neuron* **17** 991–1003
[24] Restituito S, Thompson R M and Eliet J *et al* 2000 The polyglutamine expansion in spinocerebellar ataxia type 6 causes a β subunit-specific enhanced activation of P/Q-type calcium channels in *Xenopus* oocytes *J. Neurosci.* **20** 6394–403
[25] Giunti P, Mantuano E, Frontali M and Veneziano L 2015 Molecular mechanism of spinocerebellar ataxia type 6: glutamine repeat disorder, channelopathy and transcriptional dysregulation. the multifaceted aspects of a single mutation *Front. Cell Neurosci.* **9** 36
[26] Mullner C, Broos L A, van den Maagdenberg A M and Striessnig J 2004 Familial hemiplegic migraine type 1 mutations K1336E, W1684R, and V1696I alter $Ca_V2.1$ Ca^{2+} channel gating: evidence for β-subunit isoform-specific effects. *J. Biol. Chem.* **279** 51844–50
[27] Zuhlke R D and Reuter H 1998 Ca^{2+}-sensitive inactivation of L-type Ca^{2+} channels depends on multiple cytoplasmic amino acid sequences of the α1c subunit *Proc. Natl Acad. Sci. USA* **95** 3287–94
[28] Peterson B Z, DeMaria C D, Adelman J P and Yue D T 1999 Calmodulin is the Ca^{2+} sensor for Ca^{2+}-dependent inactivation of L-type calcium channels *Neuron* **22** 549–58
[29] Ben-Johny M and Yue D T 2014 Calmodulin regulation (calmodulation) of voltage-gated calcium channels *J. Gen. Physiol.* **143** 679–92
[30] Perez-Reyes E, Cribbs L L and Daud A *et al* 1998 Molecular characterization of a neuronal low-voltage-activated T-type calcium channel *Nature* **391** 896–900
[31] Cribbs L L, Lee J H and Yang J *et al* 1998 Cloning and characterization of alpha1H from human heart, a member of the T-type Ca^{2+} channel gene family *Circ. Res.* **83** 103–9
[32] Lee J H, Daud A N and Cribbs L L *et al* 1999 Cloning and expression of a novel member of the low voltage-activated T-type calcium channel family *J. Neurosci.* **19** 1912–21
[33] Nowycky M C, Fox A P and Tsien R W 1985 Three types of neuronal calcium channel with different calcium agonist sensitivity *Nature* **316** 440–3

[34] Nilius B, Hess P, Lansman J B and Tsien R W 1986 A novel type of cardiac calcium channel in ventricular cells *Biomed. Biochim. Acta* **45** S167–70

[35] Nanou E and Catterall W A 2018 Calcium channels, synaptic plasticity, and neuropsychiatric disease *Neuron* **98** 466–81

[36] Perez-Reyes E 2003 Molecular physiology of low-voltage-activated t-type calcium channels *Physiol. Rev.* **83** 117–61

[37] Weiss N and Zamponi G W 2020 Genetic T-type calcium channelopathies *J. Med. Genet.* **57** 1–10

[38] Murbartian J, Arias J M, Lee J H, Gomora J C and Perez-Reyes E 2002 Alternative splicing of the rat $Ca_V3.3$ T-type calcium channel gene produces variants with distinct functional properties(1) *FEBS Lett.* **528** 272–8

[39] Murbartian J, Arias J M and Perez-Reyes E 2004 Functional impact of alternative splicing of human T-type $Ca_V3.3$ calcium channels *J. Neurophysiol.* **92** 3399–407

[40] Shcheglovitov A, Vitko I and Bidaud I *et al* 2008 Alternative splicing within the I–II loop controls surface expression of T-type $Ca_V3.1$ calcium channels *FEBS Lett.* **582** 3765–70

[41] Nie L, Zhu J and Gratton M A *et al* 2008 Molecular identity and functional properties of a novel T-type Ca^{2+} channel cloned from the sensory epithelia of the mouse inner ear *J. Neurophysiol.* **100** 2287–99

[42] Chemin J, Monteil A, Bourinet E, Nargeot J and Lory P 2001 Alternatively spliced α(1G) ($Ca_V3.1$) intracellular loops promote specific T-type Ca^{2+} channel gating properties *Biophys. J.* **80** 1238–50

[43] Latour I, Louw D F, Beedle A M, Hamid J, Sutherland G R and Zamponi G W 2004 Expression of T-type calcium channel splice variants in human glioma *Glia* **48** 112–9

[44] He M, Xu Z, Zhang Y and Hu C 2019 Splice-variant-specific effects of primary aldosteronism point mutations on human $Ca_V3.2$ calcium channels *Cell calcium* **84** 102104

[45] Tanabe T, Takeshima H and Mikami A *et al* 1987 Primary structure of the receptor for calcium channel blockers from skeletal muscle *Nature* **328** 313–8

[46] Bannister R A 2016 Bridging the myoplasmic gap II: more recent advances in skeletal muscle excitation-contraction coupling *J. Exp. Biol.* **219** 175–82

[47] Flucher B E 2020 Skeletal muscle $Ca_V1.1$ channelopathies *Pflugers Arch.* **472** 739–54

[48] Schneider M F and Chandler W K 1973 Voltage dependent charge movement of skeletal muscle: a possible step in excitation-contraction coupling *Nature* **242** 244–6

[49] Rios E and Brum G 1987 Involvement of dihydropyridine receptors in excitation-contraction coupling in skeletal muscle *Nature* **325** 717–20

[50] Tanabe T, Beam K G, Powell J A and Numa S 1988 Restoration of excitation-contraction coupling and slow calcium current in dysgenic muscle by dihydropyridine receptor complementary DNA *Nature* **336** 134–9

[51] Nakai J, Sekiguchi N, Rando T A, Allen P D and Beam K G 1998 Two regions of the ryanodine receptor involved in coupling with L-type Ca^{2+} channels. *J. Biol. Chem.* **273** 13403–6

[52] Avila G and Dirksen R T 2000 Functional impact of the ryanodine receptor on the skeletal muscle L-type Ca^{2+} channel *J. Gen. Physiol.* **115** 467–80

[53] Nakai J, Dirksen R T, Nguyen H T, Pessah I N, Beam K G and Allen P D 1996 Enhanced dihydropyridine receptor channel activity in the presence of ryanodine receptor *Nature* **380** 72–5

[54] Beam K G, Knudson C M and Powell J A 1986 A lethal mutation in mice eliminates the slow calcium current in skeletal muscle cells *Nature* **320** 168–70

[55] Elbaz A, Vale-Santos J and Jurkat-Rott K *et al* 1995 Hypokalemic periodic paralysis and the dihydropyridine receptor (*CACNL1A3*): genotype/phenotype correlations for two predominant mutations and evidence for the absence of a founder effect in 16 caucasian families *Am. J. Hum. Genet.* **56** 374–80

[56] Jurkat-Rott K, Lehmann-Horn F and Elbaz A *et al* 1994 A calcium channel mutation causing hypokalemic periodic paralysis *Hum. Mol. Genet.* **3** 1415–9

[57] Kil T H and Kim J B 2010 Severe respiratory phenotype caused by a *de novo* Arg528Gly mutation in the *CACNA1S* gene in a patient with hypokalemic periodic paralysis *Eur. J. Paediatr. Neurol.* **14** 278–81

[58] Wang Q, Liu M and Xu C *et al* 2005 Novel *CACNA1S* mutation causes autosomal dominant hypokalemic periodic paralysis in a Chinese family *J. Mol. Med. (Berl.)* **83** 203–8

[59] Yang B, Yang Y, Tu W, Shen Y and Dong Q 2014 A rare case of unilateral adrenal hyperplasia accompanied by hypokalaemic periodic paralysis caused by a novel dominant mutation in *CACNA1S*: features and prognosis after adrenalectomy *BMC Urol.* **14** 96

[60] Lapie P, Goudet C, Nargeot J, Fontaine B and Lory P 1996 Electrophysiological properties of the hypokalaemic periodic paralysis mutation (R528H) of the skeletal muscle alpha 1s subunit as expressed in mouse L cells *FEBS Lett.* **382** 244–8

[61] Chabrier S, Monnier N and Lunardi J 2008 Early onset of hypokalaemic periodic paralysis caused by a novel mutation of the *CACNA1S* gene *J. Med. Genet.* **45** 686–8

[62] Hanchard N A, Murdock D R and Magoulas P L *et al* 2013 Exploring the utility of whole-exome sequencing as a diagnostic tool in a child with atypical episodic muscle weakness *Clin. Genet.* **83** 457–61

[63] Hirano M, Kokunai Y and Nagai A *et al* 2011 A novel mutation in the calcium channel gene in a family with hypokalemic periodic paralysis *J. Neurol. Sci.* **309** 9–11

[64] Matthews E, Labrum R and Sweeney M G *et al* 2009 Voltage sensor charge loss accounts for most cases of hypokalemic periodic paralysis *Neurology* **72** 1544–7

[65] Kim J B, Lee K Y and Hur J K 2005 A Korean family of hypokalemic periodic paralysis with mutation in a voltage-gated calcium channel (R1239G) *J. Korean. Med. Sci.* **20** 162–5

[66] Ptacek L J, Tawil R and Griggs R C *et al* 1994 Dihydropyridine receptor mutations cause hypokalemic periodic paralysis *Cell* **77** 863–8

[67] Winczewska-Wiktor A, Steinborn B and Lehman-Horn F *et al* 2007 Myopathy as the first symptom of hypokalemic periodic paralysis–case report of a girl from a Polish family with *CACNA1S* (R1239G) mutation *Adv. Med. Sci.* **52** 155–7

[68] Dias da Silva M R, Cerutti J M and Tengan C H *et al* 2002 Mutations linked to familial hypokalaemic periodic paralysis in the calcium channel α1 subunit gene ($Ca_V1.1$) are not associated with thyrotoxic hypokalaemic periodic paralysis *Clin. Endocrinol. (Oxf.)* **56** 367–75

[69] Fuster C, Perrot J, Berthier C, Jacquemond V and Allard B 2017 Elevated resting H^+ current in the R1239H type 1 hypokalaemic periodic paralysis mutated Ca^{2+} channel *J. Physiol.* **595** 6417–28

[70] Houinato D, Laleye A and Adjien C *et al* 2007 Hypokalaemic periodic paralysis due to the *CACNA1S* R1239H mutation in a large African family *Neuromusc. Disord.* **17** 419–22

[71] Ke Q, Wu W P, Guo X H, Xu Q-g, Huang D-h, Mao Y-l and Huo C-n 2006 [R1239H mutation of *CACNA1S* gene in a Chinese family with hypokalaemic periodic paralysis] *Zhonghua Yi Xue Yi Chuan Xue Za Zhi* **23** 272–4

[72] Kusumi M, Kumada H, Adachi Y and Nakashima K 2001 Muscle weakness in a Japanese family of Arg1239His mutation hypokalemic periodic paralysis *Psychiatry Clin. Neurosci.* **55** 539–41

[73] Links T P and van der Hoeven J H 2000 Muscle fiber conduction velocity in arg1239his mutation in hypokalemic periodic paralysis *Muscle Nerve* **23** 296

[74] Ke T, Gomez C R, Mateus H E, Castano J A and Wang Q K 2009 Novel *CACNA1S* mutation causes autosomal dominant hypokalemic periodic paralysis in a South American family *J. Hum. Genet.* **54** 660–4

[75] Kurokawa M, Torio M and Ohkubo K *et al* 2020 The expanding phenotype of hypokalemic periodic paralysis in a Japanese family with p.Val876Glu mutation in *CACNA1S* *Mol. Genet. Genomic Med.* **8** e1175

[76] Yang H, Zhang H and Xing X 2015 V876E mutation in *CACNA1S* gene associated with severe hypokalemic periodic paralysis in a Chinese woman *J. Formos. Med. Assoc.* **114** 377–8

[77] Fuster C, Perrot J, Berthier C, Jacquemond V, Charnet P and Allard B 2017 Na leak with gating pore properties in hypokalemic periodic paralysis V876E mutant muscle Ca channel *J. Gen. Physiol.* **149** 1139–48

[78] Li F F, Li Q Q and Tan Z X *et al* 2012 A novel mutation in *CACNA1S* gene associated with hypokalemic periodic paralysis which has a gender difference in the penetrance *J. Mol. Neurosci.* **46** 378–83

[79] Fan C, Lehmann-Horn F and Weber M A *et al* 2013 Transient compartment-like syndrome and normokalaemic periodic paralysis due to a $Ca_V1.1$ mutation. *Brain* **136** 3775–86

[80] Edizadeh M, Vazehan R and Javadi F *et al* 2017 De novo mutation in *CACNA1S* gene in a 20-year-old man diagnosed with metabolic myopathy *Arch. Iran. Med.* **20** 617–20

[81] Toppin P J, Chandy T T, Ghanekar A, Kraeva N, Beattie W S and Riazi S 2010 A report of fulminant malignant hyperthermia in a patient with a novel mutation of the *CACNA1S* gene *Can. J. Anaesth.* **57** 689–93

[82] Weiss R G, O'Connell K M, Flucher B E, Allen P D, Grabner M and Dirksen R T 2004 Functional analysis of the R1086H malignant hyperthermia mutation in the DHPR reveals an unexpected influence of the III–IV loop on skeletal muscle EC coupling *Am. J. Physiol. Cell Physiol.* **287** C1094–102

[83] Jurkat-Rott K, McCarthy T and Lehmann-Horn F 2000 Genetics and pathogenesis of malignant hyperthermia *Muscle Nerve* **23** 4–17

[84] Monnier N, Procaccio V, Stieglitz P and Lunardi J 1997 Malignant-hyperthermia susceptibility is associated with a mutation of the α1-subunit of the human dihydropyridine-sensitive L-type voltage-dependent calcium-channel receptor in skeletal muscle *Am. J. Hum. Genet.* **60** 1316–25

[85] Miller D M, Daly C and Aboelsaod E M *et al* 2018 Genetic epidemiology of malignant hyperthermia in the UK *Br. J. Anaesth.* **121** 944–52

[86] Pirone A, Schredelseker J and Tuluc P *et al* 2010 Identification and functional characterization of malignant hyperthermia mutation T1354S in the outer pore of the $Ca_V\alpha1S$-subunit *Am. J. Physiol. Cell Physiol.* **299** C1345–54

[87] Eltit J M, Bannister R A and Moua O *et al* 2012 Malignant hyperthermia susceptibility arising from altered resting coupling between the skeletal muscle L-type Ca^{2+} channel and the type 1 ryanodine receptor. *Proc. Natl Acad. Sci. USA* **109** 7923–8

[88] Schartner V, Romero N B and Donkervoort S *et al* 2017 Dihydropyridine receptor (DHPR, *CACNA1S*) congenital myopathy *Acta Neuropathol.* **133** 517–33

[89] Hunter J M, Ahearn M E and Balak C D *et al* 2015 Novel pathogenic variants and genes for myopathies identified by whole exome sequencing *Mol. Genet. Genomic Med.* **3** 283–301

[90] Dayal A, Schrotter K, Pan Y, Fohr K, Melzer W and Grabner M 2017 The Ca^{2+} influx through the mammalian skeletal muscle dihydropyridine receptor is irrelevant for muscle performance *Nat. Commun.* **8** 475

[91] Lee C S, Dagnino-Acosta A and Yarotskyy V *et al* 2015 Ca^{2+} permeation and/or binding to $Ca_V1.1$ fine-tunes skeletal muscle Ca^{2+} signaling to sustain muscle function *Skelet. Muscle* **5** 4

[92] Georgiou D K, Dagnino-Acosta A and Lee C S *et al* 2015 Ca^{2+} Binding/permeation via calcium channel, $Ca_V1.1$, regulates the intracellular distribution of the fatty acid transport protein, CD36, and fatty acid metabolism *J. Biol. Chem.* **290** 23751–65

[93] Benedetti B, Tuluc P, Mastrolia V, Dlaska C and Flucher B E 2015 Physiological and pharmacological modulation of the embryonic skeletal muscle calcium channel splice variant $Ca_V1.1e$ *Biophys. J.* **108** 1072–80

[94] Sultana N, Dienes B and Benedetti A *et al* 2016 Restricting calcium currents is required for correct fiber type specification in skeletal muscle *Development* **143** 1547–59

[95] Alhasan K A, Abdallah M S, Kari J A and Bashiri F A 2019 Hypokalemic periodic paralysis due to *CACNA1S* gene mutation *Neurosciences (Riyadh)* **24** 225–30

[96] Striessnig J, Bolz H J and Koschak A 2010 Channelopathies in $Ca_V1.1$, $Ca_V1.3$, and $Ca_V1.4$ voltage-gated L-type Ca^{2+} channels *Pflugers Arch.* **460** 361–74

[97] Morrill J A, Brown R H Jr. and Cannon S C 1998 Gating of the L-type Ca channel in human skeletal myotubes: an activation defect caused by the hypokalemic periodic paralysis mutation R528H *J. Neurosci.* **18** 10320–34

[98] Chanda B and Bezanilla F 2008 A common pathway for charge transport through voltage-sensing domains *Neuron* **57** 345–51

[99] Cannon S C 2010 Voltage-sensor mutations in channelopathies of skeletal muscle *J. Physiol.* **588** 1887–95

[100] Struyk A F and Cannon S C 2007 A Na^+ channel mutation linked to hypokalemic periodic paralysis exposes a proton-selective gating pore *J. Gen. Physiol.* **130** 11–20

[101] Sokolov S, Scheuer T and Catterall W A 2007 Gating pore current in an inherited ion channelopathy *Nature* **446** 76–8

[102] Jurkat-Rott K, Weber M A and Fauler M *et al* 2009 K^+-dependent paradoxical membrane depolarization and Na^+ overload, major and reversible contributors to weakness by ion channel leaks *Proc. Natl Acad. Sci. USA* **106** 4036–41

[103] Wu F, Mi W and Hernandez-Ochoa E O *et al* 2012 A calcium channel mutant mouse model of hypokalemic periodic paralysis *J. Clin. Invest.* **122** 4580–91

[104] Wu F, Quinonez M, DiFranco M and Cannon S C 2018 Stac3 enhances expression of human $Ca_V1.1$ in *Xenopus* oocytes and reveals gating pore currents in HypoPP mutant channels *J. Gen. Physiol.* **150** 475–89

[105] Kubota T, Wu F and Vicart S *et al* 2020 Hypokalaemic periodic paralysis with a charge-retaining substitution in the voltage sensor *Brain Commun.* **2** fcaa103

[106] Tang Z Z, Yarotskyy V and Wei L *et al* 2012 Muscle weakness in myotonic dystrophy associated with misregulated splicing and altered gating of $Ca_V1.1$ calcium channel *Hum. Mol. Genet.* **21** 1312–24
[107] Gupta P K, Bilmen J G and Hopkins P M 2021 Anaesthetic management of a known or suspected malignant hyperthermia susceptible patient *BJA Educ.* **21** 218–24
[108] Treves S, Anderson A A and Ducreux S *et al* 2005 Ryanodine receptor 1 mutations, dysregulation of calcium homeostasis and neuromuscular disorders *Neuromusc. Disord.* **15** 577–87
[109] Rios E, Figueroa L, Manno C, Kraeva N and Riazi S 2015 The couplonopathies: a comparative approach to a class of diseases of skeletal and cardiac muscle *J. Gen. Physiol.* **145** 459–74
[110] Bannister R A and Beam K G 2013 Impaired gating of an L-Type Ca^{2+} channel carrying a mutation linked to malignant hyperthermia *Biophys. J.* **104** 1917–22
[111] Kugler G, Weiss R G, Flucher B E and Grabner M 2004 Structural requirements of the dihydropyridine receptor α1S II–III loop for skeletal-type excitation-contraction coupling *J. Biol. Chem.* **279** 4721–8
[112] Specht D, Wu S B and Turner P *et al* 2009 Effects of presynaptic mutations on a postsynaptic Cacna1s calcium channel colocalized with mGluR6 at mouse photoreceptor ribbon synapses *Invest. Ophthalmol. Vis. Sci.* **50** 505–15
[113] Torrico B, Shaw A D and Mosca R *et al* 2019 Truncating variant burden in high-functioning autism and pleiotropic effects of LRP1 across psychiatric phenotypes *J. Psychiatry Neurosci.* **44** 350–9
[114] Iossifov I, O'Roak B J and Sanders S J *et al* 2014 The contribution of *de novo* coding mutations to autism spectrum disorder *Nature* **515** 216–21
[115] Fromer M, Pocklington A J and Kavanagh D H *et al* 2014 *De novo* mutations in schizophrenia implicate synaptic networks *Nature* **506** 179–84
[116] Purcell S M, Moran J L and Fromer M *et al* 2014 A polygenic burden of rare disruptive mutations in schizophrenia *Nature* **506** 185–90
[117] Laugel-Haushalter V, Morkmued S and Stoetzel C *et al* 2018 Genetic evidence supporting the role of the calcium channel, *CACNA1S*, in tooth cusp and root patterning *Front. Physiol.* **9** 1329
[118] Mikami A, Imoto K and Tanabe T *et al* 1989 Primary structure and functional expression of the cardiac dihydropyridine-sensitive calcium channel *Nature* **340** 230–3
[119] Hofmann F, Flockerzi V, Kahl S and Wegener J W 2014 L-type $Ca_V1.2$ calcium channels: from *in vitro* findings to *in vivo* function *Physiol. Rev.* **94** 303–26
[120] Berger S M and Bartsch D 2014 The role of L-type voltage-gated calcium channels $Ca_V1.2$ and $Ca_V1.3$ in normal and pathological brain function *Cell Tissue Res.* **357** 463–76
[121] Bhat S, Dao D T and Terrillion C E *et al* 2012 *CACNA1C* (Cav1.2) in the pathophysiology of psychiatric disease *Prog. Neurobiol.* **99** 1–14
[122] Kabir Z D, Martinez-Rivera A and Rajadhyaksha A M 2017 From gene to behavior: L-type calcium channel mechanisms underlying neuropsychiatric symptoms *Neurotherapeutics* **14** 588–613
[123] Fabiato A 1983 Calcium-induced release of calcium from the cardiac sarcoplasmic reticulum *Am. J. Physiol.* **245** C1–14

[124] Goonasekera S A, Hammer K and Auger-Messier M *et al* 2012 Decreased cardiac L-type Ca^{2+} channel activity induces hypertrophy and heart failure in mice *J. Clin. Invest.* **122** 280–90
[125] Splawski I, Timothy K W and Sharpe L M *et al* 2004 $Ca_V1.2$ calcium channel dysfunction causes a multisystem disorder including arrhythmia and autism *Cell* **119** 19–31
[126] D'Gama A M, Pochareddy S and Li M *et al* 2015 Targeted DNA sequencing from autism spectrum disorder brains implicates multiple genetic mechanisms *Neuron* **88** 910–7
[127] Antzelevitch C, Pollevick G D and Cordeiro J M *et al* 2007 Loss-of-function mutations in the cardiac calcium channel underlie a new clinical entity characterized by ST-segment elevation, short QT intervals, and sudden cardiac death *Circulation* **115** 442–9
[128] Blancard M, Debbiche A and Kato K *et al* 2018 An African loss-of-function *CACNA1C* variant p.T1787M associated with a risk of ventricular fibrillation *Sci. Rep.* **8** 14619
[129] Burashnikov E, Pfeiffer R and Barajas-Martinez H *et al* 2010 Mutations in the cardiac L-type calcium channel associated with inherited J-wave syndromes and sudden cardiac death *Heart Rhythm* **7** 1872–82
[130] Campuzano O, Beltran-Alvarez P, Iglesias A, Scornik F, Perez G and Brugada R 2010 Genetics and cardiac channelopathies *Genet. Med.* **12** 260–7
[131] Chen C J, Lu T P and Lin L Y *et al* 2018 Impact of ancestral differences and reassessment of the classification of previously reported pathogenic variants in patients with brugada syndrome in the genomic era: a SADS-TW BrS registry *Front. Genet.* **9** 680
[132] Fukuyama M, Ohno S and Wang Q *et al* 2013 L-type calcium channel mutations in Japanese patients with inherited arrhythmias *Circ. J.* **77** 1799–806
[133] Fukuyama M, Wang Q and Kato K *et al* 2014 Long QT syndrome type 8: novel *CACNA1C* mutations causing QT prolongation and variant phenotypes *Europace* **16** 1828–37
[134] Hata Y, Kinoshita K and Mizumaki K *et al* 2016 Postmortem genetic analysis of sudden unexplained death syndrome under 50 years of age: a next-generation sequencing study *Heart Rhythm* **13** 1544–51
[135] Kosmicki J A, Samocha K E and Howrigan D P *et al* 2017 Refining the role of *de novo* protein-truncating variants in neurodevelopmental disorders by using population reference samples *Nat. Genet.* **49** 504–10
[136] Lim E T, Uddin M and De Rubeis S *et al* 2017 Rates, distribution and implications of postzygotic mosaic mutations in autism spectrum disorder *Nat. Neurosci.* **20** 1217–24
[137] Maltese P E, Aldanova E and Kriuchkova N *et al* 2019 Putative role of brugada syndrome genes in familial atrial fibrillation *Eur. Rev. Med. Pharmacol. Sci.* **23** 7582–98
[138] Nieto-Marin P, Jimenez-Jaimez J and Tinaquero D *et al* 2019 Digenic heterozigosity in *SCN5A* and *CACNA1C* explains the variable expressivity of the long QT phenotype in a Spanish family *Rev. Esp. Cardiol. (Engl. Ed.)* **72** 324–32
[139] Splawski I, Timothy K W and Decher N *et al* 2005 Severe arrhythmia disorder caused by cardiac L-type calcium channel mutations *Proc. Natl Acad. Sci. USA* **102** 8089–96 discussion 6–8
[140] Wemhoner K, Friedrich C and Stallmeyer B *et al* 2015 Gain-of-function mutations in the calcium channel *CACNA1C* ($Ca_V1.2$) cause non-syndromic long-QT but not Timothy syndrome *J. Mol. Cell. Cardiol.* **80** 186–95

[141] Boczek N J, Best J M and Tester D J *et al* 2013 Exome sequencing and systems biology converge to identify novel mutations in the L-type calcium channel, *CACNA1C*, linked to autosomal dominant long QT syndrome *Circ. Cardio. Genet.* **6** 279–89

[142] Boczek N J, Miller E M and Ye D *et al* 2015 Novel Timothy syndrome mutation leading to increase in *CACNA1C* window current *Heart Rhythm* **12** 211–9

[143] Boczek N J, Ye D and Jin F *et al* 2015 Identification and functional characterization of a novel *CACNA1C*-mediated cardiac disorder characterized by prolonged QT intervals with hypertrophic cardiomyopathy, congenital heart defects, and sudden cardiac death *Circ. Arrhythm. Electrophysiol.* **8** 1122–32

[144] Landstrom A P, Boczek N J and Ye D *et al* 2016 Novel long QT syndrome-associated missense mutation, L762F, in *CACNA1C*-encoded L-type calcium channel imparts a slower inactivation tau and increased sustained and window current *Int. J. Cardiol.* **220** 290–8

[145] Mellor G J, Panwar P and Lee A K *et al* 2019 Type 8 long QT syndrome: pathogenic variants in *CACNA1C*-encoded $Ca_V1.2$ cluster in STAC protein binding site *Europace* **21** 1725–32

[146] Mellor G, Laksman Z W M and Tadros R *et al* 2017 Genetic testing in the evaluation of unexplained cardiac arrest: from the CASPER (cardiac arrest survivors with preserved ejection fraction registry) *Circ. Cardiol. Genet.* **10** e001686

[147] Kojima A, Shikata F and Okamura T *et al* 2017 Refractory ventricular fibrillations after surgical repair of atrial septal defects in a patient with *CACNA1C* gene mutation – case report *J. Cardiothor. Surg.* **12** 118

[148] Chang S L, Chang C T, Hung W T and Chen L K 2019 A case of congenital long QT syndrome, type 8, undergoing laparoscopic hysterectomy with general anesthesia *Taiwan J. Obstet. Gynecol.* **58** 552–6

[149] Dufendach K A, Timothy K and Ackerman M J *et al* 2018 Clinical outcomes and modes of death in Timothy syndrome: a multicenter international study of a rare disorder *JACC Clin. Electrophysiol.* **4** 459–66

[150] Burns C, Ingles J and Davis A M *et al* 2016 Clinical and genetic features of australian families with long QT syndrome: a registry-based study *J. Arrhythm.* **32** 456–61

[151] Colson C, Mittre H and Busson A *et al* 2019 Unusual clinical description of adult with Timothy syndrome, carrier of a new heterozygote mutation of *CACNA1C Eur. J. Med. Genet.* **62** 103648

[152] Fukuyama M, Ohno S and Ozawa J *et al* 2020 High prevalence of late-appearing T-wave in patients with long QT syndrome type 8 *Circ. J.* **84** 559–68

[153] Gillis J, Burashnikov E and Antzelevitch C *et al* 2012 Long QT, syndactyly, joint contractures, stroke and novel *CACNA1C* mutation: expanding the spectrum of Timothy syndrome *Am. J. Med. Genet. Part A* **158A** 182–7

[154] Hennessey J A, Boczek N J and Jiang Y H *et al* 2014 A *CACNA1C* variant associated with reduced voltage-dependent inactivation, increased $Ca_V1.2$ channel window current, and arrhythmogenesis *PLoS One* **9** e106982

[155] Kosaki R, Ono H, Terashima H and Kosaki K 2018 Timothy syndrome-like condition with syndactyly but without prolongation of the QT interval *Am. J. Med. Genet. Part A* **176** 1657–61

[156] Liu X, Shen Y and Xie J *et al* 2017 A mutation in the *CACNA1C* gene leads to early repolarization syndrome with incomplete penetrance: a Chinese family study *PLoS One* **12** e0177532

[157] Narula N, Tester D J, Paulmichl A, Maleszewski J J and Ackerman M J 2015 Post-mortem whole exome sequencing with gene-specific analysis for autopsy-negative sudden unexplained death in the young: a case series *Pediatr. Cardiol.* **36** 768–78

[158] Ozawa J, Ohno S, Saito H, Saitoh A, Matsuura H and Horie M 2018 A novel CACNA1C mutation identified in a patient with Timothy syndrome without syndactyly exerts both marked loss- and gain-of-function effects *HeartRhythm Case Rep.* **4** 273–7

[159] Sutphin B S, Boczek N J and Barajas-Martinez H *et al* 2016 Molecular and functional characterization of rare CACNA1C variants in sudden unexplained death in the young *Congenit. Heart Dis.* **11** 683–92

[160] Endres D, Decher N and Rohr I *et al* 2020 New Cav1.2 channelopathy with high-functioning autism, affective disorder, severe dental enamel defects, a short QT interval, and a novel *CACNA1C* loss-of-function mutation *Int. J. Mol. Sci.* **21** 8611

[161] Pitt G S, Matsui M and Cao C 2021 Voltage-gated calcium channels in nonexcitable tissues *Annu. Rev. Physiol.* **83** 183–203

[162] Napolitano C, Splawski I, Timothy K W, Bloise R and Priori S G 1993 CACNA1C-related disorders *GeneReviews(R)* ed R A Pagon, M P Adam and H H Ardinger *et al* (Seattle, WA: University of Washington)

[163] Marks M L, Trippel D L and Keating M T 1995 Long QT syndrome associated with syndactyly identified in females *Am. J. Cardiol.* **76** 744–5

[164] Jacobs A, Knight B P, McDonald K T and Burke M C 2006 Verapamil decreases ventricular tachyarrhythmias in a patient with Timothy syndrome (LQT8) *Heart Rhythm* **3** 967–70

[165] Marks M L, Whisler S L, Clericuzio C and Keating M 1995 A new form of long QT syndrome associated with syndactyly *J. Am.Coll. Cardiol.* **25** 59–64

[166] Shah D P, Baez-Escudero J L, Weisberg I L, Beshai J F and Burke M C 2010 Ranolazine safely decreases ventricular and atrial fibrillation in timothy syndrome (LQT8) *Pacing Clin. Electrophysiol.* **35** e62–4

[167] Soldatov N M 1992 Molecular diversity of L-type Ca^{2+} channel transcripts in human fibroblasts *Proc. Natl Acad. Sci. USA* **89** 4628–32

[168] Schultz D, Mikala G and Yatani A *et al* 1993 Cloning, chromosomal localization, and functional expression of the alpha 1 subunit of the L-type voltage-dependent calcium channel from normal human heart *Proc. Natl Acad. Sci. USA* **90** 6228–32

[169] Liao P, Yong T F, Liang M C, Yue D T and Soong T W 2005 Splicing for alternative structures of $Ca_V1.2$ Ca^{2+} channels in cardiac and smooth muscles *Cardiovasc. Res.* **68** 197–203

[170] Barrett C F and Tsien R W 2008 The Timothy syndrome mutation differentially affects voltage- and calcium-dependent inactivation of $Ca_V1.2$ L-type calcium channels *Proc. Natl Acad. Sci. USA* **105** 2157–62

[171] Raybaud A, Dodier Y and Bissonnette P *et al* 2006 The role of the GX9GX3G motif in the gating of high voltage-activated Ca^{2+} channels *J. Biol. Chem.* **281** 39424–36

[172] Dick I E, Joshi-Mukherjee R, Yang W and Yue D T 2016 Arrhythmogenesis in Timothy syndrome is associated with defects in Ca^{2+}-dependent inactivation *Nat. Commun.* **7** 10370

[173] Bader P L, Faizi M and Kim L H *et al* 2011 Mouse model of Timothy syndrome recapitulates triad of autistic traits *Proc. Natl Acad. Sci. USA* **108** 15432–7

[174] Cheli V T, Santiago Gonzalez D A and Zamora N N *et al* 2018 Enhanced oligodendrocyte maturation and myelination in a mouse model of Timothy syndrome *Glia* **66** 2324–39

[175] Ehlinger D G and Commons K G 2017 Altered Cav1.2 function in the Timothy syndrome mouse model produces ascending serotonergic abnormalities *Eur. J. Neurosci.* **46** 2416–25

[176] Pasca S P, Portmann T and Voineagu I *et al* 2011 Using iPSC-derived neurons to uncover cellular phenotypes associated with Timothy syndrome *Nat. Med.* **17** 1657–62

[177] Yazawa M and Dolmetsch R E 2013 Modeling Timothy syndrome with iPS cells *J. Cardiovasc. Transl. Res.* **6** 1–9

[178] Yazawa M, Hsueh B and Jia X *et al* 2011 Using induced pluripotent stem cells to investigate cardiac phenotypes in Timothy syndrome *Nature* **471** 230–4

[179] Cheng E P, Yuan C and Navedo M F *et al* 2011 Restoration of normal L-type Ca^{2+} channel function during Timothy syndrome by ablation of an anchoring protein *Circ. Res.* **109** 255–61

[180] Drum B M, Dixon R E, Yuan C, Cheng E P and Santana L F 2014 Cellular mechanisms of ventricular arrhythmias in a mouse model of Timothy syndrome (long QT syndrome 8) *J. Mol. Cell. Cardiol.* **66** 63–71

[181] Ramachandran K V, Hennessey J A and Barnett A S *et al* 2013 Calcium influx through L-type $Ca_V1.2$ Ca^{2+} channels regulates mandibular development *J. Clin. Invest.* **123** 1638–46

[182] Panagiotakos G, Haveles C and Arjun A *et al* 2019 Aberrant calcium channel splicing drives defects in cortical differentiation in Timothy syndrome *Elife* **8** e51037

[183] Papineau S D and Wilson S 2014 Dentition abnormalities in a Timothy syndrome patient with a novel genetic mutation: a case report *Pediatr. Dent.* **36** 245–9

[184] Bamgboye M A, Herold K G, Vieira D C O, Traficante M K, Rogers P J, Ben-Johny M and Dick I E 2022 $Ca_V1.2$ channelopathic mutations evoke diverse pathophysiological mechanisms *J. Gen. Physiol.* **154** e202213209

[185] Kosaki R, Kubota M, Uehara T, Suzuki H, Takenouchi T and Kosaki K 2020 Consecutive medical exome analysis at a tertiary center: diagnostic and health-economic outcomes *Am. J. Med. Genet.* A **182** 1601–7

[186] Ye D, Tester D J, Zhou W, Papagiannis J and Ackerman M J 2019 A pore-localizing CACNA1C-E1115K missense mutation, identified in a patient with idiopathic QT prolongation, bradycardia, and autism spectrum disorder, converts the L-type calcium channel into a hybrid nonselective monovalent cation channel *Heart Rhythm* **16** 270–8

[187] Campuzano O, Sarquella-Brugada G and Fernandez-Falgueras A *et al* 2019 Genetic interpretation and clinical translation of minor genes related to Brugada syndrome *Hum. Mutat.* **40** 749–64

[188] Estes S I, Ye D and Zhou W *et al* 2019 Characterization of the CACNA1C-R518C missense mutation in the pathobiology of long-QT syndrome using human induced pluripotent stem cell cardiomyocytes shows action potential prolongation and L-type calcium channel perturbation *Circ. Genom. Precis. Med.* **12** e002534

[189] Splawski I, Yoo D S, Stotz S C, Cherry A, Clapham D E and Keating M T 2006 *CACNA1H* mutations in autism spectrum disorders *J. Biol. Chem.* **281** 22085–91

[190] Allegue C, Coll M and Mates J *et al* 2015 Genetic analysis of arrhythmogenic diseases in the Era of NGS: the complexity of clinical decision-making in Brugada Syndrome *PLoS One* **10** e0133037

[191] Niu J, Dick I E and Yang W *et al* 2018 Allosteric regulators selectively prevent Ca^{2+}-feedback of Ca_V and Na_V channels *Elife* **7** e35222

[192] Polster A, Dittmer P J, Perni S, Bichraoui H, Sather W A and Beam K G 2018 Stac proteins suppress Ca^{2+}-dependent inactivation of neuronal l-type Ca^{2+} channels *J. Neurosci.* **38** 9215–27

[193] Campiglio M, Coste de Bagneaux P, Ortner N J, Tuluc P, Van Petegem F and Flucher B E 2018 STAC proteins associate to the IQ domain of CaV1.2 and inhibit calcium-dependent inactivation *Proc. Natl Acad. Sci. USA* **115** 1376–81

[194] Wong King Yuen S M, Campiglio M, Tung C C, Flucher B E, Van and Petegem F 2017 Structural insights into binding of STAC proteins to voltage-gated calcium channels *Proc. Natl Acad. Sci. USA* **114** E9520

[195] Nelson B R, Wu F and Liu Y *et al* 2013 Skeletal muscle-specific T-tubule protein STAC3 mediates voltage-induced Ca^{2+} release and contractility *Proc. Natl Acad. Sci. USA* **110** 11881–6

[196] Garcia-Elias A and Benito B 2018 Ion channel disorders and sudden cardiac death *Int. J. Mol. Sci.* **19** 692

[197] Chen Y, Barajas-Martinez H and Zhu D *et al* 2017 Novel trigenic CACNA1C/DES/MYPN mutations in a family of hypertrophic cardiomyopathy with early repolarization and short QT syndrome *J. Transl. Med.* **15** 78

[198] Beziau D M, Barc J and O'Hara T *et al* 2014 Complex Brugada syndrome inheritance in a family harbouring compound *SCN5A* and *CACNA1C* mutations *Basic Res. Cardiol.* **109** 446

[199] Simms B A and Zamponi G W 2012 The Brugada syndrome mutation A39V does not affect surface expression of neuronal rat $Ca_V1.2$ channels *Mol. Brain* **5** 9

[200] Nunn L M, Lopes L R and Syrris P *et al* 2016 Diagnostic yield of molecular autopsy in patients with sudden arrhythmic death syndrome using targeted exome sequencing *Europace* **18** 888–96

[201] Lin Y, Williams N and Wang D *et al* 2017 Applying high-resolution variant classification to cardiac arrhythmogenic gene testing in a demographically diverse cohort of sudden unexplained deaths *Circ. Cardiovasc. Genet.* **10** e001839

[202] Jiang Y H, Yuen R K and Jin X *et al* 2013 Detection of clinically relevant genetic variants in autism spectrum disorder by whole-genome sequencing *Am. J. Hum. Genet.* **93** 249–63

[203] Hu H, Kahrizi K and Musante L *et al* 2019 Genetics of intellectual disability in consanguineous families *Mol. Psychiatry* **24** 1027–39

[204] Frohler S, Kieslich M and Langnick C *et al* 2014 Exome sequencing helped the fine diagnosis of two siblings afflicted with atypical Timothy syndrome (TS2) *BMC Med. Genet.* **15** 48

[205] Baig S M, Koschak A and Lieb A *et al* 2011 Loss of $Ca_V1.3$ (*CACNA1D*) function in a human channelopathy with bradycardia and congenital deafness *Nat. Neurosci.* **14** 77–84

[206] Platzer J, Engel J and Schrott-Fischer A *et al* 2000 Congenital deafness and sinoatrial node dysfunction in mice lacking class D L-type Ca^{2+} channels *Cell* **102** 89–97

[207] Mangoni M E, Couette B and Bourinet E *et al* 2003 Functional role of L-type $Ca_V1.3$ Ca^{2+} channels in cardiac pacemaker activity *Proc. Natl Acad. Sci. USA* **100** 5543–8

[208] Zhang Z, He Y and Tuteja D *et al* 2005 Functional roles of $Ca_V1.3(\alpha 1D)$ calcium channels in atria: insights gained from gene-targeted null mutant mice *Circulation* **112** 1936–44

[209] Azizan E A, Poulsen H and Tuluc P *et al* 2013 Somatic mutations in *ATP1A1* and *CACNA1D* underlie a common subtype of adrenal hypertension *Nat. Genet.* **45** 1055–60

[210] Brandt A, Striessnig J and Moser T 2003 $Ca_V1.3$ channels are essential for development and presynaptic activity of cochlear inner hair cells *J. Neurosci.* **23** 10832–40

[211] Ortner N J, Kaserer T, Copeland J N and Striessnig J 2020 *De novo CACNA1D* Ca^{2+} channelopathies: clinical phenotypes and molecular mechanism *Pflugers Arch.* **472** 755–73

[212] Pinggera A, Mackenroth L and Rump A *et al* 2017 New gain-of-function mutation shows *CACNA1D* as recurrently mutated gene in autism spectrum disorders and epilepsy *Hum. Mol. Genet.* **26** 2923–32

[213] De Mingo Alemany M C, Mifsud Grau L, Moreno Macian F, Ferrer Lorente B, Leon and Carinena S 2020 A *de novo CACNA1D* missense mutation in a patient with congenital hyperinsulinism, primary hyperaldosteronism and hypotonia *Channels (Austin)* **14** 175–80

[214] Flanagan S E, Vairo F and Johnson M B *et al* 2017 A *CACNA1D* mutation in a patient with persistent hyperinsulinaemic hypoglycaemia, heart defects, and severe hypotonia *Pediatr. Diabetes* **18** 320–3

[215] Garza-Lopez E, Lopez J A, Hagen J, Sheffer R, Meiner V and Lee A 2018 Role of a conserved glutamine in the function of voltage-gated Ca^{2+} channels revealed by a mutation in human *CACNA1D J. Biol. Chem.* **293** 14444–54

[216] Hofer N T, Tuluc P and Ortner N J *et al* 2020 Biophysical classification of a *CACNA1D* de novo mutation as a high-risk mutation for a severe neurodevelopmental disorder *Mol. Autism* **11** 4

[217] Liaqat K, Schrauwen I and Raza S I *et al* 2019 Identification of *CACNA1D* variants associated with sinoatrial node dysfunction and deafness in additional Pakistani families reveals a clinical significance *J. Hum. Genet.* **64** 153–60

[218] O'Roak B J, Vives L and Girirajan S *et al* 2012 Sporadic autism exomes reveal a highly interconnected protein network of *de novo* mutations *Nature* **485** 246–50

[219] Semenova N A, Ryzhkova O R, Strokova T V and Taran N N 2018 [The third case report a patient with primary aldosteronism, seizures, and neurologic abnormalities (PASNA) syndrome *de novo* variant mutations in the CACNA1D gene] *Zh. Nevrol. Psikhiatr. Im S S Korsakova* **118** 49–52

[220] Tan G C, Negro G and Pinggera A *et al* 2017 Aldosterone-producing adenomas: histopathology-genotype correlation and identification of a novel *CACNA1D* mutation *Hypertension* **70** 129–36

[221] Scholl U I, Goh G and Stolting G *et al* 2013 Somatic and germline *CACNA1D* calcium channel mutations in aldosterone-producing adenomas and primary aldosteronism *Nat. Genet.* **45** 1050–4

[222] Pinggera A, Lieb A and Benedetti B *et al* 2015 *CACNA1D de novo* mutations in autism spectrum disorders activate $Ca_V1.3$ L-type calcium channels *Biol. Psychiatry* **77** 816–22

[223] Pinggera A, Negro G, Tuluc P, Brown M J, Lieb A and Striessnig J 2018 Gating defects of disease-causing de novo mutations in $Ca_V1.3$ Ca^{2+} channels. *Channels (Austin)* **12** 388–402

[224] Monteleone S, Lieb A and Pinggera A *et al* 2017 Mechanisms responsible for omega-pore currents in Ca_V calcium channel voltage-sensing domains *Biophys. J.* **113** 1485–95

[225] Limpitikul W B, Dick I E, Ben-Johny M and Yue D T 2016 An autism-associated mutation in $Ca_V1.3$ channels has opposing effects on voltage- and Ca^{2+}-dependent regulation *Sci. Rep.* **6** 27235

[226] Jiao J, Zhang M and Yang P *et al* 2020 Identification of de novo JAK2 and MAPK7 mutations related to autism spectrum disorder using whole-exome sequencing in a chinese child and adolescent trio-based sample *J. Mol. Neurosci.* **70** 219–29

[227] Long S, Zhou H and Li S *et al* 2019 The clinical and genetic features of co-occurring epilepsy and autism spectrum disorder in Chinese children *Front. Neurol.* **10** 505

[228] Satterstrom F K, Kosmicki J A and Wang J *et al* 2020 Large-scale exome sequencing study implicates both developmental and functional changes in the neurobiology of autism *Cell* **180** 568–84

[229] Ross J, Gedvilaite E and Badner J A *et al* 2016 A rare variant in *CACNA1D* segregates with 7 bipolar I disorder cases in a large pedigree *Mol. Neuropsychiatry* **2** 145–50

[230] Bock G, Gebhart M and Scharinger A *et al* 2011 Functional properties of a newly identified C-terminal splice variant of $Ca_V1.3$ L-type Ca^{2+} channels *J. Biol. Chem.* **286** 42736–48

[231] Wahl-Schott C, Baumann L, Cuny H, Eckert C, Griessmeier K and Biel M 2006 Switching off calcium-dependent inactivation in L-type calcium channels by an autoinhibitory domain *Proc. Natl Acad. Sci. USA* **103** 15657–62

[232] Mansergh F, Orton N C and Vessey J P *et al* 2005 Mutation of the calcium channel gene *Cacna1f* disrupts calcium signaling, synaptic transmission and cellular organization in mouse retina *Hum. Mol. Genet.* **14** 3035–46

[233] Maddox J W, Randall K L and Yadav R P *et al* 2020 A dual role for $Ca_V1.4$ Ca^{2+} channels in the molecular and structural organization of the rod photoreceptor synapse *Elife* **9** e62184

[234] Bech-Hansen N T, Naylor M J and Maybaum T A *et al* 1998 Loss-of-function mutations in a calcium-channel alpha1-subunit gene in Xp11.23 cause incomplete X-linked congenital stationary night blindness *Nat. Genet.* **19** 264–7

[235] Hoda J C, Zaghetto F, Koschak A and Striessnig J 2005 Congenital stationary night blindness type 2 mutations S229P, G369D, L1068P, and W1440X alter channel gating or functional expression of $Ca_V1.4$ L-type Ca^{2+} channels *J. Neurosci.* **25** 252–9

[236] Koschak A, Fernandez-Quintero M L, Heigl T, Ruzza M, Seitter H and Zanetti L 2021 $Ca_V1.4$ dysfunction and congenital stationary night blindness type 2 *Pflugers Arch.* **473** 1437–54

[237] Zeitz C, Michiels C and Neuille M *et al* 2019 Where are the missing gene defects in inherited retinal disorders? Intronic and synonymous variants contribute at least to 4% of CACNA1F-mediated inherited retinal disorders *Hum. Mutat.* **40** 765–87

[238] Zeitz C, Robson A G and Audo I 2015 Congenital stationary night blindness: an analysis and update of genotype-phenotype correlations and pathogenic mechanisms *Prog. Retin. Eye Res.* **45** 58–110

[239] Gill J S, Georgiou M, Kalitzeos A, Moore A T and Michaelides M 2019 Progressive cone and cone-rod dystrophies: clinical features, molecular genetics and prospects for therapy *Br. J. Ophthalmol.* **103** 711–20

[240] Hauke J, Schild A and Neugebauer A *et al* 2013 A novel large in-frame deletion within the CACNA1F gene associates with a cone-rod dystrophy 3-like phenotype *PLoS One* **8** e76414

[241] Huang L, Zhang Q and Li S *et al* 2013 Exome sequencing of 47 chinese families with cone-rod dystrophy: mutations in 25 known causative genes *PLoS One* **8** e65546

[242] Jalkanen R, Mantyjarvi M and Tobias R *et al* 2006 X linked cone-rod dystrophy, CORDX3, is caused by a mutation in the *CACNA1F* gene *J. Med. Genet.* **43** 699–704

[243] Kim M S, Joo K and Seong M W *et al* 2019 Genetic mutation profiles in Korean patients with inherited retinal diseases *J. Korean. Med. Sci.* **34** e161

[244] Taylor R L, Parry N R A and Barton S J *et al* 2017 Panel-based clinical genetic testing in 85 children with inherited retinal disease *Ophthalmology* **124** 985–91

[245] Jalkanen R, Bech-Hansen N T and Tobias R *et al* 2007 A novel *CACNA1F* gene mutation causes Aland Island eye disease *Invest. Ophthalmol. Vis. Sci.* **48** 2498–502

[246] Vincent A, Wright T, Day M A, Westall C A and Heon E 2011 A novel p.Gly603Arg mutation in *CACNA1F* causes Aland island eye disease and incomplete congenital stationary night blindness phenotypes in a family *Mol. Vis.* **17** 3262–70

[247] Boycott K M, Pearce W G and Bech-Hansen N T 2000 Clinical variability among patients with incomplete X-linked congenital stationary night blindness and a founder mutation in *CACNA1F Can. J. Ophthalmol.* **35** 204–13

[248] Bijveld M M, Florijn R J and Bergen A A *et al* 2013 Genotype and phenotype of 101 dutch patients with congenital stationary night blindness *Ophthalmology* **120** 2072–81

[249] Burtscher V, Schicker K and Novikova E *et al* 2014 Spectrum of $Ca_V1.4$ dysfunction in congenital stationary night blindness type 2 *Biochim. Biophys. Acta* **1838** 2053–65

[250] Hoda J C, Zaghetto F, Singh A, Koschak A and Striessnig J 2006 Effects of congenital stationary night blindness type 2 mutations R508Q and L1364H on $Ca_V1.4$ L-type Ca^{2+} channel function and expression *J. Neurochem.* **96** 1648–58

[251] Zeitz C, Labs S and Lorenz B *et al* 2009 Genotyping microarray for CSNB-associated genes *Invest. Ophthalmol. Vis. Sci.* **50** 5919–26

[252] Peloquin J B, Rehak R, Doering C J and McRory J E 2007 Functional analysis of congenital stationary night blindness type-2 *CACNA1F* mutations F742C, G1007R, and R1049W *Neuroscience* **150** 335–45

[253] McRory J E, Hamid J and Doering C J *et al* 2004 The *CACNA1F* gene encodes an L-type calcium channel with unique biophysical properties and tissue distribution *J. Neurosci.* **24** 1707–18

[254] Hope C I, Sharp D M and Hemara-Wahanui A *et al* 2005 Clinical manifestations of a unique X-linked retinal disorder in a large New Zealand family with a novel mutation in *CACNA1F*, the gene responsible for CSNB2 *Clin. Exp. Ophthalmol.* **33** 129–36

[255] Hemara-Wahanui A, Berjukow S and Hope C I *et al* 2005 A *CACNA1F* mutation identified in an X-linked retinal disorder shifts the voltage dependence of $Ca_V1.4$ channel activation *Proc. Natl Acad. Sci. USA* **102** 7553–8

[256] Williams B, Lopez J A, Maddox J W and Lee A 2020 Functional impact of a congenital stationary night blindness type 2 mutation depends on subunit composition of $Ca_V1.4$ Ca^{2+} channels *J. Biol. Chem.* **295** 17215–26

[257] Knoflach D, Kerov V and Sartori S B *et al* 2013 $Ca_V1.4$ IT mouse as model for vision impairment in human congenital stationary night blindness type 2 *Channels (Austin)* **7** 503–13

[258] Bourinet E, Soong T W and Sutton K *et al* 1999 Splicing of α1A subunit gene generates phenotypic variants of P- and Q-type calcium channels *Nat. Neurosci.* **2** 407–15

[259] Starr T V, Prystay W and Snutch T P 1991 Primary structure of a calcium channel that is highly expressed in the rat cerebellum *Proc. Natl Acad. Sci. USA* **88** 5621–5

[260] Mori Y, Friedrich T and Kim M S *et al* 1991 Primary structure and functional expression from complementary DNA of a brain calcium channel *Nature* **350** 398–402

[261] Moreno H, Rudy B and Llinas R 1997 β subunits influence the biophysical and pharmacological differences between P- and Q-type calcium currents expressed in a mammalian cell line *Proc. Natl Acad. Sci. USA* **94** 14042–7

[262] Mintz I M, Adams M E and Bean B P 1992 P-type calcium channels in rat central and peripheral neurons *Neuron* **9** 85–95

[263] McDonough S I, Swartz K J, Mintz I M, Boland L M and Bean B P 1996 Inhibition of calcium channels in rat central and peripheral neurons by omega-conotoxin MVIIC *J. Neurosci.* **16** 2612–23

[264] Vacher H, Mohapatra D P and Trimmer J S 2008 Localization and targeting of voltage-dependent ion channels in mammalian central neurons *Physiol. Rev.* **88** 1407–47

[265] Westenbroek R E, Hoskins L and Catterall W A 1998 Localization of Ca^{2+} channel subtypes on rat spinal motor neurons, interneurons, and nerve terminals *J. Neurosci.* **18** 6319–30

[266] Jaudon F, Baldassari S, Musante I, Thalhammer A, Zara F and Cingolani L A 2020 Targeting alternative splicing as a potential therapy for episodic ataxia type 2 *Biomedicines* **8** 332

[267] Thalhammer A, Jaudon F and Cingolani L A 2020 Emerging roles of activity-dependent alternative splicing in homeostatic plasticity *Front. Cell Neurosci.* **14** 104

[268] Lipscombe D, Allen S E and Toro C P 2013 Control of neuronal voltage-gated calcium ion channels from RNA to protein *Trends Neurosci.* **36** 598–609

[269] Aikawa T, Watanabe T and Miyazaki T *et al* 2017 Alternative splicing in the C-terminal tail of $Ca_V2.1$ is essential for preventing a neurological disease in mice *Hum. Mol. Genet.* **26** 3094–104

[270] Soong T W, DeMaria C D and Alvania R S *et al* 2002 Systematic identification of splice variants in human P/Q-type channel $\alpha1(2.1)$ subunits: implications for current density and Ca^{2+}-dependent inactivation *J. Neurosci.* **22** 10142–52

[271] Catterall W A and Few A P 2008 Calcium channel regulation and presynaptic plasticity *Neuron* **59** 882–901

[272] Luebke J I, Dunlap K and Turner T 1993 Multiple calcium channel types control glutamatergic synaptic transmission in the hippocampus *Neuron* **11** 895–902

[273] Wheeler D B, Randall A and Tsien R W 1994 Roles of N-type and Q-type Ca^{2+} channels in supporting hippocampal synaptic transmission *Science* **264** 107–11

[274] Miyazaki T, Hashimoto K, Shin H S, Kano M and Watanabe M 2004 P/Q-type Ca^{2+} channel $\alpha1A$ regulates synaptic competition on developing cerebellar Purkinje cells *J. Neurosci.* **24** 1734–43

[275] Jun K, Piedras-Renteria E S and Smith S M *et al* 1999 Ablation of P/Q-type Ca^{2+} channel currents, altered synaptic transmission, and progressive ataxia in mice lacking the $\alpha(1A)$-subunit *Proc. Natl Acad. Sci. USA* **96** 15245–50

[276] Ophoff R A, Terwindt G M and Vergouwe M N *et al* 1996 Familial hemiplegic migraine and episodic ataxia type-2 are caused by mutations in the Ca^{2+} channel gene *CACNL1A4* *Cell* **87** 543–52

[277] Zhuchenko O, Bailey J and Bonnen P *et al* 1997 Autosomal dominant cerebellar ataxia (SCA6) associated with small polyglutamine expansions in the $\alpha1A$-voltage-dependent calcium channel *Nat. Genet.* **15** 62–9

[278] Jiang X, Raju P K and D'Avanzo N *et al* 2019 Both gain-of-function and loss-of-function *de novo CACNA1A* mutations cause severe developmental epileptic encephalopathies in the spectrum of Lennox–Gastaut syndrome *Epilepsia* **60** 1881–94

[279] Tyagi S, Bendrick T R, Filipova D, Papadopoulos S and Bannister R A 2019 A mutation in $Ca_V2.1$ linked to a severe neurodevelopmental disorder impairs channel gating *J. Gen. Physiol.* **151** 850–9

[280] Luo X, Rosenfeld J A and Yamamoto S *et al* 2017 Clinically severe *CACNA1A* alleles affect synaptic function and neurodegeneration differentially *PLoS Genet.* **13** e1006905

[281] Romaniello R, Zucca C and Tonelli A *et al* 2010 A wide spectrum of clinical, neurophysiological and neuroradiological abnormalities in a family with a novel *CACNA1A* mutation *J. Neurol. Neurosurg. Psychiatry* **81** 840–3

[282] Blumkin L, Michelson M, Leshinsky-Silver E, Kivity S, Lev D and Lerman-Sagie T 2010 Congenital ataxia, mental retardation, and dyskinesia associated with a novel *CACNA1A* mutation *J. Child Neurol.* **25** 892–7

[283] Hamdan F F, Myers C T and Cossette P *et al* 2017 High rate of recurrent *de novo* mutations in developmental and epileptic encephalopathies *Am. J. Hum. Genet.* **101** 664–85

[284] Zhang L, Wen Y and Zhang Q *et al* 2020 CACNA1A gene variants in eight chinese patients with a wide range of phenotypes *Front. Pediatr.* **8** 577544

[285] Ducros A, Denier C and Joutel A *et al* 2001 The clinical spectrum of familial hemiplegic migraine associated with mutations in a neuronal calcium channel *N. Engl. J. Med.* **345** 17–24

[286] Kors E E, Terwindt G M and Vermeulen F L *et al* 2001 Delayed cerebral edema and fatal coma after minor head trauma: role of the *CACNA1A* calcium channel subunit gene and relationship with familial hemiplegic migraine *Ann. Neurol.* **49** 753–60

[287] Stuart S, Roy B, Davies G, Maksemous N, Smith R and Griffiths L R 2012 Detection of a novel mutation in the *CACNA1A* gene *Twin Res. Hum. Genet.* **15** 120–5

[288] Riant F, Lescoat C and Vahedi K *et al* 2010 Identification of *CACNA1A* large deletions in four patients with episodic ataxia *Neurogenetics* **11** 101–6

[289] Mantuano E, Romano S and Veneziano L *et al* 2010 Identification of novel and recurrent *CACNA1A* gene mutations in fifteen patients with episodic ataxia type 2 *J. Neurol. Sci.* **291** 30–6

[290] Freilinger T, Ackl N and Ebert A *et al* 2011 A novel mutation in *CACNA1A* associated with hemiplegic migraine, cerebellar dysfunction and late-onset cognitive decline *J. Neurol. Sci.* **300** 160–3

[291] Cuenca-Leon E, Corominas R and Fernandez-Castillo N *et al* 2008 Genetic analysis of 27 Spanish patients with hemiplegic migraine, basilar-type migraine and childhood periodic syndromes *Cephalalgia* **28** 1039–47

[292] Battistini S, Stenirri S and Piatti M *et al* 1999 A new *CACNA1A* gene mutation in acetazolamide-responsive familial hemiplegic migraine and ataxia *Neurology* **53** 38–43

[293] Tantsis E M, Gill D and Griffiths L *et al* 2016 Eye movement disorders are an early manifestation of *CACNA1A* mutations in children *Dev. Med. Child Neurol.* **58** 639–44

[294] Condliffe S B, Fratangeli A and Munasinghe N R *et al* 2013 The E1015K variant in the synprint region of the $Ca_V2.1$ channel alters channel function and is associated with different migraine phenotypes *J. Biol. Chem.* **288** 33873–83

[295] Alonso I, Barros J and Tuna A *et al* 2004 A novel R1347Q mutation in the predicted voltage sensor segment of the P/Q-type calcium-channel alpha-subunit in a family with progressive cerebellar ataxia and hemiplegic migraine *Clin. Genet.* **65** 70–2

[296] Thomsen L L, Kirchmann M and Bjornsson A *et al* 2007 The genetic spectrum of a population-based sample of familial hemiplegic migraine *Brain* **130** 346–56

[297] Vahedi K, Denier C and Ducros A *et al* 2000 *CACNA1A* gene de novo mutation causing hemiplegic migraine, coma, and cerebellar atrophy *Neurology* **55** 1040–2

[298] Nardello R, Plicato G and Mangano G D *et al* 2020 Two distinct phenotypes, hemiplegic migraine and episodic Ataxia type 2, caused by a novel common *CACNA1A* variant *BMC Neurol.* **20** 155

[299] Carrera P, Piatti M and Stenirri S *et al* 1999 Genetic heterogeneity in Italian families with familial hemiplegic migraine *Neurology* **53** 26–33

[300] Pelzer N, Haan J and Stam A H *et al* 2018 Clinical spectrum of hemiplegic migraine and chances of finding a pathogenic mutation *Neurology* **90** e575–e82

[301] Grieco G S, Gagliardi S and Ricca I *et al* 2018 New *CACNA1A* deletions are associated to migraine phenotypes *J. Headache Pain* **19** 75

[302] Dichgans M, Herzog J, Freilinger T, Wilke M and Auer D P 2005 1H-MRS alterations in the cerebellum of patients with familial hemiplegic migraine type 1 *Neurology* **64** 608–13

[303] Weiss N, Tournier-Lasserve E and De Waard M 2007 Role of P/Q calcium channel in familial hemiplegic migraine *Med. Sci. (Paris)* **23** 53–63

[304] de Vries B, Stam A H and Beker F *et al* 2008 *CACNA1A* mutation linking hemiplegic migraine and alternating hemiplegia of childhood *Cephalalgia* **28** 887–91

[305] Kors E E, Melberg A and Vanmolkot K R *et al* 2004 Childhood epilepsy, familial hemiplegic migraine, cerebellar ataxia, and a new *CACNA1A* mutation *Neurology* **63** 1136–7

[306] Wilson G N 2014 Exome analysis of connective tissue dysplasia: death and rebirth of clinical genetics? *Am. J. Med. Genet.* A **164** 1209–12

[307] Imbrici P, Jaffe S L and Eunson L H *et al* 2004 Dysfunction of the brain calcium channel $Ca_V2.1$ in absence epilepsy and episodic ataxia *Brain* **127** 2682–92

[308] Maksemous N, Roy B, Smith R A and Griffiths L R 2016 Next-generation sequencing identifies novel *CACNA1A* gene mutations in episodic ataxia type 2 *Mol. Genet. Genomic Med.* **4** 211–22

[309] Soden S E, Saunders C J and Willig L K *et al* 2014 Effectiveness of exome and genome sequencing guided by acuity of illness for diagnosis of neurodevelopmental disorders *Sci. Transl. Med.* **6** 265ra168

[310] Indelicato E, Nachbauer W and Karner E *et al* 2019 The neuropsychiatric phenotype in *CACNA1A* mutations: a retrospective single center study and review of the literature *Eur. J. Neurol.* **26** 66–e7

[311] Zafeiriou D I, Lehmann-Horn F, Vargiami E, Teflioudi E, Ververi A and Jurkat-Rott K 2009 Episodic ataxia type 2 showing ictal hyperhidrosis with hypothermia and interictal chronic diarrhea due to a novel *CACNA1A* mutation *Eur. J. Paediatr. Neurol.* **13** 191–3

[312] Choi K D, Kim J S and Kim H J *et al* 2017 Genetic variants associated with episodic ataxia in Korea *Sci. Rep.* **7** 13855

[313] Sintas C, Carreno O and Fernandez-Castillo N *et al* 2017 Mutation spectrum in the *CACNA1A* gene in 49 patients with episodic ataxia *Sci. Rep.* **7** 2514

[314] van den Maagdenberg A M, Kors E E and Brunt E R *et al* 2002 Episodic ataxia type 2. Three novel truncating mutations and one novel missense mutation in the *CACNA1A* gene *J. Neurol.* **249** 1515–9

[315] Jen J, Kim G W and Baloh R W 2004 Clinical spectrum of episodic ataxia type 2 *Neurology* **62** 17–22

[316] Yue Q, Jen J C, Nelson S F and Baloh R W 1997 Progressive ataxia due to a missense mutation in a calcium-channel gene *Am. J. Hum. Genet.* **61** 1078–87

[317] Burk K, Kaiser F J and Tennstedt S *et al* 2014 A novel missense mutation in *CACNA1A* evaluated by in silico protein modeling is associated with non-episodic spinocerebellar ataxia with slow progression *Eur. J. Med. Genet.* **57** 207–11
[318] Nikaido K, Tachi N, Ohya K, Wada T and Tsutsumi H 2011 New mutation of *CACNA1A* gene in episodic ataxia type 2 *Pediatr. Int* **53** 415–6
[319] Cricchi F, Di Lorenzo C and Grieco G S *et al* 2007 Early-onset progressive ataxia associated with the first *CACNA1A* mutation identified within the I–II loop *J. Neurol. Sci.* **254** 69–71
[320] Isaacs D A, Bradshaw M J, Brown K and Hedera P 2017 Case report of novel *CACNA1A* gene mutation causing episodic ataxia type 2 *SAGE Open Med. Case Rep.* **5** 2050313X17706044
[321] Scoggan K A, Friedman J H and Bulman D E 2006 *CACNA1A* mutation in a EA-2 patient responsive to acetazolamide and valproic acid *Can. J. Neurol. Sci.* **33** 68–72
[322] Rajakulendran S, Graves T D and Labrum R W *et al* 2010 Genetic and functional characterisation of the P/Q calcium channel in episodic ataxia with epilepsy *J. Physiol.* **588** 1905–13
[323] Robbins M S, Lipton R B, Laureta E C and Grosberg B M 2009 *CACNA1A* nonsense mutation is associated with basilar-type migraine and episodic ataxia type 2 *Headache* **49** 1042–6
[324] Guerin A A, Feigenbaum A, Donner E J and Yoon G 2008 Stepwise developmental regression associated with novel *CACNA1A* mutation *Pediatr. Neurol.* **39** 363–4
[325] Roubertie A, Echenne B and Leydet J *et al* 2008 Benign paroxysmal tonic upgaze, benign paroxysmal torticollis, episodic ataxia and CACNA1A mutation in a family *J. Neurol.* **255** 1600–2
[326] Damaj L, Lupien-Meilleur A and Lortie A *et al* 2015 *CACNA1A* haploinsufficiency causes cognitive impairment, autism and epileptic encephalopathy with mild cerebellar symptoms *Eur. J. Hum. Genet.* **23** 1505–12
[327] Ohmori I, Ouchida M and Kobayashi K *et al* 2013 *CACNA1A* variants may modify the epileptic phenotype of Dravet syndrome *Neurobiol. Dis.* **50** 209–17
[328] Imbrici P, Eunson L H and Graves T D *et al* 2005 Late-onset episodic ataxia type 2 due to an in-frame insertion in *CACNA1A Neurology* **65** 944–6
[329] Bertholon P, Chabrier S, Riant F, Tournier-Lasserve E and Peyron R 2009 Episodic ataxia type 2: unusual aspects in clinical and genetic presentation. special emphasis in childhood *J. Neurol. Neurosurg. Psychiatry* **80** 1289–92
[330] Page K M, Heblich F and Davies A *et al* 2004 Dominant-negative calcium channel suppression by truncated constructs involves a kinase implicated in the unfolded protein response *J. Neurosci.* **24** 5400–9
[331] Yue Q, Jen J C, Thwe M M, Nelson S F and Baloh R W 1998 *De novo* mutation in *CACNA1A* caused acetazolamide-responsive episodic ataxia *Am. J. Med. Genet.* **77** 298–301
[332] Jen J, Wan J and Graves M *et al* 2001 Loss-of-function EA2 mutations are associated with impaired neuromuscular transmission *Neurology* **57** 1843–8
[333] Balck A, Tunc S, Schmitz J, Hollstein R, Kaiser F J and Bruggemann N 2018 A novel frameshift *CACNA1A* mutation causing episodic ataxia type 2 *Cerebellum* **17** 504–6
[334] Yugrakh M S and Levy O A 2012 Clinical reasoning: a middle-aged man with episodes of gait imbalance and a newly found genetic mutation *Neurology* **79** e135–9

[335] Ohba C, Osaka H and Iai M *et al* 2013 Diagnostic utility of whole exome sequencing in patients showing cerebellar and/or vermis atrophy in childhood *Neurogenetics* **14** 225–32
[336] Pietrobon D 2010 $Ca_V2.1$ channelopathies *Pflugers Arch.* **460** 375–93
[337] Denier C, Ducros A and Vahedi K *et al* 1999 High prevalence of *CACNA1A* truncations and broader clinical spectrum in episodic ataxia type 2 *Neurology* **52** 1816–21
[338] Jen J, Yue Q and Nelson S F *et al* 1999 A novel nonsense mutation in *CACNA1A* causes episodic ataxia and hemiplegia *Neurology* **53** 34–7
[339] Spacey S D, Hildebrand M E, Materek L A, Bird T D and Snutch T P 2004 Functional implications of a novel EA2 mutation in the P/Q-type calcium channel *Ann. Neurol.* **56** 213–20
[340] Spacey S D, Materek L A, Szczygielski B I and Bird T D 2005 Two novel *CACNA1A* gene mutations associated with episodic ataxia type 2 and interictal dystonia *Arch. Neurol.* **62** 314–6
[341] Wappl E, Koschak A and Poteser M *et al* 2002 Functional consequences of P/Q-type Ca^{2+} channel $Ca_V2.1$ missense mutations associated with episodic ataxia type 2 and progressive ataxia *J. Biol. Chem.* **277** 6960–6
[342] Tonelli A, D'Angelo M G and Salati R *et al* 2006 Early onset, non fluctuating spinocerebellar ataxia and a novel missense mutation in *CACNA1A* gene *J. Neurol. Sci.* **241** 13–7
[343] Friend K L, Crimmins D and Phan T G *et al* 1999 Detection of a novel missense mutation and second recurrent mutation in the *CACNA1A* gene in individuals with EA-2 and FHM *Hum. Genet.* **105** 261–5
[344] Petrovicova A, Brozman M and Kurca E *et al* 2017 Novel missense variant of *CACNA1A* gene in a Slovak family with episodic ataxia type 2 *Biomed. Pap. Med. Fac. Univ. Palacky Olomouc Czech Repub.* **161** 107–10
[345] Jouvenceau A, Eunson L H and Spauschus A *et al* 2001 Human epilepsy associated with dysfunction of the brain P/Q-type calcium channel *Lancet* **358** 801–7
[346] Graves T D, Imbrici P and Kors E E *et al* 2008 Premature stop codons in a facilitating EF-hand splice variant of $Ca_V2.1$ cause episodic ataxia type 2 *Neurobiol. Dis.* **32** 10–5
[347] Hu Y, Jiang H, Wang Q, Xie Z and Pan S 2013 Identification of a novel nonsense mutation p.Tyr1957Ter of *CACNA1A* in a Chinese family with episodic ataxia 2 *PLoS One* **8** e56362
[348] Iqbal Z, Rydning S L and Wedding I M *et al* 2017 Correction: targeted high throughput sequencing in hereditary ataxia and spastic paraplegia *PLoS One* **12** e0186571
[349] Melzer N, Classen J, Reiners K and Buttmann M 2010 Fluctuating neuromuscular transmission defects and inverse acetazolamide response in episodic ataxia type 2 associated with the novel $Ca_V2.1$ single amino acid substitution R2090Q *J. Neurol. Sci.* **296** 104–6
[350] Epi K C 2016 De Novo mutations in *SLC1A2* and *CACNA1A* are important causes of epileptic encephalopathies *Am. J. Hum. Genet.* **99** 287–98
[351] Reinson K, Oiglane-Shlik E and Talvik I *et al* 2016 Biallelic *CACNA1A* mutations cause early onset epileptic encephalopathy with progressive cerebral, cerebellar, and optic nerve atrophy *Am. J. Med. Genet.* A **170** 2173–6
[352] Blumkin L, Leshinsky-Silver E and Michelson M *et al* 2015 Paroxysmal tonic upward gaze as a presentation of de-novo mutations in *CACNA1A* *Eur. J. Paediatr. Neurol.* **19** 292–7
[353] Klassen T, Davis C and Goldman A *et al* 2011 Exome sequencing of ion channel genes reveals complex profiles confounding personal risk assessment in epilepsy *Cell* **145** 1036–48

[354] Molloy A, Kimmich O, Martindale J, Moore H, Hutchinson M and O'Riordan S 2013 A novel *CACNA1A* mutation associated with adult-onset, paroxysmal head tremor *Mov. Disord.* **28** 842–3

[355] Geerlings R P, Koehler P J and Haane D Y *et al* 2011 Head tremor related to *CACNA1A* mutations *Cephalalgia* **31** 1315–9

[356] Coutelier M, Coarelli G and Monin M L *et al* 2017 A panel study on patients with dominant cerebellar ataxia highlights the frequency of channelopathies *Brain* **140** 1579–94

[357] Ghoshal A, Uygun D S and Yang L *et al* 2020 Effects of a patient-derived de novo coding alteration of *CACNA1I* in mice connect a schizophrenia risk gene with sleep spindle deficits *Transl Psychiatry* **10** 29

[358] Travaglini L, Nardella M and Bellacchio E *et al* 2017 Missense mutations of *CACNA1A* are a frequent cause of autosomal dominant nonprogressive congenital ataxia *Eur. J. Paediatr. Neurol.* **21** 450–6

[359] Jiang H S, Wang D M and Wang Q *et al* 2016 Missense mutation R1345Q in *CACNA1A* gene causes a new type of ataxia with episodic tremor: clinical features, genetic analysis and treatment in a familial case *Nan Fang Yi Ke Da Xue Xue Bao* **36** 883–6

[360] Valence S, Cochet E and Rougeot C *et al* 2019 Exome sequencing in congenital ataxia identifies two new candidate genes and highlights a pathophysiological link between some congenital ataxias and early infantile epileptic encephalopathies *Genet. Med.* **21** 553–63

[361] Weyhrauch D L, Ye D and Boczek N J *et al* 2016 Whole exome sequencing and heterologous cellular electrophysiology studies elucidate a novel loss-of-function mutation in the *CACNA1A*-encoded neuronal P/Q-type calcium channel in a child with congenital hypotonia and developmental delay *Pediatr. Neurol.* **55** 46–51

[362] Self J, Mercer C and Boon E M *et al* 2009 Infantile nystagmus and late onset ataxia associated with a *CACNA1A* mutation in the intracellular loop between s4 and s5 of domain 3 *Eye (Lond.)* **23** 2251–5

[363] Kinder S, Ossig C and Wienecke M *et al* 2015 Novel frameshift mutation in the *CACNA1A* gene causing a mixed phenotype of episodic ataxia and familiar hemiplegic migraine *Eur. J. Paediatr. Neurol.* **19** 72–4

[364] Byers H M, Beatty C W, Hahn S H and Gospe S M Jr. 2016 Dramatic response after lamotrigine in a patient with epileptic encephalopathy and a *De Novo CACNA1A* variant *Pediatr. Neurol.* **60** 79–82

[365] Bahamonde M I, Serra S A and Drechsel O *et al* 2015 A single amino acid deletion (δF1502) in the S6 segment of $Ca_V2.1$ domain III associated with congenital ataxia increases channel activity and promotes Ca^{2+} influx *PLoS One* **10** e0146035

[366] Eldomery M K, Coban-Akdemir Z and Harel T *et al* 2017 Lessons learned from additional research analyses of unsolved clinical exome cases *Genome Med.* **9** 26

[367] Giffin N J, Benton S and Goadsby P J 2002 Benign paroxysmal torticollis of infancy: four new cases and linkage to *CACNA1A* mutation *Dev. Med. Child Neurol.* **44** 490–3

[368] Gambeta E, Gandini M A, Souza I A, Ferron L and Zamponi G W 2021 A *CACNA1A* variant associated with trigeminal neuralgia alters the gating of $Ca_V2.1$ channels *Mol. Brain* **14** 4

[369] Jodice C, Mantuano E and Veneziano L *et al* 1997 Episodic ataxia type 2 (EA2) and spinocerebellar ataxia type 6 (SCA6) due to CAG repeat expansion in the *CACNA1A* gene on chromosome 19p *Hum. Mol. Genet.* **6** 1973–8

[370] Pietrobon D 2007 Familial hemiplegic migraine *Neurotherapeutics* **4** 274–84

[371] Uchitel O D, Inchauspe C G, Urbano F J and Di Guilmi M N 2012 $Ca_V2.1$ voltage activated calcium channels and synaptic transmission in familial hemiplegic migraine pathogenesis *J. Physiol. Paris* **106** 12–22

[372] Charles A C and Baca S M 2013 Cortical spreading depression and migraine *Nat. Rev. Neurol.* **9** 637–44

[373] Lauritzen M 1994 Pathophysiology of the migraine aura. The spreading depression theory *Brain* **117** 199–210

[374] Hadjikhani N, Sanchez Del Rio M and Wu O *et al* 2001 Mechanisms of migraine aura revealed by functional MRI in human visual cortex *Proc. Natl Acad. Sci. USA* **98** 4687–92

[375] Leao A A 1947 Further observations on the spreading depression of activity in the cerebral cortex *J. Neurophysiol.* **10** 409–14

[376] Charles A and Brennan K 2009 Cortical spreading depression-new insights and persistent questions *Cephalalgia* **29** 1115–24

[377] Pietrobon D and Moskowitz M A 2014 Chaos and commotion in the wake of cortical spreading depression and spreading depolarizations *Nat. Rev. Neurosci.* **15** 379–93

[378] van den Maagdenberg A M, Pietrobon D and Pizzorusso T *et al* 2004 A *cacna1a* knockin migraine mouse model with increased susceptibility to cortical spreading depression *Neuron* **41** 701–10

[379] Kraus R L, Sinnegger M J, Glossmann H, Hering S and Striessnig J 1998 Familial hemiplegic migraine mutations change α1A Ca^{2+} channel kinetics *J. Biol. Chem.* **273** 5586–90

[380] Hans M, Luvisetto S and Williams M E *et al* 1999 Functional consequences of mutations in the human α1A calcium channel subunit linked to familial hemiplegic migraine *J. Neurosci.* **19** 1610–9

[381] Melliti K, Grabner M and Seabrook G R 2003 The familial hemiplegic migraine mutation R192Q reduces G-protein-mediated inhibition of P/Q-type ($Ca_V2.1$) calcium channels expressed in human embryonic kidney cells *J. Physiol.* **546** 337–47

[382] Cao Y Q and Tsien R W 2005 Effects of familial hemiplegic migraine type 1 mutations on neuronal P/Q-type Ca^{2+} channel activity and inhibitory synaptic transmission *Proc. Natl Acad. Sci. USA* **102** 2590–5

[383] Tottene A, Fellin T and Pagnutti S *et al* 2002 Familial hemiplegic migraine mutations increase Ca^{2+} influx through single human $Ca_V2.1$ channels and decrease maximal $Ca_V2.1$ current density in neurons *Proc. Natl Acad. Sci. USA* **99** 13284–9

[384] Kaja S, van de Ven R C and Broos L A *et al* 2005 Gene dosage-dependent transmitter release changes at neuromuscular synapses of *CACNA1A* R192Q knockin mice are non-progressive and do not lead to morphological changes or muscle weakness *Neuroscience* **135** 81–95

[385] van den Maagdenberg A M, Pizzorusso T and Kaja S *et al* 2010 High cortical spreading depression susceptibility and migraine-associated symptoms in $Ca_V2.1$ S218L mice *Ann. Neurol.* **67** 85–98

[386] Tottene A, Conti R and Fabbro A *et al* 2009 Enhanced excitatory transmission at cortical synapses as the basis for facilitated spreading depression in $Ca_V2.1$ knockin migraine mice *Neuron* **61** 762–73

[387] Tottene A, Pivotto F, Fellin T, Cesetti T, van den Maagdenberg A M and Pietrobon D 2005 Specific kinetic alterations of human $Ca_V2.1$ calcium channels produced by mutation

S218L causing familial hemiplegic migraine and delayed cerebral edema and coma after minor head trauma *J. Biol. Chem.* **280** 17678–86
[388] Pietrobon D 2010 Insights into migraine mechanisms and $Ca_V2.1$ calcium channel function from mouse models of familial hemiplegic migraine *J. Physiol.* **588** 1871–8
[389] Adams P J, Garcia E, David L S, Mulatz K J, Spacey S D and Snutch T P 2009 $Ca_V2.1$ P/Q-type calcium channel alternative splicing affects the functional impact of familial hemiplegic migraine mutations: implications for calcium channelopathies *Channels (Austin)* **3** 110–21
[390] Serra S A, Fernandez-Castillo N, Macaya A, Cormand B, Valverde M A and Fernandez-Fernandez J M 2009 The hemiplegic migraine-associated Y1245C mutation in *CACNA1A* results in a gain of channel function due to its effect on the voltage sensor and G-protein-mediated inhibition *Pflugers Arch.* **458** 489–502
[391] Ducros A, Denier C and Joutel A *et al* 1999 Recurrence of the T666M calcium channel *CACNA1A* gene mutation in familial hemiplegic migraine with progressive cerebellar ataxia *Am. J. Hum. Genet.* **64** 89–98
[392] Riant F, Ducros A, Ploton C, Barbance C, Depienne C and Tournier-Lasserve E 2010 *De novo* mutations in *ATP1A2* and *CACNA1A* are frequent in early-onset sporadic hemiplegic migraine *Neurology* **75** 967–72
[393] Striessnig J 2021 Voltage-gated Ca^{2+}-channel α1-subunit de novo missense mutations: gain or loss of function – implications for potential therapies *Front. Synaptic Neurosci.* **13** 634760
[394] Di Guilmi M N, Wang T and Inchauspe C G *et al* 2014 Synaptic gain-of-function effects of mutant $Ca_V2.1$ channels in a mouse model of familial hemiplegic migraine are due to increased basal $[Ca^{2+}]i$ *J. Neurosci.* **34** 7047–58
[395] Inagaki A, Frank C A, Usachev Y M, Benveniste M and Lee A 2014 Pharmacological correction of gating defects in the voltage-gated $Ca_V2.1$ Ca^{2+} channel due to a familial hemiplegic migraine mutation *Neuron* **81** 91–102
[396] Rettig J, Sheng Z H, Kim D K, Hodson C D, Snutch T P and Catterall W A 1996 Isoform-specific interaction of the α1A subunits of brain Ca^{2+} channels with the presynaptic proteins syntaxin and SNAP-25 *Proc. Natl Acad. Sci. USA* **93** 7363–8
[397] Sheng Z H, Rettig J, Cook T and Catterall W A 1996 Calcium-dependent interaction of N-type calcium channels with the synaptic core complex *Nature* **379** 451–4
[398] Cohen-Kutner M, Nachmanni D and Atlas D 2010 $Ca_V2.1$ (P/Q channel) interaction with synaptic proteins is essential for depolarization-evoked release *Channels (Austin)* **4** 266–77
[399] Yokoyama C T, Myers S J, Fu J, Mockus S M, Scheuer T and Catterall W A 2005 Mechanism of SNARE protein binding and regulation of Ca_V2 channels by phosphorylation of the synaptic protein interaction site *Mol. Cell Neurosci.* **28** 1–17
[400] Adams P J, Rungta R L, Garcia E, van den Maagdenberg A M, MacVicar B A and Snutch T P 2010 Contribution of calcium-dependent facilitation to synaptic plasticity revealed by migraine mutations in the P/Q-type calcium channel *Proc. Natl Acad. Sci. USA* **107** 18694–9
[401] Forsythe I D, Tsujimoto T, Barnes-Davies M, Cuttle M F and Takahashi T 1998 Inactivation of presynaptic calcium current contributes to synaptic depression at a fast central synapse *Neuron* **20** 797–807
[402] Mochida S, Few A P, Scheuer T and Catterall W A 2008 Regulation of presynaptic $Ca_V2.1$ channels by Ca^{2+} sensor proteins mediates short-term synaptic plasticity *Neuron* **57** 210–6

[403] Diaz-Rojas F, Sakaba T and Kawaguchi S Y 2015 Ca^{2+} current facilitation determines short-term facilitation at inhibitory synapses between cerebellar Purkinje cells *J. Physiol.* **593** 4889–904
[404] Benton M D and Raman I M 2009 Stabilization of Ca current in Purkinje neurons during high-frequency firing by a balance of Ca-dependent facilitation and inactivation *Channels (Austin)* **3** 393–401
[405] Barrett C F, Cao Y Q and Tsien R W 2005 Gating deficiency in a familial hemiplegic migraine type 1 mutant P/Q-type calcium channel *J. Biol. Chem.* **280** 24064–71
[406] Cain S M and Snutch T P 2011 Voltage-gated calcium channels and disease *Biofactors* **37** 197–205
[407] Pietrobon D 2013 Calcium channels and migraine *Biochim. Biophys. Acta* **1828** 1655–65
[408] Tao J, Liu P, Xiao Z, Zhao H, Gerber B R and Cao Y Q 2012 Effects of familial hemiplegic migraine type 1 mutation T666M on voltage-gated calcium channel activities in trigeminal ganglion neurons *J. Neurophysiol.* **107** 1666–80
[409] Inchauspe C G, Urbano F J and Di Guilmi M N *et al* 2010 Gain of function in FHM-1 $Ca_V2.1$ knock-in mice is related to the shape of the action potential *J. Neurophysiol.* **104** 291–9
[410] Cao Y Q, Piedras-Renteria E S, Smith G B, Chen G, Harata N C and Tsien R W 2004 Presynaptic Ca^{2+} channels compete for channel type-preferring slots in altered neurotransmission arising from Ca^{2+} channelopathy *Neuron* **43** 387–400
[411] Tyagi S, Ribera A B and Bannister R A 2019 Zebrafish as a model system for the study of severe $Ca_V2.1$ (α1A) channelopathies *Front Mol Neurosci.* **12** 329
[412] Wan J, Mamsa H and Johnston J L *et al* 2011 Large genomic deletions in *CACNA1A* cause episodic ataxia type 2 *Front. Neurol.* **2** 51
[413] Page K M, Heblich F and Margas W *et al* 2010 N terminus is key to the dominant negative suppression of Ca_V2 calcium channels: implications for episodic ataxia type 2 *J. Biol. Chem.* **285** 835–44
[414] Jeng C J, Chen Y T, Chen Y W and Tang C Y 2006 Dominant-negative effects of human P/Q-type Ca^{2+} channel mutations associated with episodic ataxia type 2 *Am. J. Physiol. Cell Physiol.* **290** C1209–20
[415] Cao Y Q and Tsien R W 2010 Different relationship of N- and P/Q-type Ca^{2+} channels to channel-interacting slots in controlling neurotransmission at cultured hippocampal synapses *J. Neurosci.* **30** 4536–46
[416] Kaunisto M A, Harno H and Kallela M *et al* 2004 Novel splice site *CACNA1A* mutation causing episodic ataxia type 2 *Neurogenetics* **5** 69–73
[417] Chaudhuri D, Chang S Y, DeMaria C D, Alvania R S, Soong T W and Yue D T 2004 Alternative splicing as a molecular switch for Ca^{2+}/calmodulin-dependent facilitation of P/Q-type Ca^{2+} channels *J. Neurosci.* **24** 6334–42
[418] Christel C and Lee A 2012 Ca^{2+}-dependent modulation of voltage-gated Ca^{2+} channels *Biochim. Biophys. Acta* **1820** 1243–52
[419] Thalhammer A, Contestabile A and Ermolyuk Y S *et al* 2017 Alternative splicing of P/Q-Type Ca^{2+} channels shapes presynaptic plasticity *Cell Rep.* **20** 333–43
[420] Campbell D B, North J B and Hess E J 1999 Tottering mouse motor dysfunction is abolished on the Purkinje cell degeneration (*pcd*) mutant background *Exp. Neurol.* **160** 268–78
[421] Cook A A, Fields E and Watt A J 2021 Losing the beat: contribution of Purkinje cell firing dysfunction to disease, and its reversal *Neuroscience* **462** 247–61

[422] Rose S J, Kriener L H and Heinzer A K *et al* 2014 The first knockin mouse model of episodic ataxia type 2 *Exp. Neurol.* **261** 553–62

[423] Li M, Zheng X and Zhong R *et al* 2019 Familial hemiplegic migraine with progressive cerebellar ataxia caused by a p.Thr666Met *CACNA1A* gene mutation in a Chinese family *Front. Neurol.* **10** 1221

[424] Gao Z, Todorov B and Barrett C F *et al* 2012 Cerebellar ataxia by enhanced $Ca_V2.1$ currents is alleviated by Ca^{2+}-dependent K^+-channel activators in *Cacna1a*(S218L) mutant mice *J. Neurosci.* **32** 15533–46

[425] Yabe I, Kitagawa M and Suzuki Y *et al* 2008 Downbeat positioning nystagmus is a common clinical feature despite variable phenotypes in an FHM1 family *J. Neurol.* **255** 1541–4

[426] Carreno O, Corominas R and Serra S A *et al* 2013 Screening of *CACNA1A* and *ATP1A2* genes in hemiplegic migraine: clinical, genetic, and functional studies *Mol. Genet. Genomic Med.* **1** 206–22

[427] Du X and Gomez C M 2018 Spinocerebellar [corrected] ataxia type 6: molecular mechanisms and calcium channel genetics *Adv. Exp. Med. Biol.* **1049** 147–73

[428] Kordasiewicz H B and Gomez C M 2007 Molecular pathogenesis of spinocerebellar ataxia type 6 *Neurotherapeutics* **4** 285–94

[429] Riess O, Schols L and Bottger H *et al* 1997 SCA6 is caused by moderate CAG expansion in the α_{1A}-voltage-dependent calcium channel gene *Hum. Mol. Genet.* **6** 1289–93

[430] Watase K, Barrett C F and Miyazaki T *et al* 2008 Spinocerebellar ataxia type 6 knockin mice develop a progressive neuronal dysfunction with age-dependent accumulation of mutant $Ca_V2.1$ channels *Proc. Natl Acad. Sci. USA* **105** 11987–92

[431] Tsuchiya K, Oda T and Yoshida M *et al* 2005 Degeneration of the inferior olive in spinocerebellar ataxia 6 may depend on disease duration: report of two autopsy cases and statistical analysis of autopsy cases reported to date *Neuropathology* **25** 125–35

[432] Tsunemi T, Saegusa H and Ishikawa K *et al* 2002 Novel $Ca_V2.1$ splice variants isolated from Purkinje cells do not generate P-type Ca^{2+} current *J. Biol. Chem.* **277** 7214–21

[433] Paulson H L, Shakkottai V G, Clark H B and Orr H T 2017 Polyglutamine spinocerebellar ataxias – from genes to potential treatments *Nat. Rev. Neurosci.* **18** 613–26

[434] Kordasiewicz H B, Thompson R M, Clark H B and Gomez C M 2006 C-termini of P/Q-type Ca^{2+} channel α_{1A} subunits translocate to nuclei and promote polyglutamine-mediated toxicity *Hum. Mol. Genet.* **15** 1587–99

[435] Du X, Wang J and Zhu H *et al* 2013 Second cistron in *CACNA1A* gene encodes a transcription factor mediating cerebellar development and SCA6 *Cell* **154** 118–33

[436] Saegusa H, Wakamori M and Matsuda Y *et al* 2007 Properties of human $Ca_V2.1$ channel with a spinocerebellar ataxia type 6 mutation expressed in Purkinje cells *Mol. Cell Neurosci.* **34** 261–70

[437] Matsuyama Z, Wakamori M, Mori Y, Kawakami H, Nakamura S and Imoto K 1999 Direct alteration of the P/Q-type Ca^{2+} channel property by polyglutamine expansion in spinocerebellar ataxia 6. *J. Neurosci.* **19** RC14

[438] Toru S, Murakoshi T and Ishikawa K *et al* 2000 Spinocerebellar ataxia type 6 mutation alters P-type calcium channel function *J. Biol. Chem.* **275** 10893–8

[439] Piedras-Renteria E S, Watase K and Harata N *et al* 2001 Increased expression of α_{1A} Ca^{2+} channel currents arising from expanded trinucleotide repeats in spinocerebellar ataxia type 6 *J. Neurosci.* **21** 9185–93

[440] DiFiglia M, Sapp E and Chase K O *et al* 1997 Aggregation of huntingtin in neuronal intranuclear inclusions and dystrophic neurites in brain *Science* **277** 1990–3

[441] Verhoef L G, Lindsten K, Masucci M G and Dantuma N P 2002 Aggregate formation inhibits proteasomal degradation of polyglutamine proteins *Hum. Mol. Genet.* **11** 2689–700

[442] Goellner G M and Rechsteiner M 2003 Are Huntington's and polyglutamine-based ataxias proteasome storage diseases? *Int. J. Biochem. Cell Biol.* **35** 562–71

[443] Bence N F, Sampat R M and Kopito R R 2001 Impairment of the ubiquitin-proteasome system by protein aggregation *Science* **292** 1552–5

[444] Chen H and Piedras-Renteria E S 2007 Altered frequency-dependent inactivation and steady-state inactivation of polyglutamine-expanded α_{1A} in SCA6 *Am. J. Physiol. Cell Physiol.* **292** C1078–86

[445] Llinas R and Sugimori M 1980 Electrophysiological properties of *in vitro* Purkinje cell somata in mammalian cerebellar slices *J. Physiol.* **305** 171–95

[446] Raman I M and Bean B P 1999 Ionic currents underlying spontaneous action potentials in isolated cerebellar Purkinje neurons *J. Neurosci.* **19** 1663–74

[447] Unno T, Wakamori M and Koike M *et al* 2012 Development of Purkinje cell degeneration in a knockin mouse model reveals lysosomal involvement in the pathogenesis of SCA6 *Proc. Natl Acad. Sci. USA* **109** 17693–8

[448] Orr H T 2012 Cell biology of spinocerebellar ataxia *J. Cell Biol.* **197** 167–77

[449] Sakurai T, Westenbroek R E, Rettig J, Hell J and Catterall W A 1996 Biochemical properties and subcellular distribution of the BI and rbA isoforms of α_{1A} subunits of brain calcium channels *J. Cell Biol.* **134** 511–28

[450] Kubodera T, Yokota T and Ohwada K *et al* 2003 Proteolytic cleavage and cellular toxicity of the human α_{1A} calcium channel in spinocerebellar ataxia type 6 *Neurosci. Lett.* **341** 74–8

[451] Miyazaki Y, Du X, Muramatsu S and Gomez C M 2016 An miRNA-mediated therapy for SCA6 blocks IRES-driven translation of the *CACNA1A* second cistron *Sci. Transl. Med.* **8** 347ra94

[452] Lv Y, Wang Z, Liu C and Cui L 2017 Identification of a novel *CACNA1A* mutation in a Chinese family with autosomal recessive progressive myoclonic epilepsy *Neuropsychiatr. Dis. Treat.* **13** 2631–6

[453] Sun J, Sun X, Li Z, Ma D and Lv Y 2020 An elongated tract of polyQ in the carboxylterminus of human α_{1A} calcium channel induces cell apoptosis by nuclear translocation *Oncol. Rep.* **44** 156–64

[454] Di Stefano G, Yuan J H, Cruccu G, Waxman S G, Dib-Hajj S D and Truini A 2020 Familial trigeminal neuralgia – a systematic clinical study with a genomic screen of the neuronal electrogenisome *Cephalalgia* **40** 767–77

[455] Jones M R, Urits I and Ehrhardt K P *et al* 2019 A comprehensive review of trigeminal neuralgia *Curr. Pain Headache Rep.* **23** 74

[456] Dravet C, Bureau M, Oguni H, Fukuyama Y and Cokar O 2005 Severe myoclonic epilepsy in infancy: Dravet syndrome *Adv. Neurol.* **95** 71–102

[457] Serra S A, Cuenca-Leon E and Llobet A *et al* 2010 A mutation in the first intracellular loop of *CACNA1A* prevents P/Q channel modulation by SNARE proteins and lowers exocytosis *Proc. Natl Acad. Sci. USA* **107** 1672–7

[458] Serra S A, Gene G G, Elorza-Vidal X and Fernandez-Fernandez J M 2018 Cross talk between beta subunits, intracellular Ca^{2+} signaling, and SNAREs in the modulation of $Ca_V2.1$ channel steady-state inactivation *Physiol. Rep.* **6** e13557

[459] Pragnell M, De Waard M, Mori Y, Tanabe T, Snutch T P and Campbell K P 1994 Calcium channel beta-subunit binds to a conserved motif in the I–II cytoplasmic linker of the alpha 1-subunit *Nature* **368** 67–70

[460] Catterall W A 2010 Ion channel voltage sensors: structure, function, and pathophysiology *Neuron* **67** 915–28

[461] Miki T, Zwingman T A and Wakamori M *et al* 2008 Two novel alleles of tottering with distinct $Ca_V2.1$ calcium channel neuropathologies *Neuroscience* **155** 31–44

[462] Gorman K M, Meyer E and Grozeva D *et al* 2019 Bi-allelic loss-of-function *CACNA1B* mutations in progressive epilepsy-dyskinesia *Am. J. Hum. Genet.* **104** 948–56

[463] Williams M E, Brust P F and Feldman D H *et al* 1992 Structure and functional expression of an omega-conotoxin-sensitive human N-type calcium channel *Science* **257** 389–95

[464] Chaplan S R, Pogrel J W and Yaksh T L 1994 Role of voltage-dependent calcium channel subtypes in experimental tactile allodynia *J. Pharmacol. Exp. Ther.* **269** 1117–23

[465] Bowersox S S, Gadbois T, Singh T, Pettus M, Wang Y X and Luther R R 1996 Selective N-type neuronal voltage-sensitive calcium channel blocker, SNX-111, produces spinal antinociception in rat models of acute, persistent and neuropathic pain *J. Pharmacol. Exp. Ther.* **279** 1243–9

[466] Sheng Z H, Rettig J, Takahashi M and Catterall W A 1994 Identification of a syntaxin-binding site on N-type calcium channels *Neuron* **13** 1303–13

[467] Bezprozvanny I, Scheller R H and Tsien R W 1995 Functional impact of syntaxin on gating of N-type and Q-type calcium channels *Nature* **378** 623–6

[468] Simms B A and Zamponi G W 2014 Neuronal voltage-gated calcium channels: structure, function, and dysfunction *Neuron* **82** 24–45

[469] Komuro H and Rakic P 1992 Selective role of N-type calcium channels in neuronal migration *Science* **257** 806–9

[470] Iwasaki S, Momiyama A, Uchitel O D and Takahashi T 2000 Developmental changes in calcium channel types mediating central synaptic transmission *J. Neurosci.* **20** 59–65

[471] Simms B A, Souza I A and Zamponi G W 2014 Effect of the Brugada syndrome mutation A39V on calmodulin regulation of $Ca_V1.2$ channels *Mol. Brain* **7** 34

[472] Hatakeyama S, Wakamori M and Ino M *et al* 2001 Differential nociceptive responses in mice lacking the α_{1B} subunit of N-type Ca^{2+} channels *Neuroreport* **12** 2423–7

[473] Kim C, Jun K and Lee T *et al* 2001 Altered nociceptive response in mice deficient in the α_{1B} subunit of the voltage-dependent calcium channel *Mol. Cell Neurosci.* **18** 235–45

[474] Saegusa H, Kurihara T and Zong S *et al* 2001 Suppression of inflammatory and neuropathic pain symptoms in mice lacking the N-type Ca^{2+} channel *EMBO J.* **20** 2349–56

[475] Newton P M, Orr C J, Wallace M J, Kim C, Shin H S and Messing R O 2004 Deletion of N-type calcium channels alters ethanol reward and reduces ethanol consumption in mice *J. Neurosci.* **24** 9862–9

[476] Mori Y, Nishida M and Shimizu S *et al* 2002 Ca^{2+} channel α_{1B} subunit ($Ca_V2.2$) knockout mouse reveals a predominant role of N-type channels in the sympathetic regulation of the circulatory system *Trends Cardiovasc. Med.* **12** 270–5

[477] Beuckmann C T, Sinton C M, Miyamoto N, Ino M and Yanagisawa M 2003 N-type calcium channel α_{1B} subunit ($Ca_V2.2$) knock-out mice display hyperactivity and vigilance state differences *J. Neurosci.* **23** 6793–7

[478] Groen J L, Andrade A and Ritz K *et al* 2015 *CACNA1B* mutation is linked to unique myoclonus-dystonia syndrome *Hum. Mol. Genet.* **24** 987–93
[479] Weiss N 2008 The N-type voltage-gated calcium channel: when a neuron reads a map *J. Neurosci.* **28** 5621–2
[480] Klein C and Ozelius L J 2002 Dystonia: clinical features, genetics, and treatment *Curr. Opin. Neurol.* **15** 491–7
[481] Weissbach A, Werner E and Bally J F *et al* 2017 Alcohol improves cerebellar learning deficit in myoclonus-dystonia: a clinical and electrophysiological investigation *Ann. Neurol.* **82** 543–53
[482] Mencacci N E, R'Bibo L and Bandres-Ciga S *et al* 2015 The *CACNA1B* R1389H variant is not associated with myoclonus-dystonia in a large European multicentric cohort *Hum. Mol. Genet.* **24** 5326–9
[483] Cocos R, Raicu F, Bajenaru O L, Olaru I, Dumitrescu L and Popescu B O 2021 *CACNA1B* gene variants in adult-onset isolated focal dystonia *Neurol Sci.* **42** 1113–7
[484] Wormuth C, Lundt A and Henseler C *et al* 2016 Review: $Ca_V2.3$ R-type voltage-gated Ca^{2+} channels – functional implications in convulsive and non-convulsive seizure activity *Open Neurol. J.* **10** 99–126
[485] Jing X, Li D Q and Olofsson C S *et al* 2005 $Ca_V2.3$ calcium channels control second-phase insulin release *J. Clin. Invest.* **115** 146–54
[486] Pereverzev A, Mikhna M and Vajna R *et al* 2002 Disturbances in glucose-tolerance, insulin-release, and stress-induced hyperglycemia upon disruption of the $Ca_V2.3$ (α_{1E}) subunit of voltage-gated Ca^{2+} channels *Mol Endocrinol* **16** 884–95
[487] Lu Z J, Pereverzev A and Liu H L *et al* 2004 Arrhythmia in isolated prenatal hearts after ablation of the $Ca_V2.3$ (α_{1E}) subunit of voltage-gated Ca^{2+} channels *Cell. Physiol. Biochem.* **14** 11–22
[488] Brennan S C, Finney B A and Lazarou M *et al* 2013 Fetal calcium regulates branching morphogenesis in the developing human and mouse lung: involvement of voltage-gated calcium channels *PLoS One* **8** e80294
[489] Nishiyama M, Togashi K and von Schimmelmann M J *et al* 2011 Semaphorin 3A induces $Ca_V2.3$ channel-dependent conversion of axons to dendrites *Nat. Cell Biol.* **13** 676–85
[490] Sochivko D, Pereverzev A, Smyth N, Gissel C, Schneider T and Beck H 2002 The $Ca_V2.3$ Ca^{2+} channel subunit contributes to R-type Ca^{2+} currents in murine hippocampal and neocortical neurones *J. Physiol.* **542** 699–710
[491] Day N C, Shaw P J and McCormack A L *et al* 1996 Distribution of α_{1A}, α_{1B} and α_{1E} voltage-dependent calcium channel subunits in the human hippocampus and parahippocampal gyrus *Neuroscience* **71** 1013–24
[492] Hanson J E and Smith Y 2002 Subcellular distribution of high-voltage-activated calcium channel subtypes in rat globus pallidus neurons *J. Comp. Neurol.* **442** 89–98
[493] Day N C, Wood S J and Ince P G *et al* 1997 Differential localization of voltage-dependent calcium channel α_1 subunits at the human and rat neuromuscular junction *J. Neurosci.* **17** 6226–35
[494] Wu L G, Westenbroek R E, Borst J G, Catterall W A and Sakmann B 1999 Calcium channel types with distinct presynaptic localization couple differentially to transmitter release in single calyx-type synapses *J. Neurosci.* **19** 726–36
[495] Dietrich D, Kirschstein T and Kukley M *et al* 2003 Functional specialization of presynaptic $Ca_V2.3$ Ca^{2+} channels *Neuron* **39** 483–96

[496] Westenbroek R E, Sakurai T and Elliott E M *et al* 1995 Immunochemical identification and subcellular distribution of the α_{1A} subunits of brain calcium channels *J. Neurosci.* **15** 6403–18
[497] Benkert J, Hess S and Roy S *et al* 2019 $Ca_V2.3$ channels contribute to dopaminergic neuron loss in a model of Parkinson's disease *Nat. Commun.* **10** 5094
[498] Weiergraber M, Henry M, Radhakrishnan K, Hescheler J and Schneider T 2007 Hippocampal seizure resistance and reduced neuronal excitotoxicity in mice lacking the $Ca_V2.3$ E/R-type voltage-gated calcium channel *J. Neurophysiol.* **97** 3660–9
[499] Matsuda Y, Saegusa H, Zong S, Noda T and Tanabe T 2001 Mice lacking $Ca_V2.3$ (α_{1E}) calcium channel exhibit hyperglycemia *Biochem. Biophys. Res. Commun.* **289** 791–5
[500] Saegusa H, Kurihara T and Zong S *et al* 2000 Altered pain responses in mice lacking α_{1E} subunit of the voltage-dependent Ca^{2+} channel *Proc. Natl Acad. Sci. USA* **97** 6132–7
[501] Heyne H O, Singh T and Stamberger H *et al* 2018 De novo variants in neurodevelopmental disorders with epilepsy *Nat. Genet.* **50** 1048–53
[502] Helbig K L, Lauerer R J and Bahr J C *et al* 2018 *De Novo* pathogenic variants in *CACNA1E* cause developmental and epileptic encephalopathy with contractures, macrocephaly, and dyskinesias *Am. J. Hum. Genet.* **103** 666–78
[503] Helbig K L, Farwell Hagman K D and Shinde D N *et al* 2016 Diagnostic exome sequencing provides a molecular diagnosis for a significant proportion of patients with epilepsy *Genet. Med.* **18** 898–905
[504] Mangoni M E, Traboulsie A and Leoni A L *et al* 2006 Bradycardia and slowing of the atrioventricular conduction in mice lacking $Ca_V3.1/\alpha_{1G}$ T-type calcium channels *Circ. Res.* **98** 1422–30
[505] Choi S, Yu E, Hwang E and Llinas R R 2016 Pathophysiological implication of $Ca_V3.1$ T-type Ca^{2+} channels in trigeminal neuropathic pain *Proc. Natl Acad. Sci. USA* **113** 2270–5
[506] Na H S, Choi S, Kim J, Park J and Shin H S 2008 Attenuated neuropathic pain in $Ca_V3.1$ null mice *Mol. Cells* **25** 242–6
[507] Thuesen A D, Andersen K and Lyngso K S *et al* 2018 Deletion of T-type calcium channels $Ca_V3.1$ or $Ca_V3.2$ attenuates endothelial dysfunction in aging mice *Pflugers Arch.* **470** 355–65
[508] Kim D, Song I and Keum S *et al* 2001 Lack of the burst firing of thalamocortical relay neurons and resistance to absence seizures in mice lacking α_{1G} T-type Ca^{2+} channels *Neuron* **31** 35–45
[509] Mancuso M, Orsucci D, Siciliano G and Bonuccelli U 2014 The genetics of ataxia: through the labyrinth of the minotaur, looking for Ariadne's thread *J. Neurol.* **261** S528–41
[510] Berecki G, Helbig K L and Ware T L *et al* 2020 Novel missense *CACNA1G* mutations associated with infantile-onset developmental and epileptic encephalopathy *Int. J. Mol. Sci.* **21** 6333
[511] Chemin J, Siquier-Pernet K and Nicouleau M *et al* 2018 De novo mutation screening in childhood-onset cerebellar atrophy identifies gain-of-function mutations in the CACNA1G calcium channel gene *Brain* **141** 1998–2013
[512] Kunii M, Doi H and Hashiguchi S *et al* 2020 De novo *CACNA1G* variants in developmental delay and early-onset epileptic encephalopathies *J. Neurol. Sci.* **416** 117047
[513] Coutelier M, Blesneac I and Monteil A *et al* 2015 A recurrent mutation in *CACNA1G* alters $Ca_V3.1$ T-type calcium-channel conduction and causes autosomal-dominant cerebellar ataxia *Am. J. Hum. Genet.* **97** 726–37

[514] Morino H, Matsuda Y and Muguruma K *et al* 2015 A mutation in the low voltage-gated calcium channel *CACNA1G* alters the physiological properties of the channel, causing spinocerebellar ataxia *Mol. Brain* **8** 89
[515] Kimura M, Yabe I and Hama Y *et al* 2017 SCA42 mutation analysis in a case series of Japanese patients with spinocerebellar ataxia *J. Hum. Genet.* **62** 857–9
[516] Ngo K, Aker M and Petty L E *et al* 2018 Expanding the global prevalence of spinocerebellar ataxia type 42 *Neurol. Genet.* **4** e232
[517] Li X, Zhou C and Cui L *et al* 2018 A case of a novel *CACNA1G* mutation from a Chinese family with SCA42: a case report and literature review *Medicine (Baltimore)* **97** e12148
[518] Singh B, Monteil A and Bidaud I *et al* 2007 Mutational analysis of *CACNA1G* in idiopathic generalized epilepsy. Mutation in brief #962. online *Hum. Mutat.* **28** 524–5
[519] Odgerel Z, Sonti S and Hernandez N *et al* 2019 Whole genome sequencing and rare variant analysis in essential tremor families *PLoS One* **14** e0220512
[520] Ernst W L, Zhang Y, Yoo J W, Ernst S J and Noebels J L 2009 Genetic enhancement of thalamocortical network activity by elevating α1G-mediated low-voltage-activated calcium current induces pure absence epilepsy *J. Neurosci.* **29** 1615–25
[521] Tringham E, Powell K L and Cain S M *et al* 2012 T-type calcium channel blockers that attenuate thalamic burst firing and suppress absence seizures *Sci. Transl. Med.* **4** 121ra19
[522] Powell K L, Cain S M, Snutch T P and O'Brien T J 2014 Low threshold T-type calcium channels as targets for novel epilepsy treatments *Br. J. Clin. Pharmacol.* **77** 729–39
[523] Casillas-Espinosa P M, Hicks A, Jeffreys A, Snutch T P, O'Brien T J and Powell K L 2015 Z944, a novel selective T-type calcium channel antagonist delays the progression of seizures in the amygdala kindling model *PLoS One* **10** e0130012
[524] Calhoun J D, Hawkins N A, Zachwieja N J and Kearney J A 2016 *Cacna1g* is a genetic modifier of epilepsy caused by mutation of voltage-gated sodium channel Scn2a *Epilepsia* **57** e103–7
[525] Calhoun J D, Hawkins N A, Zachwieja N J and Kearney J A 2017 *Cacna1g* is a genetic modifier of epilepsy in a mouse model of Dravet syndrome *Epilepsia* **58** e111–e5
[526] Louis E D and Ferreira J J 2010 How common is the most common adult movement disorder? update on the worldwide prevalence of essential tremor *Mov. Disord.* **25** 534–41
[527] Monteil A, Chausson P and Boutourlinsky K *et al* 2015 Inhibition of $Ca_V3.2$ T-type calcium channels by Its intracellular I–II Loop *J. Biol. Chem.* **290** 16168–76
[528] Chen C C, Lamping K G and Nuno D W *et al* 2003 Abnormal coronary function in mice deficient in α_{1H} T-type Ca^{2+} channels *Science* **302** 1416–8
[529] Hansen P B 2015 Functional importance of T-type voltage-gated calcium channels in the cardiovascular and renal system: news from the world of knockout mice *Am. J. Physiol. Regul. Integr. Comp. Physiol.* **308** R227–37
[530] Carmines P K, Fowler B C and Bell P D 1993 Segmentally distinct effects of depolarization on intracellular $[Ca^{2+}]$ in renal arterioles *Am. J. Physiol.* **265** F677–85
[531] Feng M G, Li M and Navar L G 2004 T-type calcium channels in the regulation of afferent and efferent arterioles in rats *Am. J. Physiol. Renal Physiol.* **286** F331–7
[532] Harada K, Nomura M, Nishikado A, Uehara K, Nakaya Y and Ito S 2003 Clinical efficacy of efonidipine hydrochloride, a T-type calcium channel inhibitor, on sympathetic activities *Circ. J.* **67** 139–45

[533] Hofmann F, Lacinova L and Klugbauer N 1999 Voltage-dependent calcium channels: from structure to function *Rev. Physiol., Biochem. Pharmacol.* **139** 33–87

[534] Thuesen A D, Andersen H and Cardel M *et al* 2014 Differential effect of T-type voltage-gated Ca^{2+} channel disruption on renal plasma flow and glomerular filtration rate *in vivo Am. J. Physiol. Renal Physiol.* **307** F445–52

[535] Chiang C S, Huang C H and Chieng H *et al* 2009 The $Ca_V3.2$ T-type Ca^{2+} channel is required for pressure overload-induced cardiac hypertrophy in mice *Circ. Res.* **104** 522–30

[536] Choi S, Na H S and Kim J *et al* 2007 Attenuated pain responses in mice lacking $Ca_V3.2$ T-type channels *Genes Brain Behav.* **6** 425–31

[537] Gangarossa G, Laffray S, Bourinet E and Valjent E 2014 T-type calcium channel $Ca_V3.2$ deficient mice show elevated anxiety, impaired memory and reduced sensitivity to psychostimulants *Front. Behav. Neurosci.* **8** 92

[538] Calhoun J D, Huffman A M and Bellinski I *et al* 2020 *CACNA1H* variants are not a cause of monogenic epilepsy *Hum. Mutat.* **41** 1138–44

[539] Tsakiridou E, Bertollini L, de Curtis M, Avanzini G and Pape H C 1995 Selective increase in T-type calcium conductance of reticular thalamic neurons in a rat model of absence epilepsy *J. Neurosci.* **15** 3110–7

[540] Zhang Y, Mori M, Burgess D L and Noebels J L 2002 Mutations in high-voltage-activated calcium channel genes stimulate low-voltage-activated currents in mouse thalamic relay neurons *J. Neurosci.* **22** 6362–71

[541] Zhang Y, Vilaythong A P, Yoshor D and Noebels J L 2004 Elevated thalamic low-voltage-activated currents precede the onset of absence epilepsy in the SNAP25-deficient mouse mutant coloboma *J. Neurosci.* **24** 5239–48

[542] Marescaux C, Micheletti G, Vergnes M, Depaulis A, Rumbach L and Warter J M 1984 A model of chronic spontaneous petit mal-like seizures in the rat: comparison with pentylenetetrazol-induced seizures *Epilepsia* **25** 326–31

[543] Vitko I, Chen Y, Arias J M, Shen Y, Wu X R and Perez-Reyes E 2005 Functional characterization and neuronal modeling of the effects of childhood absence epilepsy variants of *CACNA1H*, a T-type calcium channel *J. Neurosci.* **25** 4844–55

[544] Eckle V S, Shcheglovitov A and Vitko I *et al* 2014 Mechanisms by which a *CACNA1H* mutation in epilepsy patients increases seizure susceptibility *J. Physiol.* **592** 795–809

[545] Khosravani H, Altier C and Simms B *et al* 2004 Gating effects of mutations in the $Ca_V3.2$ T-type calcium channel associated with childhood absence epilepsy *J. Biol. Chem.* **279** 9681–4

[546] Khosravani H, Bladen C, Parker D B, Snutch T P, McRory J E and Zamponi G W 2005 Effects of $Ca_V3.2$ channel mutations linked to idiopathic generalized epilepsy *Ann. Neurol.* **57** 745–9

[547] Heron S E, Khosravani H and Varela D *et al* 2007 Extended spectrum of idiopathic generalized epilepsies associated with *CACNA1H* functional variants *Ann. Neurol.* **62** 560–8

[548] Peloquin J B, Khosravani H and Barr W *et al* 2006 Functional analysis of $Ca_V3.2$ T-type calcium channel mutations linked to childhood absence epilepsy *Epilepsia* **47** 655–8

[549] Daniil G, Fernandes-Rosa F L and Chemin J *et al* 2016 *CACNA1H* mutations are associated with different forms of primary aldosteronism *EBioMedicine* **13** 225–36

[550] Scholl U I, Stolting G and Nelson-Williams C *et al* 2015 Recurrent gain of function mutation in calcium channel *CACNA1H* causes early-onset hypertension with primary aldosteronism *Elife* **4** e06315
[551] Gurtler F, Jordan K and Tegtmeier I *et al* 2020 Cellular pathophysiology of mutant voltage-dependent Ca^{2+} channel *CACNA1H* in primary aldosteronism *Endocrinology* **161**
[552] Wulczyn K, Perez-Reyes E, Nussbaum R L and Park M 2019 Primary aldosteronism associated with a germline variant in *CACNA1H BMJ Case Rep.* **12** e229031
[553] Daniil G, Phedonos A A and Holleboom A G *et al* 2011 Characterization of antioxidant/anti-inflammatory properties and apoA-I-containing subpopulations of HDL from family subjects with monogenic low HDL disorders *Clin. Chim. Acta* **412** 1213–20
[554] Nanba K, Blinder A R and Rege J *et al* 2020 Somatic *CACNA1H* mutation as a cause of aldosterone-producing adenoma *Hypertension* **75** 645–9
[555] Souza I A, Gandini M A, Wan M M and Zamponi G W 2016 Two heterozygous $Ca_V3.2$ channel mutations in a pediatric chronic pain patient: recording condition-dependent biophysical effects *Pflugers Arch.* **468** 635–42
[556] Rzhepetskyy Y, Lazniewska J, Blesneac I, Pamphlett R and Weiss N 2016 *CACNA1H* missense mutations associated with amyotrophic lateral sclerosis alter $Ca_V3.2$ T-type calcium channel activity and reticular thalamic neuron firing *Channels (Austin)* **10** 466–77
[557] Stringer R N, Jurkovicova-Tarabova B and Huang S *et al* 2020 A rare *CACNA1H* variant associated with amyotrophic lateral sclerosis causes complete loss of $Ca_V3.2$ T-type channel activity *Mol. Brain* **13** 33
[558] Zhong X, Liu J R, Kyle J W, Hanck D A and Agnew W S 2006 A profile of alternative RNA splicing and transcript variation of *CACNA1H*, a human T-channel gene candidate for idiopathic generalized epilepsies *Hum. Mol. Genet.* **15** 1497–512
[559] Duzhyy D E, Viatchenko-Karpinski V Y, Khomula E V, Voitenko N V and Belan P V 2015 Upregulation of T-type Ca^{2+} channels in long-term diabetes determines increased excitability of a specific type of capsaicin-insensitive DRG neurons *Mol. Pain* **11**
[560] Jagodic M M, Pathirathna S and Joksovic P M *et al* 2008 Upregulation of the T-type calcium current in small rat sensory neurons after chronic constrictive injury of the sciatic nerve *J. Neurophysiol.* **99** 3151–6
[561] Marger F, Gelot A and Alloui A *et al* 2011 T-type calcium channels contribute to colonic hypersensitivity in a rat model of irritable bowel syndrome *Proc. Natl Acad. Sci. USA* **108** 11268–73
[562] Steinberg K M, Yu B, Koboldt D C, Mardis E R and Pamphlett R 2015 Exome sequencing of case-unaffected-parents trios reveals recessive and de novo genetic variants in sporadic ALS *Sci Rep.* **5** 9124
[563] Etkin A and Wager T D 2013 Neurodevelopmental disorders *Diagnostic and Statistical Manual of Mental Disorders* (Washington, DC: American Psychiatric Association) pp 1476–88
[564] Talley E M, Cribbs L L, Lee J H, Daud A, Perez-Reyes E and Bayliss D A 1999 Differential distribution of three members of a gene family encoding low voltage-activated (T-type) calcium channels *J. Neurosci.* **19** 1895–911
[565] Seitter H and Koschak A 2018 Relevance of tissue specific subunit expression in channelopathies *Neuropharmacology* **132** 58–70
[566] Astori S, Wimmer R D and Prosser H M *et al* 2011 The $Ca_V3.3$ calcium channel is the major sleep spindle pacemaker in thalamus *Proc. Natl Acad. Sci. USA* **108** 13823–8

[567] Pellegrini C, Lecci S, Luthi A and Astori S 2016 Suppression of sleep spindle rhythmogenesis in mice with deletion of $Ca_V3.2$ and $Ca_V3.3$ T-type Ca^{2+} channels *Sleep* **39** 875–85

[568] Lee S E, Lee J and Latchoumane C *et al* 2014 Rebound burst firing in the reticular thalamus is not essential for pharmacological absence seizures in mice *Proc. Natl Acad. Sci. USA* **111** 11828–33

[569] Harraz O F, Visser F and Brett S E *et al* 2015 $Ca_V1.2/Ca_V3.x$ channels mediate divergent vasomotor responses in human cerebral arteries *J. Gen. Physiol.* **145** 405–18

[570] Lu A T, Dai X, Martinez-Agosto J A and Cantor R M 2012 Support for calcium channel gene defects in autism spectrum disorders *Mol. Autism* **3** 18

[571] Sanchez-Roige S, Fontanillas P and Elson S L *et al* 2019 Genome-Wide association studies of impulsive personality traits (BIS-11 and UPPS-P) and drug experimentation in up to 22,861 adult research participants identify loci in the *CACNA1I* and *CADM2* genes *J. Neurosci.* **39** 2562–72

[572] Henriksen M G, Nordgaard J and Jansson L B 2017 Genetics of schizophrenia: overview of methods, findings and limitations *Front. Hum. Neurosci.* **11** 322

[573] Gulsuner S, Walsh T and Watts A C *et al* 2013 Spatial and temporal mapping of *de novo* mutations in schizophrenia to a fetal prefrontal cortical network *Cell* **154** 518–29

[574] Andrade A, Hope J, Allen A, Yorgan V, Lipscombe D and Pan J Q 2016 A rare schizophrenia risk variant of *CACNA1I* disrupts $Ca_V3.3$ channel activity *Sci Rep.* **6** 34233

[575] El Ghaleb Y, Schneeberger P E and Fernandez-Quintero M L *et al* 2021 *CACNA1I* gain-of-function mutations differentially affect channel gating and cause neurodevelopmental disorders *Brain* **144** 2092–106

[576] Wang J, Zhang Y and Liang J *et al* 2006 *CACNA1I* is not associated with childhood absence epilepsy in the Chinese Han population *Pediatr. Neurol.* **35** 187–90

[577] Qin N, Olcese R, Bransby M, Lin T and Birnbaumer L 1999 Ca^{2+}-induced inhibition of the cardiac Ca^{2+} channel depends on calmodulin *Proc. Natl Acad.* **96** 2435–8

[578] Zuhlke R D, Pitt G S, Deisseroth K, Tsien R W and Reuter H 1999 Calmodulin supports both inactivation and facilitation of L-type calcium channels *Nature* **399** 159–62

[579] Calorio C *et al* 2019 Impaired chromaffin cell excitability and exocytosis in autistic Timothy syndrome TS2-neo mouse rescued by L-type calcium channel blockers *J. Physiol.* **597** 1705–33

[580] Bozarth X, Dines J N, Cong, Mirzaa G M, Foss K, Meritt J L II, Thies J, Mefford H C and Novotny E 2018 Expanding clinical phenotype in CACNA1C related disorders: From neonatal onset severe epileptic encephalopathy to late-onset epilepsy *Am. J. Med. Genet.* A **176** 2733–9

[581] Rodan L H *et al* 2021 Phenotypic expansion of CACNA1C-associated disorders to include isolated neurological manifestations *Genet. Med.* **23** 1922–32

Chapter 14

Role of dysregulated calcium in neuropsychiatric diseases

Herie Sun and Anjali M Rajadhyaksha

Abbreviations

PTSD	Post-traumatic stress disorder
BD	Bipolar disorder
SCZ	Schizophrenia
ASD	Autism spectrum disorder
TS	Timothy syndrome
MDD	Major depressive disorder
LTCC	L-type calcium channel
GWAS	Genome-wide association study
SNP	Single-nucleotide polymorphism
CACNA1C/CACNA1D	Calcium voltage-gated channel subunit alpha-1-C/D

The L-type calcium channel (LTCC) subunits $Ca_V1.2$ and $Ca_V1.3$—encoded by the genes *CACNA1C* and *CACNA1D*, respectively—form voltage-gated ion channels that are critical in regulating calcium influx into excitable cells in the brain, heart and a variety of other important systems. Importantly, LTCCs have been shown to be essential in normal brain development and plasticity, and their genetic and functional dysregulation have been implicated in a broad range of disease processes. This chapter will examine the role of LTCCs in neuropsychiatric disease from bench to bedside, from the identification of candidate genes in large-population genetic studies, to characterizing how LTCCs underlie neuropsychiatric pathophysiology and behavioral correlates in preclinical animal models, to human studies that shed light on neuroanatomic correlates of channel-level dysfunction and provide novel ways to think about the clinical management of patients with neuropsychiatric disease.

doi:10.1088/978-0-7503-2009-2ch14

14.1 Introduction to L-type calcium channels

Evidence from both human and animal studies have unequivocally established a critical contribution of calcium signaling pathways to the pathophysiology of neurological disorders, with the L-type Ca^{2+} channel (LTCC) family of voltage-gated calcium channels serving as one major calcium source in the brain (Striessnig *et al* 2014, Simms and Zamponi 2014). The LTCC subunits $Ca_V1.2$ and $Ca_V1.3$—encoded by the genes *CACNA1C* and *CACNA1D*, respectively—are critical in regulating calcium influx into excitable cells in the brain, heart and a variety of other important systems. While their structural makeup is highly similar (Zamponi *et al* 2015), these isoforms have overlapping patterns of genetic expression present at differential levels (table 14.1, Kabir *et al* 2017b) and have been shown to contribute differently to neuronal function and behavior through distinct physiologic mechanisms (Xu and Lipscombe 2001, Lipscombe 2002, Koschak *et al* 2001).

In the nervous system, these channels are primarily present at the postsynaptic membrane (Di Biase *et al* 2008, Jenkins *et al* 2010), existing in signaling complexes that dictate regulation of calcium signal transduction pathways and interact with downstream second messenger pathways, such as Ca^{2+}/CaM-dependent protein kinases (CaMKs; Deisseroth *et al* 1998, Ma *et al* 2014) and the Ras/mitogen-activated protein kinase (MAPK) pathway (Wu *et al* 2001, Dolmetsch *et al* 2001). Importantly, LTCCs and the successful coordination of their involved pathways have been shown to be essential in normal brain development, plasticity and neuronal function (Calin-Jageman and Lee 2008, Zamponi *et al* 2015, Nanou and Catterall 2018), and their genetic and functional dysregulation have been implicated in a broad range of disease processes (Casamassima *et al* 2010, Bhat *et al* 2012, Berger and Bartsch 2014, Kabir *et al* 2016, Kabir *et al* 2017b). This chapter will examine the role of LTCCs in neuropsychiatric disorders and a number of translational applications highlighted by this area of investigation, including the use of genetic animal models for studying neuropsychiatric disease, human subject studies for the elucidation of neuroanatomic correlates of disease, and the repurposing of existing drugs for neuropsychiatric treatment.

14.2 Genetic studies and disease associations linking L-type calcium channels to neuropsychiatric disease

It is becoming increasingly apparent that no single gene underlies the pathogenesis of complex, heterogeneous neuropsychiatric diseases such as schizophrenia, bipolar disorder and autism spectrum disorder. Rather, the genetic risk for these patients appears to be polygenic, with some prominent genes conferring higher risks (Gottesman and Shields 1967, International Schizophrenia Consortium 2009, Gandal *et al* 2018). In addition, given commonalities in underlying neurocircuitry and neurobiological mechanisms between diseases, genetic risk factors and the dysfunctional neuronal processes that they are associated with are likely to contribute to clusters of symptoms shared among diseases, rather than directly to the diseases themselves. An understanding of the genetic bases of neuropsychiatric

Table 14.1. Summary of key behavioral tests in mouse models.

Trait Tested	*Test*	*Simplified Protocol*	*Example Measurements*	*Protocol Reference*
Anxiety	Open-field test	Open exploration of a bare chamber.	Measurement of total ambulatory distance, thigmotaxis (time spent near walls versus chamber center).	Seibenhener and Wooten (2015)
	Light-dark conflict test	Exploration of a chamber with light and dark components.	Measurement of approach versus avoidance behavior between light and dark compartments.	Arrant *et al* (2013)
	Elevated plus maze test	Exploration of an elevated plus-shaped maze, where two opposing arms are dark and closed, and two opposing arms are open.	Measurement of preference for open versus closed arms.	Sousa *et al* (2006)
Depression	Forced swim test	Placement of the animal in a deep, water-filled chamber.	Measurement of effort (swimming, escape attempts) versus behavioral despair (floating, immobility).	Powell *et al* (2012)
	Tail suspension test	Suspension of animal by its tail.	Measurement of effort (struggling, rearing) versus behavioral despair (immobility).	Powell *et al* (2012)
	Sucrose preference test	Choice between a bottle filled with plain water and a bottle filled with sucrose solution.	Measurement of anhedonia (lack of preference for normally appealing stimuli).	Powell *et al* (2012)
Sociability	3-chamber social interaction	Placement of mouse in a central chamber, choosing between one chamber containing a stranger mouse (social stimuli) and another containing an object.	Measurement of preference between social and non-social stimuli.	Kaidanovich-Beilin *et al* (2011)

(*Continued*)

Table 14.1. (*Continued*)

Trait Tested	*Test*	*Simplified Protocol*	*Example Measurements*	*Protocol Reference*
Learning and Memory (Spatial Learning, Working Memory)	Morris water maze	Placement of a mouse in a four-quadrant, deep, water-filled chamber, with one chamber containing a platform.	Time to swim to platform.	Sousa *et al* (2006)
	Y-maze	Placement of a mouse in a Y-shaped maze.	Spontaneous alternation (drive to explore previously unvisited areas). Preference for a novel, previously blocked off arm.	Kraeuter *et al* (2019)
Addiction	Conditioned place preference	Daily injection with a drug is paired with a distinguishable compartment. Then, this association can be extinguished (extinction paradigm) and reinstated (reinstatement paradigm).	Measurement of preference for the drug-paired chamber.	Spanagel (2017)
	Self-administration	Choice between a lever or bottle containing the addictive substance or non-addictive substance, with administration either freely available (non-operant) or tied to certain schedules, stimuli or tasks (operant).	Number of self-administration events, other addictive behavioral features (e.g. licking, lever pressing with increase in effort to deliver reward).	Spanagel (2017)

diseases is essential to furthering efforts to characterize and treat these diseases and their underlying pathophysiology.

Much of the recent light shed on the genetic basis of neuropsychiatric diseases has come from genome-wide association studies (GWAS)—large-population studies that identify genetic variants associated with a trait or disorder. GWAS can identify single-nucleotide polymorphisms (SNPs)—key elements of genetic variation that represent substitutions of a single nucleotide within the genome found in an appreciable minority of the population (Syvänen 2001).

A wealth of GWAS data and smaller-scale genetic studies in human subjects have implicated the voltage-gated calcium channel genes *CACNA1C* and *CACNA1D* as major candidate risk genes in a number of neuropsychiatric diseases. Among these, the SNP rs1006737 is perhaps the most well-known and well-studied variant of *CACNA1C* implicated in neuropsychiatric disease risk (Heyes *et al* 2015, Moon *et al* 2018, Hamshere *et al* 2013), but many SNPs have been linked across various patient populations to neuropsychiatric disease (reviewed in Heyes *et al* 2015, Moon *et al* 2018). Variants in the *CACNA1C* gene have been associated with schizophrenia (SCZ; Schizophrenia Working Group of the Psychiatric Genomics Consortium 2014, Green *et al* 2010, Ripke *et al* 2013, Hamshere *et al* 2013), bipolar disorder (BD; Ferreira *et al* 2008, Sklar *et al* 2008), drug dependence associated with BD (Mosheva *et al* 2020), major depressive disorder (MDD; Green *et al* 2010), autism spectrum disorder (ASD; Li *et al* 2015) and associated syndromic diseases such as Timothy Syndrome (TS; Splawski *et al* 2004, Splawski *et al* 2005, Bader *et al* 2011) and post-traumatic stress disorder (PTSD; Dedic *et al* 2018, Krzyzewska *et al* 2018). In addition, associations have been found between *CACNA1C* SNPs and common symptomatology across disease groups (Cross-Disorder Group of the Psychiatric Genomics Consortium 2013), suggesting that they may contribute to psychiatric disease risk by promoting particular symptom clusters rather than the diseases themselves.

CACNA1D has also been identified as a candidate risk gene for the development of neuropsychiatric disorders, including bipolar disorder, schizophrenia, ADHD, major depressive disorder and ASD (Cross-Disorder Group of the Psychiatric Genomics Consortium 2013, Ament *et al* 2015, Ross *et al* 2016, Pinggera *et al* 2015). While human genetic studies examining *CACNA1D* in drug-using individuals have been sparse, one study has shown a direct genetic association between three *CACNA1D* SNPs and cocaine dependence in human subjects (Martínez-Rivera *et al* 2017). However, work in animal models—detailed in the following sections—has elaborated on the link between $Ca_V1.3$ channels and responses to and dependence on drugs of abuse.

Most of the genetic variants of *CACNA1C* explored in the literature occur in non-coding (i.e. intronic) parts of the gene that are not translated directly into protein products but may regulate gene expression in other ways, such as by interacting with nuclear binding proteins (Roussos *et al* 2014, Eckart *et al* 2016, Won *et al* 2016). These intronic variants have been associated with both increased and decreased *CACNA1C* gene expression (Eckart *et al* 2016, Yoshimizu *et al* 2015, Roussos *et al* 2014, Gershon *et al* 2014), in a manner likely dependent on interacting factors specific to cell type and brain region. Other variants have been found to be splicing isoforms of coding (i.e. exonic) regions of the *CACNA1C* gene, most notably a gain-of-function mutation in $Ca_V1.2$ found to cause TS, a multisystem disorder characterized by ASD in 80% of patients, cardiac deficits and hand, foot and facial abnormalities (Splawski *et al* 2004, Splawski *et al* 2005, Bader *et al* 2011). Both coding (Ross *et al* 2016) and non-coding (Ament *et al* 2015) variants of *CACNA1D* have been identified.

14.3 Preclinical animal models for the study of L-type calcium channels in neuropsychiatric disease

Though neuropsychiatric disorders have historically been considered distinct clinical entities, it is becoming clearer that diseases such as bipolar disorder, schizophrenia, major depressive disorder and autism spectrum disorder share common phenotypic features, likely due to overlapping brain circuits and molecular mechanisms (Nestler and Carlezon 2006, Lüthi and Luscher 2014, Russo and Nestler 2013). Genetic rodent models that harbor mutations causing altered calcium-channel signaling analogous to naturally occurring genetic variants in humans offer a useful tool for studying the pathophysiology of these diseases and modeling associated behavioral variation. These genetic mutant rodent models can include global knockout of gene function or, in combination with viral vectors and selective breeding of mouse lines (through a technique reliant on the enzyme Cre recombinase and target gene sequences called *Lox* sequences), conditional knockouts that are restricted to specific brain regions, timepoints in development or cell types. Genetic rodent models can also harbor coding variants such as the ones that cause TS (Bader *et al* 2011). Functionally, these genetic alterations can result in either a loss or gain of function in LTCCs, affecting calcium signaling in either case. Finally, selective inhibition or activation of these channels can be achieved through non-endogenous pharmacologic means, including repurposed drugs such as dihydropyridine LTCC blockers and experimental drugs such as the $Ca_V1.3$ channel-activating drug BayK8644 (Sinnegger-Brauns *et al* 2004). Various mouse model lines used to study LTCCs, and the behavioral endophenotypes they display, are reviewed in Kabir *et al* (2017a) and Kabir *et al* (2017b), Moon *et al* (2018).

14.3.1 Behavior

In humans, neuropsychiatric endophenotypes can be extremely difficult to tie to their neural correlates. The use of preclinical animal models facilitates the controlled modeling of analogous human behaviors and the investigation of underlying pathophysiology. Anxiety-like behaviors, depressive-like behaviors, social behaviors, and cognition represent four of the most important phenotypic domains that researchers examine in these animal models. Major behavioral tests mentioned in this chapter are summarized in table 14.1 and a review of behavioral analysis in preclinical rodent models and the multitude of tests used to simulate human behaviors is provided in Sousa *et al* (2006).

14.3.1.1 Anxiety

Anxiety is one of the most prevalent symptoms in neuropsychiatric disorders (Mineka *et al* 1998, Kaufman and Charney 2000). To study anxiety in rodents, investigators use behavioral assays such as the open-field test, light-dark conflict test and elevated plus maze test that measure responses to anxiety-inducing stimuli. Both male and female mice harboring global heterozygous $Ca_V1.2$ knockout (i.e. lacking 50% of their $Ca_V1.2$ channels; note: homozygous knockout of $Ca_V1.2$ channel function is lethal in mice due to the critical role of $Ca_V1.2$ in heart development, Seisenberger *et al* 2000) display heightened anxiety-like behaviors (Dao *et al* 2010, Lee *et al* 2012, Kabir *et al* 2017a, Dedic *et al* 2018, Kabitzke *et al* 2018, Jaric *et al* 2019), with specific knockout experiments implicating excitatory glutamatergic neuron populations in the forebrain and adult prefrontal cortex as being especially important sites linking $Ca_V1.2$ function and anxiety (Lee *et al* 2012, Kabir *et al* 2017a).

Mice with global knockout of $Ca_V1.3$ channels exhibit decreased anxiety-like behaviors (Busquet *et al* 2010). As opposed to $Ca_V1.2$ channels, viral vector knockdown of $Ca_V1.3$ channels using short hairpin RNAs in the prefrontal cortex does not alter anxiety-like behavior (Lee *et al* 2012).

14.3.1.2 Depression

Depression is a comorbidity seen with many neuropsychiatric diseases (Buckley *et al* 2009, Gorman 1996), and depression-related behaviors can be examined in rodents with tests such as the forced swim test (FST), tail suspension test (TST) and the sucrose preference test (SPT), modeling key components of depressive-like behaviors such as despair (FST and TST) and anhedonia (SPT). Loss of $Ca_V1.2$ channels (Dao *et al* 2010, Kabir *et al* 2017a) or $Ca_V1.3$ channels (Busquet *et al* 2010) resulted in an antidepressant-like effect, particularly when $Ca_V1.2$ loss was induced in the prefrontal cortex (Kabir *et al* 2017a). The role of $Ca_V1.3$ channels in the pathogenesis of depressive behaviors is further supported by the induction of depressive-like behaviors by selective pharmacologic stimulation of $Ca_V1.3$ channels (Sinnegger-Brauns *et al* 2004).

14.3.1.3 Sociability

Deficits in social communication and related social behaviors are prominently seen in many neuropsychiatric disorders and are a core symptom in autism spectrum disorder (Kennedy and Adolphs 2012, DSM-V). Knockdown of $Ca_V1.2$ in excitatory glutamatergic cells in the mouse forebrain elicited social deficits (Kabir *et al* 2017a). Mouse models of TS, harboring a gain-of-function mutation in $Ca_V1.2$, recapitulate the core social deficits seen in ASD (Bader *et al* 2011). Heterozygous rats with 50% loss of $Ca_V1.2$ channels display deficits in the sending and receiving of ultrasonic vocalizations (Kisko *et al* 2018), an important rodent modality for social communication and a marker of social reward (Knutson *et al* 1998, Engelhardt *et al* 2018), as well as in reciprocal social interaction and attraction to social stimuli (Kabitzke *et al* 2018).

While less evidence exists tying dysregulation of $Ca_V1.3$ channels to altered social behaviors, enhancing $Ca_V1.3$ channel function through pharmacologic means has been shown to induce deficits in social behavior (Martínez-Rivera *et al* 2017). This is supported by the discovery that several of the mutations in the $Ca_V1.3$ gene, *CACNA1D*, in humans with ASD result in a gain of channel function (Pinggera *et al* 2015, Limpitikul *et al* 2016, Hofer *et al* 2020, Pinggera *et al* 2018, Pinggera *et al* 2017).

14.3.1.4 Cognition

An often-overlooked component of neuropsychiatric disorders are their associated cognitive symptoms. For example, while positive and negative symptoms of schizophrenia make up core DSM-V criteria for diagnosis, impairments in attention, executive function and working memory have an immense impact on functioning in settings such as work, school and home life (Bowie and Harvey 2006). Additionally, co-morbid impairments in social and cognitive processing often precede mood-related symptoms (Lieberman *et al* 2001, Heinrichs 2005). In preclinical animal models, cognitive processes can be tested through tasks measuring learning and memory such as the Morris water maze test, Y-maze test and fear-based protocols.

Experimentally induced manipulation of $Ca_V1.2$ function (through either loss-of-function knockout or gain-of-function mutations) in mice leads to deficits in hippocampal-dependent spatial memory tasks in the Morris water maze (Temme *et al* 2016, Moosmang *et al* 2005, Bader *et al* 2011) and fear-associated context discrimination tasks (Temme *et al* 2016). Many of these deficits are not primarily rooted in the initial acquisition of the task, but rather in subsequent adaptation to experimenter-prompted changes in the behavioral task, such as changing the location of the target platform. These adaptations require a degree of cognitive flexibility, and the inability to successfully alter behavior in response to these changes within an appropriate window of time demonstrates a degree of cognitive inflexibility and perseverance, a key finding in patients with schizophrenia (Waltz 2017) and ASD (Geurts *et al* 2009, D'Cruz *et al* 2013). Finally, these mice also demonstrate enhanced fear responses in response to cues associated with the initial fear stimulus (Kabir *et al* 2017b, Bader *et al* 2011).

14.3.1.5 Addiction

There is a clear link between neuropsychiatric disease and substance abuse. Substance use disorders are often co-morbid with mood disorders such as bipolar disorder (Post and Kalivas 2013) and major depressive disorder (Kessler *et al* 2009), and there is a growing body of evidence that common neural substrates exist that underlie the pathophysiology of both (Lüthi and Lüscher 2014, Nestler and Carlezon 2006, Lobo *et al* 2013, Sun *et al* 2016). Significant work has been performed in preclinical animal models of addiction examining the link between calcium signaling through LTCCs and addiction-related behaviors. For example, a variety of studies testing the effects of LTCC-blocking agents on behavioral responses to drug administration (reviewed in table 3 of Kabir *et al* 2017b) have shown a variety of altered behavioral responses to addictive substances such as cocaine, morphine, alcohol and nicotine, including decreased self-administration of substances (Kuzmin *et al* 1992, Uhrig *et al* 2017), decreased sensitization to substance effects (Reimer and Martin-Iverson 1994, Pierce *et al* 1998, Biala and Langwiński 1996, Zhang *et al* 2003) and reduced withdrawal symptoms following chronic intake (Zharkovsky *et al* 1993, Whittington *et al* 1991, Gatch 2002).

A recent study has linked *CACNA1C* SNP rs1034936 with co-morbid alcohol abuse and bipolar disorder (Mosheva *et al* 2020). The same SNP is also associated with lifetime cocaine use in these patients. Animal models harboring loss of $Ca_V1.2$ channels have identified a key role of these channels in driving drug addiction-related behaviors. $Ca_V1.2$ channels have been found to be critical in the brain region, hippocampus, in the contextual aspects of drug-associated memories, as seen for the psychostimulant cocaine (Burgdorf *et al* 2017, 2020). Human and animal studies have identified the hippocampus as a critical brain region in maintaining drug-context associations that drive relapse to addictive behavior (Koob and Volkow 2010, Kutlu and Gould 2016). Animal studies have also identified a role for $Ca_V1.2$ channels in the prefrontal cortex in driving relapse to cocaine addictive behavior following stressful stimuli and to a priming injection of cocaine (Bavley *et al* 2020).

A link between *CACNA1D* and cocaine dependence has also been established (Martínez-Rivera *et al* 2017). In rodent models with targeted deficits in $Ca_V1.3$ channel signaling, a critical role has been shown for calcium signaling through $Ca_V1.3$ lTCCs in dopaminergic neurons of the ventral tegmental area (VTA) for the development of cocaine behavioral sensitization (Schierberl *et al* 2011) and the acquisition of conditioned place preference for cocaine (Martínez-Rivera *et al* 2017). Exposure to these substances has also been shown to alter the levels of LTCC mRNA and translated protein (Shibasaki *et al* 2010, 2011, Bernardi *et al* 2014, Katsura *et al* 2002, Hayashida *et al* 2005).

In the context of co-morbid substance abuse and neuropsychiatric disease commonly occurring in patients, the hypothesis that LTCCs are a common pathophysiologic substrate in both substance abuse and neuropsychiatric disease is bolstered by recent lines of evidence from preclinical animal models. The repeated activation of $Ca_V1.3$ channels in the VTA, a brain area linked to addiction (Yap and Miczek 2008, Kauer 2003) and the modulation of the effects of addictive substances like cocaine (Pascoli *et al* 2015), was sufficient to induce social deficits, depressive-like behaviors and cocaine-related behaviors (Martínez-Rivera *et al* 2017). The presence of both neuropsychiatric symptoms and addictive behaviors in the context of experimental manipulation of $Ca_V1.3$ channels suggests a potential role for LTCCs in the pathogenesis of both.

Finally, in considering behavioral traits and their links to the genetic causes of neuropsychiatric disease, it is important to not view these behaviors in isolation, but rather in the context of other behaviors and physiologic functions. For example, deficits in social function are commonly co-morbid with heightened anxiety, making the two symptom clusters difficult to study in isolation. As another example, decreased anxiety-like behaviors in mice with $Ca_V1.3$ channel knockout mice may be partially explained by congenital deafness observed in these mice (Platzer *et al* 2000).

14.3.2 Molecular and cellular pathways

Researchers have examined the molecular and cellular pathways associated with $Ca_V1.2$ and $Ca_V1.3$ channels in genetically modified preclinical animal models to better understand the role of calcium signaling in the pathophysiology of neuropsychiatric disease.

14.3.2.1 LTCCs and mRNA translation

One method for examining how genetic variance can lead to disease states is through proteomics analyses, which characterize and quantify the protein products translated from genes. The dysregulation of mRNA translation and protein synthesis in dendrites of neurons have been hypothesized to be an important pathophysiologic explanation for behavioral impairments in neuropsychiatric diseases (Penzes *et al* 2011, Iasevoli *et al* 2013, Costa-Mattioli and Monteggia 2013, Forrest *et al* 2018). Two notable pathways involved in regulating protein synthesis and implicated in the pathogenesis of neuropsychiatric symptomatology are the mammalian target of rapamycin (mTOR) pathway (Nandagopal and Roux 2015, Costa-Mattioli and Monteggia 2013) and the eukaryotic initiation factor 2α (eIF2α) signaling pathway (Wek and Cavener 2007). For example, proteomic pathway analyses have shown that the eIF2α pathway is enriched in the postsynaptic density fractions of neuronal dendrites of patients with bipolar disorder (Föcking *et al* 2016).

Evidence in preclinical animal models has provided evidence that LTCCs are important regulators of these key pathways involved in the modulation of mRNA translation and protein synthesis. For example, the selective loss of $Ca_V1.2$ in excitatory glutamatergic neurons in the mouse forebrain results in a significant decrease in levels of general protein synthesis in this brain region, concurrent with lower activity of the mTOR complex 1 (mTORC1, a constituent of the mTOR pathway; Kabir *et al* 2017a) and higher levels of phosphorylated eIF2α, a constituent of the eIF2α pathway that represses translation when phosphorylated (Kabir *et al* 2017a). In $Ca_V1.2$-deficient mice, pharmacologic reversal of these dysregulated pathways and their associated behavioral deficits has also been achieved, such as through inhibition of phosphorylated eIF2α with the experimental drug ISRIB (Kabir *et al* 2017a), providing insight into potential future avenues for novel pharmacotherapeutics for neuropsychiatric disease.

14.3.2.2 LTCCs and neurogenesis

Adult neurogenesis in the hippocampus has been shown to be important for learning and memory processes (Yau *et al* 2015, Deng *et al* 2010), such as the formation of complex spatial representations (Dupret *et al* 2008), and dysfunctional adult neurogenesis has been hypothesized to underlie cognitive symptomatology present in neuropsychiatric disorders such as schizophrenia (Apple *et al* 2017). The link between adult neurogenesis and neuropsychiatric disease is further bolstered by the finding that the administration of various psychiatric medications—such as SSRIs, tricyclic antidepressants and SNRIs—can lead to increased neurogenesis in rodent models (Malberg 2004).

$Ca_V1.2$ channels have been shown to be important in key neuronal processes in the hippocampus necessary for the consolidation of learning and memory processes, including long-term potentiation (Moosmang *et al* 2005) and adult hippocampal neurogenesis (Lee *et al* 2016, Temme *et al* 2016). These mouse models harboring dysfunctional $Ca_V1.2$ exhibit behavior in hippocampal-dependent learning and memory tasks, such as discriminatory water-maze tasks or Y-maze tasks, that is repetitive, restrictive and perseverative in nature (Moosmang *et al* 2005, Bader *et al* 2011, Kabir *et al* 2017b). These behaviors can be viewed as indicative of a lack of cognitive flexibility, a key behavioral deficit present in neurodevelopmental diseases such as ASD (Geurts *et al* 2009, D'Cruz *et al* 2013) and in neuropsychiatric diseases such as schizophrenia (Waltz 2017).

The mechanisms by which LTCCs underlie adult hippocampal neurogenesis are unclear. However, there are many processes regulated by LTCCs that have been prominently linked to learning and memory processes, including the modulation of synaptic plasticity (Weisskopf *et al* 1999, Degoulet *et al* 2016, Moosmang *et al* 2005, Striessnig *et al* 2006), regulation of the bidirectional transition between adult hippocampal neural precursors and immature neurons (Deisseroth *et al* 2004, Lee *et al* 2016, Völkening *et al* 2017), interactions with stress and stress-related growth factors (Apple *et al* 2017), and pathways involving cAMP response element-binding protein (CREB), Ca^2/calmodulin-dependent protein kinase II (CaMKII) and brain-derived neurotrophic factor (BDNF) (Moosmang *et al* 2005, Striessnig *et al* 2006).

Similarly to $Ca_V1.2$, $Ca_V1.3$ channels have been demonstrated to be necessary for hippocampus neurogenesis with a profound effect of loss of $Ca_V1.3$ on adult hippocampal neurogenesis (Marschallinger *et al* 2015). $Ca_V1.3$ knockout results in deficit in both proliferation of neural progenitor cells as well as survival of newborn hippocampal neurons (Marschallinger *et al* 2015), an effect not observed following loss of $Ca_V1.2$ channels that impact only survival (Lee *et al* 2016). The contribution of reduced adult hippocampal neurogenesis secondary to loss of $Ca_V1.3$ channels to cognitive function and mood-related behaviors remains currently unknown. However, a neuropathological mechanism may exist underlying some of the neuropsychiatric symptomology seen in patients harboring *CACNA1D* SNPs.

14.3.2.3 LTCCs and excitatory/inhibitory synaptic balance

Neuronal signaling in the brain relies on a dense network of interconnected neurons within and across brain regions, which excite and inhibit one another through the release of excitatory and inhibitory neurotransmitters that modulate ion flow across the neuronal membrane. The nature of this delicate balance means that small perturbations in the excitatory/inhibitory balance can lead to significant downstream changes at the molecular and neurocircuitry level, with potential implications for behavior. Modeling neuropsychiatric disease in preclinical animal models has facilitated an emerging hypothesis that an imbalance between excitatory and inhibitory cells in the brain may underlie many of the behavioral deficits seen in neuropsychiatric disorders (Mullins *et al* 2016, Gao and Penzes 2015, Kabir *et al* 2017a, Santini *et al* 2013, Nelson and Valakh 2015). Dysregulated calcium signaling may play an important role in this imbalance. Treatment of neurons with LTCC-blocking agents in *ex vivo* studies led to an increase in excitatory postsynaptic currents (Gong *et al* 2007) and a decrease in receptors that mediate inhibitory neurotransmission (Saliba *et al* 2009). In live animal studies, knockout of $Ca_V1.2$ in excitatory cells in the mouse forebrain was associated with behavioral deficits and increased excitatory postsynaptic currents in the prefrontal cortex (Kabir *et al* 2017a). Targeting this brain region with a novel pharmacologic agent, ISRIB, reversed this excitatory-inhibitory imbalance and led to the rescue of behavioral deficits, further implicating calcium-related excitatory-inhibitory imbalances in neuropsychiatric symptomatology and providing a potential target for novel therapies.

14.4 Clinical applications for variation in L-type calcium channel signaling in neuropsychiatric disease

14.4.1 Neuroanatomic correlates of genetic variation in calcium signaling

In investigating the neural correlates of neuropsychiatric diseases and the role of LTCCs, one area in which human subject studies excel over preclinical animal model studies is in investigating neuroanatomic correlates. Here, a vast array of functional neuroimaging modalities—such as functional magnetic resonance imaging (fMRI), positive emission tomography (PET) and diffusion tensor imaging (DTI)—enable researchers to examine the functional (fMRI, PET) and structural (DTI) differences in neuroanatomy that correspond to genetic variation in the *CACNA1C* gene. No studies to date have been performed directly examining the link between genetic variation in *CACNA1D* and altered neuroanatomic structure.

Neuroimaging studies in carriers of *CACNA1C* SNPs have demonstrated alterations in the gross structure of brain regions as well as the functional connections within and across those regions. One prominent brain region implicated as a structure correlate of *CACNA1C* genotype in human subjects is the prefrontal cortex (PFC), a finding that has been recapitulated in preclinical $Ca_V1.2$ loss of function animal models as described above. The PFC is heavily implicated in the mediation of higher-order social and cognitive processing (Miller 2000) and as a common site of structural, functional and neurophysiologic dysfunction in neuropsychiatric diseases (Tekin and Cummings 2002, Weinberger *et al* 1986, Manoach 2003). For example, at a gross structural level, carriers of *CACNA1C* SNPs have been found to have reduced cortical surface area in parts of the right dorsolateral PFC and left superior parietal cortex (Zheng *et al* 2016) and increased gray matter volume within a bilateral corticolimbic frontotemporal system that included the PFC (Wang *et al* 2011). Functional differences in carriers of *CACNA1C* SNPs include altered connectivity between the PFC and various other parts of the brain during behavioral tasks such as facial affect processing (Wang *et al* 2011, Dima *et al* 2013) and working memory tasks (Takeuchi *et al* 2019). These differences have also been studied in patient populations with neuropsychiatric diseases such as schizophrenia and bipolar disorder, with significant neurofunctional and neuroanatomic differences seen between carriers of *CACNA1C* SNPs and their non-carrier counterparts (Soeiro-de-Souza *et al* 2017, Zhang *et al* 2019, Cosgrove *et al* 2017, Jogia *et al* 2011, Dima *et al* 2013).

14.4.2 lTCCs as a repurposed target for the treatment of neuropsychiatric diseases

The clinical implications for a better understanding of LTCCs in neuropsychiatric disease are far-ranging. Among them are the promise of a novel target for new pharmacotherapeutics and the ability for clinicians to utilize genetic information to better inform and personalize their approach to neuropsychiatric patient care.

LTCC-blocking compounds—in the form of dihydropyridines (e.g. nifedipine, amlodipine, isradipine, nicardipine) and non-dihydropyridines (e.g. verapamil, diltiazem)—have been a mainstay in the treatment of clinical hypertension and cardiac arrhythmias for decades; however, recent evidence implicating LTCCs in neuropsychiatric disease has generated interest in the possibility of repurposing these drugs for a neuropsychiatric context. Drug repurposing pipelines, wherein commonly used pharmaceutics with known mechanisms are investigated in novel disease contexts, offer higher rates of success, lower years to impact, lower costs and better known safety profiles when compared to discovering novel compounds through traditional pipelines (Pushpakom *et al* 2019, Oprea and Mestres 2012). While $Ca_V1.2$ and $Ca_V1.3$ clearly have differing downstream physiologic pathways and associations with neuropsychiatric symptom clusters, they are similar enough in sequence and structure that drugs targeting LTCCs are non-selective between these subunits (Zamponi *et al* 2015).

To date, LTCC-blocking agents have been tested in a number of neuropsychiatric contexts, for both the alleviation of symptoms secondary to neuropsychiatric disease and for the treatment of substance dependence (reviewed in Kabir *et al* 2017b). An initial surge of interest in the use of LTCC-blocking agents as repurposed drugs in neuropsychiatric disease occurred in the 1980s and 1990s, but waned due to a mixed landscape of results. Recently, as researchers have developed a better understanding of the role of LTCCs in neuropsychiatric disease and a more nuanced understanding of neuropsychiatric symptom clusters, interest in the repurposing of these agents has again risen.

In bipolar disorder, mixed effects have been demonstrated on the ability of drugs such as verapamil to attenuate episodes of acute mania (e.g. Cipriani *et al* 2016, Goodnick 1996, Gitlin and Weiss 1984, Lenzi *et al* 1995, Höschl and Kozený 1989, Pazzaglia *et al* 1998), and positive findings have been shown for the utilization of LTCC blockers as combination therapy with mood stabilizers (Mallinger *et al* 2008, Grunze *et al* 1996) or antipsychotics (Lenzi *et al* 1995). However, a recent study, part of the resurging interest in repurposing LTCC antagonists for neuropsychiatric disease, found that exposure to LTCC was associated with reduced rates of psychiatric hospitalization and self-harm in patients with bipolar disorder (Hayes *et al* 2019).

In patients with schizophrenia, treatment with LTCC blockers such as verapamil had promising results in a number of symptom domains, including decreased psychotic symptoms (Price and Pascarzi 1987, Price 1987), improvements in positive and negative symptoms (Bartko *et al* 1991) and co-morbid anxiety and depression (Bartko *et al* 1991).

Finally, consistent with the implied role of LTCCs in drug dependence, studies of LTCC-blocking agents in cocaine and morphine dependence have shown mixed results, with some trials showing a reduction in cocaine craving (Rosse *et al* 1994, Malcolm *et al* 1999, Johnson *et al* 2004) and in the subjective effect of cocaine (Muntaner *et al* 1991) and morphine (Vaupel *et al* 1993), while other studies have shown no effect of treatment (Kosten *et al* 1999, Sofuoglu *et al* 2003, Hasegawa and Zacny 1997). No effects have been shown in the treatment of alcohol dependence with LTCC-blocking agents (Zacny and Yajnik 1993, Rush and Pazzaglia 1998, Perez-Reyes *et al* 1992).

While the repurposing of these drugs in novel neuropsychiatric contexts holds great promise, it is important to understand the caveats behind some of the positive results to date. Perhaps most importantly, some anti-hypertensive and anti-arrhythmic drugs such as verapamil have relatively poorly understood mechanisms, with nonspecific effects on other channels (Bergson *et al* 2011, Zhang *et al* 1999, Motulsky *et al* 1983) and unclear, potentially toxic pharmacokinetic mechanisms in the context of the higher doses necessary for central nervous system delivery (Raderer and Scheithauer 1993).

Overall, the use of repurposed pharmacologic agents that target LTCCs for the treatment of neuropsychiatric diseases has presented mixed results in the clinical domain. However, as interest in this topic is renewed in the context of an improved understanding of the role of LTCCs in neuropsychiatric disease, there is hope that these drugs, modified versions of these drugs or isoform ($Ca_V1.2$ or $Ca_V1.3$)-specific LTCC-blocking agents, can provide a novel modality of substrate-specific therapy in a field that is sorely in need of new targeted treatments.

14.5 Conclusion

In conclusion, the LTCCs $Ca_V1.2$ and $Ca_V1.3$—encoded by the genes *CACNA1C* and *CACNA1D*, respectively—represent promising new substrates in scientific efforts to explore the pathogenesis of neuropsychiatric disease. Further, the study of these voltage-gated calcium channels serves as an excellent case study on the full spectrum of translational investigation in medical research. This process begins with the identification of putative risk genes in humans through large-scale genome-wide association studies. The identified risk genes can then be manipulated in preclinical animal models to assess their role in molecular signaling and behavior relevant to human disease. Finally, a return to human subject studies that combine behavioral metrics, functional neuroimaging technologies and genetic information provide neuroanatomic correlates of disease and the opportunity to harness this diversity of information to guide novel pharmacotherapeutics and clinician decision-making. While it is clear from a mounting body of evidence that there is a link between calcium signaling through LTCCs and the pathogenesis of neuropsychiatric disease, there is still much to be done to further parse these findings and translate them into concrete tools for the bedside.

References

Ament S A, Szelinger S, Glusman G, Ashworth J, Hou L and Akula N *et al* 2015 Rare variants in neuronal excitability genes influence risk for bipolar disorder *Proc. Natl Acad. Sci. USA* **112** 3576–81

American Psychiatric Association 2013 *Diagnostic and Statistical Manual of Mental Disorders (DSM-5®)* (Washington, DC: American Psychiatric Pub.)

Apple D M, Fonseca R S and Kokovay E 2017 The role of adult neurogenesis in psychiatric and cognitive disorders *Brain Res.* **1655** 270–6

Arrant A E, Schramm-Sapyta N L and Kuhn C M 2013 Use of the light/dark test for anxiety in adult and adolescent male rats *Behav. Brain Res.* **256** 119–27

Bader P L, Faizi M, Kim L H, Owen S F, Tadross M R and Alfa R W *et al* 2011 Mouse model of Timothy syndrome recapitulates triad of autistic traits *Proc. Natl Acad. Sci. USA* **108** 15432–7

Bartko G, Horvath S, Zador G and Frecska E 1991 Effects of adjunctive verapamil administration in chronic schizophrenic patients *Prog. Neuropsychopharmacol. Biol. Psychiatry* **15** 343–49

Bavley C C, Fetcho R N, Burgdorf C E, Walsh A P, Fischer D K and Hall B S *et al* 2020 Correction: cocaine- and stress-primed reinstatement of drug-associated memories elicit differential behavioral and frontostriatal circuit activity patterns via recruitment of L-type Ca *Mol. Psychiatry* **25** 2373–91

Berger S M and Bartsch D 2014 The role of L-type voltage-gated calcium channels $Ca_V1.2$ and $Ca_V1.3$ in normal and pathological brain function *Cell Tissue Res.* **357** 463–76

Bergson P, Lipkind G, Lee S P, Duban M E and Hanck D A 2011 Verapamil block of T-type calcium channels *Mol. Pharmacol.* **79** 411–9

Bernardi R E, Uhrig S, Spanagel R and Hansson A C 2014 Transcriptional regulation of L-type calcium channel subtypes $Ca_V1.2$ and $Ca_V1.3$ by nicotine and their potential role in nicotine sensitization *Nicotine Tobacco Res.* **16** 774–85

Bhat S, Dao D T, Terrillion C E, Arad M, Smith R J and Soldatov N M *et al* 2012 *CACNA1C* ($Ca_V1.2$) in the pathophysiology of psychiatric disease *Prog. Neurobiol.* **99** 1–14

Biała G and Langwiński R 1996 Effects of calcium channel antagonists on the reinforcing properties of morphine, ethanol and cocaine as measured by place conditioning *J .Physiol. Pharmacol.* **47** 497–502

Bowie C R and Harvey P D 2006 Cognitive deficits and functional outcome in schizophrenia *Neuropsychiatr. Dis. Treat.* **2** 531–6

Buckley P F, Miller B J, Lehrer D S and Castle D J 2009 Psychiatric comorbidities and schizophrenia *Schizophr. Bull.* **35** 383–402

Burgdorf C E, Bavley C C, Fischer D K, Walsh A P, Martinez-Rivera A and Hackett J E *et al* 2020 Contribution of D1R-expressing neurons of the dorsal dentate gyrus and Ca *Neuropsychopharmacology* **45** 1506–17

Burgdorf C E, Schierberl K C, Lee A S, Fischer D K, Van Kempen T A and Mudragel V *et al* 2017 Extinction of contextual cocaine memories requires Ca *J. Neurosci.* **37** 11894–911

Busquet P, Nguyen N K, Schmid E, Tanimoto N, Seeliger M W and Ben-Yosef T *et al* 2010 $Ca_V1.3$ L-type Ca^{2+} channels modulate depression-like behaviour in mice independent of deaf phenotype *Int. J. Neuropsychopharmacol.* **13** 499–513

Calin-Jageman I and Lee A 2008 Ca_V1 L-type Ca^{2+} channel signaling complexes in neurons *J. Neurochem.* **105** 573–83

Casamassima F, Hay A C, Benedetti A, Lattanzi L, Cassano G B and Perlis R H 2010 L-type calcium channels and psychiatric disorders: a brief review *Am. J. Med. Genet. B Neuropsychiatr. Genet.* **153B** 1373–90

Cipriani A, Saunders K, Attenburrow M J, Stefaniak J, Panchal P and Stockton S *et al* 2016 A systematic review of calcium channel antagonists in bipolar disorder and some considerations for their future development *Mol. Psychiatry* **21** 1324–32

Cosgrove D, Mothersill O, Kendall K, Konte B, Harold D and Giegling I *et al* 2017 Cognitive characterization of schizophrenia risk variants involved in synaptic transmission: evidence of *CACNA1C*'s role in working memory *Neuropsychopharmacology* **42** 2612–22

Costa-Mattioli M and Monteggia L M 2013 mTOR complexes in neurodevelopmental and neuropsychiatric disorders *Nat. Neurosci.* **16** 1537–43

Consortium C-DGotPG 2013 Identification of risk loci with shared effects on five major psychiatric disorders: a genome-wide analysis *Lancet* **381** 1371–9

D'Cruz A M, Ragozzino M E, Mosconi M W, Shrestha S, Cook E H and Sweeney J A 2013 Reduced behavioral flexibility in autism spectrum disorders *Neuropsychology* **27** 152–60

Dao D T, Mahon P B, Cai X, Kovacsics C E, Blackwell R A and Arad M *et al* 2010 Mood disorder susceptibility gene *CACNA1C* modifies mood-related behaviors in mice and interacts with sex to influence behavior in mice and diagnosis in humans *Biol. Psychiatry* **68** 801–10

Dedic N, Pöhlmann M L, Richter J S, Mehta D, Czamara D and Metzger M W *et al* 2018 Cross-disorder risk gene *CACNA1C* differentially modulates susceptibility to psychiatric disorders during development and adulthood *Mol. Psychiatry* **23** 533–43

Degoulet M, Stelly C E, Ahn K C and Morikawa H 2016 L-type Ca^{2+} channel blockade with antihypertensive medication disrupts VTA synaptic plasticity and drug-associated contextual memory *Mol. Psychiatry* **21** 394–402

Deisseroth K, Heist E K and Tsien R W 1998 Translocation of calmodulin to the nucleus supports CREB phosphorylation in hippocampal neurons *Nature* **392** 198–202

Deisseroth K, Singla S, Toda H, Monje M, Palmer T D and Malenka R C 2004 Excitation-neurogenesis coupling in adult neural stem/progenitor cells *Neuron* **42** 535–52

Deng W, Aimone J B and Gage F H 2010 New neurons and new memories: how does adult hippocampal neurogenesis affect learning and memory? *Nat. Rev. Neurosci.* **11** 339–50

Di Biase V, Obermair G J, Szabo Z, Altier C, Sanguesa J and Bourinet E *et al* 2008 Stable membrane expression of postsynaptic $Ca_V1.2$ calcium channel clusters is independent of interactions with AKAP79/150 and PDZ proteins *J. Neurosci.* **28** 13845–55

Dima D, Jogia J, Collier D, Vassos E, Burdick K E and Frangou S 2013 Independent modulation of engagement and connectivity of the facial network during affect processing by *CACNA1C* and *ANK3* risk genes for bipolar disorder *JAMA Psychiatry* **70** 1303–11

Dolmetsch R E, Pajvani U, Fife K, Spotts J M and Greenberg M E 2001 Signaling to the nucleus by an L-type calcium channel-calmodulin complex through the MAP kinase pathway *Science* **294** 333–9

Dupret D, Revest J M, Koehl M, Ichas F, De Giorgi F and Costet P *et al* 2008 Spatial relational memory requires hippocampal adult neurogenesis *PLoS One* **3** e1959

Eckart N, Song Q, Yang R, Wang R, Zhu H and McCallion A S *et al* 2016 Functional characterization of schizophrenia-associated variation in CACNA1C *PLoS One* **11** e0157086

Engelhardt K, Schwarting R K W and Wöhr M 2018 Mapping trait-like socio-affective phenotypes in rats through 50-kHz ultrasonic vocalizations *Psychopharmacology (Berl)* **235** 83–98

Ferreira M A, O'Donovan M C, Meng Y A, Jones I R, Ruderfer D M and Jones L *et al* 2008 Collaborative genome-wide association analysis supports a role for *ANK3* and *CACNA1C* in bipolar disorder *Nat. Genet.* **40** 1056–58

Föcking M, Dicker P, Lopez L M, Hryniewiecka M, Wynne K and English J A *et al* 2016 Proteomic analysis of the postsynaptic density implicates synaptic function and energy pathways in bipolar disorder *Transl. Psychiatry* **6** e959

Forrest M P, Parnell E and Penzes P 2018 Dendritic structural plasticity and neuropsychiatric disease *Nat. Rev. Neurosci.* **19** 215–34

Gandal M J, Haney J R, Parikshak N N, Leppa V, Ramaswami G and Hartl C *et al* 2018 Shared molecular neuropathology across major psychiatric disorders parallels polygenic overlap *Science* **359** 693–97

Gao R and Penzes P 2015 Common mechanisms of excitatory and inhibitory imbalance in schizophrenia and autism spectrum disorders *Curr. Mol. Med.* **15** 146–67

Gatch M B 2002 Nitrendipine blocks the nociceptive effects of chronically administered ethanol *Alcohol Clin. Exp. Res.* **26** 1181–7

Gershon E S, Grennan K, Busnello J, Badner J A, Ovsiew F and Memon S *et al* 2014 A rare mutation of *CACNA1C* in a patient with bipolar disorder, and decreased gene expression associated with a bipolar-associated common SNP of *CACNA1C* in brain *Mol. Psychiatry* **19** 890–94

Geurts H M, Corbett B and Solomon M 2009 The paradox of cognitive flexibility in autism *Trends Cogn. Sci.* **13** 74–82

Gitlin M J and Weiss J 1984 Verapamil as maintenance treatment in bipolar illness: a case report *J. Clin. Psychopharmacol.* **4** 341–3

Gong B, Wang H, Gu S, Heximer S P and Zhuo M 2007 Genetic evidence for the requirement of adenylyl cyclase 1 in synaptic scaling of forebrain cortical neurons *Eur. J. Neurosci.* **26** 275–88

Goodnick P J 1996 Treatment of mania: relationship between response to verapamil and changes in plasma calcium and magnesium levels *South Med. J.* **89** 225–26

Gorman J M 1996 Comorbid depression and anxiety spectrum disorders *Depress. Anxiety* **4** 160–68

Gottesman I I and Shields J 1967 A polygenic theory of schizophrenia *Proc. Natl Acad. Sci. USA* **58** 199–205

Green E K, Grozeva D, Jones I, Jones L, Kirov G and Caesar S *et al* 2010 The bipolar disorder risk allele at *CACNA1C* also confers risk of recurrent major depression and of schizophrenia *Mol. Psychiatry* **15** 1016–22

Grunze H, Walden J, Wolf R and Berger M 1996 Combined treatment with lithium and nimodipine in a bipolar I manic syndrome *Prog. Neuropsychopharmacol. Biol. Psychiatry* **20** 419–26

Hamshere M L, Walters J T, Smith R, Richards A L, Green E and Grozeva D *et al* 2013 Genome-wide significant associations in schizophrenia to *ITIH3/4*, *CACNA1C* and *SDCCAG8*, and extensive replication of associations reported by the schizophrenia PGC *Mol. Psychiatry* **18** 708–12

Hasegawa A E and Zacny J P 1997 The influence of three L-type calcium channel blockers on morphine effects in healthy volunteers *Anesth. Analg.* **85** 633–8

Hayashida S, Katsura M, Torigoe F, Tsujimura A and Ohkuma S 2005 Increased expression of L-type high voltage-gated calcium channel alpha1 and α2/δ subunits in mouse brain after chronic nicotine administration *Brain Res. Mol. Brain Res.* **135** 280–84

Hayes J F, Lundin A, Wicks S, Lewis G, Wong I C K and Osborn D P J *et al* 2019 Association of hydroxylmethyl glutaryl coenzyme a reductase inhibitors, L-type calcium channel antagonists, and biguanides with rates of psychiatric hospitalization and self-harm in individuals with serious mental illness *JAMA Psychiatry* **76** 382–90

Heinrichs R W 2005 The primacy of cognition in schizophrenia *Am. Psychol.* **60** 229–42

Heyes S, Pratt W S, Rees E, Dahimene S, Ferron L and Owen M J *et al* 2015 Genetic disruption of voltage-gated calcium channels in psychiatric and neurological disorders *Prog. Neurobiol.* **134** 36–54

Hofer N T, Tuluc P, Ortner N J, Nikonishyna Y V, Fernándes-Quintero M L and Liedl K R *et al* 2020 Biophysical classification of a *CACNA1D* de novo mutation as a high-risk mutation for a severe neurodevelopmental disorder *Mol. Autism* **11** 4

Höschl C and Kozený J 1989 Verapamil in affective disorders: a controlled, double-blind study *Biol. Psychiatry* **25** 128–40

Iasevoli F, Tomasetti C and de Bartolomeis A 2013 Scaffolding proteins of the post-synaptic density contribute to synaptic plasticity by regulating receptor localization and distribution: relevance for neuropsychiatric diseases *Neurochem. Res.* **38** 1–22

Jaric I, Rocks D, Cham H, Herchek A and Kundakovic M 2019 Sex and estrous cycle effects on anxiety- and depression-related phenotypes in a two-hit developmental stress model *Front. Mol. Neurosci.* **12** 74

Jenkins M A, Christel C J, Jiao Y, Abiria S, Kim K Y and Usachev Y M *et al* 2010 Ca^{2+}-dependent facilitation of $Ca_V1.3$ Ca^{2+} channels by densin and Ca^{2+}/calmodulin-dependent protein kinase II *J. Neurosci.* **30** 5125–35

Jogia J, Ruberto G, Lelli-Chiesa G, Vassos E, Maierú M and Tatarelli R *et al* 2011 The impact of the *CACNA1C* gene polymorphism on frontolimbic function in bipolar disorder *Mol. Psychiatry* **16** 1070–71

Johnson B A, Roache J D, Ait-Daoud N, Wells L T and Mauldin J B 2004 Effects of isradipine on cocaine-induced subjective mood *J. Clin. Psychopharmacol.* **24** 180–91

Kabir Z D, Che A, Fischer D K, Rice R C, Rizzo B K and Byrne M *et al* 2017a Rescue of impaired sociability and anxiety-like behavior in adult *cacna1c*-deficient mice by pharmacologically targeting eIF2α *Mol. Psychiatry* **22** 1096–109

Kabir Z D, Lee A S and Rajadhyaksha A M 2016 L-type Ca *J. Physiol.* **594** 5823–37

Kabir Z D, Martínez-Rivera A and Rajadhyaksha A M 2017b From gene to behavior: L-type calcium channel mechanisms underlying neuropsychiatric symptoms *Neurotherapeutics* **14** 588–613

Kabitzke P A, Brunner D, He D, Fazio P A, Cox K and Sutphen J *et al* 2018 Comprehensive analysis of two *Shank3* and the Cacna1c mouse models of autism spectrum disorder *Genes Brain Behav.* **17** 4–22

Kaidanovich-Beilin O, Lipina T, Vukobradovic I, Roder J and Woodgett J R 2011 Assessment of social interaction behaviors *J. Vis. Exp.* **48** 2473

Katsura M, Mohri Y, Shuto K, Hai-Du Y, Amano T and Tsujimura A *et al* 2002 Up-regulation of L-type voltage-dependent calcium channels after long term exposure to nicotine in cerebral cortical neurons *J. Biol. Chem.* **277** 7979–88

Kauer J A 2003 Addictive drugs and stress trigger a common change at VTA synapses *Neuron* **37** 549–50

Kaufman J and Charney D 2000 Comorbidity of mood and anxiety disorders *Depress. Anxiety* **12** 69–76

Kennedy D P and Adolphs R 2012 The social brain in psychiatric and neurological disorders *Trends Cogn. Sci.* **16** 559–72

Kessler R C, Aguilar-Gaxiola S, Alonso J, Chatterji S, Lee S and Ormel J *et al* 2009 The global burden of mental disorders: an update from the WHO World Mental Health (WMH) surveys *Epidemiol. Psichiatr. Soc.* **18** 23–33

Kisko T M, Braun M D, Michels S, Witt S H, Rietschel M and Culmsee C *et al* 2018 *Cacna1c* haploinsufficiency leads to pro-social 50-kHz ultrasonic communication deficits in rats *Dis. Model. Mech.* **11** dmm034116

Knutson B, Burgdorf J and Panksepp J 1998 Anticipation of play elicits high-frequency ultrasonic vocalizations in young rats *J. Comp. Psychol.* **112** 65–73

Koob G F and Volkow N D 2010 Neurocircuitry of addiction *Neuropsychopharmacology* **35** 217–38

Koschak A, Reimer D, Huber I, Grabner M, Glossmann H and Engel J *et al* 2001 α1D (Cav1.3) subunits can form l-type Ca^{2+} channels activating at negative voltages *J. Biol. Chem.* **276** 22100–6

Kosten T R, Woods S W, Rosen M I and Pearsall H R 1999 Interactions of cocaine with nimodipine: a brief report *Am. J. Addict.* **8** 77–81

Kraeuter A K, Guest P C and Sarnyai Z 2019 The Y-maze for assessment of spatial working and reference memory in mice *Methods Mol. Biol.* **1916** 105–11

Krzyzewska I M, Ensink J B M, Nawijn L, Mul A N, Koch S B and Venema A *et al* 2018 Genetic variant in *CACNA1C* is associated with PTSD in traumatized police officers *Eur. J. Hum. Genet.* **26** 247–57

Kutlu M G and Gould T J 2016 Effects of drugs of abuse on hippocampal plasticity and hippocampus-dependent learning and memory: contributions to development and maintenance of addiction *Learn. Mem.* **23** 515–33

Kuzmin A, Zvartau E, Gessa G L, Martellotta M C and Fratta W 1992 Calcium antagonists isradipine and nimodipine suppress cocaine and morphine intravenous self-administration in drug-naive mice *Pharmacol. Biochem. Behav.* **41** 497–500

Lee A S, De Jesús-Cortés H, Kabir Z D, Knobbe W, Orr M and Burgdorf C *et al* 2016 The neuropsychiatric disease-associated gene cacna1c mediates survival of young hippocampal neurons *eNeuro* **3** ENEURO.0006-16.2016

Lee A S, Ra S, Rajadhyaksha A M, Britt J K, De Jesus-Cortes H and Gonzales K L *et al* 2012 Forebrain elimination of cacna1c mediates anxiety-like behavior in mice *Mol. Psychiatry* **17** 1054–5

Lenzi A, Marazziti D, Raffaelli S and Cassano G B 1995 Effectiveness of the combination verapamil and chlorpromazine in the treatment of severe manic or mixed patients *Prog. Neuropsychopharmacol. Biol. Psychiatry* **19** 519–28

Li J, Zhao L, You Y, Lu T, Jia M and Yu H *et al* 2015 Schizophrenia related variants in *CACNA1C* also confer risk of autism *PLoS One* **10** e0133247

Lieberman J A, Perkins D, Belger A, Chakos M, Jarskog F and Boteva K *et al* 2001 The early stages of schizophrenia: speculations on pathogenesis, pathophysiology, and therapeutic approaches *Biol. Psychiatry* **50** 884–97

Limpitikul W B, Dick I E, Ben-Johny M and Yue D T 2016 An autism-associated mutation in $Ca_V1.3$ channels has opposing effects on voltage- and Ca^{2+}-dependent regulation *Sci. Rep.* **6** 27235

Lipscombe D 2002 L-type calcium channels: highs and new lows *Circ. Res.* **90** 933–35

Lobo M K, Zaman S, Damez-Werno D M, Koo J W, Bagot R C and DiNieri J A *et al* 2013 ΔFosB induction in striatal medium spiny neuron subtypes in response to chronic pharmacological, emotional, and optogenetic stimuli *J. Neurosci.* **33** 18381–95

Lüthi A and Lüscher C 2014 Pathological circuit function underlying addiction and anxiety disorders *Nat. Neurosci.* **17** 1635–43

Ma H, Groth R D, Cohen S M, Emery J F, Li B and Hoedt E *et al* 2014 γCaMKII shuttles Ca^{2+}/CaM to the nucleus to trigger CREB phosphorylation and gene expression *Cell* **159** 281–94

Malberg J E 2004 Implications of adult hippocampal neurogenesis in antidepressant action *J. Psychiatry Neurosci.* **29** 196–205

Malcolm R, Brady K T, Moore J and Kajdasz D 1999 Amlodipine treatment of cocaine dependence *J. Psychoact. Drugs* **31** 117–20

Mallinger A G, Thase M E, Haskett R, Buttenfield J, Luckenbaugh D A and Frank E *et al* 2008 Verapamil augmentation of lithium treatment improves outcome in mania unresponsive to lithium alone: preliminary findings and a discussion of therapeutic mechanisms *Bipolar Disord.* **10** 856–66

Manoach D S 2003 Prefrontal cortex dysfunction during working memory performance in schizophrenia: reconciling discrepant findings *Schizophr. Res.* **60** 285–98

Marschallinger J, Sah A, Schmuckermair C, Unger M, Rotheneichner P and Kharitonova M *et al* 2015 The L-type calcium channel $Ca_V1.3$ is required for proper hippocampal neurogenesis and cognitive functions *Cell Calcium* **58** 606–16

Martínez-Rivera A, Hao J, Tropea T F, Giordano T P, Kosovsky M and Rice R C *et al* 2017 Enhancing VTA Ca *Mol. Psychiatry* **22** 1735–45

Miller E K 2000 The prefrontal cortex and cognitive control *Nat. Rev. Neurosci.* **1** 59–65

Mineka S, Watson D and Clark L A 1998 Comorbidity of anxiety and unipolar mood disorders *Annu. Rev. Psychol.* **49** 377–412

Moon A L, Haan N, Wilkinson L S, Thomas K L and Hall J 2018 *CACNA1C*: association with psychiatric disorders, behavior, and neurogenesis *Schizophr. Bull.* **44** 958–65

Moosmang S, Haider N, Klugbauer N, Adelsberger H, Langwieser N and Müller J *et al* 2005 Role of hippocampal $Ca_V1.2$ Ca^{2+} channels in NMDA receptor-independent synaptic plasticity and spatial memory *J. Neurosci.* **25** 9883–92

Mosheva M, Serretti A, Stukalin Y, Fabbri C, Hagin M and Horev S *et al* 2020 Association between *CANCA1C* gene rs1034936 polymorphism and alcohol dependence in bipolar disorder *J. Affect. Disord.* **261** 181–6

Motulsky H J, Snavely M D, Hughes R J and Insel P A 1983 Interaction of verapamil and other calcium channel blockers with alpha 1- and alpha 2-adrenergic receptors *Circ. Res.* **52** 226–31

Mullins C, Fishell G and Tsien R W 2016 Unifying views of autism spectrum disorders: a consideration of autoregulatory feedback loops *Neuron* **89** 1131–56

Muntaner C, Kumor K M, Nagoshi C and Jaffe J H 1991 Effects of nifedipine pretreatment on subjective and cardiovascular responses to intravenous cocaine in humans *Psychopharmacology (Berl.)* **105** 37–41

Nandagopal N and Roux P P 2015 Regulation of global and specific mRNA translation by the mTOR signaling pathway *Translation (Austin)* **3** e983402

Nanou E and Catterall W A 2018 Calcium channels, synaptic plasticity, and neuropsychiatric disease *Neuron* **98** 466–81

Nelson S B and Valakh V 2015 Excitatory/inhibitory balance and circuit homeostasis in autism spectrum disorders *Neuron* **87** 684–98

Nestler E J and Carlezon W A 2006 The mesolimbic dopamine reward circuit in depression *Biol. Psychiatry* **59** 1151–59

Oprea T I and Mestres J 2012 Drug repurposing: far beyond new targets for old drugs *AAPS J.* **14** 759–63

Pascoli V, Terrier J, Hiver A and Lüscher C 2015 Sufficiency of mesolimbic dopamine neuron stimulation for the progression to addiction *Neuron* **88** 1054–66

Pazzaglia P J, Post R M, Ketter T A, Callahan A M, Marangell L B and Frye M A *et al* 1998 Nimodipine monotherapy and carbamazepine augmentation in patients with refractory recurrent affective illness *J. Clin. Psychopharmacol.* **18** 404–13

Penzes P, Cahill M E, Jones K A, VanLeeuwen J E and Woolfrey K M 2011 Dendritic spine pathology in neuropsychiatric disorders *Nat. Neurosci.* **14** 285–93

Perez-Reyes M, White W R and Hicks R E 1992 Interaction between ethanol and calcium channel blockers in humans *Alcohol Clin. Exp. Res.* **16** 769–75

Pierce R C, Quick E A, Reeder D C, Morgan Z R and Kalivas P W 1998 Calcium-mediated second messengers modulate the expression of behavioral sensitization to cocaine *J. Pharmacol. Exp. Ther.* **286** 1171–6

Pinggera A, Lieb A, Benedetti B, Lampert M, Monteleone S and Liedl K R *et al* 2015 *CACNA1D* de novo mutations in autism spectrum disorders activate $Ca_V1.3$ L-type calcium channels *Biol. Psychiatry* **77** 816–22

Pinggera A, Mackenroth L, Rump A, Schallner J, Beleggia F and Wollnik B *et al* 2017 New gain-of-function mutation shows *CACNA1D* as recurrently mutated gene in autism spectrum disorders and epilepsy *Hum. Mol. Genet.* **26** 2923–32

Pinggera A, Negro G, Tuluc P, Brown M J, Lieb A and Striessnig J 2018 Gating defects of disease-causing de novo mutations in Ca *Channels (Austin)* **12** 388–402

Platzer J, Engel J, Schrott-Fischer A, Stephan K, Bova S and Chen H *et al* 2000 Congenital deafness and sinoatrial node dysfunction in mice lacking class D L-type Ca^{2+} channels *Cell* **102** 89–97

Post R M and Kalivas P 2013 Bipolar disorder and substance misuse: pathological and therapeutic implications of their comorbidity and cross-sensitisation *Br. J. Psychiatry* **202** 172–76

Powell T R, Fernandes C and Schalkwyk L C 2012 Depression-related behavioral tests *Curr. Protoc. Mouse Biol.* **2** 119–27

Price W A 1987 Antipsychotic effects of verapamil in schizophrenia *Hillside J. Clin. Psychiatry* **9** 225–30

Price W A and Pascarzi G A 1987 Use of verapamil to treat negative symptoms in schizophrenia *J. Clin. Psychopharmacol.* **7** 357

Purcell S M, Wray N R, Stone J L, Visscher P M, O'Donovan M C and Sullivan P F *et al* 2009 Common polygenic variation contributes to risk of schizophrenia and bipolar disorder *Nature* **460** 748–52

Purcell S M, Wray N R, Stone J L, Visscher P M, O'Donovan M C, Sullivan P F and Sklar PT (International Schizophrenia Consortium) 2009 Common polygenic variation contributes to risk of schizophrenia and bipolar disorder *Nature* **460** 748–52

Pushpakom S, Iorio F, Eyers P A, Escott K J, Hopper S and Wells A *et al* 2019 Drug repurposing: progress, challenges and recommendations *Nat. Rev. Drug Discov.* **18** 41–58

Raderer M and Scheithauer W 1993 Clinical trials of agents that reverse multidrug resistance. a literature review *Cancer* **72** 3553–63

Reimer A R and Martin-Iverson M T 1994 Nimodipine and haloperidol attenuate behavioural sensitization to cocaine but only nimodipine blocks the establishment of conditioned locomotion induced by cocaine *Psychopharmacology (Berl.)* **113** 404–10

Ripke S, O'Dushlaine C, Chambert K, Moran J L, Kähler A K and Akterin S *et al* 2013 Genome-wide association analysis identifies 13 new risk loci for schizophrenia *Nat. Genet.* **45** 1150–59

Ross J, Gedvilaite E, Badner J A, Erdman C, Baird L and Matsunami N *et al* 2016 A rare variant in *Mol. Neuropsychiatry* **2** 145–50

Rosse R B, Alim T N, Fay-McCarthy M, Collins J P, Vocci F J and Lindquist T *et al* 1994 Nimodipine pharmacotherapeutic adjuvant therapy for inpatient treatment of cocaine dependence *Clin. Neuropharmacol.* **17** 348–58

Roussos P, Mitchell A C, Voloudakis G, Fullard J F, Pothula V M and Tsang J *et al* 2014 A role for noncoding variation in schizophrenia *Cell Rep.* **9** 1417–29

Rush C R and Pazzaglia P J 1998 Pretreatment with isradipine, a calcium-channel blocker, does not attenuate the acute behavioral effects of ethanol in humans *Alcohol Clin. Exp. Res.* **22** 539–47

Russo S J and Nestler E J 2013 The brain reward circuitry in mood disorders *Nat. Rev. Neurosci.* **14** 609–25

Saliba R S, Gu Z, Yan Z and Moss S J 2009 Blocking L-type voltage-gated Ca^{2+} channels with dihydropyridines reduces gamma-aminobutyric acid type a receptor expression and synaptic inhibition *J. Biol. Chem.* **284** 32544–50

Santini E, Huynh T N, MacAskill A F, Carter A G, Pierre P and Ruggero D *et al* 2013 Exaggerated translation causes synaptic and behavioural aberrations associated with autism *Nature* **493** 411–5

Schierberl K, Hao J, Tropea T F, Ra S, Giordano T P and Xu Q *et al* 2011 $Ca_V1.2$ L-type Ca^{2+} channels mediate cocaine-induced GluA1 trafficking in the nucleus accumbens, a long-term adaptation dependent on ventral tegmental area $Ca_V1.3$ channels *J. Neurosci.* **31** 13562–75

Schizophrenia Working Group of the Psychiatric Genomics Consortium 2014 Biological insights from 108 schizophrenia-associated genetic loci *Nature* **511** 421–7

Seibenhener M L and Wooten M C 2015 Use of the open field maze to measure locomotor and anxiety-like behavior in mice *J. Vis. Exp.* e52434

Seisenberger C, Specht V, Welling A, Platzer J, Pfeifer A and Kühbandner S *et al* 2000 Functional embryonic cardiomyocytes after disruption of the L-type α_{1C} ($Ca_V1.2$) calcium channel gene in the mouse *J. Biol. Chem.* **275** 39193–99

Shibasaki M, Kurokawa K, Mizuno K and Ohkuma S 2011 Up-regulation of $Ca_V1.2$ subunit via facilitating trafficking induced by Vps34 on morphine-induced place preference in mice *Eur. J. Pharmacol.* **651** 137–45

Shibasaki M, Kurokawa K and Ohkuma S 2010 Upregulation of L-type Ca_V1 channels in the development of psychological dependence *Synapse* **64** 440–44

Simms B A and Zamponi G W 2014 Neuronal voltage-gated calcium channels: structure, function, and dysfunction *Neuron* **82** 24–45

Sinnegger-Brauns M J, Hetzenauer A, Huber I G, Renström E, Wietzorrek G and Berjukov S *et al* 2004 Isoform-specific regulation of mood behavior and pancreatic beta cell and cardiovascular function by L-type Ca^{2+} channels *J. Clin. Invest.* **113** 1430–39

Sklar P, Smoller J W, Fan J, Ferreira M A, Perlis R H and Chambert K *et al* 2008 Whole-genome association study of bipolar disorder *Mol. Psychiatry* **13** 558–69

Soeiro-de-Souza M G, Lafer B, Moreno R A, Nery F G, Chile T and Chaim K *et al* 2017 The *CACNA1C* risk allele rs1006737 is associated with age-related prefrontal cortical thinning in bipolar I disorder *Transl. Psychiatry* **7** e1086

Sofuoglu M, Singha A, Kosten T R, McCance-Katz F E, Petrakis I and Oliveto A 2003 Effects of naltrexone and isradipine, alone or in combination, on cocaine responses in humans *Pharmacol. Biochem. Behav.* **75** 801–8

Sousa N, Almeida O F and Wotjak C T 2006 A hitchhiker's guide to behavioral analysis in laboratory rodents *Genes Brain Behav.* **5** 5–24

Spanagel R 2017 Animal models of addiction *Dialogues Clin. Neurosci.* **19** 247–58

Splawski I, Timothy K W, Decher N, Kumar P, Sachse F B and Beggs A H *et al* 2005 Severe arrhythmia disorder caused by cardiac L-type calcium channel mutations *Proc. Natl Acad. Sci. USA* **102** 8089–96 (discussion 8086–88)

Splawski I, Timothy K W, Sharpe L M, Decher N, Kumar P and Bloise R *et al* 2004 $Ca_V1.2$ calcium channel dysfunction causes a multisystem disorder including arrhythmia and autism *Cell* **119** 19–31

Striessnig J, Koschak A, Sinnegger-Brauns M J, Hetzenauer A, Nguyen N K and Busquet P *et al* 2006 Role of voltage-gated L-type Ca^{2+} channel isoforms for brain function *Biochem. Soc. Trans.* **34** 903–9

Striessnig J, Pinggera A, Kaur G, Bock G and Tuluc P 2014 L-type Ca *Wiley Interdiscip. Rev. Membr. Transp. Signal.* **3** 15–38

Sun H, Martin J A, Werner C T, Wang Z J, Damez-Werno D M and Scobie K N *et al* 2016 BAZ1B in nucleus accumbens regulates reward-related behaviors in response to distinct emotional stimuli *J. Neurosci.* **36** 3954–61

Syvänen A C 2001 Accessing genetic variation: genotyping single nucleotide polymorphisms *Nat. Rev. Genet.* **2** 930–42

Takeuchi H, Tomita H, Taki Y, Kikuchi Y, Ono C and Yu Z *et al* 2019 A common *CACNA1C* gene risk variant has sex-dependent effects on behavioral traits and brain functional activity *Cereb. Cortex* **29** 3211–9

Tekin S and Cummings J L 2002 Frontal-subcortical neuronal circuits and clinical neuropsychiatry: an update *J. Psychosom. Res.* **53** 647–54

Temme S J, Bell R Z, Fisher G L and Murphy G G 2016 Deletion of the mouse homolog of *CACNA1C* disrupts discrete forms of hippocampal-dependent memory and neurogenesis within the dentate gyrus *eNeuro* **3** ENEURO.0118-16.2016

Uhrig S, Vandael D, Marcantoni A, Dedic N, Bilbao A and Vogt M A *et al* 2017 Differential roles for L-type calcium channel subtypes in alcohol dependence *Neuropsychopharmacology* **42** 1058–69

Vaupel D B, Lange W R and London E D 1993 Effects of verapamil on morphine-induced euphoria, analgesia and respiratory depression in humans *J. Pharmacol. Exp. Ther.* **267** 1386–94

Völkening B, Schönig K, Kronenberg G, Bartsch D and Weber T 2017 Deletion of psychiatric risk gene *Cacna1c* impairs hippocampal neurogenesis in cell-autonomous fashion *Glia* **65** 817–27

Waltz J A 2017 The neural underpinnings of cognitive flexibility and their disruption in psychotic illness *Neuroscience* **345** 203–17

Wang F, McIntosh A M, He Y, Gelernter J and Blumberg H P 2011 The association of genetic variation in *CACNA1C* with structure and function of a frontotemporal system *Bipolar Disord.* **13** 696–700

Weinberger D R and Berman K F 1996 Prefrontal function in schizophrenia: confounds and controversies *Philos. Trans. R. Soc. Lond. B Biol. Sci.* **351** 1495–503

Weinberger D R, Berman K F and Zec R F 1986 Physiologic dysfunction of dorsolateral prefrontal cortex in schizophrenia. I. Regional cerebral blood flow evidence *Arch. Gen. Psychiatry* **43** 114–24

Weisskopf M G, Bauer E P and LeDoux J E 1999 L-type voltage-gated calcium channels mediate NMDA-independent associative long-term potentiation at thalamic input synapses to the amygdala *J. Neurosci.* **19** 10512–9

Wek R C and Cavener D R 2007 Translational control and the unfolded protein response *Antioxid. Redox Signal.* **9** 2357–71

Whittington M A, Dolin S J, Patch T L, Siarey R J, Butterworth A R and Little H J 1991 Chronic dihydropyridine treatment can reverse the behavioural consequences of and prevent adaptations to, chronic ethanol treatment *Br. J. Pharmacol.* **103** 1669–76

Won H, de la Torre-Ubieta L, Stein J L, Parikshak N N, Huang J and Opland C K *et al* 2016 Chromosome conformation elucidates regulatory relationships in developing human brain *Nature* **538** 523–7

Wu G Y, Deisseroth K and Tsien R W 2001 Spaced stimuli stabilize MAPK pathway activation and its effects on dendritic morphology *Nat. Neurosci.* **4** 151–8

Xu W and Lipscombe D 2001 Neuronal $Ca_V1.3\alpha_1$ L-type channels activate at relatively hyperpolarized membrane potentials and are incompletely inhibited by dihydropyridines *J. Neurosci.* **21** 5944–51

Yap J J and Miczek K A 2008 Stress and rodent models of drug addiction: role of VTA-Accumbens-PFC-Amygdala circuit *Drug Discov. Today Dis. Models* **5** 259–70

Yau S Y, Li A and So K F 2015 Involvement of adult hippocampal neurogenesis in learning and forgetting *Neural. Plast.* **2015** 717958

Yoshimizu T, Pan J Q, Mungenast A E, Madison J M, Su S and Ketterman J *et al* 2015 Functional implications of a psychiatric risk variant within *CACNA1C* in induced human neurons *Mol. Psychiatry* **20** 162–69

Zacny J P and Yajnik S 1993 Effects of calcium channel inhibitors on ethanol effects and pharmacokinetics in healthy volunteers *Alcohol* **10** 505–9

Zamponi G W, Striessnig J, Koschak A and Dolphin A C 2015 The physiology, pathology, and pharmacology of voltage-gated calcium channels and their future therapeutic potential *Pharmacol. Rev.* **67** 821–70

Zhang Q, Li J X, Zheng J W, Liu R K and Liang J H 2003 L-type Ca^{2+} channel blockers inhibit the development but not the expression of sensitization to morphine in mice *Eur. J. Pharmacol.* **467** 145–50

Zhang S, Zhou Z, Gong Q, Makielski J C and January C T 1999 Mechanism of block and identification of the verapamil binding domain to HERG potassium channels *Circ. Res.* **84** 989–98

Zhang Z, Wang Y, Zhang Q, Zhao W, Chen X and Zhai J *et al* 2019 The effects of *CACNA1C* gene polymorphism on prefrontal cortex in both schizophrenia patients and healthy controls *Schizophr. Res.* **204** 193–200

Zharkovsky A, Tötterman A M, Moisio J and Ahtee L 1993 Concurrent nimodipine attenuates the withdrawal signs and the increase of cerebral dihydropyridine binding after chronic morphine treatment in rats *Naunyn Schmiedebergs Arch. Pharmacol.* **347** 483–86

Zheng F, Cui Y, Yan H, Liu B and Jiang T 2016 The effects of a genome-wide supported variant in the *CACNA1C* gene on cortical morphology in schizophrenia patients and healthy subjects *Sci. Rep.* **6** 34298

www.ingramcontent.com/pod-product-compliance
Ingram Content Group UK Ltd.
Pitfield, Milton Keynes, MK11 3LW, UK
UKHW051906100926
8556IPUK00005B/788

9 780750 320078